Partial Differential and Integral Equations

International Society for Analysis, Applications and Computation

Volume 2

The titles published in this series are listed at the end of this volume.

Partial Differential and Integral Equations

Edited by

Heinrich G. W. Begehr

Freie Universität Berlin,
Berlin, Germany

Robert P. Gilbert

University of Delaware,
Newark, Delaware, U.S.A.

and

Guo-Chun Wen

Peking University,
Beijing, P.R. China

KLUWER ACADEMIC PUBLISHERS
DORDRECHT / BOSTON / LONDON

A C.I.P. Catalogue record for this book is available from the Library of Congress.

ISBN-13: 978-1-4613-3278-7 e-ISBN-13: 978-1-4613-3276-3
DOI: 10.1007/978-1-4613-3276-3

Published by Kluwer Academic Publishers,
P.O. Box 17, 3300 AA Dordrecht, The Netherlands.

Sold and distributed in North, Central and South America
by Kluwer Academic Publishers,
101 Philip Drive, Norwell, MA 02061, U.S.A.

In all other countries, sold and distributed
by Kluwer Academic Publishers,
P.O. Box 322, 3300 AH Dordrecht, The Netherlands.

Printed on acid-free paper

Contents

Preface

This volume of the Proceedings of the congress ISAAC '97 collects the contributions of the four sections

1. Function theoretic and functional analytic methods for pde,
2. Applications of function theory of several complex variables to pde,
3. Integral equations and boundary value problems,
4. Partial differential equations.

Most but not all of the authors have participated in the congress. Unfortunately some from Eastern Europe and Asia have not managed to come because of lack of financial support. Nevertheless their manuscripts of the proposed talks are included in this volume.
The majority of the papers deal with complex methods. Among them boundary value problems in particular the Riemann–Hilbert, the Riemann (Hilbert) and related problems are treated. Boundary behaviour of vector–valued functions are studied too. The Riemann–Hilbert problem is solved for elliptic complex equations, for mixed complex equations, and for several complex variables. It is considered in a general topological setting for mappings into $\mathbb{C}^n$ and related to Toeplitz operators. Convolution operators are investigated for nilpotent Lie groups leading to some consequences for the null space of the tangential Cauchy Riemann operator. Some boundary value problems for overdetermined systems in balls of $\mathbb{C}^n$ are solved explicitly. A survey is given for the Gauss–Manin connection associated with deformations of curve singularities.

Several papers deal with generalizations of analytic functions with various applications to mathematical physics. Singular integrals in quaternionic analysis are studied which are applied to the time–harmonic Maxwell equations. The theory of first order elliptic systems in the plane are described by generalized analytic and M– (or J–) analytic functions. On one hand this theory is applied to functions of several variables on the other hand to some mixed problems in plane elasiticity. The Riemann–Hilbert problem is also studied for these functions. Some boundary value problems are studied for composite type first order systems in a similar way.
There are two kinds of bianalytic functions. One is just a particular case of the concept of polyanalytic functions which are related to polyharmonic functions as analytic are to harmonic. The other arises in the theory of second order systems in the plane. The Riemann (Hilbert) problem which inversely

formulated is solved for analytic functions is also studied for bi–(poly–) analytic functions. Bianalytic functions in connection with second order systems are used in plane elasticity to solve boundary value problems. Plane contact problems often can be reduced to the Riemann problem. Here some periodic problem is considered. Some crack problems are reduced to singular integral equations with Cauchy kernels. Singular integral equations with Hilbert kernels are investigated too. Existence and uniqueness results are given for boundary value problems of parabolic and hyperbolic equations. Some boundary–contact problems are considered having applications in accoustic scattering. A survey on weakly hyperbolic equations is provided. Eigenvalue problems with a spectral parameter linearly involved in the boundary conditions as well as indefinite nonlinear Neumann problems are studied for elliptic equations. Also higher order ordinary differential equations with super–singularities and the minimizing estimate in some linear progression model are investigated.

The editors are indepted to Mrs Barbara G. Wengel from the I. Math. Inst. of FU Berlin for unifying the manuscripts according to the publishers demand, producing the figure in one contribution, and preparing the camera–ready copy.

1 GAUSS-MANIN CONNEXION OF CURVE SINGULARITIES

A.G. Aleksandrov †

Russian Academy of Sciences
Institute of Control Sciences
Profsojuznaja ul. 65,
GSP-7, MOSCOW, 117806,
Russian Federation
alexandr@ipu.rssi.ru

Abstract: The paper is a brief survey of recent works where a computational approach to the study of Gauss-Manin connexion associated with deformations of curve singularities is developed. Related questions concerning the theory of logarithmic differential forms, Picard-Fuchs systems, hypergeometric functions and their generalizations are discussed.

AMS Classification: Primary 32S30, 32S40, 33C20, 14D05; Secondary 32S25, 14M10, 14F40
Key words and phrases: Gauss-Manin connexion, curve singularities, simple singularities, versal deformations, discriminants, free divisors, logarithmic differential forms, torsion differentials, hypergeometric functions, Picard-Fuchs systems, Fuchsian equations

1.1 INTRODUCTION

In the most general context the notion of Gauss-Manin connexion was introduced in works of B. Malgrange, P. Deligne and F. Pham. However, probably E. Brieskorn was the first who has begun to adopt efficiently this tool to the study of hypersurface *singularities* and related topics. His approach was slightly modified and extended to the case of complete intersections with isolated sin-

†Partially supported by Grant 96-0713 from the International Association INTAS and by Ruprecht-Karls-Universität Heidelberg

H.G.W. Begehr et al. (eds.), Partial Differential and Integral Equations, 1–22.

gularities by G.-M. Greuel [16], and to the general case of complex space germs with non-isolated singularities by H. Hamm [18].

Meanwhile K. Saito [24], calculating the connexion associated with the 3-parameter versal deformation of the hypersurface A_3-singularity, has obtained an elegant representation of the corresponding Gauss-Manin system of differential equations using logarithmic differential forms. His ideas have led not only to the opening of the important results in the theory of connexions but also have given a strong impetus to the development of the quite intrigued theory of logarithmic differential forms (see [25], [2]).

In a different manner the study of Gauss-Manin systems for versal integrals of types A, D, E (close related with the study of generalized functions) has been developed by V.P. Palamodov with H. Malikov [21], [23]. In particular, the corresponding systems of differential equations of the first order have been represented as local Picard-Fuchs systems. Later Gauss-Manin systems associated with 2-parameter *principal* deformations of the simple space curve singularities from the list [15] has been computed by S. Guzev [17]. However, in all works mentioned above their solutions were not determined.

The concept of Gauss-Manin connexion was also investigated by S. Ishiura and M. Noumi [19] from another point of view. They have described Gauss-Manin systems for the A_μ-singularities with the help of the hamiltonian representation. The Gauss-Manin systems naturally appear also in works of I.M. Gelfand, K. Aomoto, A.N. Varchenko and their followers in connection with the study of integrals of the type

$$J(t) = \int F_0^{\lambda_0}(z,t) F_1^{\lambda_1}(z,t) \dots F_k^{\lambda_k}(z,t)\, dz.$$

Thus, the case where $F_j(z,t)$, $1 \leq j \leq k$, are linear functions in the variables $z = (z_1, \dots, z_m)$ is considered in [7], [29], some special cases where $F_0(z,t)$ is a linear deformation of the Pham singularities were studied in [22], the case where all $F_j(z,t)$ are quasihomogeneous was investigated in [13], and so on. Moreover, in some cases an explicit representation of the solutions of Gauss-Manin systems in terms of the known types of generalized hypergeometric functions have been obtained.

To conclude our brief excursus to the history we shall note that the famous proof of Yu.I. Manin of the Mordell conjectures for algebraic curves over functional fields bases essentially on properties of the differential equations to which *period integrals* satisfy. During the time when these works have made so strong an impression on the mathematicians, it seems just an easy hand of A. Grothendieck naming the whole class of the systems of differential equations considered in Manin's papers by the *Gauss-Manin systems* or *Gauss-Manin connexions.*

In the present survey we will shortly discuss recent works where we develop a computational approach to the study of Gauss-Manin connexion associated with deformations of simple isolated singularities of hypersurfaces and with deformations of curve singularities that are complete or, more generally, non-complete intersections.

1.2 THE NOTION OF GAUSS–MANIN CONNECTION

We shall begin with an unformal discussion of the basic ideas of the theory. Let us consider the system of algebraic or analytic equations

$$\begin{cases} f_1(z_1, z_2, \ldots, z_m) & = & 0 \\ \vdots & & \\ f_k(z_1, z_2, \ldots, z_m) & = & 0 \end{cases} \tag{1.1}$$

given in a neighborhood U of the origin in $\mathbf{C}^m$. For simplicity we shall assume that $k = m - 1$ and the set X_0 of the solutions of our system in the neighborhood U is one-dimensional. We will say that a point, laying on the curve X_0, is *nonsingular* if the differential form $df_1 \wedge \ldots \wedge df_{m-1}$ does not vanish at it. Otherwise, this point (and the curve) refers to as *singular*, or shortly a *singularity*. Of course, without loss of generality one can suppose that X_0 has the only singularity at the origin $\{0\} \in X_0 \subset \mathbf{C}^m$, that is, X_0 is the germ of a reduced *curve*.

We shall assume now that the equations of system (1.1) can be *perturbed*

$$\begin{cases} f_1(z_1, z_2, \ldots, z_m) & = & t_1, \\ \vdots & & \\ f_{m-1}(z_1, z_2, \ldots, z_m) & = & t_{m-1}, \end{cases} \tag{1.2}$$

in such a manner that at each sufficiently small values of parameters $t = (t_1, \ldots, t_{m-1}) \in \mathbf{C}^{m-1}$ in the chosen neighborhood U the set X_t of the solutions of the system (1.2) is also *one-dimensional*. In other words, we shall consider the *principal deformation* of the curve singularity X_0 given by the holomorphic map

$$f : \mathcal{X} \longrightarrow \mathbf{C}^{m-1}. \tag{1.3}$$

Let X be the intersection of a ball of a small radius $\epsilon > 0$ centered at the distinguished point $\{0\} \in X_0$ with $f^{-1}(T)$, where $T \subset \mathbf{C}^{m-1}$ is a punctured ball of a radius $0 < \delta \ll \epsilon$ centered at the origin $0 = f(0)$. Consider the natural restriction of the mapping (1.3) :

$$f : X \longrightarrow T.$$

For some values of parameters $t \in T$ the fibers X_t will be non-singular curve-germs, i.e. 1-dimensional germs with no singularities, for other ones the fibers can have singular points which are called the critical points of the mapping f. Denote by $C \subset X$ the set of critical points of f and by D its image $f(C) \subset T$. Thus, parameters, corresponding to the fibers with singularities, are contained in the set D which refers to as the discriminant set, or the *discriminant* of the principal deformation X_0. In many important cases the discriminant is the zero-set of the only equation $h(t_1, \ldots, t_k) = 0$, i.e. D is a *hypersurface*. Set

$$T' = T - D, \quad X' = X - C.$$

Then the restriction $f : X' \to T'$ is a local trivial differentiable fiber bundle called the *Milnor fibration* of f, that is, the fibers $X_t = f^{-1}(t) \cap X$ (of *real* dimension two) form the smooth fiber bundle over T'. Fix a point $t_0 \in T'$. Then for each *smooth closed path* $\gamma_0 \subset X_{t_0}$, corresponding to the 1-cycle in $H^1(X_{t_0}, \mathbf{C})$, it is possible to construct a family of 1-cycles $\gamma(t) \subset X_t$, $t \in T'$ such that $\gamma(t_0) = \gamma_0$. If one takes a holomorphic differential form $\omega = g(z)dz_1 \wedge \ldots \wedge dz_m$ of the maximal degree in a neighborhood of the origin in $\mathbf{C}^m$ then using the identity $df_1 \wedge \ldots \wedge df_{m-1} \wedge \psi = \omega$ one can find a differential form ψ which is the result of the division of ω by $df_1 \wedge \ldots \wedge df_{m-1}$. Certainly, the form ψ is not determined uniquely, but only up to the summands containing differentials of the functions $f_1, \ldots, f_{m-1}$. Nevertheless, it is easy to prove, that for all parameters t, rather close to zero, the integral

$$I(t) = \int_{\gamma(t)} \psi = \int_{\gamma(t)} \frac{\omega}{df_1 \wedge \ldots \wedge df_{m-1}} \tag{1.4}$$

is determined correctly. Moreover, the integral $I(t)$ is not only continuous, but also an *analytic* function in t. The integrals of such type are called the *period integrals.* Replacing the differential form ω with any other one, the integral (1.4), generally speaking, will also change. However, it is possible to prove that the set of all such integrals contains a finite number of elements $I_1(t), \ldots, I_\mu(t)$ such that any integral (1.4) may be expressed by means of these generators as a linear combination with *holomorphic* coefficients. In the present context μ is nothing but the *Milnor number* which is a topological invariant of the singularity X_0. The same observation remains valid if one fixes the form ω and takes various families $\gamma(t)$. For definiteness, we shall fix a family of vanishing cycles $\gamma(t)$ and consider μ independent *period integrals* of the following type

$$I_j(t) = \int_{\gamma(t)} \frac{\omega_j}{df_1 \wedge \ldots \wedge df_{m-1}}, \qquad 1 \leq j \leq \mu. \tag{1.5}$$

Now it should be noticed that the period integrals $I_j(t)$ can be differentiated on parameters. Thus, between integrals and their derivatives there arise linear relations *(syzygies)* with polynomial coefficients in t. As a result, these relations generate a system of differential equations for the integrals $I_j(t)$ consisting of an infinite number of equations that are expressed through a finite number of independent ones.

In such a way to the singularity X_0 given by the system (1.1) there corresponds a system of differential equations, depending on parameters t, and determined correctly outside of the discriminant. This system refers to as a *connexion,* or *the Gauss-Manin system,* associated with the principal deformation of X_0. The main problem is to determine a system of differential equations defined on the *whole* space of parameters, which is equivalent to the initial one outside of the discriminant (or, in other words, *to prolong* the initial system to the discriminant set). It is possible to show that the solution of this problem depends mainly on the properties of fibers over the discriminant.

1.3 HYPERSURFACE A_μ-SINGULARITIES

The basic properties of the Gauss-Manin connexion of hypersurface singularities have been stated for the first time by E. Brieskorn. In the famous work [10] he proved that the connexion associated with the 1-parameter principal deformation of an isolated hypersurface singularity can be represented as a system of ordinary differential equations with regular singularities. Moreover, it follows from his results that after a matrix transformation with meromorphic entries such an equation is reduced to one having poles of first order. In fact, an analogue of his result remains valid for the *versal* deformations of A_μ-singularity [19]. However, the latter proof is based essentially on properties of the special flat coordinate system. On the contrary, in the earlier work [21] the Gauss-Manin systems for the versal deformations of simple hypersurface singularities were written down explicitly in the standard coordinate representation. Based on this result (see also [23]) and following [4] let us describe the case of A_μ-singularity in more details.

Consider the minimal versal deformation of the singularity A_μ, given by the equation $F(z,s) = z^{\mu+1}+s_{\mu-1}z^{\mu-1}+\ldots+s_1 z+s_0 = 0$. Due to V.P.Palamodov [21] the *versal integrals*

$$K_i^\lambda(s) = \int_\Gamma z^i (z^{\mu-1}+s_{\mu-1}z^{\mu-1}+\ldots+s_1 z+s_0)^\lambda \, dz, \qquad i = 0,\ldots,\mu+1, \quad (1.6)$$

are well defined where the contour $\Gamma \subset \{z \in \mathbb{C} : F(z,s) \neq 0\}$ of the integration is homotopic to the real line $\mathbb{R}$.

Proposition 1 ([21], [23]). *The versal integrals $K_0^\lambda(s),\ldots,K_{\mu+1}^\lambda(s)$ satisfy the following overdetermined system of differential equations:*

$$\sum_{\ell=0}^{\mu-1} s_\ell \frac{\partial}{\partial s_0} K_{\ell+i}^\lambda + \frac{\partial}{\partial s_0} K_{\mu+1+i}^\lambda = \lambda K_i^\lambda, \qquad 0 \le i \le \mu-1,$$

$$\sum_{\ell=1}^{\mu-1} \ell s_\ell \frac{\partial}{\partial s_0} K_{\ell+j}^\lambda + (\mu+1)\frac{\partial}{\partial s_0} K_{\mu+1+j}^\lambda = -(j+1)K_j^\lambda, \qquad -1 \le j \le \mu-1.$$

In fact, the above system is transformed to the following one:

Proposition 2 ([3]). *The integrals $K_0(s),\cdots,K_{\mu-1}(s)$, $s = (s_0,\cdots,s_{\mu-1})$, satisfy the system of differential equations*

$$S\frac{\partial}{\partial s_0}\mathbf{K}(s) = (L + V(s_2,\cdots,s_{\mu-1}))\mathbf{K}(s), \quad (1.7)$$

where $\mathbf{K}(s) = (K_0(s),\cdots,K_{\mu-1}(s))^{tr}$ is a vector-column of the versal integrals, $S = s_0 \cdot \mathrm{id}_\mu + C(s')$, $C(s')$ is a polynomial matrix depending on $s' = (s_1,\ldots,s_{\mu-1})$, L is the diagonal matrix $\mathrm{diag}\,(\lambda + \frac{1}{\mu+1},\ldots,\lambda+\frac{\mu}{\mu+1})$, and the matrix $V(s_2,\cdots,s_{\mu-1})$ can be presented as

$$\frac{1}{(\mu+1)^2}\begin{bmatrix} 0 & 0 & 0 & 0 & \cdots & 0 & 0 \\ 0 & 0 & 0 & 0 & \cdots & 0 & 0 \\ 2s_{\mu-1} & 0 & 0 & 0 & \cdots & 0 & 0 \\ 3s_{\mu-2} & 2\cdot 2s_{\mu-1} & 0 & 0 & \cdots & \vdots & \vdots \\ 4s_{\mu-3} & 3\cdot 2s_{\mu-2} & 2\cdot 3s_{\mu-1} & 0 & \cdots & \vdots & \vdots \\ \vdots & \vdots & \vdots & \ddots & \cdots & \vdots & \vdots \\ (\mu-1)s_2 & (\mu-2)\cdot 2s_3 & (\mu-3)\cdot 3s_4 & \cdots & 2\cdot(\mu-2)s_{\mu-1} & 0 & 0 \end{bmatrix}.$$

Corollary 1 ([4]). *The solutions to the system* (1.7) *can be presented in the expansion form*

$$\mathbf{K}(s_0,s') = \left(\mathrm{id}_\mu + \sum_{k\geq 0}\frac{1}{k!}\prod_{j=0}^{k-1} D(s')(L - j\cdot\mathrm{id}_\mu + V(s'))(\frac{s_0}{\Delta_\mu})^k\right)\mathbf{K}(0,s'), \tag{1.8}$$

where $\Delta_\mu = \det|S(0,s')|$, *and* $D(s') = \Delta_\mu \cdot S^{-1}(0,s'), \quad s' = (s_1,\ldots,s_{\mu-1})$.

Proof The proof is quite elementary. First of all observe that $\partial S/\partial s_0 = \mathrm{id}_\mu$ is the identity matrix of μ-th order. Thus,

$$S\left(\frac{\partial}{\partial s_0}\right)^2\mathbf{K}(s) + \frac{\partial}{\partial s_0}\mathbf{K}(s) = (L+V(s'))\frac{\partial}{\partial s_0}\mathbf{K}(s),$$

$$\begin{aligned}\left(\frac{\partial}{\partial s_0}\right)^2\mathbf{K}(s) &= S^{-1}(L+V(s')-\mathrm{id}_\mu)\frac{\partial}{\partial s_0}\mathbf{K}(s)\\ &= S^{-1}(L+V(s')-\mathrm{id}_\mu)S^{-1}(L+V(s'))\mathbf{K}(s).\end{aligned}$$

Similarly,

$$\left(\frac{\partial}{\partial s_0}\right)^j\mathbf{K}(s) = \prod_{j=0}^{k-1}(S^{-1}(L+V(s')-j\cdot\mathrm{id}_\mu))\mathbf{K}(s).$$

It remains to apply Taylor's expansion theorem to $K(s)$ at the point that lies outside of the discriminant set and then to make an analytic continuation.

As an example one can show (see [4]) that in the case of an A_3-singularity the function in two variables $K^0_{A_\mu}(s_0,0,\ldots,0,s_i,0,\ldots,0)$ can be expressed in terms of the generalized hypergeometric functions in Pochhammer's sense. Namely, $K^0_{A_3}(s_0,s_1,0)$ is equal to the expression

$$2^{12}s_0^2\cdot {}_3F_2\left(\frac{17}{12}-\frac{\lambda}{3},\frac{13}{6}-\frac{\lambda}{3},\frac{23}{12}-\frac{\lambda}{3};\frac{5}{3},\frac{4}{3}\mid\frac{2^{20}s_0^3}{3^3s_1^4}\right)K^0_{A_3}(0,s_1,0)$$

$$-(\frac{5}{4}-\lambda)(\frac{1}{2}-\lambda)\frac{9s_1^5}{4}\cdot {}_3F_2\left(\frac{17}{12}-\frac{\lambda}{3},\frac{7}{6}-\frac{\lambda}{3},\frac{11}{12}-\frac{\lambda}{3};\frac{1}{3},\frac{2}{3}\mid\frac{2^{20}s_0^3}{3^3s_1^4}\right)K^1_{A_3}(0,s_1,0)$$

$$-2^8(\lambda+\frac{3}{4})\frac{s_0}{s_1^2}\cdot {}_3F_2\left(\frac{1}{2}-\frac{\lambda}{3},\frac{1}{4}-\frac{\lambda}{3},\frac{3}{4}-\frac{\lambda}{3};\frac{4}{3},\frac{2}{3}\mid \frac{2^{20}s_0^3}{3^3s_1^4}\right) K^2_{A_3}(0,s_1,0),$$

where ${}_3F_2(\alpha_1,\alpha_2,\alpha_3;\beta_1,\beta_2 \mid z)$ denotes the generalized hypergeometric function of Pochhammer type (see [8],[20])

$${}_3F_2(\alpha_1,\alpha_2,\alpha_3;\beta_1,\beta_2 \mid x)=\sum_{j\geq 0}\frac{(\alpha_1;j)(\alpha_2;j)(\alpha_3;j)}{(\beta_1;j)(\beta_2;j)j!}x^j.$$

1.4 COMPLETE INTERSECTION S_μ-SINGULARITIES

Let us consider the pair of equations

$$\begin{cases} f_1(x,y,z) &= x^2+y^2+z^{\mu-3} &= 0,\\ f_2(x,y,z) &= yz &= 0,\end{cases}$$

in $(\mathbb{C}^3,0)$ where $\mu\geq 5$. The corresponding *curve* singularity X_0 is usually called the complete intersection of type S_μ (see [15]). Set $\nu=\mu-3$. Then the principal deformation of X_0 is defined as

$$\begin{cases} f_1(x,y,z) &= x^2+y^2+z^{\nu} &= t_1,\\ f_2(x,y,z) &= yz &= t_2.\end{cases}$$

As in section 1 we see that the Gauss-Manin connexion associated with the principal deformation of X_0 is represented by a system of differential equations which, in its turn, describe nontrivial relations between the integrals

$$I_j(t)=\int_{\gamma(t)} q_j(x,y,z)\,dx\wedge dy\wedge dz/df_1\wedge df_2$$

and their derivatives. It is easy to verify (see [3]) that the functions $q_j(x,y,z)$, $j=1,\ldots,\mu$, can be represented by the set of monomials $\{1,z,z^2,\ldots,z^\nu,\ y,x\}$. Thus, we define exactly μ differential 1-forms

$$\begin{aligned} \frac{z^j dx\wedge dy\wedge dz}{df_1\wedge df_2} &= \frac{z^{j-1}dz}{2x}, \qquad 0\leq j\leq \nu,\\ \frac{y dx\wedge dy\wedge dz}{df_1\wedge df_2} &= \frac{t_2 dz}{z^2x},\\ \frac{x dx\wedge dy\wedge dz}{df_1\wedge df_2} &= -\frac{dy}{2y}.\end{aligned}$$

Evidently the latter form can be easily integrated. So *nontrivial* integrals for which the differential equation shall be calculated are given by the following set of differential forms consisting of $\mu-1$ elements:

$$\frac{z^{-2}dz}{x},\frac{z^{-1}dz}{x},\frac{dz}{x},\ldots,\frac{z^{\nu-1}dz}{x}.$$

Set

$$I_j(t) = \int_{\gamma(t)} \frac{z^{j-1}dz}{x}, \quad -1 \le j \le \nu.$$

It is not difficult to see [3] that the problem is reduced to the study of the system of equations which satisfy the *period integrals* $J_1(t), \ldots, J_{\nu+2}(t)$ defined as follows:

$$J_{j+2}(t) = -2\frac{\partial}{\partial t_1} I_j(t) = \int_{\gamma(t)} \frac{z^{j+2}\, dz}{\sqrt{(z^{\nu+2} + t_1 z^2 + t_2^2)^3}}, \qquad j = -1, \ldots, \nu.$$

The key point of the proof is the existence of evident relations between $J_i(t)$ and the versal integrals (1.6)

$$K_i^{-3/2}(t_2^2, 0, t_1, 0, \ldots, 0) = J_i(t), \qquad i = 1, \ldots, \nu + 2. \tag{1.9}$$

As remarked above we have exactly $\mu - 1$ nontrivial period integrals $J_1(t), \ldots, J_{\nu+2}(t)$ which correspond to the integrals $K_1(s_0, 0, s_2, 0, \ldots, 0), \ldots, K_{\nu+2}(s_0, 0, s_2, 0, \ldots, 0)$ in view of the relationship (1.9). For technical reasons we shall also extend formally the relation (1.9) for $i = 0$. As a result the following statement is obtained.

Proposition 3 ([3]). *The period integrals $J_0(t), \ldots, J_{\nu+2}(t)$ for the isolated complete intersection S_μ-singularity satisfy the system of differential equations*

$$\begin{aligned}
(t_2^2 \mathrm{id}_{\nu+1} + C(t_1))\frac{\partial}{\partial t_2}\mathbf{J}(t) &= 2t_2(L + V(t_1))\mathbf{J}(t),\\
J_{\nu+1}(t) &= -\frac{2t_1}{\nu+2} J_1(t),\\
\frac{\partial}{\partial t_2} J_{\nu+2}(t) &= -\frac{2}{\nu+2}\left(t_1 \frac{\partial}{\partial t_2} J_2(t) + t_2 J_0(t)\right),
\end{aligned}$$

where $\mathbf{J}(t) = (J_0(t), \ldots, J_\nu(t))^{tr}$,

$$C(t_1) = \begin{bmatrix}
0 & 0 & 0 & 0 & 0 & \cdots & 0\\
0 & 0 & 0 & \frac{\nu}{\nu+2}t_1 & 0 & \cdots & 0\\
0 & 0 & 0 & 0 & \frac{\nu}{\nu+2}t_1 & \cdots & \vdots\\
\vdots & \vdots & \vdots & \vdots & \cdots & \ddots & \vdots\\
0 & 0 & 0 & 0 & \ddots & \ddots & \frac{\nu}{\nu+2}t_1\\
0 & \frac{-2\nu}{(\nu+2)^2}t_1^2 & 0 & 0 & \cdots & \cdots & 0\\
0 & 0 & \frac{-2\nu}{(\nu+2)^2}t_1^2 & 0 & \cdots & \cdots & 0
\end{bmatrix},$$

$$L = \mathrm{diag}\left(\lambda + \tfrac{1}{\nu+2}, \ldots, \lambda + \tfrac{\nu+1}{\nu+2}\right),$$

$$V(t_1) = \begin{bmatrix}
0 & 0 & \cdots & 0\\
\vdots & \vdots & \cdots & \vdots\\
0 & 0 & \cdots & 0\\
\frac{-2\nu}{(\nu+2)^2}t_1 & 0 & \cdots & 0
\end{bmatrix}.$$

Set $m = [(\nu+1)/2]$. Then the above system can be transformed to two kinds of differential equations of order $m+1$ satisfied by the period integrals $J_{2j}(t)$ and $J_{2j+1}(t)$ for all $0 \le j \le m$. Further calculations show (see [3]) that the period integrals can be expressed in terms of the generalized hypergeometric functions. For example, for odd ν altogether $(m+1)$ solutions $\mathcal{U}_k(t)$, $0 \le k \le m$, are obtained that can be expressed through the series

$$\mathcal{U}_k(t) = t_1^{-(3\nu+4)/2\nu} \tau^{\rho_k} U_k(\tau^m, -\frac{3}{2}),$$

where

$$\begin{aligned} U_k(x,\lambda) &= \sum_{\ell \ge 0} a_{m\ell+k}(\lambda) x^\ell, \\ a_{m\ell+k}(\lambda) &= \prod_{j=0}^{m-1} \frac{((k-\lambda-1)/m - j/(m+1); \ell)}{((j+k)/m - 1; \ell)} \frac{(-\lambda + 1/(\nu+2); \ell)}{(m - (\lambda + \frac{1}{2}); \ell)}, \\ \tau &= (1/\nu)^{1/m} (-\nu t_1/\nu + 2)^{\frac{-m-1}{m}} t_2^2, \end{aligned}$$

$(\alpha;\ell) = \Gamma(\alpha+\ell)/\Gamma(\alpha)$, and $0 \le k \le m$.

Of course, an analogous result is also true for even ν (see [3]). It should be remarked that the functions $U_k(x,\lambda)$ may be considered as the generalized hypergeometric functions

$${}_{m+1}F_m(\alpha_1^{(k)}, \ldots, \alpha_{m+1}^{(k)}; \beta_1^{(k)}, \ldots, \beta_m^{(k)} | x) = \sum_{\ell \ge 0} \frac{(\alpha_1^{(k)}; \ell) \ldots (\alpha_{m+1}^{(k)}; \ell)}{(\beta_1^{(k)}; \ell) \ldots (\beta_m^{(k)}; \ell)} \frac{x^\ell}{\ell!}$$

in Pochhammer's sense (see [8],[20]).

1.5 NON–COMPLETE INTERSECTION CASE

Recall that if $X_0 \subset \mathbb{C}^m$ is the germ of a complete intersection then its defining ideal $\mathcal{I} \subset \mathcal{O}_{\mathbb{C}^m}$ can be generated by a *regular* sequence of holomorphic functions $g_1, \ldots, g_k$ such that g_i is not a zero divisor in $\mathcal{O}_{\mathbb{C}^m}/(g_1, \ldots, g_{i-1})$ for all $i = 1, \ldots, k$. Otherwise, X_0 is called a non-complete intersection.

Note that an arbitrary deformation of a hypersurface singularity can be presented as the principal deformation of a complete intersection. For example, the versal deformation of the plane curve A_2-singularity $x^3 + t_2 x + t_1 = y^2$ can be rewritten as the principal deformation of a complete intersection given be the system of two equations in $\mathbb{C}^3$

$$\begin{cases} x^3 + xz - y^2 &= -t_1, \\ z &= t_2. \end{cases}$$

However, it seems for germs which are not a complete intersection there do not exist any analogs of the principal deformation. Consider a simple example of such a type: the germ $X_0 \subset (\mathbb{C}^3, 0)$ is the union of three coordinate axes in $\mathbb{C}^3$ given by the equations

$$\left\{\begin{array}{rcl} xy & = & 0, \\ xz & = & 0, \\ yz & = & 0. \end{array}\right. \tag{1.10}$$

Then the principle deformation

$$\left\{\begin{array}{rcl} xy & = & t_1, \\ xz & = & t_2, \\ yz & = & t_3, \end{array}\right.$$

is not ***flat*** since fibers X_t over different parameters t have different complex dimensions. Indeed, $\dim_{\mathbb{C}} X_0 = 1$, while $\dim_{\mathbb{C}} X_t = 0$, $\#(X_t) = 2$, if $t_1 t_2 t_3 \neq 0$, and $\dim_{\mathbb{C}} X_t = -1$, $X_t = \emptyset$, if $t_1 = 0$, $t_2 t_3 \neq 0$, and so on. However, one can describe the Gauss-Manin connexion associated with the *minimal versal* deformation of the *smoothable* germ X_0 (that is, X_0 possesses a deformation with smooth generic fiber) similarly to section 1. In such a case the key role is played by a general notion of discriminant. Let us go into more details.

At first recall the following definition. Let $f : X \longrightarrow Y$ be a morphism of germs, $\Delta : X \longrightarrow X \times_Y X$ be the diagonal inclusion into the fiber product of X over Y, and $\mathcal{J}$ is the defining ideal of the diagonal $\Delta(X) \subset X \times_Y X$. Then $\Omega^1_{X/Y} = \Delta^*(\mathcal{J}/\mathcal{J}^2)$ is called the module of relative regular differentials on X/S.

Now consider the minimal versal deformation $f : X \longrightarrow S$ of a pure 1-dimensional smoothable complex germ $X_0 \subset \mathbb{C}^m$ with a singularity at the origin. Assume also that the base S is smooth. Following [28], (2.5), p. 587, we define the critical subgerm $C \subset X$ by the first Fitting Ideal $F_1(\Omega^1_{X/S})$ of $\Omega^1_{X/S}$, so $\mathcal{O}_C = \mathcal{O}_X / F_1(\Omega^1_{X/S})$. Then the discriminant $D \subset S$ of f is defined by $F_0(f_* \mathcal{O}_C)$. In order to calculate the latter, one has to construct a resolution

$$\ldots \longrightarrow \mathcal{O}_S^q \xrightarrow{A} \mathcal{O}_S^p \longrightarrow f_* \mathcal{O}_C \longrightarrow 0. \tag{1.11}$$

Hence, the 0-Fitting Ideal $F_0(f_* \mathcal{O}_C) \subset \mathcal{O}_S$ is generated by minors of maximal order of the matrix A and does not depend on such a presentation.

In the case where X_0 is a complete intersection K.Saito [26] has proved that $\mathcal{O}_C$ is a Cohen-Macaulay ring of Krull dimension $\tau - 1$, where $\tau = \dim S = \dim_{\mathbb{C}} T^1(X_0)$ is the Tjurina number and $T^1(X_0)$ is the vector space of infinitesimal deformations of the first order of the singularity X_0. Moreover, in such a case $p = q$ and $h = \det(A)$, that is, the ideal

$$F_0(f_* \mathcal{O}_C) \simeq (h) \mathcal{O}_S$$

is *principal,* its generator has no multiple factors and defines the *reduced* hypersurface in the τ-dimensional smooth germ S.

In general the situation is much more complicated. There is a special class of space curve singularities which can serve as a good illustration for occurring difficulties. In a certain sense these curves are very far from being complete intersections. More precisely, it is well-known [1] that curve-germs with the minimal conductor (that is, with invariants $c = \delta+1$, δ is the arithmetic genus), are non-complete intersections for $c \geq 3$, since for complete intersections the Serre's equality $c = 2\delta$ holds. We will consider in more detail two simplest space-curve germs of genus 2, that is, $c = 3$.

The first case was already mentioned before. It is the germ $X_0 \subset (\mathbb{C}^3, 0)$ of the union of three coordinate axes in $\mathbb{C}^3$ given by the system of equations (1.10). In other words, X_0 is the *bouquet* of z-axis and the A_1-singularity in the complex (x,y)-plane. The minimal versal deformation $f: (X,0) \to (S,0)$ of X_0 is described as follows.

The base deformation space is $S \simeq (\mathbb{C}^3, 0)$, while the total deformation space $X \subset \mathbb{C}^3 \times S$ is defined by equations

$$\begin{cases} xy - t_2 t_3 & = & 0, \\ xz - t_2 z - t_1 t_2 & = & 0, \\ yz + t_1 y - t_3 z & = & 0. \end{cases} \tag{1.12}$$

The morphism f is induced by the natural projection $\mathbb{C}^3 \times S \to S$. The critical set $C \subset X$ of f is defined by the Fitting Ideal $F_1(\Omega_{X/S})$, that is, by the minors of order two of the jacobian matrix

$$J = \begin{pmatrix} y & x & 0 \\ z & 0 & x - t_2 \\ 0 & z + t_1 & y - t_3 \end{pmatrix}.$$

The discriminant of the versal deformation is defined by the Fitting 0-ideal $F_0(f_*(\mathcal{O}_C))$. It is not difficult to verify that there is a resolution (1.11) with 6×5-matrix A which can be presented in the following form:

$$A = \begin{pmatrix} t_2 t_3 & t_3 & 0 & 0 \\ 0 & t_3 & -t_2 & 0 \\ t_1 t_2 & 0 & 0 & t_2 \\ 0 & 0 & 0 & t_3 \\ 0 & t_1 & 0 & 0 \\ 0 & 0 & t_1 & 0 \end{pmatrix}.$$

Hence,

$$F_0(f_*(\mathcal{O}_C)) = (t_1 t_2 t_3^3, t_1 t_2^2 t_3^2, t_1 t_2^3 t_3, t_1^2 t_2 t_3^2, t_1^2 t_2^2 t_3, t_1^3 t_2 t_3) = (t_1, t_2, t_3)^3 \cap (t_1 t_2 t_3),$$

and the ideal $(t_1, t_2, t_3)\mathcal{O}_S$ corresponds to the 0-dimensional embedded component of the discriminant D. Thus, D is not reduced. Its reduced part D_{red} is called the discriminant set and defined by the equation $t_1 t_2 t_3 = 0$. As a result D_{red} is the germ of the union of three coordinate planes in $\mathbb{C}^3$ and its singular locus Sing D_{red} is the union of three coordinate axes.

Consider a point $t = (t_1, t_2, t_3) \notin D_{\text{red}}$. It is easy to see that the fiber X_t over this point is the *complete intersection* singularity

$$\begin{cases} xy - t_2 t_3 & = & 0, \\ xz - t_2 z - t_1 t_2 & = & 0. \end{cases} \tag{1.13}$$

Let us change the variable $x := x/t_2$. Obviously, the system (1.13) defines the principal deformation of the complete intersection curve

$$\begin{cases} xy &= 0, \\ xz - z &= 0. \end{cases} \tag{1.14}$$

Evidently the curve (1.13) is isomorphic to the multigerm of two nodes (transversal intersection of two lines) and its Milnor number is equal to 2. Each node is isomorphic to the Morse point (or, equivalently to the A_1-singularity defined by the equation $x^2 - y^2 = 0$ in the plane) with Milnor number 1. The associated Gauss-Manin connexion is trivial since the 1-form $\omega = dx/\sqrt{x^2 - u}$ on the generic fiber X_u of the principal deformation $x^2 - y^2 = u$ is integrated. This implies the triviality of the Gauss-Manin connexion associated with the deformation (1.13) outside of the discriminant, and, of course, the triviality of its extension to the whole base space S.

The second example is the germ of a two-branch space curve X_0 which is the bouquet of a line and A_2-singularity. Its equations are the following:

$$\begin{cases} x^3 - y^2 &= 0, \\ xz &= 0, \\ yz &= 0. \end{cases} \tag{1.15}$$

The minimal versal deformation $f : (X, 0) \to (S, 0)$ of this germ is described analogously: $S \simeq (\mathbb{C}^4, 0)$, while $X \subset \mathbb{C}^3 \times S$ is given by three equations

$$\begin{cases} x^3 - y^2 + t_2x^2 + t_4x + t_3y &= 0, \\ xz + t_1y - t_1t_3 &= 0, \\ yz + t_1x^2 + t_1t_2x + t_1t_4 &= 0. \end{cases}$$

It is not difficult to verify that the matrix A in the resolution (1.11) can be presented in the form

$$A = \begin{pmatrix} -t_1t_4 & 0 & 0 & -t_3 & t_1 \\ 0 & 4t_4 & 3t_3 & 0 & 2t_2 \\ 2t_1t_2t_3 & 3t_1t_3 & 0 & 2t_4 & 0 \\ 0 & 2t_2t_3 & 2t_4 & 0 & 3t_3 \\ t_1t_4 & 2t_1t_2 & 0 & 0 & 3t_1 \\ -t_1t_3 & 0 & 2t_1 & 0 & 0 \end{pmatrix}.$$

Hence, $F_0(f_*(\mathcal{O}_C)) = (t_1, t_3, t_4)^3 \cap (t_1)^2 \cap (P)$, that is, it is generated by the three elements $t_1^2t_4P, t_1^2t_3P, t_1^3P$, where

$$P = 64t_4^3 - 16t_2^2t_4^2 - 72t_2t_3^2t_4 + 16t_2^3t_3^2 + 27t_3^4.$$

The germ D is non-reduced, since (t_1, t_3, t_4) corresponds to the 1-dimensional embedded component of the discriminant D. Furthermore, D_{red} is reducible and its equation is $t_1P = 0$.

Assume $t_1 \neq 0$. Consider a point $t = (t_1, \ldots, t_4) \notin D_{\text{red}}$ and make the substitution $z := z/t_1$

$$\begin{cases} x^3 - y^2 + t_2x^2 + t_4x + t_3y &= 0, \\ xz + y - t_3 &= 0, \\ yz + x^2 + t_2x + t_4 &= 0. \end{cases}$$

One can take the expression for y obtained from the second equation and substitute it to the first and third ones. Easy calculations show that the first equation is equal to the third one multiplied by x. Hence, the fiber X_t over such a point is isomorphic to the complete intersection given by the second and third equations:

$$\begin{cases} xz + y - t_3 &= 0, \\ x^2 - xz^2 + t_2x + t_3z + t_4 &= 0, \end{cases} \tag{1.16}$$

or, equivalently, X_t is contained in the 3-parameter versal deformation of the quasihomogeneous complete intersection space curve singularity

$$\begin{cases} xz + y &= 0, \\ x^2 - xz^2 &= 0. \end{cases} \tag{1.17}$$

The latter germ is nothing but the union of the coordinate axis z and the twisted cubic

$$\begin{cases} x &= s^2, \\ y &= -s^3, \\ z &= s. \end{cases}$$

More exactly, the germ (1.17) is isomorphic to the plane curve A_3-singularity $x^2 - z^4 = 0$. Hence, the fiber X_t given by the system (1.15) is isomorphic to the plane singularity

$$z^4 - 2t_2z^2 - 4t_3z - 4t_4 - 4t_2x - 4x^2 = 0$$

which is equivalent to the minimal versal deformation of A_3-singularity. Indeed, after the following changing of variables

$$t_2 = -2t_2,\ t_3 = -4t_3,\ t_4 = -4t_4,\ x = 2x,\ x = x + t_2/2,\ t_4 = t_4 + t_2^2/4$$

this equation is reduced to the form

$$z^4 + t_2z^2 + t_3z + t_4 - x^2 = 0.$$

Thus, the Gauss-Manin system associated with the versal deformation of the space curve singularity (1.15) outside of the discriminant describes relations between the derivatives of the period integrals similar to those for the A_3-singularity, see [5]. In order to explain how one can solve the problem of the extension of such system to the whole base space S we are in need of further details which will be discussed in the next section.

In conclusion of the present section it should be remarked that similar considerations are suitable for the investigation of Gauss-Manin connections associated with versal deformations of bouquets of a line and simple hypersurface singularities. For example, the bouquet of a line and A_3-singularity whose Milnor number is equal to 4 can be treated with the help of D_4-singularity, and so on (see [5]). The key role in all such examples plays the observation that the generic fibers of the minimal versal deformations are isomorphic up to a smooth factor to the generic fibers of the minimal versal deformations of hypersurface singularities which are adjacent to the non-smooth factor of bouquets.

1.6 GAUSS–MANIN SYSTEMS AND LOGARITHMIC FORMS

Here and further we shall adhere to the terminology and notations of the works [10] and [16]. Let $f : X \to S$ be a *flat* mapping of germs of complex spaces. As in section 1.1 denote by $C \subset X$ the critical set of f, and by $D \subset S$ the set of critical values of the mapping f or, equivalently, the discriminant set. Set $S' = S \backslash D$ and $X' = X \backslash f^{-1}(D)$. Let $\Omega_X^\bullet = (\Omega_X^p, d)_{p=0,1,\dots}$ be the (absolute) De Rham complex of holomorphic differential forms on X. If the germ $X \subset (\mathbb{C}^m, 0)$ is defined by an ideal $\mathcal{I} = (g_1, \dots, g_k)$ then, by definition,

$$\Omega_X^p = \Omega_{\mathbf{C}^m}^p / (\sum_{j=1}^{k} g_j \Omega_{\mathbf{C}^m}^p + dg_j \wedge \Omega_{\mathbf{C}^m}^{p-1}) \mid_X ,$$

and the differentiation $d : \Omega_X^p \to \Omega_X^{p-1}$, $p \geq 0$, is induced by the standard one acting on $\Omega_{\mathbf{C}^m}^\bullet$.

Suppose that X_0 is an isolated singularity of complete intersection and $\Omega_{X/S}^\bullet$ is the *relative* De Rham complex of X/S (see section 1.4). Let $f : X \to S$ be its deformation with *smooth* generic fiber. Then the base space S is also smooth. In such a case the discriminant *set* D is a *hypersurface* and all fibers X_t, $t \in S \backslash D$, are nonsingular. Let $\Omega_{X'/S'}^\bullet$ be the *relative* De Rham complex of X'/S' which can be also described in the following way

$$\Omega_{X'/S'}^\bullet = \Omega_{X'}^\bullet / f^*(\Omega_{S'}^1) \wedge \Omega_{X'}^{\bullet - 1}.$$

Then coherent sheaves $\mathcal{H}^p(X'/S') = H^p(f_* \Omega_{X'/S'}^\bullet, d)$ are well defined as the p-th cohomology group of the relative de Rham complex $\Omega_{X'/S'}^\bullet$ which are naturally isomorphic to the p-th cohomology group $H^p(X_t, \mathbb{C})$, $t \in S \setminus D$, of nonsingular fibers X_t for all $p \geq 0$. Since we consider the case of 1-dimensional germs then $H^1(X_t, \mathbb{C}) = 0$ for $p \geq 2$. The transference of the cohomology classes $\mathcal{H}^1(X'/S')$ along the tangent directions on the complement $S \setminus D$ induces a differential operator or a connexion

$$\nabla_{X'/S'} : H^1(f_* \Omega_{X'/S'}^\bullet) \longrightarrow H^1(f_* \Omega_{X'/S'}^\bullet) \otimes_{\mathcal{O}_{S'}} \Omega_{S'}^1 \qquad (1.18)$$

whose kernel coincides with the sheaf of horizontal sections of the cohomological bundle. One can describe this operator more precisely. Let ω be a holomorphic form representing a class of the *relative* de Rham cohomology in $f^{-1}(B)$, where B is a small enough ball in S'. Let $\gamma(s)$ be any continuous family of cycles over B. Then (cf. section 1) the integral

$$I(t) = \int_{\gamma(t)} \omega$$

is a holomorphic function on B. Using the Stokes formula one can verify

$$dI = \sum a_i dt_i, \quad a_i = \int_{\gamma(t)} \omega_i,$$

where $d\omega = \sum df_i \wedge \omega_i$, and ω_i are holomorphic in $f^{-1}(B)$. After some additional considerations our connection can be rewritten in the form

$$dI = \Theta I$$

where Θ is the matrix connection of ∇ and the entries of Θ are holomorphic 1-form on S' satisfying the integrability condition

$$d\Theta = \Theta \wedge \Theta.$$

Such a system is often called a local Picard-Fuchs system [23]. In the case where the germ X_0 is a *complete intersection* G.-M. Greuel [16] has proved that the connexion (1.18) is extended to the whole S :

$$\nabla_{X/S} : \mathcal{H}^1(f_*\Omega^\bullet_{X/S}) \longrightarrow \mathcal{H}^1(f_*\Omega^\bullet_{X/S}) \otimes \Omega^1_S(D) \tag{1.19}$$

given by the rule

$$\nabla_{X/S}[\omega] = \sum_{i=1}^{k} h[\alpha_i] \otimes dt_i/h,$$

where $h = 0$ is the defining equation of the discriminant set D, $\Omega^1_S(D)$ denotes the sheaf of meromorphic 1-forms with poles of the *first* order along D, the symbol $[\alpha]$ means the corresponding relative de Rham cohomology class of α in $f_*\Omega^1_{X/S}/d\,\mathcal{O}_S$ so that we have the decomposition $d\omega = \sum_{i=1}^k df_i \wedge \alpha_i, \quad \alpha_i \in f_*\Omega^1_X$. The formula (1.19) yields immediately that the Gauss-Manin connexion has poles of the first order in the case where h has no multiple factors, i.e. D is *reduced.*

Definition. A meromorphic form $\omega \in \Omega^1_S(D)$ is called a *logarithmic* differential form if and only if ω and its differential $d\omega$ have poles along D at worst of first order. That is, $h\omega$ as well as $hd\omega$ are *holomorphic* differential forms on S.

We use the notation $\Omega^1_S(\log D)$ for the $\mathcal{O}_S$-module of logarithmic differential forms. Of course, the module $\Omega^1_S(\log D)$ is a submodule of $\Omega^1_S(D)$. It should be remarked that the logarithmic differential forms have various beautiful analytic and algebraic properties (see [25], [2]). Moreover, Saito's considerations [24, 26] imply that the connexion (1.13) can be rewritten in the following way

$$\nabla_{X/S} : \mathcal{H}^1(f_*\Omega^\bullet_{X/S}) \to \mathcal{H}^1(f_*\Omega^\bullet_{X/S}) \otimes_{\mathcal{O}_S} \Omega^1_S(\log D). \tag{1.20}$$

There is a good illustration of this statement in the case of A_2-singularity. Let $F(z,s) = z^3 + s_1 z + s_0$ be the equation of the minimal versal deformation of A_2-singularity. Then its discriminant $D \subset \mathbb{C}^2$ is defined by $h = 27s_0^2 + 4s_1^3 = 0$. Set

$$I_1(s) = \int \frac{dz}{\sqrt{z^3 + s_1 z + s_0}}, \qquad I_2(s) = \int \frac{z\,dz}{\sqrt{z^3 + s_1 z + s_0}},$$

where the contour of integration is a homology cycle of the elliptic curve given by the equation $w^2 = F(z,s)$. Of course, these integrals are expressed in terms of the versal integrals (1.6) where $\lambda = -1/2$. Following K. Saito [24] we obtain the following representation of the Gauss-Manin system

$$d\mathbf{I} = \Omega \mathbf{I},$$

where

$$\Omega = \begin{pmatrix} -\frac{1}{12}\frac{dh}{h} & \frac{3}{2}\frac{\vartheta}{h} \\ -\frac{1}{2}s_1\frac{\vartheta}{h} & \frac{1}{12}\frac{dh}{h} \end{pmatrix},$$

$d\Omega = \Omega \wedge \Omega$, $\vartheta = 2s_1 ds_0 - 3s_0 ds_1$, and $\omega = \vartheta/h$ is a *logarithmic* differential form. Moreover, this equation is reduced to an equation of second order for the integral I_1 which is nothing but the classical Legendre equation (cf. section 1.8). Moreover, in this example the discriminant D is a *free divisor*, that is, $\Omega_S^1(\log D)$ is a free $\mathcal{O}_S$-module with generators dh/h and ω. In fact, due to K. Saito [26], this is true for discriminants of isolated hypersurface singularities as well as for discriminants of isolated complete intersection singularities.

A similar representation also exists in the case of A_3-singularity (see [24]). In fact, in this case the discriminant is a free divisor defined by the polynomial in three variables t_2, t_3, t_4 of weights $2, 3, 4$, respectively:

$$h = 256t_4^3 - 128t_2^2t_4^2 + 16t_2^4t_4 + 144t_2t_3^2t_4 - 4t_2^3t_3^2 - 27t_3^4 = 0.$$

This divisor is usually called the "swallow tail". On the other hand, an easy calculation show that the non-linear factor of the discriminant of the non-complete intersection (1.15) defined by the equation

$$P = 64t_4^3 - 16t_2^2t_4^2 - 72t_2t_3^2t_4 + 16t_2^3t_3^2 + 27t_3^4 = 0$$

is a free divisor also. That is why we can apply the technique of [24] developed in the case of A_3-singularity in order to compute an analogous Gauss-Manin system for the singularity (1.15) defined on the whole base space of the versal deformation (see details in [5]).

In conclusion, it should be remarked that the latter polynomial can be considered as the "twin" of the "swallow tail". In fact, it is not difficult to prove that there exist only two polynomials of degree 12 depending on 3 variables of weights $2, 3, 4$ which define free divisors and are not equivalent under weighted variable transformations. For example, the polynomial P can be transformed to

$$64t_4^3 + 48t_2^2t_4^2 - 72t_2t_3^2t_4 + 27t_3^4,$$

as well as to

$$t_4^3 - 18t_2t_3^2t_4 - 3t_2^4t_4 + 18t_2^3t_3^2 + 27t_3^4 + 2t_2^6.$$

For completeness it might be worthwhile to mention that the latter polynomial has been also found by M.Sato in an essentially different context [27].

1.7 GENERALIZED PICARD–FUCHS SYSTEMS

Now consider an example of another type. Let $D \subset \mathbb{C}^m$ be a divisor given by $h(t) = h_1(t) \dots h_n(t) = 0$ where $h_1(t), \dots, h_n(t) \in \mathbb{C}[t]$ are irreducible factors, that is,

$$D = \bigcup_{i=1}^{n} D_i, \ D_i = \{t \in \mathbf{C}^m : h_i(t) = 0\}.$$

Assume that $D_i, 1 \leq i \leq n$, are hypersurfaces whose singular loci have codimension at least two satisfying the following conditions:

1) D_i have normal crossing with D_j outside of some analytic subset D_{ij} contained in $D_i \cup D_j$ such that the codimension of D_{ij} is at least two,

2) $\dim_{\mathbf{C}} D_i \cap D_j \cap D_k \leq m-3$ for $i \neq j \neq k \neq i, \ i, j, k = 1, \dots, n.$

Then the module $\Omega^1_{\mathbb{C}^m}(\log D)$ of logarithmic differential 1-form is *generated* by

$$\frac{dh_1}{h_1}, \dots, \frac{dh_n}{h_n} \tag{1.21}$$

as an $\mathcal{O}_{\mathbb{C}^m}$-module (see [24], [25], (2.9)). In such a case, "Pfaffian system of Fuchsian type" has been defined in a quite natural manner for a set of unknown functions $\mathbf{I} = (I_1, \dots, I_k)^{tr}$:

$$d\mathbf{I} = (\sum_{j=1}^{n} A_j \frac{dh_j}{h_j})\, \mathbf{I}, \tag{1.22}$$

where $A_j \in \mathrm{End}(\mathbf{C}^k) \otimes \mathcal{O}_{\mathbf{C}^m}$. Pfaffian systems of this type were studied in [14]. However, in general, if $n > 1$ and the divisor D has irreducible components which are not hyperplanes with mutually transversal intersections or if $n = 1$ and $D = D_1$ is an irreducible hypersurface having a singular locus of codimension one, there will appear logarithmic differential forms in addition to those of principal type (1.21). More generally, there is the following exact sequence of $\mathcal{O}_{\mathbb{C}^m}$-modules [2]:

$$0 \longrightarrow \frac{dh}{h}\mathcal{O}_{\mathbb{C}^m} + \Omega^1_{\mathbb{C}^m} \longrightarrow \Omega^1_{\mathbb{C}^m}(\log D) \xrightarrow{\cdot h} \mathrm{Tors}\,\Omega^1_D \longrightarrow 0,$$

where $\Omega^1_D = \Omega^1_{\mathbb{C}^m}/(h\Omega^1_{\mathbb{C}^m} + dh \wedge \mathcal{O}_{\mathbb{C}^m})$, and $\mathrm{Tors}\,\Omega^1_D$ is the torsion submodule whose rank is equal to $m-1$. For instance, the form $\vartheta = 2s_1ds_0 - 3s_0ds_1$ of the example from the preceding section is contained in $\mathrm{Tors}\,\Omega^1_D$, where $D \subset \mathbb{C}^2$ is given by the equation $27s_0^2 + 4s_1^3 = 0$.

Thus, it is quite natural to consider the following class of Pfaffian systems with coefficients from $\Omega^1_{\mathbf{C}^m}(\log D)$:

$$d\mathbf{I} = (A\frac{dh}{h} + \sum_{j=1}^{m-1} B_j \frac{\vartheta_j}{h})\,\mathbf{I}, \tag{1.23}$$

where $\mathbf{I} = (I_1, \ldots, I_k)^{tr}$ is a vector-column of unknown functions, $\vartheta_j \in \operatorname{Tors}\Omega^1_D$ are generators of the torsion module, and $A, B_j \in \operatorname{End}(\mathbb{C}^k) \otimes \mathcal{O}_{\mathbb{C}^m}$, $j = 1, \ldots, m-1$, are coefficient matrices with holomorphic entries satisfying the standard integrability condition

$$dA \wedge \frac{dh}{h} + \sum_{j=1}^{m-1} dB_j \wedge \frac{\vartheta_j}{h} + B_j d(\frac{\vartheta_j}{h}) = \sum_{j=1}^{m-1} (B_j A - AB_j) \frac{\vartheta_j \wedge dh}{h^2}.$$

Remark that in the case where the divisor D is the union of normally crossing divisors D_i that satisfy conditions 1) and 2) mentioned above, $\mathcal{O}_D$-module of torsion differentials $\operatorname{Tors}\Omega^1_D$ is generated by $(n-1)$ differential 1-forms

$$h\frac{dh_1}{h_1}, \ldots, \widehat{h\frac{dh_i}{h_i}}, \ldots, h\frac{dh_n}{h_n},$$

where the i–th form hdh_i/h_i, $1 \le i \le m$, is to be omitted. Therefore, in this situation system (1.22) can be considered as a special case of (1.23). Thus, one may regard the system of type (1.23) as a natural generalization of "the Pfaffian system of Fuchsian type" to the case of arbitrary divisors without multiple components.

Indeed, the class of generalized Picard-Fuchs systems (1.23) contains all classical Fuchsian systems. For instance, in the case where the divisor D has normal crossing the system (1.23) is reduced to a system of Fuchsian type in the sense of R. Gerard and A.H.M. Levelt [14]. Another important example where D is the union of coordinate hyperplanes was investigated by A.A. Bolibrukh [9] in full detail.

1.8 HYPERGEOMETRIC FUNCTIONS

As we already have remarked in the Introduction in several cases representations of solutions of the Gauss-Manin systems in terms of the known types of hypergeometric functions were obtained. In connection with these results recall that the hypergeometric series

$$1 + \frac{\alpha\beta}{1\gamma}z + \frac{\alpha(\alpha+1)\beta(\beta+1)}{1\cdot 2\gamma(\gamma+1)}z^2 + \frac{\alpha(\alpha+1)(\alpha+2)\beta(\beta+1)(\beta+2)}{1\cdot 2\cdot 3\gamma(\gamma+1)(\gamma+2)}z^3 + \ldots \tag{1.24}$$

has for the first time appeared in a little bit more general form in 1769 in the second volume of the famous memoir of L. Euler [11], where the decomposition in power series of integrals of the following type

$$\int_0^1 u^{\beta-1}(1-u)^{\gamma-\beta-1}(1-zu)^{-\alpha}du$$

was investigated. As it turned out, the series (1.24) is the solution of the following differential equation of the second order

$$z(1-z)u'' + [\gamma - (\alpha+\beta+1)z]u' - \alpha\beta u = 0, \tag{1.25}$$

which is called the *hypergeometric* equation. Its second fundamental solution is expressed through the series (1.24) also. One of the special case of the equation (1.25)

$$(z^2 - 1/4)u'' + (\alpha+\beta+1)zu' + \alpha\beta u = 0 \tag{1.26}$$

is well known as the classical *Legendre equation.* On the other hand, it should be noted that the hypergeometric equation (1.25) is possible to be considered as a special case of a *Fuchsian equation* which, in its turn, belong to a class of the differential equations with *regular singularities* (see [6]).

Recall that in section 1.5 the Gauss-Manin connexion associated with the *versal* deformation $x^3 + s_1x + s_0 - y^2 = 0$ of A_2-singularity or, equivalently, the *"cusp"*-singularity is described. The discriminant set D is given by the equation $h(s) = 27s_0^2 + 4s_1^3 = 0$, and one can verify that the corresponding Gauss-Manin system (1.7) is reduced to the Legendre equation (1.26) with $\alpha = 1/6$, $\beta = 5/6$:

$$(t^2 - \frac{1}{4})u'' + 2tu' + \frac{5}{36}u = 0, \tag{1.27}$$

where the parameter t is equal to $s_0 s_1^{-3/2}$ up to a constant multiple.

On the other hand, in the first volume of his fundamental work [11] L. Euler has developed the graceful theory of integrals of the type

$$\Pi(z) = \int \frac{ez^4 + 2dz^3 + cz^2 + 2bz + a}{\sqrt{Ez^4 + 2Dz^3 + Cz^2 + 2Bz + A}} dz, \tag{1.28}$$

using only elementary (though sometimes too bulky) computational methods. Thus, it was found out that the transcendental function (1.28), depending on 10 parameters $A, B, \dots, d, e$, at some special values of these parameters turns in various known functions, such as $\log(z)$, $\arcsin(z)$ and many others. Such type of the integrals under the name *elliptic* were well known already to the mathematicians of the seventeenth century as examples of integrals which are not expressed through elementary functions and led to new transcendental functions.

After that time a huge amount of scientific works were devoted to the study of hypergeometric functions and equations, and of various generalizations of these notions. Thus, these problems were investigated by the whole galaxy of outstanding mathematicians of the following generation, such as A. Legendre, C. Gauss, N.H. Abel and C. Jacobi. In particular, C. Jacobi introduced the notion of *period integrals* when he studied integrals similar to (1.28) that contain in the denominator of the integrated expression a polynomial of fifth or higher degree (such integrals are called *ultra-elliptic,* or *hyperbolic*). At the end of the nineteenth century L. Pochhammer [20], Ju. Mellin and others investigated the integral representation of hypergeometric functions, G. Lauricella and P. Appell

[8] have found out some new classes and types of hypergeometric functions depending in several complex variables. At the present time the significant progress in this range of problems has been achieved due to the works of I.M. Gelfand and his school (see [12], [13]), K. Aomoto [7], A.N. Varchenko [29] and many others.

However, the interesting problem of the existence of new types of hypergeometric functions remains still open. In fact, the preceding example as well as the ones in sections 2,3,4 show that the period integrals can be expressed in terms of known hypergeometric functions. Hence, these integrals satisfy systems of differential equations of hypergeometric type. Nevertheless, we hope that the following *question* might be a subject of much current interest: does there exist a Gauss-Manin system associated with the versal deformation of an isolated singularity whose solutions can not be expressed in terms of hypergeometric functions of the known types and their generalizations ?

References

[1] Aleksandrov, A.G.: Deformations of one-dimensional singularities with invariants $c = \delta + 1$. Uspekhi Mat. Nauk **33** (1978), No. 3, pp. 157-158 (Russian).

[2] Aleksandrov, A.G.: Nonisolated hypersurface singularities. Advances in Soviet Math. **1**, Amer. Math. Soc., Providence, RI, 1990, 211-246.

[3] Aleksandrov, A.G., Tanabé, S.: Computing Gauss-Manin systems for complete intersection singularities S_μ. Georgian Math. J. **3** (1996), no. 5, 401-422.

[4] Aleksandrov, A.G., Tanabé, S.: Gauss-Manin connexions, logarithmic forms and hypergeometric functions. In: Geometry from the Pacific Rim (ed. J.Berrick et al.). Walter de Gruyter, Berlin – New York, 1997, 1-21.

[5] Aleksandrov, A.G., Martin, B.: Gauss-Manin connexion associated with bouquets of curve singularities. Technische Universität Cottbus, Preprint, 1998.

[6] Anosov, D.V., Bolibruch, A.A.: The Riemann-Hilbert problem. Aspects of Mathematics: E **22**, Vieweg, Braunschweig, Wiesbaden, 1994.

[7] Aomoto, K.: Configurations and invariant Gauss-Manin connections of integrals I. Tokyo J.Math. **5** (1982), 249-287.

[8] Appell, P., Kampé de Fériet, J.: Fonctions hypergeometriques et hypersphériques. Gauthier–Villars, Paris, 1926.

[9] Bolibrukh, A.A.: On the fundamental matrix of a Pfaffian system of Fuchsian type. Izv. AN SSSR, Ser. Mat. **41** (1977), No.5, 1084-1109 (Russian).

[10] Brieskorn, E.: Die Monodromie der isolierten Singularitäten von Hyperflächen. Manuscripta Math. **2** (1970), 103-161.

[11] Euler, L.: Institutionum calculi integralis, Vol. primum, secundum, tertium. Academiae Imperialis Scientiarum, Petropoli, 1768-1769.

[12] Gelfand, I.M.: General theory of hypergeometric functions. Dokl. Acad. Nauk SSSR, Ser. Mat. **291** (1986), 573-577 (Russian).

[13] Gelfand, I.M., Kapranov, M.M., Zelevinski, A.V.: Generalized Euler integrals and A-hypergeometric functions. Adv. Math. **84** (1990), No.2, 255-271.

[14] Gerard, R., Levelt, A.H.M.: Étude d'une classe particulière de système de Pfaff du type de Fuchs sur l'espace projectif complexe. J. Math. Pures Appl. **51** (1972), 189-217.

[15] Giusti, M.: Classification des singularités isolées simples d'intersections completès. Singularities, Proc. Symp. Pure Math. **40**, 1, Amer. Math. Soc., Providence, RI, 1983, 457-494.

[16] Greuel, G.–M.: Der Gauß-Manin-Zusammenhang isolierter Singularitäten von vollständigen Durchschnitten. Math. Ann. **214** (1975), 235-266.

[17] Guzev, S.: Picard-Fuchs systems for deformations of simple complete intersections. Unpublished notes. Moscow State University, Moscow, 1983 (Russian).

[18] Hamm, H.A.: Zur analytischen und algebraishen Beschreibung der Picard-Lefschetz-Monodromie. Habilitationschrift. Göttingen, 1974.

[19] Ishiura, S., Noumi, M.: A calculus of the Gauss-Manin system of type A_ℓ, I, II. Proc. Japan Acad. Ser. A Math. Sci. **58** (1982), 13-16, 62-65.

[20] Levelt, A.H.M.: Hypergeometric functions. Proc. Konikl. Nederl. Akad. Wet., Ser. A, Math. Sci. **64** (1961), 190-211.

[21] Malikov, H.M.: Redefinition of the differential systems for versal integrals of type A, D, E. Diff. Urav. **18** (1982), 1394-1400 (Russian).

[22] Noumi, M.: Expansion of the solutions of a Gauss-Manin system at a point of infinity. Tokyo J. Math. **7** (1984), 1-60.

[23] Palamodov, V.P.: Deformations of complex spaces. Itogi Nauki i Tekhn., Fund. Naprav. **10** (1985), 123–221, VINITI AN SSSR (Russian).

[24] Saito, K.: On the uniformization of complements of discriminant loci. In: Hyperfunctions and linear part. diff. eq. RIMS Kōkyūroku **287** (1977), 117-137.

[25] Saito, K.: Theory of logarithmic differential forms. J. Fac. Sci. Univ. Tokyo, Ser. IA **27** (1980), 265-291.

[26] Saito, K.: On the periods of primitive integrals I, II. Preprint, RIMS, Kyoto Univ., 1982.

[27] Sekiguchi, J.: Some topics related with discriminant polynomials. In: Daisukaisekigaku (Symposium on algebraic analysis). Abstracts, RIMS, Kyoto Univ., 1992.

[28] Tessier, D.: The hunting of invariants in the geometry of discriminants. Real and complex singularities, Oslo-76. Sijthoff-Noordhoff Publ., Alphen aan den Rijn, 1977, 565-678.

[29] Varchenko, A.N.: The Euler beta-function, the Vandermonde determinant, Legendre's equation, and critical values of linear functions on a configuration of hyperplanes, I. Izv. Akad. Nauk SSSR, Ser. Mat. **53**(1989), 1206-1235 (Russian).

2 GROUP CONVOLUTION OPERATORS ON STEP OF TWO NILPOTENT LIE GROUPS

N.L. Vasilevski

Departamento de Matemáticas
CINVESTAV del I.P.N.
México D.F.
México
nvasilev@math.cinvestav.mx

Abstract: We introduce a new technique for the investigation of group convolution operators on step of two nilpotent Lie groups, based on the direct integral construction, Weyl symbolic calculus for pseudodifferential operators and symplectic changes of variables. It allows to study at the same time not only both the right and the left convolution operators, but it turns out to be useful to consider the so-called two-sided convolutions. The structure of these group convolution operators is described. Some applications to complex analysis are given.

2.1 INTRODUCTION

The classical approach to the investigation of convolution operators on the real line is based on the use of the Fourier transformation. The usual step under consideration of group convolution operators on the nilpotent Lie groups is to generalize the classical Fourier transformation up to appropriate group Fourier transformation and to use afterwards the machinery of non-commutative harmonic analysis.

The investigation of group convolution operators on step of two nilpotent Lie groups and pseudodifferential operators connected with them has a sufficiently long history by this time. We mention here the works [1, 2, 3, 5, 7, 8, 10, 11, 12] only, which have extensive bibliography.

H.G.W. Begehr et al. (eds.), Partial Differential and Integral Equations, 23–35.

But the classical Fourier transformation gives an example of symplectic change of variables in $T^*\mathbb{R}$ as well. Our aim is to introduce a new technique for the investigation of group convolution operators on step two nilpotent Lie groups, originated on the symplectic nature of the classical Fourier transformation. This technique is based on the direct integral construction, Weyl symbolic calculus for pseudodifferential operators and symplectic changes of variables. It allows to consider at the same time not only both the right and the left convolution operators (which is impossible following non-commutative harmonic analysis line), but it turns out to be useful to consider the so-called two-sided convolutions.

The structure of these group convolution operators is described. The boundedness results extend the known ones onto the case of noncontinuous Fourier image of the convolution operator kernel with respect to the "real" variable $\lambda \in \mathbb{R}^m$.

The groups under consideration appear naturally in complex analysis. They can be identified with the so-called standard *CR*-manifolds, and in some cases with the distinguished boundary of type II Siegel domains. The simple and more evident than, for example, in [4,6,9] isomorphic description of the Hardy space and of the Szegö projection associated with the type II Siegel domains is given. For the case of the Heisenberg group the algebra generated by Toeplitz operators associated with the corresponding Siegel domain is described as well.

This work is an expanded version of the talk given at the ISAAC Congress, Newark, Delaware, June 3–7, 1997, and presents a condensed survey of the results of [13, 14].

2.2 MAIN DEFINITIONS

Denote by $\mathbb{H}_F$ the space $\mathbb{R}^N = \mathbb{R}^m \times \mathbb{C}^n$, $N = m + 2n$, with the following group operation

$$(t_1, z_1)(t_2, z_2) = (t_1 + t_2 + 2\,\mathrm{Im}\, F(z_1, z_2), z_1 + z_2),$$

where $(t_k, z_k) \in \mathbb{R}^m \times \mathbb{C}^n$, $k = 1, 2$, and $F(z_1, z_2)$ is a Hermitian form defined on $\mathbb{C}^n$ and valued in $\mathbb{C}^m$.

It is known that the group $\mathbb{H}_F$ is a step two nilpotent Lie group and its (two-sided) Haar measure coincides with the Lebesque measure on $\mathbb{R}^N$. For $m = 1$ and $F(z_1, z_2) = z_1\overline{z}_2$ $\mathbb{H}_F$ is just the Heisenberg group $\mathbb{H}_n$.

The noncommutativity of the group operation induces (in contrast to the standard summation operation in $\mathbb{R}^N$) the existence of two shift operators: the right and the left shift operators that are defined on $L_2(\mathbb{H}_n)$ by the rule

$$\begin{aligned}(\pi_r(g)\varphi)(h) &= \varphi(hg),\\ (\pi_l(g)\varphi)(h) &= \varphi(g^{-1}h).\end{aligned}$$

Integrating these shift operators with some kernel functions we arrive at the right and the left group convolution operators

$$B = (2\pi)^{-\frac{N}{2}} \int_{\mathbb{H}_F} b(g)\pi_r(g)dg,$$

$$C = (2\pi)^{-\frac{N}{2}} \int_{\mathbb{H}_F} c(g)\pi_l(g)dg.$$

Moreover, it turns out to be useful to consider the so-called two-sided convolution operator

$$K = (2\pi)^{-N} \int_{\mathbb{H}_F \times \mathbb{H}_F} k(g_1, g_2)\pi_r(g_2)\pi_l(g_2)dg_1dg_2.$$

We will use the following notations for the direct and the dual coordinates of the points of $\mathbb{H}_F = \mathbb{R}^N = \mathbb{R}^m \times \mathbb{C}^n = \mathbb{R}^m \times \mathbb{R}^{2n} = \mathbb{R}^m \times \mathbb{R}^n \times \mathbb{R}^n$:

$$g = (t, z) = (t, \tilde{x}) = (t, x, y)$$

and

$$\gamma = (\lambda, \zeta) = (\lambda, \tilde{\xi}) = (\lambda, \xi, \eta).$$

Here, for example, $z = x + iy \in \mathbb{C}^n$, $\tilde{x} = (x, y) \in \mathbb{R}^{2n}$, $t \in \mathbb{R}^m$.

2.3 PSEUDODIFFERENTIAL FORM OF CONVOLUTION OPERATORS

The center of $\mathbb{H}_F$ has the form

$$\mathcal{Z} = \{(t, z) \in \mathbb{H}_F : z = 0\}.$$

All the operators $\pi_l(g_0)$, with $g_0 \in \mathcal{Z}$, commute with both the right and the left shift operators. And so they commute with all bounded on $L_2(\mathbb{H}_F) = L_2(\mathbb{R}^N)$ right, left and two-sided convolution operators.

We have

$$(\pi_l(g_0)f)(t, z) = f(t - t_0, z) = (\tau_{t_0} \otimes I)f,$$

where $(\tau_{t_0}\varphi)(t) = \varphi(t - to)$, is acting on the space $L_2(\mathbb{R}^m)$ shift operator, $L_2(\mathbb{H}_F) = L_2(\mathbb{R}^m) \otimes L_2(\mathbb{C}^n)$.

Commuting with $\pi_l(g_0)$ operators B, C and K should commute with all elements of the von Neumann algebra $\mathfrak{A}$ generated by these shift operators π_l. Hence these convolution operators belong to the commutant $\mathfrak{A}'$ of the algebra $\mathfrak{A}$.

We have

$$\mathfrak{A} = F^{-1}L_\infty(\mathbb{R}^m)F \otimes \mathbb{C} \cdot I$$

and

$$\begin{aligned} \mathfrak{A}' &= F^{-1}L_\infty(\mathbb{R}^m)F \otimes \mathfrak{B}(L_2(\mathbb{C}^n)) \\ &= (F^{-1} \otimes I)(L_\infty(\mathbb{R}^m) \otimes \mathfrak{B}(L_2(\mathbb{C}^n)))(F \otimes I), \end{aligned}$$

where $\mathfrak{B}(X)$ is the algebra of all bounded linear operators acting on a Hilbert space X.

Lemma 1 *Let $K \in \mathfrak{B}(L_2(\mathbb{H}_F))$ be an operator commuting with all the operators $\pi_l(g_0)$, for $g_0 \in \mathcal{Z}$. Then:*

(i) $K \in (F^{-1} \otimes I)(L_\infty(\mathbb{R}^m) \otimes \mathfrak{B}(L_2(\mathbb{C}^n)))(F \otimes I)$;

(ii) there exists an essentially bounded measurable operator–function

$$K(\cdot) : \lambda \in \mathbb{R}^m \longmapsto K(\lambda) \in \mathfrak{B}(L_2(\mathbb{C}^n)),$$

which is determined by the operator K with respect to a zero–measure set and which determines the following acting on the space $L_2(\mathbb{H}_F) = L_2(\mathbb{R}^m) \otimes L_2(\mathbb{C}^n)$ operator:

$$\tilde{K} f(\lambda, x) = (I \otimes K(\lambda)) f(\lambda, x);$$

(iii) $K = (F^{-1} \otimes I)\tilde{K}(F \otimes I)$;

(iv) $\|K\| = \text{ess-sup}_{\lambda \in \mathbb{R}^m} \|K(\lambda)\|$.

Remark 2 The values of an operator–function $K(\lambda)$ almost everywhere in $\mathbb{R}^m$ are obtained by applying the partial Fourier transformation with respect to the variable t $(t \to \lambda)$ and by fixing then the value of λ i.e.

$$I \otimes K(\lambda) = ((F \otimes I)K(F^{-1} \otimes I))(\lambda).$$

Let $(F_1, F_2, \ldots, F_m)$ be the coordinate notation of the vector form F. Here F_ν, $\nu = 1, \ldots, m$, are the scalar Hermitian forms in $\mathbb{C}^n$:

$$F_\nu(z_1, z_2) = z_1' F_\nu \bar{z}_2, \quad \nu = 1, \ldots, m.$$

For $\lambda \in \mathbb{R}^m$ denote by $L(\lambda)$ the following scalar Hermitian form

$$L(\lambda)(z_1, z_2) = \lambda F(z_1, z_2) = \sum_{\nu=1}^{m} \lambda_\nu F_\nu(z_1, z_2).$$

Introduce the $2n \times 2n$ (real) matrix

$$\Lambda(\lambda) = \begin{pmatrix} \text{Im } L(\lambda) & -\text{Re } L(\lambda) \\ \text{Re } L(\lambda) & \text{Im } L(\lambda) \end{pmatrix}.$$

Calculations show that for almost all $\lambda \in \mathbb{R}^m$

$$K(\lambda)\varphi(\tilde{x}) = (2\pi)^{-2n} \int_{\mathbb{R}^{4n}} \hat{k}\left(-\lambda, -\tilde{\xi} + 2\Lambda\left(\frac{\tilde{x}+\tilde{y}}{2}\right), \lambda, \tilde{\xi} + 2\Lambda\left(\frac{\tilde{x}+\tilde{y}}{2}\right)\right)$$

$$\times \; e^{i(\tilde{x}-\tilde{y})\cdot\tilde{\xi}} \varphi(\tilde{y})\, d\tilde{y}\, d\tilde{\xi},$$

where $\hat{k}$ is the (Euclidean) Fourier transformation of k.

Lemma 3 *The Weyl symbols of the operators $B(\lambda)$, $C(\lambda)$ and $K(\lambda)$ (corresponding to the right, left and two-sided convolution operators on $\mathbb{H}_F$) have the forms*

$$\begin{aligned} \omega_{B(\lambda)} &= \hat{b}(-\lambda, -\xi + 2\text{Im } L(\lambda)\bar{z}, -\eta + 2\text{Re } L(\lambda)\bar{z}), \\ \omega_{C(\lambda)} &= \hat{c}(\lambda, \xi + 2\text{Im } L(\lambda)\bar{z}, \eta + 2\text{Re } L(\lambda)\bar{z}), \\ \omega_{K(\lambda)} &= \hat{k}(-\lambda, -\xi + 2\text{Im } L(\lambda)\bar{z}, -\eta + 2\text{Re } L(\lambda)\bar{z}, \\ &\qquad \lambda, \xi + 2\text{Im } L(\lambda)\bar{z}, \eta + 2\text{Re } L(\lambda)\bar{z}). \end{aligned}$$

Theorem 4 *The right, left and the two-sided convolution operators are pseudodifferential operators on $L_2(\mathbb{H}_F) = L_2(\mathbb{R}^N)$ having the symbols*

$$\begin{aligned}
\omega_B(t,\tilde{x},\lambda,\tilde{\xi}) &= \hat{b}(-\lambda,-\tilde{\xi}+2\Lambda(\lambda)\tilde{x}),\\
\omega_C(t,\tilde{x},\lambda,\tilde{\xi}) &= \hat{c}(\lambda,\tilde{\xi}+2\Lambda(\lambda)\tilde{x}),\\
\omega_K(t,\tilde{x},\lambda,\tilde{\xi}) &= \hat{k}(-\lambda,-\tilde{\xi}+2\Lambda(\lambda)\tilde{x},\lambda,\tilde{\xi}+2\Lambda(\lambda)\tilde{x}).
\end{aligned}$$

2.4 STRUCTURE OF OPERATORS $K(\lambda)$

Let a point $\lambda \in \mathbb{R}^m \setminus \{0\}$ be fixed. Consider the operator

$$K(\lambda)\varphi(\tilde{x}) = (2\pi)^{-2n} \int_{\mathbb{R}^{4n}} e^{i(\tilde{x}-\tilde{y})\cdot\tilde{\xi}} \omega_{K(\lambda)}(\frac{1}{2}(\tilde{x}+\tilde{y}),\,\tilde{\xi})\,\varphi(\tilde{y})\,d\tilde{y}\,d\tilde{\xi}$$

bounded on $L_2(\mathbb{R}^{2n}) = L_2(\mathbb{C}^n)$ with the Weyl symbol

$$\omega_{K(\lambda)}(\tilde{x},\tilde{\xi}) = \hat{k}(-\lambda,\ -\tilde{\xi}+2\Lambda(\lambda)\tilde{x};\ \lambda,\ \tilde{\xi}+2\Lambda(\lambda)\tilde{x}).$$

The Hermitian matrix $L(\lambda)$ is homogeneous of the first order

$$L(\lambda) = |\lambda| L\left(\frac{\lambda}{|\lambda|}\right),$$

using some unitary matrix $U = U\left(\frac{\lambda}{|\lambda|}\right)$ it can be reduced to a diagonal form

$$U(\frac{\lambda}{|\lambda|})\,L(\frac{\lambda}{|\lambda|})\,U(\frac{\lambda}{|\lambda|})^* = \operatorname{diag}\left(\varepsilon_1^2,\dots,\varepsilon_{p,}^2\,-\varepsilon_{p+1}^2,\dots,-\varepsilon_{n-k,}^2\underbrace{0,\dots,0}_{k}\right).$$

Introduce in $T^*(\mathbb{C}^n) = T^*(\mathbb{R}^{2n})$ the following symplectic changes of variables

$$z = U'z_1, \qquad \zeta = U'\zeta_1,$$

or

$$\begin{aligned}
x &= U_1'x_1 - U_2'y_1, & y &= U_2'x_1 + U_1'y_1,\\
\xi &= U_1'\xi_1 - U_2'\eta_1, & \eta &= U_2'\xi_1 + U_1'\eta_1,
\end{aligned}$$

where $U = U_1 + iU_2$, and for $j = 1,\dots,p$

$$\begin{aligned}
x_j &= -\frac{1}{2\sqrt{|\lambda|}\varepsilon_j}\eta_{1,j} + \sqrt{|\lambda|}\varepsilon_j x_{1,j},\\
y_j &= \frac{1}{2\sqrt{|\lambda|}\varepsilon_j}\eta_{1,j} + \sqrt{|\lambda|}\varepsilon_j x_{1,j},\\
\xi_j &= \frac{1}{2\sqrt{|\lambda|}\varepsilon_j}\xi_{1,j} + \sqrt{|\lambda|}\varepsilon_j y_{1,j},\\
\eta_j &= \frac{1}{2\sqrt{|\lambda|}\varepsilon_j}\xi_{1,j} - \sqrt{|\lambda|}\varepsilon_j y_{1,j},
\end{aligned}$$

for $j = p+1, \ldots, n-k$

$$\begin{aligned} x_j &= -\frac{1}{2\sqrt{|\lambda|}\varepsilon_j}\eta_{1,j} - \sqrt{|\lambda|}\varepsilon_j x_{1,j}, \\ y_j &= \frac{1}{2\sqrt{|\lambda|}\varepsilon_j}\eta_{1,j} - \sqrt{|\lambda|}\varepsilon_j x_{1,j}, \\ \xi_j &= -\frac{1}{2\sqrt{|\lambda|}\varepsilon_j}\xi_{1,j} + \sqrt{|\lambda|}\varepsilon_j y_{1,j}, \\ \eta_j &= -\frac{1}{2\sqrt{|\lambda|}\varepsilon_j}\xi_{1,j} - \sqrt{|\lambda|}\varepsilon_j y_{1,j}, \end{aligned}$$

for $j = n-k+1, \ldots, n$

$$x_j = \frac{1}{2\sqrt{|\lambda|}}\xi_{1,j}, \quad y_j = \frac{1}{2\sqrt{|\lambda|}}\eta_{1,j},$$

$$\xi_j = -2\sqrt{|\lambda|}x_{1,j}, \quad \eta_j = -2\sqrt{|\lambda|}y_{1,j}.$$

For $z = x + iy \in \mathbb{C}^n$, introduce $x = (x', x'')$, $y = (y', y'') \in \mathbb{R}^{n-k} \times \mathbb{R}^k$ and denote by $\xi', \eta' \in \mathbb{R}^{n-k}$ the dual coordinates for x' and y'. Now

$$L_2(\mathbb{C}^n) = L_2(\mathbb{R}^{n-k}) \otimes L_2(\mathbb{R}^{2k}) \otimes L_2(\mathbb{R}^{n-k}),$$

and the point

$$(x', x'', y'', y') \in \mathbb{R}^{n-k} \times \mathbb{R}^k \times \mathbb{R}^k \times \mathbb{R}^{n-k} = \mathbb{R}^{n-k} \times \mathbb{R}^{2k} \times \mathbb{R}^{n-k}$$

corresponds to the point $z = x + iy \in \mathbb{C}^n$.

Introduce n–dimensional complex vectors $\hat{z}_r$ and $\hat{z}_l$ by

$$\hat{z}_{r,j} = \begin{cases} x_j + i\xi_j & = x'_j + i\xi'_j, \\ x_j - i\xi_j & = x'_j - i\xi'_j, \\ x_j + iy_j & = x''_{j-(n-k)} + iy''_{j-(n-k)}, \end{cases}$$

$$\hat{z}_{l,j} = \begin{cases} y_i - i\eta_j & = y'_j - i\eta'_j, \\ y_i + i\eta_j & = y'_j + i\eta'_j, \\ x_j + iy_j & = x''_{j-(n-k)} + iy''_{j-(n-k)}, \end{cases}$$

and two $n \times n$ complex matrices

$$\begin{aligned} \hat{U}_r &= \hat{U}_r\left(\frac{\lambda}{|\lambda|}\right) = 2iU'\mathrm{diag}\,(\varepsilon_1, \ldots, \varepsilon_{n-k}, i, \ldots, i), \\ \hat{U}_l &= \hat{U}_l\left(\frac{\lambda}{|\lambda|}\right) = 2iU'\mathrm{diag}\,(\varepsilon_1, \ldots, \varepsilon_{n-k}, -i, \ldots, -i). \end{aligned}$$

Theorem 5 *Let $\lambda \in \mathbb{R}^m \setminus \{0\}$. Then there exists a unitary operator ν on $L_2(\mathbb{C}^n)$ such that the Weyl symbol of the operator $\nu\, K(\lambda)\, \nu^{-1}$ is calculated by the formula*

$$\omega_{\nu\, K(\lambda)\, \nu^{-1}} = \hat{k}\left(-\lambda, \sqrt{|\lambda|}\hat{U}_r\left(\frac{\lambda}{|\lambda|}\right)\hat{z}_r, \lambda, \sqrt{|\lambda|}\hat{U}_l\left(\frac{\lambda}{|\lambda|}\right)\hat{z}_l\right).$$

Corollary 6 *In the case of the Heisenberg group $\mathbb{H}_n$ we have $k = 0$,*

$$L_2(\mathbb{C}^n) = L_2(\mathbb{R}^n) \otimes L_2(\mathbb{R}^n)$$

and

$$\begin{aligned} \nu\, B(\lambda)\, \nu^{-1} &= B_\lambda \otimes I, \\ \nu\, C(\lambda)\, \nu^{-1} &= I \otimes C_\lambda, \end{aligned}$$

where the Weyl symbols of the operators B_λ and C_λ acting on $L_2(\mathbb{R}^n)$ are calculated by the formulas

$$\begin{aligned} \omega_{B_\lambda}(x,\xi) &= \hat{b}\,(-\lambda,\, -2\sqrt{|\lambda|}x,\, -2(\operatorname{sign}\lambda)\sqrt{|\lambda|}\xi), \\ \omega_{C_\lambda}(y,\eta) &= \hat{c}\,(\lambda,\, 2(\operatorname{sign}\lambda)\sqrt{|\lambda|}\eta,\, 2\sqrt{|\lambda|}y), \end{aligned}$$

and where

$$\begin{aligned} \omega_B &= \hat{b}\,(-\lambda,\, -\xi - 2\lambda y,\, -\eta + 2\lambda x), \\ \omega_C &= \hat{c}\,(\lambda,\, \xi - 2\lambda y,\, \eta + 2\lambda x) \end{aligned}$$

are the Weyl symbols of the right B and left C convolution operators on the Heisenberg group $\mathbb{H}_n$.

2.5 BOUNDEDNESS OF CONVOLUTION OPERATORS

Introduce on $\mathbb{H}_F = \mathbb{R}^m \times \mathbb{C}^n = \mathbb{R}^m \times \mathbb{R}^{2n}$ the one–parameter group of the nonisotropic dilations $\delta_\tau, \tau > 0$,

$$\delta_\tau(\tau,\zeta) = (\tau\lambda, \tau^{\frac{1}{2}}\zeta), \quad (\lambda,\zeta) \in \mathbb{R}^m \times \mathbb{C}^n,$$

or

$$\delta_\tau(\lambda,\tilde{\xi}) = (\tau\lambda, \tau^{\frac{1}{2}}\tilde{\xi}), \quad (\lambda,\tilde{\xi}) \in \mathbb{R}^m \times \mathbb{R}^{2n}.$$

A function ψ on $\mathbb{H}_F$ is called a δ_τ–homogeneous function of degree $s \in \mathbb{R}$ if

$$\psi \circ \delta_\tau = \tau^s \psi.$$

The values of an arbitrary δ_τ–homogeneous function ψ is determined uniquely by its values on the nonisotropic unit sphere Ω in $\mathbb{H}_F$

$$\Omega = \{(\lambda,\tilde{\xi}) \in \mathbb{H}_F = \mathbb{R}^m \times \mathbb{R}^{2n} : \quad ||(\lambda,\tilde{\xi})|| = 1\},$$

where $||(\lambda,\tilde{\xi})||^2=\sum_{\nu=1}^{m}|\lambda_\nu|^2+\sum_{k=1}^{n}(|\xi_k|^4+|\eta_k|^4)$.

For an algebra $\mathcal{A}$ ($\subset L_\infty(\Omega)$) of functions on Ω denote by $H_\delta^s(\mathcal{A})$ the algebra of all functions in $\mathbb{H}_F$ δ_τ-homogeneous of degree s and having $\mathcal{A}$ restrictions onto Ω.

Theorem 7 *Let E be a closed subset in $\mathbb{R}^m$ with the measure zero and with the "cone" property*

$$\forall\lambda\in E \qquad \forall\tau>0 \qquad \tau\lambda\in E,$$

and let $E_\Omega=\Omega\cap(E\times\mathbb{R}^{2n})$. Further let $\tilde{b}(-\lambda,\tilde{\xi})$, $\tilde{c}(\lambda,\tilde{\xi})$ be functions from $H_\delta^0(C^\infty(\Omega\backslash E_\Omega))$, and $\tilde{k}(-\lambda,\tilde{\xi},\lambda,\tilde{\xi})\in H_\delta^0(C^\infty(\Omega\backslash E_\Omega))\otimes H_\delta^0(C^\infty(\Omega\backslash E_\Omega))$. Then pseudodifferential operators B,C and K with the Weyl symbols

$$\begin{aligned}\omega_B(\tilde{x},\lambda,\tilde{\xi}) &= \tilde{b}(-\lambda,-\tilde{\xi}+2\Lambda(\lambda)\tilde{x}),\\ \omega_C(\tilde{x},\lambda,\tilde{\xi}) &= \tilde{c}(\lambda,\tilde{\xi}+2\Lambda(\lambda)\tilde{x})\end{aligned}$$

and

$$\omega_K(\tilde{x},\lambda,\tilde{\xi})=\tilde{k}(-\lambda,-\tilde{\xi}+2\Lambda(\lambda)\tilde{x};\lambda,\tilde{\xi}+2\Lambda(\lambda)\tilde{x}),$$

respectively, are bounded on $L_2(\mathbb{H}_F)$.

Each of the operators, say $A\in\{B,C,K\}$, generates an operator–function

$$A(\cdot):\quad \mathbb{R}^m\to\mathfrak{B}(L_2(\mathbb{R}^{2n}))$$

which is essentially bounded and measurable in $\mathbb{R}^m$ and is strongly continuous in $\mathbb{R}^m\backslash E$. The operator

$$\tilde{A}f(\lambda,\tilde{\xi})=(I\otimes A(\lambda))f(\lambda,\tilde{\xi})$$

defined by $A(\cdot)$ is bounded on $L_2(\mathbb{R}^m\times\mathbb{R}^{2n})$ and is unitary equivalent to the operator A:

$$A=(F^{-1}\otimes I)\tilde{A}(F\otimes I),$$

where F and F^{-1} are direct and inverse Fourier transformations on $L_2(\mathbb{R}^m)$.

And

$$||A||=\sup_{\lambda\in\mathbb{R}^m\backslash E}||A(\lambda)||.$$

Remark 8 The Weyl symbol of the operator $A(\lambda)$ for $\lambda\in\mathbb{R}^m\backslash E$ is calculated by the formula

$$\omega_{A(\lambda)}(\tilde{x},\tilde{\xi})=\omega_A(\tilde{x},\lambda,\tilde{\xi}).$$

Remark 9 For this important case, when the functions $\tilde{b}$, $\tilde{c}$ and $\tilde{k}$ are δ_τ-homogeneous of zero order, the formulas of Theorem 5 and Corollary 6 acquire more simple and nice form

$$\begin{aligned}\omega_{B_\lambda}(x,\xi) &= \tilde{b}(-\operatorname{sign}\lambda,\ -2x,\ -2(\operatorname{sign}\lambda)\xi),\\ \omega_{C_\lambda}(y,\eta) &= \tilde{c}(\operatorname{sign}\lambda,\ 2(\operatorname{sign}\lambda)\eta,\ 2y),\\ \omega_{\nu K(\lambda)\nu^{-1}} &= \tilde{k}\left(-\frac{\lambda}{|\lambda|},\hat{U}_r\left(\frac{\lambda}{|\lambda|}\right)\hat{z}_r,\frac{\lambda}{|\lambda|},\hat{U}_l\left(\frac{\lambda}{|\lambda|}\right)\hat{z}_l\right).\end{aligned}$$

2.6 HARDY SPACE AND SZEGÖ PROJECTION ASSOCIATED WITH SIEGEL DOMAINS

The distinguished boundary of the type II Siegel domain

$$D = \{(w,z) \in \mathbb{C}^{m+n} : \ \mathrm{Im}\, w - F(z,z) \in V\},$$

has the form

$$B = \{(w,z) \in \mathbb{C}^{m+n} : \ \mathrm{Im}\, w - F(z,z) = 0\}.$$

The mapping

$$\kappa_0 : \ (t,z) \in \mathbb{R}^m \times \mathbb{C}^n \ \longmapsto \ (t + iF(z,z),\ z) \in B$$

identifies $\mathbb{R}^m \times \mathbb{C}^n = \mathbb{H}_F$ with B, and the corresponding mapping

$$\kappa : \ L_2(B) \ \longrightarrow \ L_2(\mathbb{R}^m \times \mathbb{C}^n) = L_2(\mathbb{H}_F),$$

where $(\kappa f)(t,z) = f(\kappa_0(t,z))$, transfers isometrically the Hardy space $H^2(D)$ onto some subspace $H^2(\mathbb{H}_F) = \kappa(H^2(D))$ of $L_2(\mathbb{H}_F)$.

The space $H^2(\mathbb{H}_F)$ is the closure in $L_2(\mathbb{H}_F)$ of the linear vector space of functions which satisfy the tangential Cauchy–Riemann equations on $\mathbb{H}_F$:

$$\overline{Z} f_0 = \left(\frac{\partial}{\partial \overline{z}} - i\, \overline{L(\frac{\partial}{\partial t})\, z} \right) f_0 = 0.$$

The orthogonal projection $P : \ L_2(\mathbb{H}_F) \ \longrightarrow \ H^2(\mathbb{H}_F)$ is a right group convolution operator on $\mathbb{H}_F$

$$P = (2\pi)^{-\frac{N}{2}} \int_{\mathbb{H}_F} p(g)\, \pi_r(g)\, dg$$

with the Weyl symbol

$$\omega_P(\tilde{x}, \lambda, \tilde{\xi}) = \hat{p}(-\lambda,\ -\tilde{\xi} + 2\Lambda(\lambda)\tilde{x}),$$

where $\hat{p}$ is a Fourier transformation of p, or (in the complex form)

$$\omega_P(z, \lambda, \zeta) = \hat{p}(-\lambda,\ 2i\,\omega_{\overline{Z}}(z, \lambda, \zeta)),$$

where $\omega_{\overline{Z}}(z,\lambda,\zeta) = \frac{1}{2} i\zeta + \overline{L(\lambda)} z$ is a Weyl symbol of the tangential Cauchy–Riemann operator $\overline{Z}$.

The operator P generates an essentially bounded measurable operator–function

$$P(\cdot) : \ \mathbb{R}^m \ \longrightarrow \ \mathfrak{B}(L_2(\mathbb{C}^n))$$

such that

$$P = (F^{-1} \otimes I)\,(I \otimes P(\lambda))\,(F \otimes I),$$

and for all $\lambda \in \overline{\mathbb{R}^m \setminus V^*}$ $P(\lambda) = 0$, for $\lambda \in \operatorname{Int} V^*$ the Weyl symbol of the operator $P(\lambda)$ has the form

$$\omega_{P(\lambda)}(z, \zeta) = \hat{p}(-\lambda, 2i\,\omega_{\overline{Z}(\lambda)}(z, \zeta)),$$

where $\omega_{\overline{Z}(\lambda)}(z, \zeta)$ is a Weyl symbol of the operator

$$\overline{Z}(\lambda) = \frac{\partial}{\partial \overline{z}} + \overline{L(\lambda)} z,$$

and V^* is the dual cone to the cone V.

The operator $P(\lambda)$ is an orthogonal projection of $L_2(\mathbb{C}^n)$ onto the space $H^2(\lambda)$ which is a closure in $L_2(\mathbb{C}^n)$ of a linear vector space of functions satisfying to the equation $\overline{Z}(\lambda)\varphi = 0$.

We have

$$\nu \circ \overline{Z}(\lambda) \circ \nu^{-1} = \sqrt{|\lambda|}\, E \left(\frac{\partial}{\partial x} + x \right),$$

and the operator ν maps isometrically $H^2(\lambda)$ onto the space $\nu_1(\lambda)\,(H^2(\lambda)) = H_1^2(\lambda)$ which is the closure in $L_2(\mathbb{C}^n)$ of a linear vector space of functions satisfying to the equation

$$\left(\frac{\partial}{\partial x} + x \right) \varphi(x, y) = 0.$$

Its general solution has the form

$$\pi^{-\frac{n}{4}} e^{-\frac{x^2}{2}} \cdot g(y),$$

where $g(y) \in L_2(\mathbb{R}^n)$. The function $l(x) = \pi^{-n/4} e^{-x^2/2} \in L_2(\mathbb{R}^n)$ and $\|l(x)\| = 1$.

Denote by L_0 the one-dimensional subspace of $L_2(\mathbb{R}^n)$ generated by $l(x)$. Then $H_1^2(\lambda) = L_0 \otimes L_2(\mathbb{R}^n)$ and $P_1(\lambda) = P_0 \otimes I$, where

$$(P_0 f)(x) = \pi^{-n/2} \int_{\mathbb{R}^n} e^{-(x^2+u^2)/2} f(u)\, du$$

is the orthogonal projection of $L_2(\mathbb{R}^n)$ onto L_0.

Theorem 10 *There exists a unitary operator ν_0 which gives an isometrical isomorphism of the space*

$$L_2(\mathbb{R}^m \times \mathbb{C}^n) = L_2(\mathbb{R}^m) \otimes L_2(\mathbb{R}^n) \otimes L_2(\mathbb{R}^n)$$

under which

(i) the space $H^2(\mathbb{H}_F)$ is mapped onto

$$L_2(V^*) \otimes L_0 \otimes L_2(\mathbb{R}^n);$$

(ii) $\nu P \nu^{-1} = \chi_{V^*}(\lambda) I \otimes P_0 \otimes I$, *where* $\chi_{V^*}(\lambda)$ *is the characteristic function of the dual cone* V^*.

Corollary 11 *The operator* $\nu_D = \nu_0 \circ \kappa$ *maps isometrically* $L_2(B)$ *onto*

$$L_2(\mathbb{R}^m) \otimes L_2(\mathbb{R}^n) \otimes L_2(\mathbb{R}^n)$$

and under this mapping we have

(i) $\nu_D(H^2(D)) = L_2(V^*) \otimes L_0 \otimes L_2(\mathbb{R}^n)$;

(ii) $\nu_D P_D \nu_D^{-1} = \chi_{V^*}(\lambda) I \otimes P_0 \otimes I$.

2.7 TOEPLITZ OPERATORS ON THE HEISENBERG GROUP

Consider

$$D = \{(w, z) \in \mathbb{C} \times \mathbb{C}^n : \mathrm{Im}\, w - z \cdot \overline{z} > 0\},$$

its distinguished boundary

$$B = \{(w, z) \in \mathbb{C} \times \mathbb{C}^n : \mathrm{Im}\, w - z \cdot \overline{z} = 0\}$$

after identification with $\mathbb{R} \times \mathbb{C}^n$ coincides with the Heisenberg group $\mathbb{H}_n$.

Denote by χ_+ the characteristic function of the "upper half sphere", i.e.

$$\chi_+(\lambda, \xi, \eta) = \begin{cases} 1, & \lambda \geq 0 \\ 0, & \lambda < 0, \end{cases}$$

where $(\lambda, \xi, \eta) \in \Omega$, and by $PC^\infty(\Omega)$ the algebra generated by χ_+ and functions from $C^\infty(\Omega)$.

And let $\mathcal{R} = \mathcal{R}(H^0_\delta(PC(\Omega)))$ the C^*–algebra generated by all the operators B and C with

$$\begin{aligned} \omega_B(\tilde{x}, \lambda, \tilde{\xi}) &= \tilde{b}(-\lambda, -\tilde{\xi} + 2\Lambda(\lambda)\tilde{x}), \\ \omega_C(\tilde{x}, \lambda, \tilde{\xi}) &= \tilde{c}(\lambda, \tilde{\xi} + 2\Lambda(\lambda)\tilde{x}), \end{aligned}$$

where $\tilde{b}, \tilde{c} \in H^0_\delta(PC^\infty(\Omega))$.

The algebra $\mathcal{R}$ is generated also by all operators K with

$$\tilde{k} \in H^0_\delta(PC^\infty(\Omega)) \otimes H^0_\delta(PC^\infty(\Omega)).$$

Denote by $CS(\mathbb{R}^k)$ the algebra of functions in $\mathbb{R}^k$ generated by

$$s_0(x) = \frac{1}{\sqrt{1+x^2}}, \quad s_j = \frac{x_j}{\sqrt{1+x^2}}, \quad j = 1, \ldots, k, \quad x \in \mathbb{R}^k,$$

and let $\Psi(CS(\mathbb{R}^{2n}))$ be the C^*–algebra generated by all operators with (sufficiently smooth) Weyl symbols $a(x, \xi) \in CS(\mathbb{R}^{2n})$ acting on the space $L_2(\mathbb{R}^n)$.

Denote by $\mathcal{T}(\mathcal{R})$ the algebra generated by all Toeplitz operators

$$T_K : \quad f \in H^2(\mathbb{H}_n) \longmapsto PKf \in H^2(\mathbb{H}_n)$$

with presymbols K from the algebra $\mathcal{R} = \mathcal{R}(H^0_\delta(PC(\Omega)))$.

Theorem 12 *The algebra $\mathcal{T}(\mathcal{R})$ is isometrically isomorphic to the algebra $\Psi(CS(\mathbb{R}^{2n}))$. The isomorphism*

$$\tau : \quad \mathcal{T}(\mathcal{R}) \to \Psi(CS(\mathbb{R}^{2n}))$$

is generated by the following mapping of generators of $\mathcal{T}(\mathcal{R})$:

$$\tau : \quad T_K \longmapsto \hat{K},$$

where the Weyl symbol of the operator $\hat{K} \in \Psi(CS(\mathbb{R}^{2n}))$ is calculated by the formula

$$\omega_{\hat{K}}(y,\eta) = \pi^{-n} \int_{\mathbb{R}^{2n}} e^{-(x^2+\xi^2)} \tilde{k}(-1, -2x, -2\xi; 1, 2\eta, 2y) dx d\xi,$$

where

$$\omega_K(\tilde{x}, \lambda, \tilde{\xi}) = \tilde{k}(-\lambda, -\xi - 2\lambda y, -\eta + 2\lambda x; \lambda, \xi - 2\lambda y, \eta + 2\lambda x)$$

is the Weyl symbol of the operator $K \in \mathcal{R}(H^0_\delta(PC(\Omega)))$.

References

[1] Beals, R., Greiner, P.: Calculus on Heisenberg manifolds. Princeton Univ. Press, Princeton, 1988.

[2] Folland, G.: Harmonic analysis is phase space. Princeton Univ. Press, Princeton, 1989.

[3] Geller, D.: Analytic pseudodifferential operators for the Heisenberg group and local solvability. Princeton Univ. Press, Princeton, 1990.

[4] Gindikin, S.: Analysis on homogeneous domains. Russian Math. Surveys 19 (1969), 1–89.

[5] Goodman, R.: Nilpotent Lie groups: Structure and applications to analysis. Lecture Notes in Math. No. 562, Springer, Berlin–Heidelberg–New York, 1976.

[6] Korányi, A., Stein, E.: H^2 spaces of generalized half–spaces. Studia Math. 44 (1972), 379–388.

[7] Korányi, A., Vagi, S.: Singular integrals is homogeneous spaces and some problems of classical analysis. Ann. Scuola Norm. Sup. Pisa. 25 (1971), 575–648.

[8] Nagel, A., Stein, E.: Lectures on pseudodifferential operators: Regularity theorems and application to non–elliptic problems. Princeton Univ. Press, Princeton, 1979.

[9] Ogden, R., Vági, S.: Harmonic analysis of a nilpotent group and function theory on Siegel domains of type II. Adv. Math. 33 (1979), 31–92.

[10] Schempp, W.: Harmonic Analysis on the Heisenberg nilpotent Lie group. Pitman, London, 1986.

[11] Taylor, M.: Noncommutative microlocal analysis, Part 1. Memoirs Amer. Math. Soc. 313, Providence, R.I., 1984.

[12] Taylor, M. : Noncommutative harmonic analysis. Amer. Math. Soc., Providence, R.I., 1986.

[13] Vasilevski, N.L., Trujillo, R.: Group convolution operators on standard Cauchy–Riemann manifolds. I. Structural properties. Integral Eq. Oper. Theory. 19 (1994), 65–105.

[14] Vasilevski, N.L.: Group convolution operators on standard Cauchy–Riemann manifolds. II. Algebras of convolution operators on the Heisenberg group. Integral Eq. Oper. Theory. 19 (1994), 327–348.

3 SOME BOUNDARY VALUE PROBLEMS FOR SECOND ORDER OVERDETERMINED ELLIPTIC SYSTEMS IN THE UNIT BALL OF $\mathbb{C}^N$

Abduhamid Dzhuraev

Mathematical Institute of Tajik Academy of Science
Prospekt Rudaki 33
734025 Dushanbe
Tajikistan

ajuraev@td.silk.glas.apc.org

Abstract: It is well known that an application of the techniques of a single complex variable allows to reach some exhaustive results on PDEs in particular in the theory of boundary value problems for second order elliptic partial differential equations which are not necessarily strongly elliptic [4]. This paper is an attempt to apply the techniques of several complex variables to boundary value problems for second order overdetermined elliptic systems of partial differential equations in the unit ball of $\mathbb{C}^n$, $n > 1$.

3.1 DIRICHLET PROBLEMS

Let $B^n = \left\{ z = (z_1, \cdots, z_n) \in \mathbb{C}^n : |z| = \sqrt{|z_1|^2 + \cdots + |z_n|^2} < 1 \right\}$ be the unit ball of $\mathbb{C}^n$. We consider the following two Dirichlet problems:

Problem D_1 *Find a function $u(z) : \mathbb{C}^n \to \mathbb{C}$ which is a $C^2(B^n)$ solution in B^n to the following second order overdetermined elliptic system of n^2 equations*

$$u_{\bar{z}_k z_\ell} = 0 \,, \quad 1 \le k, \ell \le n \,, \tag{3.1}$$

continuous on the closure $\overline{B^n}$, satisfying the condition

$$u(\zeta) = \gamma(\zeta) \tag{3.2}$$

H.G.W. Begehr et al. (eds.), *Partial Differential and Integral Equations*, 37–57.

on $S^{2n-1} = \{\zeta = (\zeta_1, \cdots, \zeta_n) \in \mathbb{C}^n : |\zeta| = 1\}$ *where* $\gamma(\zeta)$*is a continuous function given on* S^{2n-1}.

Problem D_2 *Find a function* $u(z) : \mathbb{C}^n \to \mathbb{C}$ *which is a* $C^2(B^n)$ *solution in* B^n *to the following second order overdetermined elliptic system of* $n(n+1)/2$ *independent equations*

$$u_{\bar{z}_k \bar{z}_\ell} = 0, \ 1 \le k, \ell \le n \,, \tag{3.3}$$

satisfying the condition (1.2) on S^{2n-1}.

When $\gamma(\zeta) \equiv 0$ these problems will be called the homogeneous Dirichlet problems and denoted by D_1^0 and D_2^0, respectively. The functions, satisfying (3.1) are called pluriharmonic and the functions, satisfying (3.3) are called pluriholomorphic while the functions, satisfying the second order overdetermined elliptic system of only n equations in (3.3) with $k = \ell$,

$$u_{\bar{z}_k \bar{z}_k} = 0 \,, \quad 1 \le k \le n \,, \tag{3.3'}$$

are called bianalytic, see[1].
Since every pluriharmonic function in B^n is apparently harmonic in B^n, then the homogeneous Dirichlet problem for pluriharmonic functions has no nontrivial solutions, while the homogeneous Dirichlet problem for pluriholomorphic functions has in B^n infinitely many linearly independent solutions

$$u_k(z) = (1 - |z|^2) z^k = \left(1 - |z_1|^2 - \cdots - |z_n|^2\right) z_1^{k_1} \cdots z_n^{k_n} \,,$$

$|k| = k_1 + \cdots + k_n \ge 0$. This distinction is the result of the essential differences between the systems (3.1) and (3.3): In contrast to (3.1) system (3.3) cannot be splitted into separated equations for the real and imaginary parts of $u(z)$ and is not reducible to the system (3.1). The same distinction between Problems D_1^0 and D_2^0 is observed also in case $n = 1$ (see [2]), but in case $n > 1$ new conditions arise for the inhomogeneous Dirichlet problem to be solvable. To describe the solvability of the inhomogeneous Dirichlet problems let $\mathcal{H}^{p,q}$ be the space consisting of all restrictions to the sphere S^{2n-1} of the harmonic polynomials that are homogeneous of degree p in z and of degree q in $\bar{z}$. It is well known [7] that $L^2(S^{2n-1}) = \cup_{p \ge 0, q \ge 0} \mathcal{H}^{p,q}$, so that $L^2(S^{2n-1}) = \cup_{p \ge 1, q \ge 1} \mathcal{H}^{p,q} \cup P_h$, where $P_h = \cup_{p \ge 0} \mathcal{H}^{p,0} \cup \cup_{q \ge 1} \mathcal{H}^{0,q}$ is the orthogonal complement of the subspace $\cup_{p \ge 1, q \ge 1} \mathcal{H}^{p,q}$ and consists of the boundary values on S^{2n-1} of pluriharmonic functions in B^n. Using the Cauchy integral formula

$$\phi(z) = C(\phi) = \int_{S^{2n-1}} \phi(\zeta) \frac{d\sigma(\zeta)}{(1 - \langle z, \zeta \rangle)^n} \,, \quad \langle z, \zeta \rangle = \sum_{k=1}^{n} z_k \bar{\zeta}_k \tag{3.4}$$

for holomorphic functions $\phi(z)$ on $\overline{B^n}$ and the Plemelj formula for the boundary value of the integral in (3.4) (see for example [6]) we conclude:

Proposition 1 *The function* $f(\zeta) \in C(S^{2n-1}) \cap L^2(S^{2n-1})$ *is the boundary value on* S^{2n-1} *of a holomorphic function in* B^n *if and only if*

$$\int_{S^{2n-1}} \frac{f(\eta) d\sigma(\eta)}{(1 - \langle \zeta, \eta \rangle)^n} = \frac{1}{2} f(\zeta) \tag{3.5}$$

for any $\zeta \in S^{2n-1}$, where the integral is understood as the Cauchy principal value (see, for example [5]).

Proposition 2 *The function $f(\zeta) \in C(S^{2n-1}) \cap L^2(S^{2n-1})$ belongs to P_h if and only if*

$$\int_{S^{2n-1}} f(\eta)\, \mathrm{Re}\, \frac{1}{(1-\langle\zeta,\eta\rangle)^n} d\sigma(\eta) = \frac{1}{2}\int_{S^{2n-1}} f(\eta) d\sigma(\eta) \tag{3.6}$$

for any $\zeta \in S^{2n-1}$ (see [5]).

Theorem 1 *If $n > 1$, the problem D_1 is solvable if and only if its right-hand side $\gamma(\zeta)$ satisfies the condition* (3.6)*. In this case the unique solution of D_1 is represented by*

$$u(z) = \int_{S^{2n-1}} \gamma(\eta) \left(\frac{1}{(1-\langle z,\eta\rangle)^n} + \frac{1}{(1-\langle \eta,z\rangle)^n} - 1 \right) d\sigma(\eta) \tag{3.7}$$

Proof From (3.1) it follows that any pluriharmonic function is a sum of holomorphic and antiholomorphic functions $u(z) = \phi(z) + \overline{\psi(z)}$, so that the Dirichlet condition (3.2) is

$$\phi(\zeta) + \overline{\psi(\zeta)} = \gamma(\zeta)\ ,\quad \zeta \in S^{2n-1}\ . \tag{3.8}$$

From this relation and the Cauchy formula (3.4) we obtain

$$\phi(z) \;=\; \int_{S^{2n-1}} \frac{\gamma(\zeta)d\sigma(\zeta)}{(1-\langle z,\zeta\rangle)^n} - \overline{C}\ ,$$

$$\psi(z) \;=\; \int_{S^{2n-1}} \overline{\gamma(\zeta)} \left(\frac{1}{(1-\langle z,\zeta\rangle)^n} - 1 \right) d\sigma(\zeta) + C$$

with $C = \psi(0)$, because

$$\int_{S^{2n-1}} \frac{\overline{\psi(\zeta)}d\sigma(\zeta)}{(1-\langle z,\zeta\rangle)^n} = \overline{\int_{S^{2n-1}} \frac{\psi(\zeta)d\sigma(\zeta)}{(1-\langle \zeta,z\rangle)^n\,(1-\langle w,\zeta\rangle)^n}\bigg|_{w=0}} = \overline{\psi(0)} = \overline{C}\ ,$$

$$\int_{S^{2n-1}} \frac{\overline{\phi(\zeta)}d\sigma(\zeta)}{(1-\langle z,\zeta\rangle)^n} = \overline{\phi(0)} = \int_{S^{2n-1}} \overline{\gamma(\zeta)}d\sigma(\zeta) - C\ .$$

Now the condition (3.6) for $\gamma(\zeta)$ is the consequence of the Plemelj formula.

Theorem 2 *Problem D_2 is solvable if its right-hand side $\gamma(\zeta)$ satisfies the conditions*

$$\int_{S^{2n-1}} \frac{\eta_j\gamma(\eta)d\sigma(\eta)}{(1-\langle\zeta,\eta\rangle)^n} = \frac{\zeta_j}{2}\gamma(\zeta)\ ,\quad 1 \le j \le n, \tag{3.9}$$

for any $\zeta \in S^{2n-1}$. In this case all solutions of problem D_2 are given by

$$u(z) = \int_{S^{2n-1}} \frac{\langle\eta,z\rangle\gamma(\eta)d\sigma(\eta)}{(1-\langle z,\eta\rangle)^n} + \sum_{|k|\ge 0} a_k z^k (1-|z|^2)\ , \tag{3.10}$$

where a_k are arbitrary constants.

Proof From (3.3) it follows that a pluriholomorphic function is represented as (see for example [3])

$$u(z) = \phi_0(z) + \langle \phi(z), z \rangle = \phi_0(z) + \sum_{j=1}^{n} \phi_j(z)\bar{z}_j \,, \tag{3.11}$$

with arbitrary functions $\phi_k(z)$, $0 \leq k \leq n$, holomorphic in $\overline{B^n}$, so that the Dirichlet condition (3.2) is

$$\sum_{j=1}^{n} \bar{\zeta}_j \left(\phi_j(\zeta) + \zeta_j \phi_0(\zeta) \right) = \sum_{j=1}^{n} \bar{\zeta}_j \zeta_j \gamma(\zeta) \,, \quad |\zeta| = 1 \,,$$

which is fulfilled if

$$\zeta_j \gamma(\zeta) = \phi_j(\zeta) + \zeta_j \phi_0(\zeta)$$

for any $\zeta \in S^{2n-1}$ and $1 \leq j \leq n$. These equations mean that the functions $\zeta_j \gamma(\zeta)$ are boundary values of holomorphic functions in B^n, *i.e.* by (3.5):

$$\int_{S^{2n-1}} \frac{\eta_j \gamma(\eta) d\sigma(\eta)}{(1 - \langle \zeta, \eta \rangle)^n} = \frac{1}{2} \zeta_j \gamma(\zeta) \tag{3.12}$$

for any $\zeta \in S^{2n-1}$ and $1 \leq j \leq n$. Note that from these conditions after multiplication by $\bar{\zeta}_j$ and summation we obtain the following integral equation for $\gamma(\zeta)$:

$$\frac{1}{2}\gamma(\zeta) - \int_{S^{2n-1}} \frac{\langle \eta, \zeta \rangle \gamma(\eta) d\sigma(\eta)}{(1 - \langle \zeta, \eta \rangle)^n} = 0 \,.$$

3.2 NEUMANN PROBLEMS

It is natural to consider also the following Neumann problems.

Problem N_1 *Find a function $u(z) : \mathbb{C}^n \to \mathbb{C}$, which is a $C^2(B^n)$ solution in B^n to system (3.1), belonging to $C^1(\overline{B^n})$ and satisfying the condition*

$$\frac{\partial u}{\partial \nu_\zeta} = \gamma(\zeta) \tag{3.13}$$

on S^{2n-1} and normalized by the condition

$$\int_{S^{2n-1}} u(\zeta) d\sigma(\zeta) = 0 \,. \tag{3.14}$$

Problem N_2 *Find a function $u(z) : \mathbb{C}^n \to \mathbb{C}$, which is a $C^2(B^n)$ solution in B^n to system (3.3), belonging to $C^1(\overline{B^n})$ and satisfying condition (3.13) on S^{2n-1} and normalized by condition (3.14).*

Again the homogeneous problem N_1^0 corresponding to the problem N_1 has no nontrivial solutions, because condition (3.13), with $\gamma \equiv 0$ is

$$\sum_{j=1}^{n}\left[\zeta_j\phi_{z_j}(\zeta)+\overline{\zeta_j\psi_{z_j}(\zeta)}\right]=0\,,\quad \zeta\in S^{2n-1}\,,$$

and hence

$$\sum_{j=1}^{n} z_j\phi_{z_j}(z)=0\,,\quad \sum_{j=1}^{n} z_j\psi_{z_j}(z)=0\,,\quad |z|<1\,.$$

From these equations it follows that ϕ and ψ are constants and as a consequence $u(z)$ is a constant which is only zero for (3.14).
The homogeneous problem N_2^0 corresponding to N_2 has infinitely many linearly independent solutions

$$u_k(z)=\left(1-\frac{|k|}{|k|+2}|z|^2\right)z^k\,.$$

Theorem 3 *If $n>1$, then Problem N_1 is solvable if and only if its right-hand side $\gamma(\zeta)$ satisfies the conditions* (3.6) *and* (3.15′). *Under these conditions the solution is unique.*

Proof Neumann boundary condition (3.13) for holomorphic functions ϕ, ψ is

$$\sum_{j=1}^{n}\zeta_j\phi_{\zeta_j}(\zeta)+\overline{\zeta_j\psi_{\zeta_j}(\zeta)}=\gamma(\zeta)\,,\quad \zeta\in S^{2n-1},$$

from which it follows that ϕ, ψ are solutions to the following holomorphic partial differential equations in B^n:

$$\begin{aligned}\sum_{j=1}^{n} z_j\phi_{z_j}(z) &= \int_{S^{2n-1}}\gamma(\eta)\frac{d\sigma(\eta)}{(1-\langle z,\eta\rangle)^n}\,,\\ \sum_{j=1}^{n} z_j\psi_{z_j}(z) &= \int_{S^{2n-1}}\overline{\gamma(\eta)}\left(\frac{1}{(1-\langle z,\eta\rangle)^n}-1\right)d\sigma(\eta)\,.\end{aligned}\tag{3.15}$$

From the first equation in (3.15) we obtain

$$\int_{S^{2n-1}}\gamma(\eta)d\sigma(\eta)=0\,,\tag{3.15′}$$

so that solving (3.15) we obtain

$$\begin{aligned}\phi(z) &= \int_{S^{2n-1}}\gamma(\eta)\left(\sum_{j=1}^{n-1}\frac{1}{j(1-\langle z,\eta\rangle)^j}+\log\frac{1}{1-\langle z,\eta\rangle}\right)d\sigma(\eta)+\phi(0)\,,\\ \psi(z) &= \int_{S^{2n-1}}\overline{\gamma(\eta)}\left(\sum_{j=1}^{n-1}\frac{1}{j(1-\langle z,\eta\rangle)^j}+\log\frac{1}{1-\langle z,\eta\rangle}\right)d\sigma(\eta)+\psi(0)\,.\end{aligned}$$

Hence, the desired pluriharmonic function is represented as

$$\begin{aligned} u(z) &= \phi(z) + \overline{\psi(z)} \\ &= \int_{S^{2n-1}} \gamma(\eta) \left(\sum_{j=1}^{n-1} \operatorname{Re} \frac{2}{j(1-\langle z,\eta\rangle)^j} + \log \frac{1}{|1-\langle z,\eta\rangle|^2} \right) d\sigma(\eta) + C \end{aligned}$$

where C is a constant, which is uniquely determined by using the normalization (3.14),

$$C = -\int_{S^{2n-1}} d\sigma(\zeta) \int_{S^{2n-1}} \gamma(\eta) \left(\sum_{j=1}^{n-1} \operatorname{Re} \frac{2}{j(1-\langle \zeta,\eta\rangle)^j} + \log \frac{1}{|1-\langle \zeta,\eta\rangle|^2} \right) d\sigma(\eta).$$

Theorem 4 *If $n > 1$, then Problem N_2 is solvable if its right-hand side $\gamma(\zeta)$ satisfies the conditions* (3.9) *and*

$$\sum_{\ell=1}^{n-1} \int_{S^{2n-1}} d\sigma(\zeta) \int_{S^{2n-1}} \frac{\langle \eta,\zeta\rangle \gamma(\eta) d\sigma(\eta)}{(1-\langle \zeta,\eta\rangle)^{n-\ell}} = 0\,. \tag{3.16}$$

In this case the general solution to Problem N_2 is given by

$$u(z) = \frac{1}{n-1} \sum_{\ell=1}^{n-1} \int_{S^{2n-1}} \frac{\langle \eta,z\rangle \gamma(\eta) d\sigma(\eta)}{(1-\langle z,\eta\rangle)^{n-\ell}} + \sum_{|k|>0} a_k z^k \left(1 - \frac{|k|}{|k|+2}|z|^2\right), \tag{3.17}$$

where a_k are arbitrary constants.

Proof Substituting (3.11) into (3.13) we obtain the equality

$$\sum_{k=1}^{n} \bar{\zeta}_k \left(\sum_{j=1}^{n} \zeta_j \frac{\partial \phi_k}{\partial \zeta_j} + \phi_k(\zeta) + \zeta_k \sum_{j=1}^{n} \zeta_j \frac{\partial \phi_0}{\partial \zeta_j} \right) = \sum_{k=1}^{n} \bar{\zeta}_k \zeta_k \gamma(\zeta)\,,\ \zeta \in S^{2n-1}\,,$$

which is fulfilled if the equalities

$$\sum_{j=1}^{n} \zeta_j \frac{\partial \phi_k}{\partial \zeta_j} + \phi_k(\zeta) + \zeta_k \sum_{j=1}^{n} \zeta_j \frac{\partial \phi_0}{\partial \zeta_j} = \zeta_k \gamma(\zeta)\,,\ \zeta \in S^{2n-1} \tag{3.18}$$

hold. This means that the right-hand sides must be the boundary values of holomorphic functions in the ball B^n as so the left-hand sides are, *i.e.*

$$\int_{S^{2n-1}} \frac{\eta_k \gamma(\eta) d\sigma(\eta)}{(1-\langle \zeta,\eta\rangle)^n} = \frac{1}{2}\zeta_k \gamma(\zeta)\,,\ \ \zeta \in S^{2n-1}\,.$$

Next, by application of the Cauchy formula to (3.18) we obtain the following partial differential equations for the holomorphic functions ϕ_k

$$\sum_{j=1}^{n} z_j \frac{\partial \phi_k}{\partial z_j} + \phi_k(z) = -z_k \sum_{j=1}^{n} z_j \frac{\partial \phi_0}{\partial z_j} + \int_{S^{2n-1}} \frac{\eta_k \gamma(\eta) d\sigma(\eta)}{(1-\langle z,\eta\rangle)^n}. \tag{3.19}$$

Solving these equations for ϕ_k and choosing $\phi_0(z) = z_1^{k_1} \cdots z_n^{k_n}$ we obtain (3.17). Substituting (3.17) into (3.14) and taking into account that

$$\int_{S^{2n-1}} \zeta^k d\sigma(\zeta) = 0\,, \quad |k| > 0$$

we obtain (3.16).

For pluriharmonic functions in B^n instead of problem N_1 we also consider the following problem with complex parameter α:

$$\frac{\partial u}{\partial \nu_\zeta} + \alpha u(\zeta) = \gamma(\zeta)\,, \quad \zeta \in S^{2n-1}\,. \tag{3.20}$$

As we have seen in case $\alpha = 0$ there is no nontrivial pluriharmonic function, satisfying the homogeneous condition (3.20) and the normalization condition (3.14).
If $\operatorname{Re}\alpha > 0$ and $u(z)$ is a pluriharmonic function, satisfying the homogeneous condition (3.20), then using Green's theorem we obtain

$$\begin{aligned}\sum_{j=1}^{n} \int_{B^n} \left(\left|u_{\bar z_j}\right|^2 + \left|u_{z_j}\right|^2\right) d\nu(z) &= \sum_{j=1}^{n} \int_{S^{2n-1}} u(\zeta) \left(\frac{\partial \bar u}{\partial \zeta_j}\zeta_j + \frac{\partial \bar u}{\partial \bar\zeta_j}\bar\zeta_j\right) d\sigma(\zeta) \\ &= \int_{S^{2n-1}} u(\zeta) \frac{\partial \bar u}{\partial \nu_\zeta} d\sigma(\zeta) = -\bar\alpha \int_{S^{2n-1}} |u(\zeta)|^2 d\sigma(\zeta)\end{aligned}$$

from which it follows that $u_{z_j} = u_{\bar z_j} \equiv 0$, $1 \le j \le n$, in B^n, *i.e.* $u(z) = const.$ and $u(z) \equiv 0$ for $\alpha \neq 0$.
If α coincides with a negative integer, $\alpha = -m$, $m > 0$, then there are exactly $2\left((n-1)(n-2)(m-1)/2 + m(n-1) + 1\right)^2$ linearly independent non-trivial pluriharmonic functions in B^n, which satisfy the homogeneous condition (3.20). Indeed, if $u(z) = \phi(z) + \overline{\psi(z)}$, then the last condition is equivalent to the condition

$$\sum_{k=1}^{n} \zeta_k \frac{\partial \phi}{\partial \zeta_k} - m\phi(\zeta) + \overline{\sum_{k=1}^{n} \zeta_k \frac{\partial \psi}{\partial \zeta_k} - m\psi(\zeta)} = 0$$

on S^{2n-1}, so that an application of the Cauchy formula shows that the functions ϕ and ψ are solutions of the following holomorphic partial differential equations

$$\sum_{k=1}^{n} \zeta_k \frac{\partial \phi}{\partial \zeta_k} - m\phi(\zeta) = -\overline{C}\,, \quad \sum_{k=1}^{n} \zeta_k \frac{\partial \psi}{\partial \zeta_k} - m\psi(\zeta) = C$$

in B^n, where C is an arbitrary constant. Only the constants $\phi = \overline{C}/m$ and $\psi = -C/m$ are particular solutions to these equations. Solutions to the corresponding homogeneous equations coincide with the homogeneous polynomials of degree m and hence they are linear combinations of the $(n-1)(n-2)(m-1)/2 + m(n-1) + 1$ monomials z^k of degree $|k| = k_1 + \cdots + k_n$ and corresponding pluriharmonic functions are linear combinations of monomials z^k and $\bar{z}^k$ with $|k| = m$.

We have seen that there are infinitely many pluriholomorphic functions in B^n, satisfying the homogeneous condition (3.20) with $\alpha = 0$.
If α coincides with a nonnegative integer $\alpha = m \geq 0$, then there exist also infinitely many such functions

$$u_k(z) = \left(1 - \frac{|k| + m}{|k| + m + 2}|z|^2\right) z^k \ , \ |k| \neq -m - 2$$

together with the finite number $(n-1)(n-2)(m-2)/2 + (m-1)(n-1) + 1$ of more such functions. Indeed, the homogeneous condition (3.20) is equivalent to the following conditions

$$\sum_{k=1}^{n} \zeta_k \frac{\partial \phi_j}{\partial \zeta_k} + (\alpha + 1)\phi_j(\zeta) = -\zeta_j \left(\sum_{k=1}^{n} \zeta_k \frac{\partial \phi_0}{\partial \zeta_k} + \alpha \phi_0(\zeta)\right) \ , \ \zeta \in S^{2n-1} \ ,$$

and hence in B^n, by the Cauchy formula

$$\sum_{k=1}^{n} z_k \frac{\partial \phi_j}{\partial z_k} + (\alpha + 1)\phi_j(z) = -z_j \left(\sum_{k=1}^{n} z_k \frac{\partial \phi_0}{\partial z_k} + \alpha \phi_0(z)\right) \ , \ 1 \leq j \leq n \ .$$

If $\operatorname{Re} \alpha \geq 0$, then the homogeneous equations

$$\sum_{k=1}^{n} z_k \frac{\partial \phi_j}{\partial z_k} + (\alpha + 1)\phi_j(z) = 0 \ , \ 1 \leq j \leq n \ ,$$

have no nontrivial solutions, continuous in $\overline{B^n}$, but if $\alpha = -m$, $m > 0$, then they have solutions, which are homogeneous polynomials of degree $m - 1$ and are combinations of $(n-1)(n-2)(m-2)/2 + (m-1)(n-1) + 1$ monomials and, moreover, there are infinitely many solutions of the corresponding inhomogeneous equations generated by $\phi_0(z) = z^k$ with $|k| = m - 1$.

For a positive integer m we consider one more problem.

Problem N_1^m *Find a pluriharmonic function $u(z)$ in B^n, belonging to $C^m(\overline{B^n})$ and satisfying the condition*

$$\frac{\partial^m u}{\partial \nu_\zeta^m} = \gamma(\zeta) \tag{3.21}$$

on S^{2n-1}.

Theorem 5 *Problem N_1^m is solvable if and only if its right-hand side $\gamma(\zeta)$ satisfies the conditions* (3.6) *and* (3.15′)*. In this case the solution to the problem is given by*

$$u(z) = \int_{S^{2n-1}} \gamma(\eta) \operatorname{Re} \frac{2z_n^{n-1}\langle z, \eta\rangle}{(m-1)!} \int_1^{z_n} \left(\log \frac{z_n}{\tau}\right)^{m-1} \frac{d\tau}{(z_n - \tau\langle z, \eta\rangle)^n} d\sigma(\eta) + C \ ,$$

with an arbitrary constant C.

Proof Since on S^{2n-1}

$$\frac{\partial}{\partial \nu_\zeta} = \sum_{k=1}^{n} \left[\zeta_k \frac{\partial}{\partial \zeta_k} + \bar{\zeta}_k \frac{\partial}{\partial \bar{\zeta}_k} \right] ,$$

condition (3.21) for the holomorphic functions ϕ and ψ in the representation $u(z) = \phi(z) + \overline{\psi(z)}$ becomes

$$D_\zeta^m \phi(\zeta) + \overline{D_\zeta^m \psi(\zeta)} = \gamma(\zeta) \, , \quad \zeta \in S^{2n-1} \, , \tag{3.22}$$

where $D_\zeta = \sum_{k=1}^{n} \zeta_k \partial/\partial \zeta_k$, $D_\zeta^m = D_\zeta(D_\zeta(\cdots(D_\zeta))$.
Since $D_z^m \phi(z)$, $D_z^m \psi(z)$ are holomorphic in B^n, we conclude that problem (3.22) has a solution if and only if $\gamma(\zeta)$ satisfies condition (3.6) and we obtain

$$\begin{aligned} D_z^m \phi(z) &= \int_{S^{2n-1}} \gamma(\eta) \frac{d\sigma(\eta)}{(1 - \langle z, \eta \rangle)^n} - \bar{C} \, , \\ D_z^m \psi(z) &= \int_{S^{2n-1}} \overline{\gamma(\eta)} \left(\frac{d\sigma(\eta)}{(1 - \langle z, \eta \rangle)^n} - 1 \right) + C \end{aligned} \tag{3.23}$$

with an arbitrary constant C. Since $D_z^m \phi|_{z=0} = 0$, $D_z^m \psi|_{z=0} = 0$, we obtain $C = 0$ and condition (3.15′). Now we take into account that the right-hand side of (3.23) vanishes at $z = 0$ and that only the constants are solutions to the corresponding homogeneous equations $D_z^m \phi(z) = 0$, $D_z^m \psi(z) = 0$. After integrating (3.23) we get

$$\begin{aligned} \phi(z) &= \int_{S^{2n-1}} \gamma(\eta) \frac{z_n^{n-1} \langle z, \eta \rangle}{(m-1)!} \\ &\quad \times \int_1^{z_n} \left(\log \frac{z_n}{\tau} \right)^{m-1} \frac{d\tau}{(z_n - \tau \langle z, \eta \rangle)^n} d\sigma(\eta) + C_1 \, , \\ \psi(z) &= \int_{S^{2n-1}} \overline{\gamma(\eta)} \frac{z_n^{n-1} \langle z, \eta \rangle}{(m-1)!} \\ &\quad \times \int_1^{z_n} \left(\log \frac{z_n}{\tau} \right)^{m-1} \frac{d\tau}{(z_n - \tau \langle z, \eta \rangle)^n} d\sigma(\eta) + C_2 \, , \end{aligned}$$

where C_1, C_2 are arbitrary constants.

3.3 OTHER PROBLEMS

As we have seen the Dirichlet and Neumann problems in B^n are well-posed for system (3.1) (the corresponding homogeneous problems have no nontrivial solutions and the inhomogeneous ones are uniquely solvable for the right-hand side belonging to the subspace P_h). But they are not so for system (3.3), the corresponding homogeneous problems have infinitely many linearly independent solutions and the inhomogeneous problems are not uniquely solvable if they are solvable. This means that, in general the Dirichlet and Neumann problems are not well-posed for all second order overdetermined elliptic systems. It is natural therefore to know whether there is any other boundary value problem which would be a good problem (in the above sense) for both systems (3.1) and (3.3). Here such problems are considered in B^n.

Problem A_1 (A_2) *Find a function $u(z) : \mathbb{C}^n \to \mathbb{C}$ which is a $C^2(B^n)$ solution to system (3.1) ((3.3)), belonging to $C^1(\overline{B^n})$, and satisfying the conditions*

$$\operatorname{Re} u(\zeta) = \gamma_0(\zeta)\,, \quad \operatorname{Re} u_{\bar{z}_k}(\zeta) = \gamma_k(\zeta)\,, \quad 1 \le k \le n\,, \tag{3.24}$$

on S^{2n-1} and normalized by the conditions

$$\operatorname{Im} \int_{S^{2n-1}} u(\zeta) d\sigma(\zeta) = 0\,, \quad \operatorname{Im} \int_{S^{2n-1}} u_{\bar{z}_k}(\zeta) d\sigma(\zeta) = 0\,, \quad 1 \le k \le n\,. \tag{3.25}$$

Theorem 6 *The homogeneous problems A_1^0 and A_2^0 have no nontrivial solutions. Problem A_1 is solvable if and only if its right-hand sides satisfy condition* (3.6) *and*

$$\frac{\partial}{\partial z_\ell} C(\gamma_k) = \frac{\partial}{\partial z_k} C(\gamma_\ell)\,.$$

In this case its unique solution is given by

$$\begin{aligned} u(z) \;=\; & \int_{S^{2n-1}} \gamma_0(\zeta) \left(\frac{2}{(1-\langle z,\zeta\rangle)^n} - 1 \right) d\sigma(\zeta) \\ & -2i \operatorname{Im} \int_{S^{2n-1}} z_k \gamma_k(\zeta) \left(\frac{1}{n-1} \sum_{\ell=1}^{n-1} \frac{2}{(1-\langle z,\zeta\rangle)^{n-\ell}} - 1 \right) d\sigma(\zeta) \\ & +4i \sum_{k=1}^{n} z_k \operatorname{Im} \int_{S^{2n-1}} d\sigma(\zeta) \int_{S^{2n-1}} \frac{\gamma_k(\eta) d\sigma(\eta)}{(1-\langle \zeta,\eta\rangle)^n} \\ & -2i \operatorname{Im} \int_{S^{2n-1}} d\sigma(\zeta) \int_{S^{2n-1}} \frac{\gamma_0(\eta) d\sigma(\eta)}{(1-\langle \zeta,\eta\rangle)^n}\,. \end{aligned} \tag{3.26}$$

Problem A_2 is solvable if and only if its right-hand side $\gamma(\zeta) = (\gamma_1(\zeta), \cdots, \gamma_n(\zeta))$ and the function

$$\begin{aligned} \gamma_{n+1}(\zeta) \;=\; & \gamma_0(\zeta) - \langle \gamma(\zeta), \operatorname{Re}\zeta\rangle - \int_{S^{2n-1}} \langle \gamma(\zeta), \operatorname{Im}\zeta\rangle \operatorname{Im} \frac{2}{(1-\langle \zeta,\eta\rangle)^n} d\sigma(\eta) \\ & + \int_{S^{2n-1}} d\sigma(\zeta) \int_{S^{2n-1}} \langle \gamma(\zeta), \operatorname{Im}\zeta\rangle \operatorname{Im} \frac{2}{(1-\langle \zeta,\eta\rangle)^n} d\sigma(\eta) \end{aligned}$$

satisfy condition (3.6) *and in this case its unique solution is given by*

$$\begin{aligned} u(z) &= \int_{S^{2n-1}} (\gamma_{n+1}(\zeta) - \langle \gamma(\zeta), z\rangle) \left(\frac{2}{(1-\langle z,\zeta\rangle)^n} - 1 \right) d\sigma(\zeta) \\ &\quad -2i \operatorname{Im} \int_{S^{2n-1}} d\sigma(\zeta) \int_{S^{2n-1}} \frac{\gamma_{n+1}(\eta) d\sigma(\eta)}{(1-\langle \zeta,\eta\rangle)^n} \\ &\quad -2i \sum_{k=1}^{n} \bar{z}_k \operatorname{Im} \int_{S^{2n-1}} d\sigma(\zeta) \int_{S^{2n-1}} \frac{\gamma_k(\eta) d\sigma(\eta)}{(1-\langle \zeta,\eta\rangle)^n} . \end{aligned} \tag{3.27}$$

Proof Using the representation $u(z) = \phi(z) + \overline{\psi(z)}$ the last n conditions in (3.24) for problem A_1 are

$$\operatorname{Re} \psi_{z_k}(\zeta) = \gamma_k(\zeta) , \quad \zeta \in S^{2n-1} , \quad 1 \le k \le n ,$$

so that we obtain condition (3.6) for $\gamma_k(\zeta)$, $1 \le k \le n$, and using the last n conditions in (3.25) we obtain

$$\begin{aligned} \psi_{z_k}(z) &= \int_{S^{2n-1}} \gamma_k(\zeta) \left(\frac{2}{(1-\langle z,\zeta\rangle)^n} - 1 \right) d\sigma(\zeta) \\ &\quad -2i \operatorname{Im} \int_{S^{2n-1}} d\sigma(\zeta) \int_{S^{2n-1}} \frac{\gamma_k(\eta) d\sigma(\eta)}{(1-\langle \zeta,\eta\rangle)^n} . \end{aligned}$$

So that through the identity

$$\psi(z) - \psi(0) = \int_0^1 \frac{d}{ds} \psi(sz) ds = \sum_{k=1}^{n} z_k \int_0^1 \psi_{\zeta_k}(sz) ds$$

we obtain

$$\begin{aligned} \psi(z) &= \int_{S^{2n-1}} \sum_{k=1}^{n} z_k \gamma_k(\zeta) \left(\frac{1}{n-1} \sum_{\ell=1}^{n-1} \frac{1}{(1-\langle z,\zeta\rangle)^{n-\ell}} - 1 \right) d\sigma(\zeta) \\ &\quad -2i \sum_{k=1}^{n} z_k \operatorname{Im} \int_{S^{2n-1}} d\sigma(\zeta) \int_{S^{2n-1}} \frac{\gamma_k(\eta) d\sigma(\eta)}{(1-\langle \zeta,\eta\rangle)^n} + \psi(0) . \end{aligned} \tag{3.28}$$

Now the first condition in (3.24) is

$$\operatorname{Re} [\phi(\zeta) + \psi(\zeta)] = \gamma_0(\zeta) , \quad \zeta \in S^{2n-1} ,$$

so that we derive condition (3.6) also for $\gamma_0(\zeta)$. Hence, using the first condition in (3.25) we find

$$\begin{aligned} \phi(z) &= -\psi(z) + \int_{S^{2n-1}} \gamma_0(\zeta) \left(\frac{2}{(1-\langle z,\zeta\rangle)^n} - 1 \right) d\sigma(\zeta) \\ &\quad +2i \operatorname{Im} \int_{S^{2n-1}} \psi(\zeta) d\sigma(\zeta) - 2i \operatorname{Im} \int_{S^{2n-1}} d\sigma(\zeta) \int_{S^{2n-1}} \frac{\gamma_0(\eta) d\sigma(\eta)}{(1-\langle \zeta,\eta\rangle)^n} . \end{aligned} \tag{3.29}$$

Inserting (3.28) and (3.29) into $u(z) = \phi(z) + \overline{\psi(z)}$ and taking into account the equality

$$\psi(0) = \int_{S^{2n-1}} \psi(\zeta) d\sigma(\zeta)$$

following from the Cauchy formula for holomorphic functions in B^n, we obtain (3.27).
Using the representation (3.11) in the last n conditions in (3.24) for Problem A_2 we derive the Schwarz conditions

$$\operatorname{Re} \phi_k(\zeta) = \gamma_k(\zeta) \, , \quad 1 \leq k \leq n \, ,$$

on S^{2n-1} for the holomorphic functions $\phi_k(z)$ and hence condition (3.6) follows for $\gamma_k(\zeta)$, $1 \leq k \leq n$, so that using the last n conditions in (3.25) we see that

$$\begin{aligned} \phi_k(z) &= \int_{S^{2n-1}} \gamma_k(\zeta) \left(\frac{2}{(1 - \langle z, \zeta \rangle)^n} - 1 \right) d\sigma(\zeta) \\ &\quad -2i \operatorname{Im} \int_{S^{2n-1}} d\sigma(\zeta) \int_{S^{2n-1}} \frac{\gamma_k(\eta) d\sigma(\eta)}{(1 - \langle \zeta, \eta \rangle)^n} \, . \end{aligned} \tag{3.30}$$

Now the first condition in (3.24) is the Schwarz condition

$$\operatorname{Re} \phi_0(\zeta) = \gamma_{n+1}(\zeta)$$

on S^{2n-1} for the holomorphic function $\phi_0(z)$. Hence by means of the first condition in (3.25)

$$\begin{aligned} \phi_0(z) &= \int_{S^{2n-1}} \gamma_{n+1}(\zeta) \left(\frac{2}{(1 - \langle z, \zeta \rangle)^n} - 1 \right) d\sigma(\zeta) \\ &\quad -2i \operatorname{Im} \int_{S^{2n-1}} d\sigma(\zeta) \int_{S^{2n-1}} \frac{\gamma_{n+1}(\eta) d\sigma(\eta)}{(1 - \langle \zeta, \eta \rangle)^n} \, . \end{aligned} \tag{3.31}$$

Inserting (3.30) and (3.31) into (3.11) we obtain (3.27).

Note that under suitable compatibility conditions on the right-hand sides all problems above can be investigated also for the inhomogeneous systems

$$w_{\bar{z}_k z_\ell} = f_{k\ell}(z) \ \text{ and } \ w_{\bar{z}_k \bar{z}_\ell} = f_{k\ell}(z) \ \text{ in } B^n \, .$$

Let us consider, for instance, problem A_2 for the last system.

Proposition 3 *If $f_{k\ell}(z)$ are $C^1(B^n)$-functions, bounded on $\overline{B^n}$, satisfying the compatibility conditions*

$$f_{k\ell}(z) = f_{\ell k}(z) \, , \quad \frac{\partial f_{k\ell}}{\partial \bar{z}_j} - \frac{\partial f_{j\ell}}{\partial \bar{z}_k} = 0 \, , \tag{3.32}$$

then the function

$$w^*(z) = (Vf)(z) = (TF)(z) + \sum_{\ell=1}^{n} \bar{z}_\ell (Tf^{(\ell)})(z) \tag{3.33}$$

is a particular solution to the inhomogeneous system

$$w^*_{\bar{z}_k \bar{z}_\ell} = f_{k\ell} , \tag{3.34}$$

where $f^{(\ell)}(z) = (f_{1\ell}(z), \cdots, f_{n\ell}(z))$, $F(z) = (F_1(z), \cdots, F_n(z))$ *with*

$$F_k(z) = \sum_{\ell=1}^{n} \bar{z}_\ell f_{k\ell}(z) ,$$

$$(TF)(z) = \int_{S^{2n-1}} \frac{1 - \langle \zeta, z \rangle}{|\zeta - z|^{2n}} (TF)(\zeta) d\sigma(\zeta) - \frac{1}{n} \int_{B^n} \frac{\langle F(\zeta), \zeta - z \rangle}{|\zeta - z|^{2n}} d\nu(\zeta)$$

for $z \in B^n$ *and*

$$(TF)(\zeta) = \frac{1}{n} \int_{B^n} \frac{\langle F(\eta), \zeta - \eta \rangle}{(1 - \langle \zeta, \eta \rangle)^n (1 - \langle \eta, \zeta \rangle)} d\nu(\eta) , \quad \zeta \in S^{2n-1} .$$

Proof Since for $g(z) = (g_1(z), \cdots, g_n(z))$, $\partial g_k / \partial \bar{z}_j - \partial g_j / \partial \bar{z}_k = 0$ we have [8]

$$\frac{\partial (Tg)(z)}{\partial \bar{z}_k} = g_k(z) , \quad 1 \le k \le n .$$

Then, taking into account that by (3.32)

$$\frac{\partial F_k}{\partial \bar{z}_j} - \frac{\partial F_j}{\partial \bar{z}_k} = 0$$

we obtain (3.34) from (3.33) by direct differentiation.
Now if $w(z)$ is a solution to problem A_2 in B^n for the system $w_{\bar{z}_k \bar{z}_\ell} = f_{k\ell}(z)$, then the function $u(z) = w(z) - w^*(z)$ will be a solution to system (3.3) in B^n, satisfying the boundary conditions

$$\operatorname{Re} u(\zeta) = \gamma_0(\zeta) - \operatorname{Re} w^*(\zeta) , \quad \operatorname{Re} u_{\bar{z}_k}(\zeta) = \gamma_k(\zeta) - \operatorname{Re} (Tf^{(k)})(\zeta) , 1 \le k \le n ,$$

on S^{2n-1}, normalized by the conditions

$$\operatorname{Im} \int_{S^{2n-1}} u(\zeta) d\sigma(\zeta) = - \operatorname{Im} \int_{S^{2n-1}} w^*(\zeta) d\sigma(\zeta) ,$$

$$\operatorname{Im} \int_{S^{2n-1}} u_{\bar{z}_k}(\zeta) d\sigma(\zeta) = - \operatorname{Im} \int_{S^{2n-1}} (Tf^{(k)})(\zeta) d\sigma(\zeta) , \quad 1 \le k \le n .$$

The elliptic system (3.3′) for bianalytic functions, obviously, is a particular case of the following more general second order overdetermined elliptic system of n independent equations

$$\sum_{k,\ell=1}^{n} a_{k\ell}^{(j)}(z) u_{\bar{z}_k \bar{z}_\ell} = 0 , \quad 1 \le j \le n .$$

In case $n = 2$ every such system may be reduced to the following system

$$u_{\bar{z}_1\bar{z}_1} + A(z)u_{\bar{z}_1\bar{z}_2} = 0\ , \quad u_{\bar{z}_2\bar{z}_2} + B(z)u_{\bar{z}_1\bar{z}_2} = 0 \tag{3.35}$$

with the ellipticity condition: $A(z)B(z) \neq 1$.

We consider now the following boundary value problem.

Problem A_{12} *Find a function $u(z_1, z_2) : \mathbb{C}^2 \to \mathbb{C}$, which is a $C^2(B^2)$ solution to system (3.24), belonging to $C^2(\overline{B^2})$ and satisfying the conditions*

$$\mathrm{Re}\, u(\zeta) = \gamma_0(\zeta)\ , \ \mathrm{Re}\, u_{\bar{z}_k}(\zeta) = \gamma_k(\zeta),\ k = 1,2\ , \ \mathrm{Re}\, u_{\bar{z}_1\bar{z}_2}(\zeta) = \gamma_{12}(\zeta) \tag{3.36}$$

on $S^3 = \partial B^2$ and normalized by the conditions

$$\begin{aligned} \mathrm{Im} \int_{S^3} u(\zeta)d\sigma(\zeta) &= \alpha_0\ , \\ \mathrm{Im} \int_{S^3} u_{\bar{z}_k}(\zeta)d\sigma(\zeta) &= \alpha_k\ ,\ k = 1,2\ , \\ \mathrm{Im} \int_{S^3} u_{\bar{z}_1\bar{z}_2}(\zeta)d\sigma(\zeta) &= \alpha_{12}\ . \end{aligned} \tag{3.37}$$

We assume that system (3.34) is complete, that is, the commutator $[L^1, L^2] = L^1L^2 - L^2L^1$ of the differential operators L^1 and L^2 of the left-hand sides of (3.34) is equal to zero. In terms of the coefficients $A(z)$, $B(z)$ this condition is expressed as

a) $B_{\bar{z}_1\bar{z}_1} - A_{\bar{z}_2\bar{z}_2} + A(z)B_{\bar{z}_1\bar{z}_2} - B(z)A_{\bar{z}_1\bar{z}_2} = 0$,

b) $A(z)B_{\bar{z}_2} - B(z)A_{\bar{z}_2} + 2B_{\bar{z}_1} = 0$,

c) $A(z)B_{\bar{z}_1} - B(z)A_{\bar{z}_1} - 2A_{\bar{z}_2} = 0$

in B^2. Differentiating the first equation in (3.34) with respect to $\bar{z}_2$ and the second with respect to $\bar{z}_1$, we get

$$u_{\bar{z}_1\bar{z}_1\bar{z}_2} + A(z)u_{\bar{z}_1\bar{z}_2\bar{z}_2} = -A_{\bar{z}_2}u_{\bar{z}_1\bar{z}_2}\ ,$$

$$B(z)u_{\bar{z}_1\bar{z}_1\bar{z}_2} + u_{\bar{z}_1\bar{z}_2\bar{z}_2} = -B_{\bar{z}_1}u_{\bar{z}_1\bar{z}_2}\ ,$$

so that the function $V(z) = u_{\bar{z}_1\bar{z}_2}$ is a solution to the following first order overdetermined system

$$V_{\bar{z}_1} + A_1V = 0\ , \quad V_{\bar{z}_2} + A_2V = 0\ , \tag{3.38}$$

where

$$A_1(z) = (1 - AB)^{-1}(A_{\bar{z}_2} - AB_{\bar{z}_1})\ , \quad A_2(z) = (1 - AB)^{-1}(B_{\bar{z}_1} - BA_{\bar{z}_2})\ .$$

Using a)–c) it is easily verified that

$$\frac{\partial A_1}{\partial \bar{z}_2} - \frac{\partial A_2}{\partial \bar{z}_1} = 0 .$$

Let $s(z)$ be a solution to the Schwarz problem $\operatorname{Im} s(\zeta) = 0$, $\zeta \in S^3$ for the inhomogeneous Cauchy-Riemann system

$$s_{\bar{z}_1} = A_1(z) \, , \quad s_{\bar{z}_2} = A_2(z) \tag{3.39}$$

in B^2 (see [5]).
Then system (3.38) is the Cauchy-Riemann system for the unknown function $V(z) \exp s(z)$ and the last condition in (3.36) is the Schwarz condition

$$\operatorname{Re} \left[V(\zeta) e^{s(\zeta)} \right] = \gamma_{12}(\zeta) e^{s(\zeta)}$$

for this function. Thus taking into account the last condition in (3.39) we uniquely determine the function $V(z)e^{s(z)}$ and hence, the function $u_{\bar{z}_1\bar{z}_2}$ in the ball B^2. This means that through (3.35) the values of $u_{\bar{z}_1\bar{z}_1}$, $u_{\bar{z}_1\bar{z}_2}$, $u_{\bar{z}_2\bar{z}_2}$ will be specified in B^2, *i.e.* we have the inhomogeneous system $u_{\bar{z}_k\bar{z}_\ell} = f_{k\ell}(z)$ in B^2 together with the boundary conditions for the values of u on S^3 and its first order derivatives $u_{\bar{z}_1}$, $u_{\bar{z}_2}$, normalized by conditions (3.37), which is just the case when $n = 2$ of Problem A_2.

Let us now consider the following problems.

Problem B_1 (B_2) *Find a $C^2(B^n) \cap C^1(\overline{B^n})$ function $u(z) : \mathbb{C}^n \to \mathbb{C}$, which is a solution to system* (3.1) ((3.3)) *in B^n, satisfying the conditions*

$$\operatorname{Re} u(\zeta) = \gamma(\zeta) \, , \quad \operatorname{Re} \langle grad_{\bar{z}} u(\zeta), \zeta \rangle = \operatorname{Re} \sum_{k=1}^{n} \bar{\zeta}_k u_{\bar{z}_k}(\zeta) = h(\zeta) \tag{3.40}$$

on S^{2n-1} and normalized by the condition

$$\operatorname{Im} \int_{S^{2n-1}} u(\zeta) d\sigma(\zeta) = 0 . \tag{3.41}$$

Theorem 7 *Problem B_1 is solvable if and only if its right-hand sides $\gamma(\zeta)$, $h(\zeta)$ satisfy the condition* (3.6) *and*

$$\int_{S^{2n-1}} h(\zeta) d\sigma(\zeta) = 0 . \tag{3.42}$$

In this case its unique solution is of the form

$$\begin{aligned} u(z) \; &= \; \int_{S^{2n-1}} \gamma(\zeta) \left(\frac{2}{(1 - \langle z, \zeta \rangle)^n} - 1 \right) d\sigma(\zeta) \\ &\quad - i \operatorname{Im} \int_{S^{2n-1}} h(\zeta) \left(\sum_{j=1}^{n-1} \frac{1}{j \, (1 - \langle z, \zeta \rangle)^j} + \log \frac{1}{1 - \langle z, \zeta \rangle} \right) d\sigma(\zeta) \end{aligned}$$

$$-i\,\mathrm{Im}\int_{S^{2n-1}} d\sigma(\zeta)\int_{S^{2n-1}}\left[\frac{2\gamma(\eta)}{(1-\langle\zeta,\eta\rangle)^n}\right.$$
$$\left.-h(\eta)\left(\sum_{j=1}^{n-1}\frac{1}{j\,(1-\langle\zeta,\eta\rangle)^j}+\log\frac{1}{1-\langle\zeta,\eta\rangle}\right)\right]d\sigma(\eta) \tag{3.43}$$

and the corresponding homogeneous problem has no nontrivial solutions.

Proof Through the representation $u(z)=\phi(z)+\overline{\psi(z)}$, for the pluriharmonic functions in B^n, conditions (3.40) can be reduced to

$$\mathrm{Re}\,(\phi(\zeta)+\psi(\zeta))=\gamma(\zeta)\,,\quad \mathrm{Re}\sum_{k=1}^{n}\zeta_k\psi_{z_k}(\zeta)=h(\zeta)\,,\quad \zeta\in S^{2n-1}\,, \tag{3.44}$$

so that condition (3.6) follows for $\gamma(\zeta)$, $h(\zeta)$ in order for problem (3.44) to be solvable. Then we obtain from (3.44) that

$$\phi(z)=-\psi(z)+\int_{S^{2n-1}}\gamma(\zeta)\left(\frac{2}{(1-\langle z,\zeta\rangle)^n}-1\right)d\sigma(\zeta)+ic_1\,, \tag{3.45}$$

where c_1 is an arbitrary real constant and $\psi(z)$ is a holomorphic solution to the following partial differential equation

$$\sum_{k=1}^{n}z_k\frac{\partial\psi}{\partial z_k}=\int_{S^{2n-1}}h(\zeta)\left(\frac{2}{(1-\langle z,\zeta\rangle)^n}-1\right)d\sigma(\zeta) \tag{3.46}$$

in B^n. The condition (3.42) follows from (3.46) at $z=0$. Hence, using the identity

$$\begin{aligned}\frac{1}{(1-\langle z,\zeta\rangle)^n}-1 &= \langle z,\zeta\rangle\sum_{j=1}^{n}\frac{1}{(1-\langle z,\zeta\rangle)^j}\\ &=\sum_{k=1}^{n}z_k\frac{\partial}{\partial z_k}\left(\sum_{j=1}^{n-1}\frac{1}{j\,(1-\langle z,\zeta\rangle)^j}+\log\frac{1}{1-\langle z,\zeta\rangle}\right)\end{aligned}$$

in equation (3.46) one gets

$$\begin{aligned}&\sum_{k=1}^{n}z_k\frac{\partial}{\partial z_k}\left[\psi(z)-2\int_{S^{2n-1}}h(\zeta)\right.\\ &\qquad\left.\times\left(\sum_{j=1}^{n-1}\frac{1}{j\,(1-\langle z,\zeta\rangle)^j}+\log\frac{1}{1-\langle z,\zeta\rangle}\right)d\sigma(\zeta)\right]=0\,.\end{aligned}$$

This means that the holomorphic function in the square brackets is continuous in $\overline{B^n}$ and homogeneous of degree zero, *i.e.* it is a constant, say C. Then

$$\psi(z)=2\int_{S^{2n-1}}h(\zeta)\left(\sum_{j=1}^{n-1}\frac{1}{j\,(1-\langle z,\zeta\rangle)^j}+\log\frac{1}{1-\langle z,\zeta\rangle}\right)d\sigma(\zeta)+C\,.$$

The desired pluriharmonic function $u(z)$ is then given by

$$\begin{aligned} u(z) &= -2i\,\mathrm{Im}\psi(z) + \int_{S^{2n-1}} \gamma(\zeta)\left(\frac{2}{(1-\langle z,\zeta\rangle)^n} - 1\right) d\sigma(\zeta) + ic_1 \\ &= \int_{S^{2n-1}} \gamma(\zeta)\left(\frac{2}{(1-\langle z,\zeta\rangle)^n} - 1\right) d\sigma(\zeta) \\ &\quad -i\,\mathrm{Im}\int_{S^{2n-1}} h(\zeta)\left(\sum_{j=1}^{n-1}\frac{1}{j\,(1-\langle z,\zeta\rangle)^j} + \log\frac{1}{1-\langle z,\zeta\rangle}\right) d\sigma(\zeta) \\ &\quad +i(c_1 - 2ImC) \end{aligned}$$

or eliminating the constant $c_1 - 2\,\mathrm{Im}\,C$ by means of (3.42) we obtain (3.43).

The Dirichlet and Neumann problems as well as probem B_2, with its two real conditions (3.40), are underdetermined for pluriholomorphic functions generated, according to (3.11), by $n+1$ independent holomorphic functions, when $n > 1$.
Moreover, if in (3.11) the functions $\phi_j(z)$ coinside with $z_j\psi(z)$ where $\psi(z)$ is an arbitrary holomorphic function in B^n, then we obtain the subclass S_p of pluriholomorphic functions, represented in B^n by

$$u(z) = \phi_0 + |z|^2\psi(z)\ , \tag{3.47}$$

which, obviously coincides with those functions, pluriholomorphic in B^n, having boundary values on S^{2n-1}, satisfying condition (3.5). Hence, problem (3.40), (3.41) is a good problem for functions from S_p.

Theorem 8 *Problem B_2 for pluriholomorphic functions from S_p is solvable if and only if its right-hand sides $\gamma(\zeta)$, $h(\zeta)$ satisfy condition* (3.6). *In this case its unique solution, satisfying the additional normalization condition*

$$\mathrm{Im}\int_{S^{2n-1}} \langle grad_{\bar z} u(\zeta), \zeta\rangle d\sigma(\zeta) = 0\ , \tag{3.48}$$

is given by

$$\begin{aligned} u(z) &= \int_{S^{2n-1}} \left(\frac{2}{(1-\langle z,\zeta\rangle)^n} - 1\right)\left[\gamma(\zeta) - (1-|z|^2)h(\zeta)\right] d\sigma(\zeta) \\ &\quad -2i\,\mathrm{Im}\int_{S^{2n-1}} d\sigma(\zeta)\int_{S^{2n-1}} \frac{\gamma(\eta) - (1-|\zeta|^2)h(\eta)}{(1-\langle\zeta,\eta\rangle)^n} d\sigma(\eta) \end{aligned} \tag{3.49}$$

and the corresponding homogeneous problem has no nontrivial solutions.

Proof By (3.47) conditions (3.40) are

$$\mathrm{Re}\,(\phi_0(\zeta) + \psi(\zeta)) = \gamma(\zeta)\ ,\quad \mathrm{Re}\ \psi(\zeta) = h(\zeta)\ ,\quad \zeta \in S^{2n-1}\ ,$$

so that condition (3.6) for $\gamma(\zeta)$ and $h(\zeta)$ follows for these problems to be solvable. Using (3.48) we obtain

$$\phi_0(z) = -\psi(z) + \int_{S^{2n-1}} \gamma(\zeta)\left(\frac{2}{(1-\langle z,\zeta\rangle)^n} - 1\right) d\sigma(\zeta) + ic_1 ,$$

in B^n, where $\psi(z)$ is given in B^n by

$$\begin{aligned}\psi(z) &= \int_{S^{2n-1}} h(\zeta)\left(\frac{2}{(1-\langle z,\zeta\rangle)^n} - 1\right) d\sigma(\zeta) \\ &\quad -2i\,\mathrm{Im} \int_{S^{2n-1}} d\sigma(\zeta) \int_{S^{2n-1}} \frac{h(\eta)d\sigma(\eta)}{(1-\langle \zeta,\eta\rangle)^n} .\end{aligned}$$

Hence, the desired solution is given by

$$\begin{aligned}u(z) &= \phi_0(z) + |z|^2\psi(z) \\ &= (|z|^2-1)\psi(z) + \int_{S^{2n-1}} \gamma(\zeta)\left(\frac{2}{(1-\langle z,\zeta\rangle)^n} - 1\right) d\sigma(\zeta) + ic_1 \\ &= \int_{S^{2n-1}} \left(\frac{2}{(1-\langle z,\zeta\rangle)^n} - 1\right)\left[\gamma(\zeta) - (1-|z|^2)h(\zeta)\right] d\sigma(\zeta) \\ &\quad +2i\mathrm{Im} \int_{S^{2n-1}} d\sigma(\zeta) \int_{S^{2n-1}} \frac{h(\eta)d\sigma(\eta)}{(1-\langle \zeta,\eta\rangle)^n} + ic_1\end{aligned}$$

and eliminating the constant c_1 by means of (3.42) we obtain (3.49).

3.4 A SYSTEM, PERTURBED BY LOWER ORDER TERMS

We consider now the second order overdetermined system

$$u_{\bar{z}_k\bar{z}_\ell} - z_k z_\ell u = 0 \ , \quad 1 \le k,\ell \le n \ , \tag{3.50}$$

which is obtained from (3.3) by adding lower order terms.

Proposition 4 *System* (3.50) *is compatible in* B^n *if and only if*

$$z_\ell u_{\bar{z}_k} - z_k u_{\bar{z}_\ell} = 0 \ , \quad 1 \le k,\ell \le n \ , \tag{3.51}$$

in B^n*. In this case all regular solutions of this system are represented by*

$$u(z) = \phi(z)e^{|z|^2-1} + \psi(z)e^{1-|z|^2} , \tag{3.52}$$

where $\phi(z)$, $\psi(z)$ *are holomorphic functions in* B^n.

Proof The equations (3.51) are a consequence of the conditions (3.32) for $f_{k\ell} = z_k z_\ell u$. If $u(z)$ is a regular solution to (3.50), then considering the two functions

$$\begin{aligned} \phi(z) &= \frac{e^{1-|z|^2}}{2}\left(u(z)+\sum_{k=1}^{n}\frac{\overline{z_k}}{|z|^2}u_{\bar z_k}\right), \\ \psi(z) &= \frac{e^{|z|^2-1}}{2}\left(u(z)-\sum_{k=1}^{n}\frac{\overline{z_k}}{|z|^2}u_{\bar z_k}\right), \end{aligned} \tag{3.53}$$

using (3.50) and (3.51) by direct differentiation, we conclude that

$$\phi_{\bar z_\ell}=0\,,\quad \psi_{\bar z_\ell}=0\,,\quad 1\le \ell\le n\,,$$

and from (3.53) we obtain (3.52).

The homogeneous Dirichlet problem for system (3.50) in B^n has infinitely many linearly independent solutions

$$u_k(z)=\left(e^{2(1-|z|^2)}-1\right)z^k e^{|z|^2-1}\,.$$

Restricted to S^{2n-1} equations (3.51) are the tangential Cauchy–Riemann equations, which means that the inhomogeneous Dirichlet problem (3.2) for system (3.50) is solvable if and only if $\gamma(\zeta)$ is the boundary value of a holomorphic function in B^n, *i.e.* satisfies condition (3.5). In this case by (3.51) it is easy to see that all solutions to this problem are given by

$$u(z)=\phi(z)e^{|z|^2-1}+e^{1-|z|^2}\left(\int_{S^{2n-1}}\frac{\gamma(\zeta)d\sigma(\zeta)}{(1-\langle z,\zeta\rangle)^n}-\phi(z)\right),$$

where $\phi(z)$ is an arbitrary holomorphic function in B^n.

Now the Neumann condition (2.1) is

$$\sum_{k=1}^{n}\zeta_k\left(\phi_{z_k}(\zeta)+\psi_{z_k}(\zeta)\right)+2(\phi(\zeta)-\psi(\zeta))=\gamma(\zeta)\,,\quad \zeta\in S^{2n-1}\,,$$

which means that the Neumann problem for system (3.50) is solvable if and only if $\gamma(\zeta)$ satisfies the condition (3.5) and from which we conclude then that

$$\sum_{\ell=1}^{n} z_\ell\frac{\partial\phi}{\partial z_\ell}+2\phi(z)=\int_{S^{2n-1}}\frac{\gamma(\zeta)d\sigma(\zeta)}{(1-\langle z,\zeta\rangle)^n}+2\psi(z)-\sum_{k=1}^{n}z_k\frac{\partial\psi}{\partial z_k}\,,\quad z\in B^n\,.$$

This is a holomorphic partial differential equation with respect to $\phi(z)$. If we choose a holomorphic function ψ arbitrarily in B^n, solving this equation we obtain all solutions to the Neumann problem which contain an arbitrary holomorphic function in B^n. In particular choosing $\psi(z)=z^k=z_1^{k_1}\cdots z_n^{k_n}$ and $\gamma(\zeta)\equiv 0$ we obtain the set of nontrivial solution to the homogeneous Neumann problem

$$u_k(z)=\left(e^{2(1-|z|^2)}-\frac{|k|-2}{|k|+2}\right)z^k e^{|z|^2-1}\,.$$

These results show that the Dirichlet and Neumann problems are not well-posed not only for system (3.3), but also for the perturbed one (3.50).

Let us finally consider problem (3.40) with the normalization conditions (3.41), (3.48).

Theorem 9 *Problem* (3.40), (3.41), (3.48) *for system* (3.50) *in* B^n *is solvable if and only if its right-hand sides* $\gamma(\zeta)$ *and* $h(\zeta)$ *satisfy condition* (3.6). *In this case its unique solution is given by*

$$\begin{aligned} u(z) = & \frac{e^{|z|^2-1}+e^{1-|z|^2}}{2}\left\{ \int_{S^{2n-1}} \gamma(\zeta)\left(\frac{2}{(1-\langle z,\zeta\rangle)^n}-1\right)d\sigma(\zeta) \right. \\ & \left. -2i\,\mathrm{Im}\int_{S^{2n-1}} d\sigma(\zeta)\int_{S^{2n-1}} \frac{\gamma(\eta)d\sigma(\eta)}{(1-\langle\zeta,\eta\rangle)^n}\right\} \\ & +\frac{e^{|z|^2-1}+e^{1-|z|^2}}{2}\left\{ \int_{S^{2n-1}} h(\zeta)\left(\frac{2}{(1-\langle z,\zeta\rangle)^n}-1\right)d\sigma(\zeta) \right. \\ & \left. -2i\,\mathrm{Im}\int_{S^{2n-1}} d\sigma(\zeta)\int_{S^{2n-1}} \frac{h(\eta)d\sigma(\eta)}{(1-\langle\zeta,\eta\rangle)^n}\right\} \end{aligned} \tag{3.54}$$

and the corresponding homogeneous problem has no nontrivial solution.

Proof By (3.51) conditions (3.40) are the following Schwarz problems

$$\mathrm{Re}\,[\phi(\zeta)+\psi(\zeta)] = \gamma(\zeta)\,, \quad \mathrm{Re}\,[\phi(\zeta)-\psi(\zeta)] = h(\zeta)\,, \quad \zeta\in S^{2n-1} \tag{3.55}$$

for the in B^n holomorphic functions $\phi(z)+\psi(z)$ and $\phi(z)-\psi(z)$ and hence, are solvable if and only if $\gamma(\zeta)$ and $h(\zeta)$ satisfy (3.6), so that using (3.41) and (3.48) we obtain

$$\begin{aligned} \phi(z)+\psi(z) = & \int_{S^{2n-1}} \gamma(\zeta)\left(\frac{2}{(1-\langle z,\zeta\rangle)^n}-1\right)d\sigma(\zeta) \\ & -2i\,\mathrm{Im}\int_{S^{2n-1}} d\sigma(\zeta)\int_{S^{2n-1}} \frac{\gamma(\eta)d\sigma(\eta)}{(1-\langle\zeta,\eta\rangle)^n}\,, \\ \phi(z)-\psi(z) = & \int_{S^{2n-1}} h(\zeta)\left(\frac{2}{(1-\langle z,\zeta\rangle)^n}-1\right)d\sigma(\zeta) \\ & -2i\,\mathrm{Im}\int_{S^{2n-1}} d\sigma(\zeta)\int_{S^{2n-1}} \frac{h(\eta)d\sigma(\eta)}{(1-\langle\zeta,\eta\rangle)^n} \end{aligned}$$

and (3.54) follows from these equalities and (3.52).

Acknowledgement

The author is grateful to INTAS for supporting this work through INTAS grant no 93–322.

References

[1] Balk, M.: Polyanalytic functions. Akademie–Verlag, Berlin, 1991.

[2] Bitsadze, A.: On the uniqueness of the Dirichlet problem for elliptic partial differential equations. Uspechi Math. Nauk 3, No 6(1948), 211–212.

[3] Detraz, J.: Probleme de Dirichlet pour le systeme $\partial^2 f/\partial \bar{z}_i \partial \bar{z}_j = 0$. Arkiv för Mathematik 26 (1988), 173–184.

[4] Dzhuraev, A.: Method of singular integral equations. Longman, Harlow, 1992.

[5] Dzhuraev, A.: On Riemann-Hilbert boundary value problems in several complex variables. Complex Variables, Theory Appl. 29 (1996), 287–303.

[6] Kerzman, N.: Singular integrals in complex analysis. Proc. Symp. Pure Math. AMS, 35 (1979), 2, 3-41.

[7] Krantz, S.: Partial differential equations and complex analysis. CRC Press, Boca Raton, 1992.

[8] Rudin, W.: Function theory in the unit ball of $\mathbb{C}^n$. Springer–Verlag, New York, 1980.

4 RIEMANN–HILBERT BOUNDARY VALUE PROBLEMS IN $\mathbb{C}^N$

Heinrich Begehr

Freie Universität Berlin
I. Mathematisches Institut
Arnimallee 3
D-14195 Berlin
Germany

begehr@math.fu-berlin.de

Abstract: The Riemann–Hilbert boundary value problem is discussed for Cauchy–Riemann systems in polydiscs. For simplicity mainly the case of two variables is handled and the differences to the case of one variable is explained.

4.1 INTRODUCTION

Boundary value problems for analytic functions of one complex variable were investigated extensively and many applications and generalizations exist, see e.g. [1, 6, 7, 10, 15, 17]. It is a phenomenon that for several complex variables almost nothing is known about boundary value problems. One reason is that there are no proper explicit integral representation formulas as for one complex variable. Moreover, boundary value problems turn out not to be universally solvable. Some solvability conditions have to be observed.

Up to now only simple domains have been considered as the ball and polydomains. Kakichev has treated both the Riemann and the Riemann–Hilbert boundary value problem for analytic functions in several complex variables, see [7, 12, 13, 16]. Recently some results on these problems were obtained for even nonlinear Cauchy–Riemann systems in $\mathbb{C}^n$, see [7, 18]. They originally were published in Chinese.

In this paper the theory of the Riemann–Hilbert problem will be developed for polydomains rather than for the ball. This case is more interesting because

H.G.W. Begehr et al. (eds.), Partial Differential and Integral Equations, 59–84.

here an index arises. The reason that this is not the case for balls in $\mathbb{C}^n$ for $1 < n$ is that the boundary of a ball is a simply connected surface of real dimension $2n - 1$ which is a sphere in $\mathbb{R}^{2n}$.

In order to get explicit solutions the problem will be studied for polydiscs. Moreover, the considerations are restricted to $n = 2$ in order to keep the formulas less involved. The general case can be treated similarly, see [5]. For comparison the classical results for $n = 1$ will be reported on.

4.2 ANALYTIC FUNCTIONS

The classical theory of boundary value problems for analytic functions of one complex variable is developed in [9, 14] see also [1] and [8, 11]. For several complex variables see [4, 5, 7, 12, 13, 18, 19].

Riemann–Hilbert Problem *For a domain $D \subset \mathbb{C}^n$ and $\lambda \in C(\partial D; \mathbb{C})$, $\lambda(\zeta) \neq 0$ on $\partial D, \gamma \in C(\partial D; \mathbb{R})$ given find an analytic function φ in D, continuous on $\overline{D} = D \cup \partial D$ such that*

$$\operatorname{Re}\{\overline{\lambda}\varphi\} = \gamma \text{ on } \partial D . \tag{4.1}$$

In the following D is assumed to be the unit polydisc

$$\mathbb{D}^n := \{z : z \in \mathbb{C}^n, \ |z_k| < 1, \ 1 \leq k \leq n\}$$

in $\mathbb{C}^n$ with the distinguished boundary

$$\partial_0 \mathbb{D}^n := \{z : z \in \mathbb{C}^n, \ |z_k| = 1, \ 1 \leq k \leq n\} .$$

The boundary condition (1.1) is considered only on $\partial_0 \mathbb{D}^n$ rather than on the entire topological boundary

$$\partial \mathbb{D}^n := \{z : z \in \mathbb{C}^n, \ |z_k| = 1 \text{ for at least one } k, \ 1 \leq k \leq n\} .$$

For the Riemann–Hilbert problem in the unit ball

$$B^n := \left\{ z : z \in \mathbb{C}^n , \ \sum_{k=1}^{n} |z_k|^2 < 1 \right\}$$

of $\mathbb{C}^n$, see [5].

4.2.1 *Schwarz problem*

For $\lambda \equiv 1$ the Riemann–Hilbert problem reduces to the Schwarz problem. Its solution can be found by modifying the Cauchy formula. For $n = 1$ we have

$$\varphi(z) = \frac{1}{2\pi i} \int_{|\zeta|=1} \varphi|\zeta| \frac{d\zeta}{\zeta - z} \quad , \quad |z| < 1 ,$$

$$0 = \frac{1}{2\pi i} \int_{|\zeta|=1} \varphi(\zeta) \frac{\overline{z} d\zeta}{1 - \overline{z}\zeta} \quad , \quad |z| < 1 .$$

Adding the complex conjugate of the latter to the first gives with $\overline{\zeta}d\zeta+\zeta d\overline{\zeta}=0$ on $|\zeta|=1$

$$\varphi(z)=\frac{1}{2\pi i}\int_{|\zeta|=1}\mathrm{Re}\,\varphi(\zeta)\left[\frac{2\zeta}{\zeta-z}-1\right]\frac{d\zeta}{\zeta}+ic\,,\ c=\mathrm{Im}\,\varphi(0),\ |z|<1\,. \quad (4.2)$$

The real part satisfies

$$\mathrm{Re}\,\varphi(z)=\frac{1}{2\pi i}\int_{|\zeta|=1}\mathrm{Re}\,\varphi(\zeta)\left[\frac{\zeta}{\zeta-z}+\frac{\overline{\zeta}}{\overline{\zeta-z}}-1\right]\frac{d\zeta}{\zeta},\ |z|<1\,. \quad (4.3)$$

If in (4.2), (4.3) the Schwarz condition

$$\mathrm{Re}\,\varphi=\gamma \ \text{ on } \ |\zeta|=1 \quad (4.4)$$

is inserted one has the solution to this Schwarz problem. This follows as the kernel in (4.3) is the Poisson kernel

$$\frac{\zeta}{\zeta-z}+\frac{\overline{\zeta}}{\overline{\zeta-z}}-1=\frac{1-|z|^2}{|\zeta-z|^2},\ |z|<|\zeta|=1\,.$$

Following this procedure for $n=2$ we have to add

$$\varphi(z)=\frac{1}{(2\pi i)^2}\int_{|\zeta_1|=1}\int_{|\zeta_2|=1}\varphi(\zeta)\frac{d\zeta_1}{\zeta_1-z_1}\frac{d\zeta_2}{\zeta_2-z_2},\ |z_1|<1,\ |z_2|<1\,,$$

and the complex conjugate of

$$0=\frac{1}{(2\pi i)^2}\int_{|\zeta_1|=1}\int_{|\zeta_2|=1}\varphi(\zeta)\frac{\overline{z_1}d\zeta_1}{1-\overline{z_1}\zeta_1}\frac{\overline{z_2}d\zeta_2}{1-\overline{z_2}\zeta_2},\ |z_1|<1,\ |z_2|<1\,.$$

This gives for $|z_1|,|z_2|<1$

$$\varphi(z)=\frac{1}{(2\pi i)^2}\int_{|\zeta_1|=1}\int_{|\zeta_2|=1}\mathrm{Re}\,\varphi(\zeta)\left(2\frac{\zeta_1}{\zeta_1-z_1}\frac{\zeta_2}{\zeta_2-z_2}-1\right)\frac{d\zeta_1}{\zeta_1}\frac{d\zeta_2}{\zeta_2}+i\mathrm{Im}\,\varphi(0)$$

if one observes

$$\frac{1}{(2\pi i)^2}\int_{|\zeta_1|=1}\int_{|\zeta_2|=1}\overline{\varphi(\zeta)}\left(\frac{z_1}{\zeta_1-z_1}+\frac{z_2}{\zeta_2-z_2}\right)\frac{d\zeta_1}{\zeta_1}\frac{d\zeta_2}{\zeta_2}=0,\ |z_1|<1,\ |z_2|<1\,.$$

Lemma 1 *For $\gamma\in C(\partial\mathbb{D}^2;\mathbb{R}), c\in\mathbb{R}$, the analytic function*

$$\varphi(z):=\frac{1}{(2\pi i)^2}\int_{\partial_0\mathbb{D}^2}\gamma(\zeta)\left[2\frac{\zeta}{\zeta-z}-1\right]\frac{d\zeta}{\zeta}+ic,\ z\in\mathbb{D}^2\,, \quad (4.5)$$

with

$$\frac{\zeta}{\zeta-z}:=\frac{\zeta_1}{\zeta_1-z_1}\frac{\zeta_2}{\zeta_2-z_2},\ \frac{d\zeta}{\zeta}:=\frac{d\zeta_1}{\zeta_1}\frac{d\zeta_2}{\zeta_2}\,,$$

satisfies the Schwarz condition

$$\operatorname{Re}\varphi = \gamma \ \ on \ \ \partial \mathbb{D}^2$$

if and only if

$$\frac{1}{(2\pi i)^2}\int_{\partial_0 \mathbb{D}^2} \gamma(\zeta)\left[\frac{z_1}{\zeta_1 - z_1}\frac{\overline{z_2}}{\overline{\zeta_2 - z_2}}\right.$$

$$\left. + \frac{\overline{z_1}}{\overline{\zeta_1 - z_1}}\frac{z_2}{\zeta_2 - z_2}\right]\frac{d\zeta_1}{\zeta_1}\frac{d\zeta_2}{\zeta_2} = 0,\ z \in \mathbb{D}^2 . \tag{4.6}$$

Proof For $z \in \mathbb{D}^2, \zeta + \partial_0 \mathbb{D}^2$ we have

$$\frac{\zeta_1}{\zeta_1 - z_1}\frac{\zeta_2}{\zeta_2 - z_2} + \frac{\overline{\zeta_1}}{\overline{\zeta_1 - z_1}}\frac{\overline{\zeta_2}}{\overline{\zeta_2 - z_2}} - 1 + \frac{z_1}{\zeta_1 - z_1}\frac{\overline{z_2}}{\overline{\zeta_2 - z_2}} + \frac{\overline{z_1}}{\overline{\zeta_1 - z_1}}\frac{z_2}{\zeta_2 - z_2}$$

$$= \frac{1 - |z_1|^2}{|\zeta_1 - z_1|^2}\frac{1 - |z_2|^2}{|\zeta_2 - z_2|^2} ,$$

the right–hand side being the Poisson kernel of $\mathbb{D}^2$.

The solvability condition (4.6) is

$$2\operatorname{Re}\frac{1}{(2\pi i)^2}\int_{\partial_0 \mathbb{D}^2} \gamma(\zeta)\frac{z_1\overline{\zeta_1}}{1 - z_1\overline{\zeta_1}}\frac{\overline{z_2}\zeta_2}{1 - \overline{z_2}\zeta_2}\frac{d\zeta_1}{\zeta_1}\frac{d\zeta_2}{\zeta_2} = 0 .$$

This equation is equivalent to the set of conditions

$$2\operatorname{Re}\frac{1}{(2\pi i)^2}\int_{\partial_0 \mathbb{D}^2} \gamma(\zeta)\overline{\zeta_1}^k\zeta_2^l\frac{\overline{d\zeta_1}}{\overline{\zeta_1}}\frac{d\zeta_2}{\zeta_2} = 0 ,\ 1 \le k, l . \tag{4.6'}$$

4.2.2 Riemann–Hilbert problem

The Riemann–Hilbert problem can in principal be reduced to a particular one by using the "Schwarz operator" from the preceding section.

Definition *Let $D^n = \times_{k=1}^n D_k$ be a polydomain in $\mathbb{C}^n, 1 \le n$, with rectifiable $\partial D_k, 1 \le k \le n$. Let $\lambda \in C(\partial_0 D^n; \mathbb{C}), \lambda \ne 0$ on $\partial_0 D^n := \times_{k=1}^{\overline{n}} \partial D_k$. Then*

$$\kappa = (\kappa_1, \ldots, \kappa_n) ,\ \kappa_k := \frac{1}{2\pi i}\int_{\partial D_k} d\log\lambda(\zeta) \in \mathbb{Z} ,\ 1 \le k \le n ,$$

is called multi–index of λ, $\kappa := ind\,\lambda$.

Assuming $0 \in D^n$ the function $\zeta^{-\kappa} = \prod_{k=1}^n \zeta_k^{-\kappa_k}$ is well–defined on $\partial_0 D^n$. For $D^n = \mathbb{D}^n$ the analytic function

$$\delta(z) := \frac{1}{(2\pi i)^n}\int_{\partial_0 \mathbb{D}^n} \arg\{\zeta^{-\kappa}\lambda(\zeta)\}\left[\frac{2\zeta}{\zeta - z} - 1\right]\frac{d\zeta}{\zeta}$$

solves the Schwarz problem

$$\operatorname{Re}\delta(\zeta) = \arg\{\zeta^{-\kappa}\lambda(\zeta)\} \text{ on } \partial_0 \mathbb{D}^n$$

for $n = 1$ without any further condition, for $n = 2$ under condition (4.6) with $\gamma(\zeta) = \arg\{\zeta^{-\kappa}\lambda(\zeta)\}$. Thus on $\partial_0 \mathbb{D}^n$ the condition

$$\operatorname{Re}\{\overline{\lambda}\varphi\} = \gamma$$

is equivalent to

$$\operatorname{Re}\left\{\overline{\zeta}^{\kappa} e^{-i\delta(\zeta)}\varphi(\zeta)\right\} = e^{\operatorname{Im}\delta(\zeta)}\gamma(\zeta) \, .$$

Hence, in the following we will concentrate on $\lambda(\zeta) = \overline{\zeta}^{\kappa}$.

For $n = 1$ the particular Riemann–Hilbert problem

$$\operatorname{Re}\{\overline{\zeta}^{\kappa}\varphi(\zeta)\} = \gamma(\zeta) \text{ on } |\zeta| = 1 \tag{4.7}$$

for $0 \leq \kappa$ obviously has the particular solution

$$\varphi_0(z) := z^{\kappa}\psi(z) \, , \; \psi(z) := \frac{1}{2\pi i}\int_{|\zeta|=1} \gamma(\zeta)\left[\frac{2\zeta}{\zeta - z} - 1\right]\frac{d\zeta}{\zeta} \, .$$

The general solution to the related homogeneous problem is of the form $\sum_{\nu=0}^{2\kappa} a_\nu z^\nu$ where the a_ν are arbitrary complex constants satisfying $a_\nu + \overline{a_{2\kappa-\nu}} = 0$, $0 \leq \nu \leq \kappa$. This can easily be seen by adjusting an arbitrary power series to the boundary condition

$$\operatorname{Re}\{\overline{\zeta}^{\kappa}\varphi(\zeta)\} = 0 \text{ on } |\zeta| = 1 \, .$$

Thus, the general solution to (4.7) is

$$\begin{gathered} \varphi(z) = \frac{z^{\kappa}}{2\pi i}\int_{|\zeta|=1} \gamma(\zeta)\left[\frac{2\zeta}{\zeta - z} - 1\right]\frac{d\zeta}{\zeta} + \sum_{\nu=0}^{2\kappa} a_\nu z^\nu \, , \\ a_\nu + \overline{a_{2\kappa-\nu}} = 0 \, , \; 0 \leq \nu \leq \kappa \, . \end{gathered} \tag{4.8}$$

In case $\kappa < 0$ a solution to (4.7) would satisfy

$$\operatorname{Re}\{\zeta^{-\kappa}\varphi(\zeta)\} = \gamma(\zeta) \text{ on } |\zeta| = 1 \, . \tag{4.7'}$$

As $z^{-\kappa}\varphi(z)$ would be analytic in $|z| < 1$ the solution to (4.7′) would be given by the Schwarz operator as

$$z^{-\kappa}\varphi(z) = \frac{1}{2\pi i}\int_{|\zeta|=1} \gamma(\zeta)\left[\frac{2\zeta}{\zeta - z} - 1\right]\frac{d\zeta}{\zeta} \, .$$

This is only the case if the right–hand side has a zero at the origin of order $-\kappa$, i.e.

$$\frac{1}{2\pi i}\int_{|\zeta|=1} \gamma(\zeta)\frac{d\zeta}{\zeta^{k+1}} = 0 \, , \; 0 \leq k \leq -\kappa - 1 \, .$$

These conditions can be expressed as

$$\frac{1}{2\pi i}\int_{|\zeta|=1}\gamma(\zeta)\left[2\frac{1-(z\overline{\zeta})^{-\kappa}}{1-z\overline{\zeta}}-1\right]\frac{d\zeta}{\zeta}=0\,. \tag{4.9}$$

These are $-2\kappa-1$ real conditions on γ in order that (4.7′) is solvable for $\kappa<0$. The solution then is

$$\varphi(z)=\frac{2}{2\pi i}\int_{|\zeta|=1}\gamma(\zeta)\frac{\zeta^{\kappa}}{\zeta-z}d\zeta\;,\;|z|<1\,. \tag{4.10}$$

The homogeneous problem in this case is only trivially solvable because $z^{-\kappa}\varphi(z)$ would be an analytic function in the unit disc vanishing at the origin and having vanishing real part on the boundary.

Lemma 2 *Let $\gamma\in C(\partial\mathbb{D}^2;\mathbb{R}\,)$ satisfy* (4.6). *Then the Riemann-Hilbert problem*

$$\operatorname{Re}\{\overline{\zeta}_1^{\kappa_1}\overline{\zeta}_2^{\kappa_2}\varphi(\zeta)\}=\gamma(\zeta)\quad on\ \partial_0\mathbb{D}^2 \tag{4.11}$$

is solvable for $0\le\kappa_1,\kappa_2$. The general solution is

$$\varphi(z)=z_1^{\kappa_1}z_2^{\kappa_2}\psi(z)+\sum_{\nu_1=0}^{2\kappa_1}\sum_{\nu_2=0}^{2\kappa_2}a_{\nu_1\nu_2}z_1^{\nu_1}z_2^{\nu_2},$$

$$a_{\nu_1\nu_2}+\overline{a_{2\kappa_1-\nu_1,2\kappa_2-\nu_2}}=0\;,\;\;0\le\nu_1\le\kappa_1,0\le\nu_2\le\kappa_2\;, \tag{4.12}$$

$$\psi(z)=\frac{1}{(2\pi i)^2}\int_{\partial_0\mathbb{D}^2}\gamma(\zeta)\left[2\frac{\zeta_1}{\zeta_1-z_1}\frac{\zeta_2}{\zeta_2-z_2}-1\right]\frac{d\zeta_1}{\zeta_1}\frac{d\zeta_2}{\zeta_2}\,.$$

As soon as κ_1 or κ_2 is negative the related homogeneous problem (4.11) *with $\gamma=0$ is only trivially solvable. The inhomogeneous problem is solvable if γ besides* (4.6) *satisfies additional conditions. These are in the case $\kappa_1<0\le\kappa_2$*

$$\begin{aligned}\frac{1}{(2\pi i)^2}\int_{\partial_0\mathbb{D}^2}\gamma(\zeta)&\Big[2(1-(z_1\overline{\zeta_1})^{-\kappa_1})\\ &\times\frac{1}{1-z_1\overline{\zeta_1}}\frac{1}{1-z_2\overline{\zeta_2}}-1\Big]\frac{d\zeta_1}{\zeta_1}\frac{d\zeta_2}{\zeta_2}=0\,,\end{aligned} \tag{4.13}$$

and for $\kappa_1,\kappa_2<0$

$$\begin{aligned}\frac{1}{(2\pi i)^2}\int_{\partial_0\mathbb{D}^2}\gamma(\zeta)&\Big[2(1-(z_1\overline{\zeta_1})^{-\kappa_1}(z_2\overline{\zeta_2})^{-\kappa_2})\\ &\times\frac{1}{1-z_1\overline{\zeta_1}}\frac{1}{1-z_2\overline{\zeta_2}}-1\Big]\frac{d\zeta_1}{\zeta_1}\frac{d\zeta_2}{\zeta_2}=0\,.\end{aligned} \tag{4.13′}$$

The solution then is given by

$$\varphi(z) = \frac{2}{(2\pi i)^2} \int_{\partial_0 \mathbb{D}^2} \gamma(\zeta) \frac{\zeta_1^{\kappa_1}}{\zeta_1 - z_1} \frac{z_2^{\kappa_2}}{\zeta_2 - z_2} d\zeta_1 d\zeta_2 \text{ for } \kappa_1 < 0 \leq \kappa_2 , \tag{4.14}$$

and

$$\varphi(z) = \frac{2}{(2\pi i)^2} \int_{\partial_0 \mathbb{D}^2} \gamma(\zeta) \frac{\zeta_1^{\kappa_1}}{\zeta_1 - z_1} \frac{\zeta_2^{\kappa_2}}{\zeta_2 - z_2} d\zeta_1 d\zeta_2 \text{ for } \kappa_1, \kappa_2 < 0 . \tag{4.14'}$$

Proof For $0 \leq \kappa_1, \kappa_2$ the function $z_1^{\kappa_1} z_2^{\kappa_2} \psi(z)$ from (4.12) obviously is a particular solution to (4.11) under condition (4.6). If φ_0 would be a solution to the related homogeneous problem

$$\operatorname{Re} \{\zeta_1^{-\kappa_1} \zeta_2^{-\kappa_2} \varphi_0(\zeta)\} = 0 \text{ on } \partial_0 \mathbb{D}^2$$

then $z_1^{-\kappa_1} z_2^{-\kappa_2} \varphi_0(z)$ would be analytic in $\mathbb{D}^2$ up to $z_1 = 0$ or $z_2 = 0$ having a pole of order κ_ν in $z_\nu = 0$ as a function of $z_\nu, 1 \leq \nu \leq 2$. Therefore this function has the form

$$z_1^{-\kappa_1} z_2^{-\kappa_2} \varphi(z) = \sum_{\nu_1 = -\kappa_1}^{+\infty} \sum_{\nu_2 = -\kappa_2}^{+\infty} a_{\nu_1 \nu_2} z_1^{\nu_1} z_2^{\nu_2} .$$

From the homogeneous Schwarz condition

$$\sum_{\nu_1 = -\kappa_1}^{+\infty} \sum_{\nu_2 = -\kappa_2}^{+\infty} \{a_{\nu_1 \nu_2} z_1^{\nu_1} z_2^{\nu_2} + \overline{a_{\nu_1 \nu_2}} z_1^{-\nu_1} z_2^{-\nu_2}\}$$

$$= \sum_{\nu_1 = -\kappa_1}^{\kappa_1} \{A_{\nu_1} z_1^{\nu_1} + \overline{A_{\nu_1}} z_1^{-\nu_1}\} + \sum_{\nu_1 = \kappa_1 + 1}^{+\infty} A_{\nu_1} z_1^{\nu_1} + \sum_{\nu_1 = -\infty}^{-\kappa_1 - 1} \overline{A_{-\nu_1}} z_1^{\nu_1} = 0$$

follows for $|z_1| = 1$ with

$$A_{\nu_1} := \sum_{\nu_2 = -\kappa_2}^{+\infty} a_{\nu_1 \nu_2} z_2^{\nu_2} , \; -\kappa_1 \leq \nu_1 .$$

Hence,

$$A_{\nu_1} + \overline{A_{-\nu_1}} = 0 \text{ for } |\nu_1| \leq \kappa_1 , \; A_{\nu_1} = 0 \text{ for } \kappa_1 < \nu_1 .$$

Similarly, for $|z_2| = 1$

$$A_{\nu_1} + \overline{A_{-\nu_1}} = \sum_{\nu_2 = -\kappa_2}^{+\infty} \{a_{\nu_1 \nu_2} z_2^{\nu_2} + \overline{a_{-\nu_1 \nu_2}} z_2^{-\nu_2}\} = 0 \text{ for } |\nu_1| \leq \kappa ,$$

implies

$$a_{\nu_1 \nu_2} + \overline{a_{-\nu_1 - \nu_2}} = 0 \text{ for } |\nu_1| \leq \kappa_1, |\nu_2| \leq \kappa_2 , \; a_{\nu_1 \nu_2} = 0 \text{ for } |\nu_1| \leq \kappa_1, \kappa_2 < \nu_2 .$$

Thus

$$z_1^{-\kappa_1} z_2^{-\kappa_2}\varphi_0(z) = \sum_{\nu_1=-\kappa_1}^{\kappa_1} \sum_{\nu_2=-\kappa_2}^{\kappa_2} a_{\nu_1\nu_2} z_1^{\nu_1} z_2^{\nu_2} ,$$

$$a_{\nu_1\nu_2} + \overline{a_{-\nu_1-\nu_2}} = 0 \ , \ 0 \le \nu_1 \le \kappa_1, 0 \le \nu_2 \le \kappa_2 ,$$

i.e.

$$\varphi_0(z) = \sum_{\nu_1=0}^{2\kappa_1} \sum_{\nu_2=0}^{2\kappa_2} \alpha_{\nu_1\nu_2} z_1^{\nu_1} z_2^{\nu_2} ,$$

$$\alpha_{\nu_1\nu_2} + \overline{\alpha_{2\kappa_1-\nu_1 2\kappa_2-\nu_2}} = 0 \ , \ 0 \le \nu_1 \le \kappa_1, 0 \le \nu_2 \le \kappa_2 .$$

For $\kappa_1 < 0 \le \kappa_2$ a solution φ would lead to a solution $\widetilde{\varphi}(z) = z_1^{-\kappa_1}\varphi(z)$ of the problem

$$\operatorname{Re}\{\zeta_2^{-\kappa_2}\widetilde{\varphi}(\zeta)\} = \gamma(\zeta) \ \text{ on } \ \partial_0 \boldsymbol{D}^2$$

which is of the type just considered where the index κ_1 is zero. Hence,

$$\widetilde{\varphi}(z) = z_2^{\kappa_2}\psi(z) + \sum_{\nu_2=0}^{2\kappa_2} a_{\nu_2} z_2^{\kappa_2} \ , \ a_{\nu_2} + \overline{a_{2\kappa_2-\nu_2}} = 0, \ 0 \le \nu_2 \le \kappa_2 ,$$

i.e.

$$\varphi(z) = z_1^{\kappa_1} z_2^{\kappa_2}\psi(z) + z_1^{\kappa_1} \sum_{\nu_2=0}^{2\kappa_2} a_{\nu_2} z_2^{\nu_2} .$$

But this function is analytic in $z_1 = 0$ as a function of z_1 if and only if $a_{\nu_2} = 0, 0 \le \nu_2 \le 2\kappa_2$, and in analogy to (4.9) equation (4.13) is satisfied. And then (4.14) is the solution which follows by adding (4.13) to $\overline{z_1}^{\kappa_1}\overline{z_2}^{\kappa_2}\varphi(z)$. A similar reasoning leads to the result for $\kappa_1, \kappa_2 < 0$. There is no principle obstade in generalizing this result to $\mathbb{C}^n$, see [4].

4.3 INHOMOGENEOUS CAUCHY–RIEMANN SYSTEMS

For one complex variable the inhomogeneous Cauchy–Riemann system in complex form is

$$w_{\overline{z}} = f \ . \tag{4.15}$$

A particular solution is provided by the Pompeiu operator

$$Tf(z) := -\frac{1}{\pi}\int_D f(\zeta)\frac{d\xi d\eta}{\zeta - z}$$

where D is the domain under consideration. The general solution to (4.15) then is with arbitrary analytic function φ in D

$$w = \varphi + Tf \ .$$

Expressing φ by the Cauchy integral

$$\varphi(z) = \frac{1}{2\pi i}\int_{\partial D} w(\zeta)\frac{d\zeta}{\zeta - z}$$

gives the Cauchy–Pompeiu representation formula for w. The Riemann–Hilbert boundary condition (4.1) thus means

$$\operatorname{Re}\{\overline{\lambda}\varphi\} = \gamma_0 := \gamma - \operatorname{Re}\{\overline{\lambda}Tf\} \text{ on } \partial D .$$

Determining φ by this condition leads to the solution of (4.15). The main problem therefore is to express φ through γ and f. In the case where D is the unit disc $\mathbb{D}$ and $\lambda = 1$ the Schwarz operator gives

$$\begin{aligned}\varphi(z) &= \frac{1}{2\pi i}\int_{|\zeta|=1}\gamma(\zeta)\left(2\frac{\zeta}{\zeta - z} - 1\right)\frac{d\zeta}{\zeta}\\ &\quad -\frac{1}{2\pi i}\int_{|\zeta|=1}(Tf(\zeta) + \overline{Tf(\zeta)})\left(\frac{\zeta}{\zeta - z} - \frac{1}{2}\right)\frac{d\zeta}{\zeta}.\end{aligned}$$

From

$$\frac{1}{2\pi i}\int_{|\zeta|=1} Tf(\zeta)\frac{d\zeta}{\zeta - z} = 0 ,$$

$$\frac{1}{2\pi i}\int_{|\zeta|=1} \overline{Tf(\zeta)}\frac{d\zeta}{\zeta - z} = \frac{1}{\pi}\int_{|\zeta|<1} \overline{f(\zeta)}\frac{z\,d\xi d\eta}{1 - z\overline{\zeta}}$$

for $|z| < 1$ then

$$\begin{aligned}w(z) &= \frac{1}{2\pi i}\int_{|\zeta|=1}\gamma(\zeta)\left[\frac{\zeta}{\zeta - z} - 1\right]\frac{d\zeta}{\zeta} + \widetilde{T}f(z) ,\\ \widetilde{T}f(z) &:= -\frac{1}{\pi}\int_{|\zeta|<1}\left[\frac{f(\zeta)}{\zeta - z} + \frac{z\overline{f(\zeta)}}{1 - z\overline{\zeta}}\right]d\xi d\eta\end{aligned}$$

follows. This $\widetilde{T}$–operator was given by Vekua, [15]. It obviously is a particular solution to (4.15) satisfying the homogeneous Schwarz condition on $\partial\mathbb{D}$.

For the particular Riemann–Hilbert problem

$$\operatorname{Re}\{\overline{\zeta}^{\kappa} w(\zeta)\} = \gamma(\zeta) \text{ on } |\zeta| = 1$$

i.e.

$$\operatorname{Re}\{\zeta^{-\kappa}\varphi(\zeta)\} = \gamma_0(\zeta) := \gamma(\zeta) - \operatorname{Re}\{\zeta^{-\kappa}Tf(\zeta)\} \text{ on } |\zeta| = 1$$

from (4.8)

$$\begin{aligned}\varphi(z) &= \frac{z^{\kappa}}{2\pi i}\int_{|\zeta|=1}\gamma(\zeta)\left[\frac{2\zeta}{\zeta - z} - 1\right]\frac{d\zeta}{\zeta} + \sum_{\nu=0}^{2\kappa} a_{\nu} z^{\nu}\\ &\quad - \frac{z^{\kappa}}{2\pi i}\int_{|\zeta|=1}\left[\zeta^{-\kappa}Tf(\zeta) + \zeta^{\kappa}\overline{Tf(\zeta)}\right]\left(\frac{\zeta}{\zeta - z} - \frac{1}{2}\right)\frac{d\zeta}{\zeta}\end{aligned}$$

follows if $0 \le \kappa$. Because then, see [5], Lemma 5.27 and Corollary 5.29

$$\frac{1}{2\pi i}\int_{|\zeta|=1} \zeta^{-\kappa} Tf(\zeta)\frac{d\zeta}{\zeta - z} = 0\,,$$

$$\frac{1}{2\pi i}\int_{|\zeta|=1} \zeta^{\kappa}\overline{Tf(\zeta)}\frac{d\zeta}{\zeta - z} = \frac{z^{\kappa}}{\pi}\int_{|\zeta|<1} \frac{z}{1 - z\overline{\zeta}}\overline{f(\zeta)}d\xi d\eta \tag{4.16}$$

we have

$$\begin{aligned} \varphi(z) &= \frac{z^{\kappa}}{2\pi i}\int_{|\zeta|=1} \gamma(\zeta)\left[\frac{2\zeta}{\zeta - 2} - 1\right]\frac{d\zeta}{\zeta} \\ &\quad - \frac{z^{2\kappa}}{\pi}\int_{|\zeta|<1} \frac{z}{1 - z\overline{\zeta}}\overline{f(\zeta)}d\xi d\eta + \sum_{\nu=0}^{2\kappa} a_{\nu} z^{\nu} \end{aligned}$$

and hence with $a_{\nu} + a_{2k-\nu} = 0,\ 0 \le \nu \le \kappa$,

$$\begin{aligned} w(z) &= \frac{z^{\kappa}}{2\pi i}\int_{|\zeta|=1} \gamma(\zeta)\left[\frac{2\zeta}{\zeta - z} - 1\right]\frac{d\zeta}{\zeta} \\ &\quad - \frac{1}{\pi}\int_{|\zeta|<1} \left(\frac{f(\zeta)}{\zeta - z} + \frac{z^{2\kappa+1}\overline{f(\zeta)}}{1 - z\overline{\zeta}}\right) d\xi d\eta + \sum_{\nu=0}^{2\kappa} a_{\nu} z^{\nu}\,. \end{aligned} \tag{4.17}$$

For $\kappa < 0$ from (4.10)

$$\varphi(z) = \frac{2}{2\pi i}\int_{|\zeta|=1} \gamma(\zeta)\frac{\zeta^{\kappa}}{\zeta - z}d\zeta - \frac{1}{2\pi i}\int_{|\zeta|=1} [Tf(\zeta) + \zeta^{2\kappa}\overline{Tf(\zeta)}]\frac{d\zeta}{\zeta - z}$$

if (4.9) is satisfied with γ_0 instead of γ. Observing

$$\frac{1}{2\pi i}\int_{|\zeta|=1} Tf(\zeta)\frac{d\zeta}{\zeta - z} = 0\,,$$

$$\frac{1}{2\pi i}\int_{|\zeta|=1} \zeta^{2\kappa}\overline{Tf(\zeta)}\frac{d\zeta}{\zeta - z} = \frac{1}{\pi}\int_{|\zeta|<1} \overline{\zeta^{-2\kappa-1}\,f(\zeta)}\frac{d\xi d\eta}{1 - z\overline{\zeta}} \tag{4.18}$$

shows

$$\begin{aligned} w(z) &= \frac{2}{2\pi i}\int_{|\zeta|=1} \gamma(\zeta)\frac{\zeta^{\kappa}}{\zeta - z}d\zeta \\ &\quad - \frac{1}{\pi}\int_{|\zeta|<1} \left(\frac{f(\zeta)}{\zeta - z} + \frac{\overline{\zeta^{-2\kappa-1}f(\zeta)}}{1 - z\overline{\zeta}}\right) d\xi d\eta\,. \end{aligned} \tag{4.19}$$

The solvability condition reads

$$\frac{1}{2\pi i}\int_{|\zeta|=1} \gamma(\zeta)\left[2\frac{1 - (z\overline{\zeta})^{-\kappa}}{1 - z\overline{\zeta}} - 1\right]\frac{d\zeta}{\zeta}$$

$$
\begin{aligned}
&= \frac{1}{2\pi i}\int_{|\zeta|=1}[\zeta^{-\kappa}Tf(\zeta)+\zeta^{\kappa}\overline{Tf(\zeta)}]\left[\frac{1-(z\overline{\zeta})^{-\kappa}}{1-z\overline{\zeta}}-\frac{1}{2}\right]\frac{d\zeta}{\zeta} \\
&= \frac{1}{\pi}\int_{|\zeta|<1}\left\{f(\zeta)\left[\frac{\zeta^{-\kappa}-z^{-\kappa}}{\zeta-z}-\frac{\zeta^{-\kappa-1}}{2}\right]\right. \\
&\qquad \left. +\overline{f(\zeta)}\,\overline{\zeta}^{-\kappa-1}\left[\frac{1-(z\overline{\zeta})^{-\kappa}}{1-z\overline{\zeta}}-\frac{1}{2}\right]\right\}d\xi d\eta\,.
\end{aligned}
\tag{4.20}
$$

Combining (4.17) with (4.18) gives

$$
\begin{aligned}
z^{-\kappa}w(z) &= \frac{1}{2\pi i}\int_{|\zeta|=1}\gamma(\zeta)\left[\frac{2\zeta}{\zeta-z}-1\right]\frac{d\zeta}{\zeta} \\
&\quad -\frac{1}{\pi}\int_{|\zeta|<1}\left\{\zeta^{-\kappa-1}f(\zeta)\frac{\zeta+z}{\zeta-z}+\overline{\zeta^{-\kappa-1}f(\zeta)}\frac{1+z\overline{\zeta}}{1-z\overline{\zeta}}\right\}d\xi d\eta
\end{aligned}
$$

from which in fact follows

$$\mathrm{Re}\,\{\overline{z}^{\kappa}w(z)\}=\gamma \ \text{ for } \ |z|=1\,.$$

The following result is proved.

Theorem 1 *The special Riemann–Hilbert problem* (4.7) *for the inhomogeneous Cauchy–Riemann equation* (4.15) *is solvable for* $0 \le \kappa$. *The general solution is given by* (4.17). *For* $\kappa < 0$ *the related homogeneous problem is only trivially solvable. Under the condition* (4.20) *the inhomogeneous problem is uniquely solvable and the solution is given by* (4.19).

The Cauchy–Riemann system in $\mathbb{C}^2$ is

$$w_{\overline{z_k}} = f_k\ ,\ 1\le k\le 2\ ;\ f_{1\overline{z_2}} = f_{2\overline{z_1}}\,. \tag{4.21}$$

A particular solution to this system in a bidomain $D^2 = D_1\times D_2, D_k\subset\mathbb{C}\,, 1\le k\le 2$, is

$$
\begin{aligned}
w_0 &= T_1f_1+T_2f_2-T_2T_1f_{1\overline{\zeta_2}}\,; \\
T_kf(z_k) &:= -\frac{1}{\pi}\int_{D_k}\frac{f(\zeta_k)}{\zeta_k-z_k}d\xi_k d\eta_k,\ \ 1\le k\le 2\,.
\end{aligned}
\tag{4.22}
$$

This can easily be verified by differentiation. The general solution to (4.21) is $w=\varphi+w_0$ with an arbitrary analytic function φ in D^2. Let in the following D^2 be the bidisc $\mathbb{D}^2=\mathbb{D}\times\mathbb{D}$.

The Riemann–Hilbert problem (4.11) for (4.21) means

$$\mathrm{Re}\left\{\overline{\zeta_1}^{\kappa_1}\overline{\zeta_2}^{\kappa_2}\varphi(\zeta)\right\}=\gamma_0(\zeta):=\gamma(\zeta)-\mathrm{Re}\left\{\overline{\zeta_1}^{\kappa_1}\overline{\zeta_2}^{\kappa_2}w_0(\zeta)\right\}\ \text{ on }\ \partial_0\mathbb{D}^2\,.$$

This problem can be handled via Lemma 2.
(1) $0 \leq \kappa_1, \kappa_2$. φ is of the form (4.12) with

$$\psi(z) = \frac{1}{(2\pi i)^2} \int_{\partial_0 \mathbb{D}^2} \gamma(\zeta) \left[2 \frac{\zeta_1}{\zeta_1 - \zeta_2} \frac{\zeta_2}{\zeta_2 - z_2} - 1 \right] \frac{d\zeta_1}{\zeta_1} \frac{d\zeta_2}{\zeta_2}$$

$$- \frac{1}{(2\pi i)^2} \int_{\partial_0 \mathbb{D}^2} \left[\overline{\zeta_1}^{\kappa_1} \overline{\zeta_2}^{\kappa_2} w_0(\zeta) + \zeta_1^{\kappa_1} \zeta_2^{\kappa_2} \overline{w_0(\zeta)} \right] \tag{4.23}$$

$$\times \left[\frac{\zeta_1}{\zeta_1 - \zeta_2} \frac{\zeta_2}{\zeta_2 - z_2} - \frac{1}{2} \right] \frac{d\zeta_1}{\zeta_1} \frac{d\zeta_2}{\zeta_2}$$

if and only if (4.6) is satisfied with γ replaced by γ_0. From (4.16) and the Cauchy–Pompeiu formula

$$\frac{1}{(2\pi i)^2} \int_{\partial_0 \mathbb{D}^2} \overline{\zeta_1}^{\kappa_1} \overline{\zeta_2}^{\kappa_2} w_0(\zeta) \left[\frac{\zeta_1}{\zeta_1 - z_1} \frac{\zeta_2}{\zeta_2 - z_2} - \frac{1}{2} \right] \frac{d\zeta_1}{\zeta_1} \frac{d\zeta_2}{\zeta_2} = 0 ,$$

$$\frac{1}{(2\pi i)^2} \int_{\partial_0 \mathbb{D}^2} \zeta_1^{\kappa_1} \zeta_2^{\kappa_2} \overline{w_0(\zeta)} \left[\frac{\zeta_1}{\zeta_1 - z_1} \frac{\zeta_2}{\zeta_2 - z_2} - \frac{1}{2} \right] \frac{d\zeta_1}{\zeta_1} \frac{d\zeta_2}{\zeta_2}$$

$$= \frac{z_1^{\kappa_1}}{2\pi^2 i} \int_{|\zeta_1|<1} \int_{|\zeta_2|=1} \zeta_2^{\kappa_2} \overline{f_1(\zeta)} \frac{d\zeta_2}{\zeta_2 - z_2} \frac{z_1 d\xi_1 d\eta_1}{1 - z_1 \overline{\zeta_1}}$$

$$+ \frac{z_2^{\kappa_2}}{2\pi^2 i} \int_{|\zeta_2|<1} \int_{|\zeta_1|=1} \zeta_1^{\kappa_1} \overline{f_2(\zeta)} \frac{d\zeta_1}{\zeta_1 - z_1} \frac{z_2 d\xi_2 d\eta_2}{1 - z_2 \overline{\zeta_2}}$$

$$- \frac{z_1^{\kappa_1} z_2^{\kappa_2}}{\pi^2} \int_{\mathbb{D}^2} \overline{f_{1\overline{\zeta_2}}(\zeta)} \frac{z_1 d\xi_1 d\eta_1}{1 - z_1 \overline{\zeta_1}} \frac{z_2 d\xi_2 d\eta_2}{1 - z_2 \overline{\zeta_2}}$$

$$= \frac{z_1^{\kappa_1} z_2^{\kappa_2}}{\pi} \int_{|\zeta_1|<1} \overline{f_1(\zeta_1, z_2)} \frac{z_1 d\xi_1 d\eta_1}{1 - z_1 \overline{\zeta_1}} + \frac{z_1^{\kappa_1} z_2^{\kappa_2}}{\pi} \int_{|\zeta_2|<1} \overline{f_2(z_1, \zeta_2)} \frac{z_2 d\xi_2 d\eta_2}{1 - z_2 \overline{\zeta_2}}$$

$$+ \frac{1}{\pi^2} \int_{\mathbb{D}^2} \left\{ z_1^{\kappa_1} \zeta_2^{\kappa_2} \overline{f_{1\zeta_2}(\zeta)} \frac{z_1}{1 - z_2 \overline{\zeta_1}} \frac{1}{\zeta_2 - z_2} + \zeta_1^{\kappa_1} z_2^{\kappa_2} \overline{f_{2\zeta_1}(\zeta)} \frac{1}{\zeta_1 - z_1} \frac{z_2}{1 - z_2 \overline{\zeta_2}} \right.$$

$$\left. - z_1^{\kappa_1} z_2^{\kappa_2} \overline{f_{1\overline{\zeta_2}}(\zeta)} \frac{z_1}{1 - z_1 \overline{\zeta_1}} \frac{z_2}{1 - z_2 \overline{\zeta_2}} \right\} d\xi_1 d\eta_1 d\xi_2 d\eta_2 .$$

Thus the solution to (4.21), (4.11) in the case $0 \leq \kappa_1, \kappa_2$ is

$$w(z) = \frac{z_1^{\kappa_1} z_2^{\kappa_2}}{(2\pi i)^2} \int_{\partial_0 \mathbb{D}^2} \gamma(\zeta) \left[2 \frac{\zeta_1}{\zeta_1 - z_1} \frac{\zeta_2}{\zeta_2 - z_2} - 1 \right] \frac{d\zeta_1}{\zeta_1} \frac{d\zeta_2}{\zeta_2} + \sum_{\nu_1=0}^{2\kappa_1} \sum_{\nu_2=0}^{2\kappa_2} a_{\nu_1 \nu_2} z_1^{\nu_1} z_2^{\nu_2}$$

$$- \frac{1}{\pi} \int_{|\zeta_1|<1} \left\{ \frac{f_1(\zeta_1, z_2)}{\zeta_1 - z_1} + \frac{z_1^{2\kappa_1} z_2^{2\kappa_2} z_1 \overline{f_1(\zeta_1, z_2)}}{1 - z_1 \overline{\zeta_1}} \right\} d\xi_1 d\eta_1 \tag{4.24}$$

$$- \frac{1}{\pi} \int_{|\zeta_2|<1} \left\{ \frac{f_2(z_1, \zeta_2)}{\zeta_2 - z_2} + \frac{z_1^{2\kappa_1} z_2^{2\kappa_2} z_2 \overline{f_2(z_1, \zeta_2)}}{1 - z_2 \overline{\zeta_2}} \right\} d\xi_2 d\eta_2$$

$$
\begin{aligned}
-\frac{1}{\pi^2}\int_{\mathbb{D}^2}\Big\{& z_1^{2\kappa_1} z_2^{\kappa_2}\zeta_2^{\kappa_2}\overline{f_{1\zeta_2}(\zeta)}\frac{z_1}{1-z_1\overline{\zeta_1}}\frac{1}{\zeta_2-z_2} \\
&+z_1^{\kappa_1}\zeta_1^{\kappa_1} z_2^{2\kappa_2}\overline{f_{2\zeta_1}(\zeta)}\frac{1}{\zeta_1-z_1}\frac{z_2}{1-z_2\overline{\zeta_2}} \\
&+\frac{f_{1\overline{\zeta_2}}(\zeta)}{(\zeta_1-z_1)(\zeta_2-z_2)}-\frac{z_1^{2\kappa_1} z_2^{2\kappa_2} z_1 z_2\overline{f_{1\overline{\zeta_2}}(\zeta)}}{(1-z_1\overline{\zeta_1})(1-z_2\overline{\zeta_2})}\Big\} d\xi_1 d\eta_1 d\xi_2 d\eta_2 .
\end{aligned}
$$

The solvability condition (4.6) for γ_0 means

$$
\begin{aligned}
I &:= \frac{1}{(2\pi i)^2}\int_{\partial_0\mathbb{D}^2}\gamma(\zeta)\left[\frac{z_1}{\zeta_1-z_1}\frac{\overline{z_2}}{\overline{\zeta_2-z_2}}+\frac{\overline{z_1}}{\overline{\zeta_1-z_1}}\frac{z_2}{\zeta_2-z_2}\right]\frac{d\zeta_1}{\zeta_1}\frac{d\zeta_2}{\zeta_2} \\
&= \frac{1}{2(2\pi i)^2}\int_{\partial_0\mathbb{D}^2}\left\{\overline{\zeta_1}^{\kappa_1}\overline{\zeta_2}^{\kappa_2} w_0(\zeta)+\zeta_1^{\kappa_1}\zeta_2^{\kappa_2}\overline{w_0(\zeta)}\right\} \qquad (4.25) \\
&\quad\times\left[\frac{z_1}{\zeta_1-z_1}\frac{\overline{z_2}}{\overline{\zeta_2-z_2}}+\frac{\overline{z_1}}{\overline{\zeta_{1-z_1}}}\frac{z_2}{\zeta_2-z_2}\right]\frac{d\zeta_1}{\zeta_1}\frac{d\zeta_2}{\zeta_2} .
\end{aligned}
$$

Treating the right–hand side as before the respective terms from (4.23) gives

$$
\begin{aligned}
I &= \mathrm{Re}\left\{\frac{1}{(2\pi i)^2}\int_{\partial_0\mathbb{D}^2}\left[\overline{\zeta_1}^{\kappa_1}\overline{\zeta_2}^{\kappa_2} w_0(\zeta)+\zeta_1^{\kappa_1}\zeta_2^{\kappa_2}\overline{w_0(\zeta)}\right]\right. \\
&\qquad\left.\times\frac{z_1}{\zeta_1-z_1}\frac{\overline{z_2}}{\overline{\zeta_2-z_2}}\frac{d\zeta_1}{\zeta_1}\frac{d\zeta_2}{\zeta_2}\right\} \\
&= \mathrm{Re}\left\{\frac{1}{(2\pi i)^2}\int_{\partial_0\mathbb{D}^2}\left[\overline{\zeta_1}^{\kappa_1}\overline{\zeta_2}^{\kappa_2} T_2 f_2(\zeta)+\zeta_1^{\kappa_1}\zeta_2^{\kappa_2}\overline{T_1 f_1(\zeta)}\right]\right. \\
&\qquad\left.\times\frac{z_1}{\zeta_1-z_1}\frac{\overline{z_2}}{\overline{\zeta_2-z_2}}\frac{d\zeta_1}{\zeta_1}\frac{d\zeta_2}{\zeta_2}\right\} \\
&= \mathrm{Re}\left\{\frac{1}{2\pi^2 i}\int_{|\zeta_2|<1}\int_{|\zeta_1|=1}\overline{\zeta_1}^{\kappa_1}\overline{z_2}^{\kappa_2} f_2(\zeta)\frac{z_1}{\zeta_1-z_1}\frac{d\zeta_1}{\zeta_1}\frac{\overline{z_2}d\xi_2 d\eta_2}{1-\overline{z_2}\zeta_2}\right. \\
&\qquad\left.+\frac{1}{2\pi^2 i}\int_{|\zeta_1|<1}\int_{|\zeta_2|=1} z_1^{\kappa_1}\zeta_2^{\kappa_2}\overline{f_1(\zeta)}\frac{\overline{z_2}}{\overline{\zeta_2-z_2}}\frac{d\zeta_2}{\zeta_2}\frac{z_1 d\xi_1 d\eta_1}{1-z_1\overline{\zeta_1}}\right\}
\end{aligned}
$$

$$= \operatorname{Re}\left\{\frac{1}{\pi^2}\int_{\mathbb{D}^2}\left[\overline{\zeta_1}^{\kappa_1}\overline{z_2}^{\kappa_2}f_{2\zeta_1}(\zeta)\frac{z_1}{1-z_1\overline{\zeta_1}}\,\frac{\overline{z_2}}{1-\overline{z_2}\zeta_2}\right.\right.$$

$$\left.\left.+z_1^{\kappa_1}\zeta_2^{\kappa_2}\overline{f_{1\zeta_2}(\zeta)}\frac{z_1}{1-z_1\overline{\zeta_1}}\,\frac{\overline{z_2}}{1-\overline{z_2}\zeta_2}\right]d\xi_1 d\eta_1 d\xi_2 d\eta_2\right\}$$

as for $0 \leq \kappa$

$$\frac{1}{2\pi i}\int_{|\zeta|=1}\overline{\zeta}^{\kappa}f(\zeta)\frac{z}{\zeta-z}\,\frac{d\zeta}{\zeta} = \frac{1}{\pi}\int_{|\zeta|<1}\overline{\zeta}^{\kappa}f_{\zeta}(\zeta)\frac{z}{1-z\overline{\zeta}}d\xi d\eta\,. \tag{4.26}$$

Hence, the solvability condition for (4.21), (4.11) for $0 \leq \kappa_1, \kappa_2$ is

$$\operatorname{Re}\left\{\frac{2}{(2\pi i)^2}\int_{\partial_0\mathbb{D}^2}\gamma(\zeta)\frac{z_1}{\zeta_1-z_1}\,\frac{\overline{z}_2}{\zeta_2-z_2}\,\frac{d\zeta_1}{\zeta_1}\,\frac{d\zeta_2}{\zeta_2}\right\}$$

$$= \operatorname{Re}\left\{\frac{1}{\pi^2}\int_{\mathbb{D}^2}\left[z_1^{\kappa_1}\zeta_2^{\kappa_2}\overline{f_{1\zeta_2}(\zeta)}+\overline{\zeta_1}^{\kappa_1}\overline{z_2}^{\kappa_2}f_{2\zeta_1}(\zeta)\right]\right. \tag{4.27}$$

$$\left.\times\frac{z_1}{1-z_1\overline{\zeta_1}}\,\frac{\overline{z_2}}{1-\overline{z_2}\zeta_2}d\xi_1 d\eta_1 d\xi_2 d\eta_2\right\}.$$

(2) $\kappa_1 < 0 \leq \kappa_2$. φ has the form (4.14) under conditions (4.13), (4.6) with γ_0 instead of γ. (4.14) gives

$$\varphi(z) = \frac{2}{(2\pi i)^2}\int_{\partial_0\mathbb{D}^2}\gamma(\zeta)\frac{\zeta_1^{\kappa_1}}{\zeta_1-z_1}\,\frac{z_2^{\kappa_2}}{\zeta_2-z_2}d\zeta_1 d\zeta_2$$

$$-\frac{z_2^{\kappa_2}}{(2\pi i)^2}\int_{\partial_0\mathbb{D}^2}\left\{\overline{\zeta_2}^{\kappa_2}w_0(\zeta)+\zeta_1^{2\kappa_1}\zeta_2^{\kappa_2}\overline{w_0(\zeta)}\right\}\frac{d\zeta_1}{\zeta_1-z_1}\,\frac{d\zeta_2}{\zeta_2-z_2}$$

$$= \frac{2}{(2\pi i)^2}\int_{\partial_0\mathbb{D}^2}\gamma(\zeta)\frac{\zeta_1^{\kappa_1}}{\zeta_1-z_1}\,\frac{z_2^{\kappa_2}}{\zeta_2-z_2}d\zeta_1 d\zeta_2$$

$$-\frac{z_2^{\kappa_2}}{(2\pi i)^2}\int_{\partial_0\mathbb{D}^2}\zeta_1^{2\kappa_1}\zeta_2^{\kappa_2}\overline{w_0(\zeta)}\frac{d\zeta_1}{\zeta_1-z_1}\,\frac{d\zeta_2}{\zeta_2-z_2}$$

$$= \frac{2}{(2\pi i)^2}\int_{\partial_0\mathbb{D}^2}\gamma(\zeta)\frac{\zeta_1^{\kappa_1}}{\zeta_1-z_1}\,\frac{z_2^{\kappa_2}}{\zeta_2-z_2}d\zeta_1 d\zeta_2$$

$$-\frac{z_2^{2\kappa_2}}{\pi}\int_{|\zeta_1|<1}\overline{\zeta_1}^{-2\kappa_1-1}\overline{w_{0\overline{\zeta_1}}(\zeta_1,z_2)}\frac{d\xi_1 d\eta_1}{1-z_1\overline{\zeta_1}}$$

$$-\frac{1}{\pi^2}\int_{\mathbb{D}^2}\overline{\zeta_1}^{-2\kappa_1-1}\zeta_2^{\kappa_2}z_2^{\kappa_2}\overline{w_{0\overline{\zeta_1}\zeta_2}(\zeta)}\frac{d\xi_1 d\eta_1}{1-z_1\overline{\zeta_1}}\,\frac{d\xi_2 d\eta_2}{\zeta_2-z_2}\,.$$

Observing $w_{0\overline{z_1}} = f_1$ then

$$\begin{aligned} w(z) &= \frac{2}{(2\pi i)^2}\int_{\partial_0 \mathbb{D}^2} \gamma(\zeta)\frac{\zeta_1^{\kappa_1}}{\zeta_1 - z_1}\frac{z_2^{\kappa_2}}{\zeta_2 - z_2} d\zeta_1 d\zeta_2 \\ &\quad -\frac{1}{\pi}\int_{|\zeta_1|<1}\left\{\frac{f_1(\zeta_1, z_2)}{\zeta_1 - z_1} + \frac{\overline{\zeta_1}^{-2\kappa_1-1} z_2^{2\kappa_2}\overline{f_1(\zeta_1, z_2)}}{1 - z_1\overline{\zeta_1}}\right\} d\xi_1 d\eta_1 \\ &\quad -\frac{1}{\pi}\int_{|\zeta_2|<1}\frac{f_2(z_1, \zeta_2)}{\zeta_2 - z_2} d\xi_2 d\eta_2 \end{aligned}$$

$$-\frac{1}{\pi^2}\int_{\mathbb{D}^2}\left\{\frac{f_{1\overline{\zeta_2}}(\zeta)}{(\zeta_1 - z_1)(\zeta_2 - z_2)} + \frac{\overline{\zeta_1}^{-2\kappa_1-1}\zeta_2^{\kappa_2} z_2^{\kappa_2}\overline{f_{1\zeta_2}(\zeta)}}{(1 - z_1\overline{\zeta_1})(\zeta_2 - z_2)}\right\} d\xi_1 d\eta_1 d\xi_2 d\eta_2 \,. \quad (4.28)$$

Condition (4.6) means

$$\mathrm{Re}\,\frac{2}{(2\pi i)^2}\int_{\partial_0 \mathbb{D}^2} \gamma(\zeta)\frac{z_1}{\zeta_1 - z_1}\frac{\overline{z_2}}{\overline{\zeta_2 - z_2}}\frac{d\zeta_1}{\zeta_1}\frac{d\zeta_2}{\zeta_2}$$

$$= \mathrm{Re}\,\frac{1}{(2\pi i)^2}\int_{\partial_0 \mathbb{D}^2}\left\{\overline{\zeta_1}^{\kappa_1}\overline{\zeta_2}^{\kappa_2} w_0(\zeta) + \zeta_1^{\kappa_1}\zeta_2^{\kappa_2}\overline{w_0(\zeta)}\right\}\frac{z_1}{\zeta_1 - z_1}\frac{\overline{z_2}}{\overline{\zeta_2 - z_2}}\frac{d\zeta_1}{\zeta_1}\frac{d\zeta_2}{\zeta_2} \,.$$

From the Gauß and the Cauchy–Pompeiu formulas then

$$\begin{aligned} &\frac{1}{(2\pi i)^2}\int_{\partial_0 \mathbb{D}^2} \zeta_1^{-\kappa_1}\overline{\zeta_2}^{\kappa_2} w_0(\zeta)\frac{z_1}{\zeta_1 - z_1}\frac{\overline{z_2}}{\overline{\zeta_2 - z_2}}\frac{d\zeta_1}{\zeta_1}\frac{d\zeta_2}{\zeta_2} \\ &= \frac{1}{(2\pi i)^2}\int_{\partial_0 \mathbb{D}^2} \zeta_1^{-\kappa_1-1}\overline{\zeta_2}^{\kappa_2} w_0(\zeta)\frac{z_1}{\zeta_1 - z_1}\frac{\overline{z_2}}{1 - \overline{z_2}\zeta_2} d\zeta_1 d\zeta_2 \\ &= \frac{\overline{z_1}^{-\kappa_1}}{\pi}\int_{|\zeta_2|<1}\frac{\partial}{\partial\overline{\zeta_2}}(\overline{\zeta_2}^{\kappa_2} w_0(\zeta))\frac{\overline{z_2}}{1 - \overline{z_2}\zeta_2} d\xi_2 d\eta_2 \\ &\quad + \frac{1}{\pi^2}\int_{\mathbb{D}^2} \zeta_1^{-\kappa_2-1}\frac{\partial}{\partial\overline{\zeta_2}}(\overline{\zeta_2}^{\kappa_2} f_1(\zeta))\frac{z_1}{\zeta_1 - z_1}\frac{\overline{z_2}}{1 - \overline{z_2}\zeta_2} d\xi_1 d\eta_1 d\xi_2 d\eta_2 \\ &= \frac{\overline{z_1}^{-\kappa_1}}{\pi}\int_{|\zeta_2|<1}\frac{\partial}{\partial\overline{\zeta_2}}(\overline{\zeta_2}^{\kappa_2} T_1 f_1(\zeta))\frac{\overline{z_2}}{1 - \overline{z_2}\zeta_2} d\xi_2 d\eta_2 \\ &\quad + \frac{\overline{z_1}^{-\kappa_1}\overline{z_2}^{\kappa_2}}{\pi}\int_{|\zeta_2|<1}(f_2(\zeta) - T_1 f_{1\overline{\zeta_2}}(\zeta))\frac{\overline{z_2}}{1 - \overline{z_2}\zeta_2} d\xi_2 d\eta_2 \\ &\quad + \frac{1}{\pi^2}\int_{\mathbb{D}^2} \zeta_1^{-\kappa_1-1}\frac{\partial}{\partial\overline{\zeta_2}}(\overline{\zeta_2}^{\kappa_2} f_1(\zeta))\frac{z_1}{\zeta_1 - z_1}\frac{\overline{z_2}}{1 - \overline{z_2}\zeta_2} d\xi_1 d\eta_1 d\xi_2 d\eta_2 \end{aligned}$$

$$= \frac{z^{-\kappa_1}\overline{z_2}^{\kappa_2}}{\pi}\int_{|\zeta_2|<1} f_2(\zeta)\frac{\overline{z_2}}{1-\overline{z_2}\zeta_2}d\xi_2 d\eta_2$$

$$+\frac{z_1^{-\kappa_1}\overline{z_2}^{\kappa_2}}{\pi^2}\int_{\mathbb{D}^2} f_{1\overline{\zeta_2}}(\zeta))\frac{1}{\zeta_1-z_1}\frac{\overline{z_2}}{1-\overline{z_2}\zeta_2}d\xi_1 d\eta_1 d\xi_2 d\eta_2$$

$$+\frac{1}{\pi^2}\int_{\mathbb{D}^2}(\zeta_1^{-\kappa_1}-z_1^{-\kappa_1})\frac{\partial}{\partial\overline{\zeta_2}}(\overline{\zeta_2}^{\kappa_2} f_1(\zeta))\frac{1}{\zeta_1-z_1}\frac{\overline{z_2}}{1-\overline{z_2}\zeta_2}d\xi_1 d\eta_1 d\xi_2 d\eta_2$$

$$-\frac{1}{\pi^2}\int_{\mathbb{D}^2}\zeta_1^{-\kappa_1-1}\frac{\partial}{\partial\overline{\zeta_2}}(\overline{\zeta_2}^{\kappa_2} f_1(\zeta))\frac{\overline{z_2}}{1-\overline{z_2}\zeta_2}d\xi_1 d\eta_1 d\xi_2 d\eta_2 \,.$$

Here in the third step for $0 \leq \kappa$

$$\frac{1}{\pi}\int_{|\zeta|<1}\frac{\partial}{\partial\overline{\zeta}}(\overline{\zeta}^{\kappa} Tf(\zeta))\frac{\overline{z}}{1-\overline{z}\zeta}d\xi d\eta = \frac{\overline{z}^{\kappa}}{\pi}\int_{|\zeta|<1} f(\zeta)\frac{\overline{z}}{1-\overline{z}\zeta}d\xi d\eta$$

is used which holds together with $T(\overline{\zeta}^{\kappa}Tf)_{\overline{\zeta}} = \overline{z}^{\kappa}Tf$, see Corollary 5.28 in [5].

Similarly

$$\frac{1}{(2\pi i)^2}\int_{\partial_0\mathbb{D}^2}\zeta_1^{-\kappa_1}\overline{\zeta_2}^{\kappa_2} w_0(\zeta)\frac{\overline{z_1}}{\overline{\zeta_1-z_1}}\frac{z_2}{\zeta_2-z_2}\frac{d\zeta_1}{\zeta_1}\frac{d\zeta_2}{\zeta_2}$$

$$= -\frac{1}{(2\pi i)^2}\int_{\partial_0\mathbb{D}^2}\zeta_1^{-\kappa_1}\overline{\zeta_2}^{\kappa_2} w_0(\zeta)\frac{\overline{z_1}}{1-\overline{z_1}\zeta_1}\frac{z_2}{1-z_2\overline{\zeta_2}}d\zeta_1 d\overline{\zeta_2}$$

$$= \frac{1}{\pi^2}\int_{\mathbb{D}^2}\zeta_1^{-\kappa_1}\overline{\zeta_2}^{\kappa_2} f_{1\zeta_2}(\zeta)\frac{\overline{z_1}}{1-\overline{z_1}\zeta_1}\frac{z_2}{1-z_2\overline{\zeta_2}}d\xi_1 d\eta_1 d\xi_2 d\eta_2 \,.$$

Thus

$$\operatorname{Re}\frac{2}{(2\pi i)^2}\int_{\partial_0\mathbb{D}^2}\gamma(\zeta)\frac{z_1}{\zeta_1-z_1}\frac{\overline{z_2}}{\overline{\zeta_2-z_2}}\frac{d\zeta_1}{\zeta_1}\frac{d\zeta_2}{\zeta_2}$$

$$= \operatorname{Re}\left\{\frac{z_1^{-\kappa_1}\overline{z_2}^{\kappa_2}}{\pi}\int_{|\zeta_2|<1} f_2\frac{\overline{z_2}}{1-\overline{z_2}\zeta_2}\, d\xi_2 d\eta_2\right.$$

$$+\frac{z_1^{-\kappa_1}\overline{z_2}^{\kappa_2}}{\pi^2}\int_{\mathbb{D}^2} f_{1\overline{\zeta_2}}(\zeta)\frac{1}{\zeta_1-z_1}\frac{\overline{z_2}}{1-\overline{z_2}\zeta_2}d\xi_1 d\eta_1 d\xi_2 d\eta_2$$

$$+\frac{1}{\pi^2}\int_{\mathbb{D}^2}(\zeta_1^{-\kappa_1}-z_1^{-\kappa_1})\frac{\partial}{\partial\overline{\zeta_2}}(\overline{\zeta_2}^{\kappa_2} f_1(\zeta))\frac{1}{\zeta_1-z_1}\frac{\overline{z_2}}{1-\overline{z_2}\zeta_2}d\xi_1 d\eta_1 d\xi_2 d\eta_2$$

$$+\frac{1}{\pi^2}\int_{\mathbb{D}^2}\left\{\zeta_1^{-\kappa_1}\overline{\zeta_2}^{\kappa_2} f_{1\zeta_2}(\zeta)\frac{\overline{z_1}}{1-\overline{z_1}\zeta_1}\frac{z_2}{1-z_2\overline{\zeta_2}}\right.$$
$$\left.-\zeta_1^{-\kappa_1-1}\frac{\partial}{\partial\overline{\zeta_2}}(\overline{\zeta_2}^{\kappa_2} f_1(\zeta))\frac{\overline{z_2}}{1-\overline{z_2}\zeta_2}\right\}d\xi_1 d\eta_1 d\xi_2 d\eta_2\Bigg\}. \tag{4.29}$$

Condition (4.13) gives

$$\frac{2}{(2\pi i)^2}\int_{\partial_0\mathbb{D}^2}\gamma(\zeta)\left[2(1-(z_1\overline{\zeta_1})^{-\kappa_1})\frac{1}{1-z_1\overline{\zeta_1}}\frac{1}{1-z_2\overline{\zeta_2}}-1\right]\frac{d\zeta_1}{\zeta_1}\frac{d\zeta_2}{\zeta_2}$$
$$=\frac{1}{(2\pi i)^2}\int_{\partial_0\mathbb{D}^2}\left\{\zeta_1^{-\kappa_1}\overline{\zeta_2}^{\kappa_2}w_0(\zeta)+\overline{\zeta_1}^{-\kappa_1}\zeta_2^{\kappa_2}\overline{w_0(\zeta)}\right\}$$
$$\times\left[2(1-(z_1\overline{\zeta_1})^{-\kappa_1})\frac{1}{1-z_1\overline{\zeta_1}}\frac{1}{1-z_2\overline{\zeta_2}}-1\right]\frac{d\zeta_1}{\zeta_1}\frac{d\zeta_2}{\zeta_2}.$$

To simplify the right–hand side the relation

$$\frac{1}{2\pi i}\int_{|\zeta|=1}f(\zeta)\frac{d\zeta}{\zeta} = \frac{1}{2\pi i}\int_{|\zeta|=1}f(\zeta)\left[\frac{1}{\zeta-z}-\frac{z}{(\zeta-z)\zeta}\right]d\zeta$$
$$= f(z)+\frac{1}{\pi}\int_{|\zeta|<1}\left[\frac{f_{\overline{\zeta}}(\zeta)}{\zeta-z}-\frac{zf_{\zeta}(\zeta)}{1-z\overline{\zeta}}\right]d\xi d\eta$$

is used. The first term can be rewritten as

$$\frac{2}{(2\pi i)^2}\int_{\partial_0\mathbb{D}^2}\zeta_1^{-\kappa_1}\overline{\zeta_2}^{\kappa_2}w_0(\zeta)(1-z_1^{-\kappa_1}\overline{\zeta_1}^{-\kappa_1})\frac{d\zeta_1}{\zeta_1-z_1}\frac{d\zeta_2}{\zeta_2-z_2}$$
$$-\frac{1}{(2\pi i)^2}\int_{\partial_0\mathbb{D}^2}\zeta_1^{-\kappa_1}\overline{\zeta_2}^{\kappa_2}w_0(\zeta)\frac{d\zeta_1}{\zeta_1}\frac{d\zeta_2}{\zeta_2}$$

$$= \frac{2}{(2\pi i)^2} \int_{\partial_0 \mathbb{D}^2} (\zeta_1^{-\kappa_1} - z_1^{-\kappa_1}) \overline{\zeta_2}^{\kappa_2} w_0(\zeta) \frac{d\zeta_1}{\zeta_1 - z_1} \frac{d\zeta_2}{\zeta_2 - z_2}$$

$$- \frac{1}{(2\pi i)^2} \int_{\partial_0 \mathbb{D}^2} \zeta_1^{-\kappa_1 - 1} \overline{\zeta_2}^{\kappa_2} w_0(\zeta) d\zeta_1 \frac{d\zeta_2}{\zeta_2}$$

$$= \frac{2\overline{z_2}^{\kappa_2}}{\pi} \int_{|\zeta_1|<1} (\zeta_1^{-\kappa_1} - z_1^{-\kappa_1}) f_1(\zeta_1, z_2) \frac{d\xi_1 d\eta_1}{\zeta_1 - z_1}$$

$$+ \frac{2}{\pi^2} \int_{\mathbb{D}^2} (\zeta_1^{-\kappa_1} - z_1^{-\kappa_1}) \frac{\partial}{\partial \overline{\zeta_2}} (\overline{\zeta_2}^{\kappa_2} f_1(\zeta)) \frac{d\xi_1 d\eta_1}{\zeta_1 - z_1} \frac{d\xi_2 d\eta_2}{\zeta_2 - z_2}$$

$$- \frac{\overline{z_2}^{\kappa_2}}{\pi} \int_{|\zeta_1|<1} \zeta_1^{-\kappa_1 - 1} f_1(\zeta_1, z_2) d\xi_1 d\eta_1$$

$$- \frac{1}{\pi^2} \int_{\mathbb{D}^2} \zeta_1^{-\kappa_1 - 1} \Bigg[\frac{\partial}{\partial \overline{\zeta_2}} (\overline{\zeta_2}^{\kappa_2} f_1(\zeta)) \frac{1}{\zeta_2 - z_2}$$

$$- \overline{\zeta_2}^{\kappa_2} f_{1\zeta_2}(\zeta) \frac{z_2}{1 - z_2 \overline{\zeta_2}} \Bigg] d\xi_1 d\eta_1 d\xi_2 d\eta_2 \,.$$

The second term becomes

$$- \frac{1}{(2\pi i)^2} \int_{\partial_0 \mathbb{D}^2} \overline{\zeta_1}^{-\kappa_1} \zeta_2^{\kappa_2} \overline{w_0(\zeta)} \left[2(1 - z_1^{-\kappa_1} \overline{\zeta_1}^{-\kappa_1}) \frac{1}{1 - z_1 \overline{\zeta_1}} \frac{1}{\zeta_2 - z_2} - \frac{1}{\zeta_2} \right] \frac{d\overline{\zeta_1}}{\overline{\zeta_1}} d\zeta_2$$

$$= \frac{1}{2\pi^2 i} \int_{|\zeta_1|<1} \int_{|\zeta_2|=1} \overline{\zeta_1}^{-\kappa_1 - 1} \zeta_2^{\kappa_2} \overline{f_1(\zeta)}$$

$$\times \left[2(1 - z_1^{-\kappa_1} \overline{\zeta_1}^{-\kappa_1}) \frac{1}{1 - z_1 \overline{\zeta_1}} \frac{1}{\zeta_2 - z_2} - \frac{1}{\zeta_2} \right] d\xi_1 d\eta_1 d\zeta_2$$

$$= \frac{2 z_2^{\kappa_2}}{\pi} \int_{|\zeta_1|<1} \overline{\zeta_1}^{-\kappa_1 - 1} \overline{f_1(\zeta_1, z_2)} (1 - z_1^{-\kappa_1} \overline{\zeta_1}^{-\kappa_1}) \frac{d\xi_1 d\eta_1}{1 - z_1 \overline{\zeta_1}}$$

$$+ \frac{2}{\pi^2} \int_{\mathbb{D}^2} \overline{\zeta_1}^{-\kappa_1 - 1} \zeta_2^{\kappa_2} \overline{f_{1\zeta_2}(\zeta)} (1 - z_1^{-\kappa_1} \overline{\zeta_1}^{-\kappa_1}) \frac{d\xi_1 d\eta_1}{1 - z_1 \overline{\zeta_1}} \frac{d\xi_2 d\eta_2}{\zeta_2 - z_2}$$

$$- \frac{z_2^{\kappa_2}}{\pi} \int_{|\zeta_1|<1} \overline{\zeta_1}^{-\kappa_1 - 1} \overline{f_1(\zeta_1, z_2)} d\xi_1 d\eta_1$$

$$- \frac{1}{\pi^2} \int_{\mathbb{D}^2} \overline{\zeta_1}^{-\kappa_1 - 1} \Bigg[\zeta_2^{\kappa_2} \overline{f_{1\zeta_2}(\zeta)} \frac{1}{\zeta_2 - z_2}$$

$$- \frac{\partial}{\partial \zeta_2} (\zeta_2^{\kappa_2} \overline{f_1(\zeta)}) \frac{z_2}{1 - z_2 \overline{\zeta_2}} \Bigg] d\xi_1 d\eta_1 d\xi_2 d\eta_2 \,.$$

Hence condition (4.13) is

$$\frac{1}{(2\pi i)^2}\int_{\partial_0 \mathbb{D}^2} \gamma(\zeta)\left[2(1-z_1^{-\kappa_1}\overline{\zeta_1}^{-\kappa_1})\frac{1}{\zeta_1-z_1}\,\frac{1}{\zeta_2-z_2}-\frac{1}{\zeta_1\zeta_2}\right]d\zeta_1 d\zeta_2 \tag{4.30}$$

$$= \frac{1}{\pi}\int_{|\zeta_1|<1}\Bigg\{(\zeta_1^{-\kappa_1}-z_1^{-\kappa_1})\overline{z_2}^{\kappa_2}\frac{f_1(\zeta_1,z_2)}{\zeta_1-z_1}+\overline{\zeta_1}^{-\kappa_1-1}z_2^{\kappa_2}(1-z_1^{-\kappa_1}\overline{\zeta_1}^{-\kappa_1})$$

$$\times\frac{\overline{f_1(\zeta_1,z_2)}}{1-z_1\overline{\zeta_1}}-\frac{1}{2}\zeta_1^{-\kappa_1-1}\overline{z_2}^{\kappa_2}f_1(\zeta_1,z_2)-\frac{1}{2}\overline{\zeta_1}^{-\kappa_1-1}z_2^{\kappa_2}\overline{f_1(\zeta_1,z_2)}\Bigg\}d\xi_1 d\eta_1$$

$$+\frac{1}{\pi^2}\int_{\mathbb{D}^2}\Bigg\{(\zeta_1^{-\kappa_1}-z_1^{-\kappa_1})\frac{\partial}{\partial\overline{\zeta_2}}(\overline{\zeta_2}^{\kappa_2}f_1(\zeta))\frac{1}{\zeta_1-z_1}\,\frac{1}{\zeta_2-z_2}$$

$$+\overline{\zeta_1}^{-\kappa_1-1}\zeta_2^{\kappa_2}\overline{f_{1\zeta_2}(\zeta)}(1-z_1^{-\kappa_1}\overline{\zeta_1}^{-\kappa_1})\frac{1}{1-z_1\overline{\zeta_1}}\,\frac{1}{\zeta_2-z_2}$$

$$-\frac{1}{2}\zeta_1^{-\kappa_1-1}\frac{\partial}{\partial\overline{\zeta_2}}(\overline{\zeta_2}^{\kappa_2}f_1(\zeta))\frac{1}{\zeta_2-z_2}$$

$$+\frac{1}{2}\overline{\zeta_1}^{-\kappa_1-1}\frac{\partial}{\partial\zeta_2}(\zeta_2^{\kappa_2}\overline{f_1(\zeta)})\frac{z_2}{1-z_2\overline{\zeta_2}}+\frac{1}{2}\zeta_1^{-\kappa_1-1}\overline{\zeta_2}^{\kappa_2}f_{1\zeta_2}(\zeta)\frac{z_2}{1-z_2\overline{\zeta_2}}$$

$$-\frac{1}{2}\overline{\zeta_1}^{-\kappa_1-1}\zeta_2^{\kappa_2}\overline{f_{1\zeta_2}(\zeta)}\frac{1}{\zeta_2-z_2}\Bigg\}d\xi_1 d\eta_1 d\xi_2 d\eta_2\,.$$

In order to check that (4.28) is the solution to (4.21), (4.11) for $\kappa_1 < 0 \le \kappa_2$ under conditions (4.29), (4.30) we proceed as follows. Multiplying (4.28) with $z_1^{-\kappa_1}\overline{z_2}^{\kappa_2}$ and adding (4.30) gives

$$z_1^{-\kappa_1}\overline{z_2}^{\kappa_2}w(z)=\frac{1}{(2\pi i)^2}\int_{\partial_0\mathbb{D}^2}\gamma(\zeta)\left(2\frac{\zeta_1}{\zeta_1-z_1}\,\frac{\zeta_2}{\zeta_2-z_2}-1\right)\frac{d\zeta_1}{\zeta_1}\,\frac{d\zeta_2}{\zeta_2}$$

$$-\frac{1}{2\pi}\int_{|\zeta_1|<1}\Bigg\{\zeta_1^{-\kappa_1-1}\overline{z_2}^{\kappa_2}f_1(\zeta_1,z_2)\frac{\zeta_1+z_1}{\zeta_1-z_1}$$

$$+\overline{\zeta_1}^{-\kappa_1-1}z_2^{\kappa_2}\overline{f_1(\zeta_1,z_2)}\frac{1+z_1\overline{\zeta_1}}{1-z_1\overline{\zeta_1}}\Bigg\}d\xi_1 d\eta_1$$

$$-\frac{z_1^{-\kappa_1}\overline{z_2}^{\kappa_2}}{\pi}\int_{|\zeta_2|<1}\frac{f_2(z_1,z_2)}{\zeta_2-z_2}d\xi_2 d\eta_2$$

$$-\frac{1}{\pi^2}\int_{\mathbb{D}^2}\left\{\frac{z_1^{-\kappa_1}\overline{z_2}^{\kappa_2}f_{1\overline{\zeta_2}}(\zeta)}{(\zeta_1-z_1)(\zeta_2-z_2)}+\frac{\overline{\zeta_1}^{\kappa_1-1}\zeta_2^{\kappa_2}\overline{f_{1\zeta_2}(\zeta)}}{(1-z_1\overline{\zeta_1})(\zeta_2-z_2)}\right.$$

$$+\frac{1}{2}\frac{\zeta_1^{-\kappa_1-1}\overline{\zeta_2}^{\kappa_2}z_2}{1-z_2\overline{\zeta_2}}f_{1\zeta_2}(\zeta)-\frac{1}{2}\frac{\overline{\zeta_1}^{-\kappa_1-1}\zeta_2^{\kappa_2}}{\zeta_2-z_2}\overline{f_{1\zeta_2}(\zeta)}$$

$$+\frac{\zeta_1^{-\kappa_1}-z_1^{-\kappa_1}}{(\zeta_1-z_1)(\zeta_2-z_2)}\frac{\partial}{\partial\overline{\zeta_2}}(\overline{\zeta_2}^{\kappa_2}f_1(\zeta))-\frac{1}{2}\frac{\zeta_1^{-\kappa_1-1}}{\zeta_2-z_2}\frac{\partial}{\partial\overline{\zeta_2}}(\overline{\zeta_2}^{\kappa_2}f_1(\zeta))$$

$$\left.+\frac{1}{2}\frac{\overline{\zeta_1}^{-\kappa_1-1}z_2}{1-z_2\overline{\zeta_2}}\frac{\partial}{\partial\zeta_2}(\zeta_2^{\kappa_2}\overline{f_1(\zeta)})\right\}d\xi_1d\eta_1d\xi_2d\eta_2$$

so that on $\partial_0\mathbb{D}^2$

$$\operatorname{Re}\{z_1^{-\kappa_1}\overline{z_2}^{\kappa_2}w(z)\}=\operatorname{Re}\frac{1}{(2\pi i)^2}\int_{\partial_0\mathbb{D}^2}\gamma(\zeta)\left(2\frac{\zeta_1}{\zeta_1-z_1}\frac{\zeta_2}{\zeta_2-z_2}-1\right)\frac{d\zeta_1}{\zeta_1}\frac{d\zeta_2}{\zeta_2}$$

$$-\operatorname{Re}\frac{z_1^{-\kappa_1}\overline{z_2}^{\kappa_2}}{\pi}\int_{|\zeta_2|<1}\frac{f_2(z_1,\zeta_2)}{\zeta_2-z_2}d\xi_2d\eta_2$$

$$-\operatorname{Re}\frac{1}{\pi^2}\int_{\mathbb{D}^2}\left\{\frac{z_1^{-\kappa_1}\overline{z_2}^{\kappa_2}f_{1\overline{\zeta_2}}(\zeta)}{(\zeta_1-z_1)(\zeta_2-z_2)}-\frac{\overline{\zeta_1}^{-\kappa_1}\zeta_2^{\kappa_2}\overline{f_{1\zeta_2}(\zeta)}}{\overline{(\zeta_1-z_1)(\zeta_2-z_2)}}\right.$$

$$\left.+\frac{\zeta_1^{-\kappa_1-1}z_1-z_1^{-\kappa_1}}{(\zeta_1-z_1)(\zeta_2-z_2)}\frac{\partial}{\partial\overline{\zeta_2}}(\overline{\zeta_2}^{\kappa_2}f_1(\zeta))\right\}d\xi_1d\eta_1d\xi_2d\eta_2\,.$$

From (4.29) this is just

$$\operatorname{Re}\{z_1^{-\kappa_1}\overline{z_2}^{\kappa_2}w(z)\}$$

$$=\frac{1}{(2\pi i)^2}\int_{\partial_0\mathbb{D}^2}\gamma(\zeta)\left(\frac{\zeta_1}{\zeta_1-z_1}+\frac{\overline{\zeta_1}}{\overline{\zeta_1-z_1}}-1\right)$$

$$\times\left(\frac{\zeta_2}{\zeta_2-z_2}+\frac{\overline{\zeta_2}}{\overline{\zeta_2-z_2}}-1\right)\frac{d\zeta_1}{\zeta_1}\frac{d\zeta_2}{\zeta_2}=\gamma(z)$$

on $\partial_0\mathbb{D}^2$. That (4.28) solves system (4.21) follows at once by differentiation and observing the compatibility condition.

(3) $\kappa_1,\kappa_2<0$. φ is given by (4.14′) under conditions (4.13′), (4.6) with γ replaced by γ_0. (4.14′) leads to

$$\varphi(z) = \frac{2}{(2\pi i)^2}\int_{\partial_0\mathbb{D}^2}\gamma(z)\frac{\zeta_1^{\kappa_1}}{\zeta_1-z_1}\frac{\zeta_2^{\kappa_2}}{\zeta_2-z_2}d\zeta_1d\zeta_2$$

$$-\frac{1}{(2\pi i)^2}\int_{\partial_0\mathbb{D}^2}\left\{w_0(\zeta)+\zeta_1^{2\kappa_1}\zeta_2^{2\kappa_2}\overline{w_0(\zeta)}\right\}\frac{d\zeta_1}{\zeta_1-z_1}\frac{d\zeta_2}{\zeta_2-z_2}\,.$$

As

$$\frac{1}{(2\pi i)^2}\int_{\partial_0 \mathbb{D}^2} w_0(\zeta)\frac{d\zeta_1}{\zeta_1 - z_1}\,\frac{d\zeta_2}{\zeta_2 - z_2} = 0\,,$$

$$\frac{1}{(2\pi i)^2}\int_{\partial_0 \mathbb{D}^2} \overline{\zeta_1}^{-2\kappa_1}\overline{\zeta_2}^{-2\kappa_2}\overline{w_0(\zeta)}\frac{d\zeta_1}{\zeta_1 - z_1}\,\frac{d\zeta_2}{\zeta_2 - z_2}$$

$$= \frac{1}{(2\pi i)^2}\int_{\partial_0 \mathbb{D}^2} \overline{\zeta_1}^{-2\kappa_1-1}\overline{\zeta_2}^{-2\kappa_2-1}\overline{w_0(\zeta)}\frac{d\overline{\zeta}_1}{\overline{1 - z_1\overline{\zeta_1}}}\,\frac{d\overline{\zeta}_2}{\overline{1 - z_2\overline{\zeta_2}}}$$

$$= \frac{1}{\pi^2}\int_{\mathbb{D}^2} \overline{\zeta_1}^{-2\kappa_1-1}\overline{\zeta_2}^{-2\kappa_2-1}\overline{w_{0\overline{\zeta_1}\,\overline{\zeta_2}}}\frac{d\xi_1 d\eta_1}{\overline{1 - z_1\overline{\zeta_1}}}\,\frac{d\xi_2 d\eta_2}{\overline{1 - z_2\overline{\zeta_2}}}$$

then

$$\begin{aligned} w(z) \;=\; & \frac{2}{(2\pi i)^2}\int_{\partial_0 \mathbb{D}^2} \gamma(\zeta)\frac{\zeta_1^{\kappa_1}}{\zeta_1 - z_1}\,\frac{\zeta_2^{\kappa_2}}{\zeta_2 - z_2}d\zeta_1 d\zeta_2 \\ & -\frac{1}{\pi}\int_{|\zeta_1|<1}\frac{f_1(\zeta_1 z_2)}{\zeta_1 - z_1}d\xi_1 d\eta_1 - \frac{1}{\pi}\int_{|\zeta_2|<1}\frac{f_2(z_1,\zeta_2)}{\zeta_2 - z_2}d\xi_2 d\eta_2 \\ & -\frac{1}{\pi^2}\int_{\mathbb{D}^2}\left\{\frac{f_{1\overline{\zeta_2}}(\zeta)}{(\zeta_1 - z_1)(\zeta_2 - z_2)}\right. \\ & \left. + \frac{\overline{\zeta_1}^{-2\kappa_1-1}\overline{\zeta_2}^{-2\kappa_2-1}\overline{f_{1\overline{\zeta_2}}(\zeta)}}{(1 - z_1\overline{\zeta_1})(1 - z_2\overline{\zeta_2})}\right\}d\xi_1 d\eta_1 d\xi_2 d\eta_2\,. \end{aligned} \tag{4.31}$$

Condition (4.6) gives as before

$$\mathrm{Re}\,\frac{2}{(2\pi i)^2}\int_{\partial_0 \mathbb{D}^2} \gamma(\zeta)\frac{z_1}{\zeta_1 - z_1}\,\frac{\overline{z_2}}{\overline{\zeta_2 - z_2}}\,\frac{d\zeta_1}{\zeta_1}\,\frac{d\zeta_2}{\zeta_2}$$

$$= \mathrm{Re}\,\frac{1}{(2\pi i)^2}\int_{\partial_0 \mathbb{D}^2}\left\{\overline{\zeta_1}^{\kappa_1}\overline{\zeta_2}^{\kappa_2}w_0(\zeta) + \zeta_1^{\kappa_1}\zeta_2^{\kappa_2}\overline{w_0(\zeta)}\right\}\frac{z_1}{\zeta_1 - z_1}\,\frac{\overline{z_2}}{\overline{\zeta_2 - z_2}}\,\frac{d\zeta_1}{\zeta_1}\,\frac{d\zeta_2}{\zeta_2}\,.$$

Here similarly as in case (2)

$$\frac{1}{(2\pi i)^2}\int_{\partial_0 \mathbb{D}^2} \zeta_1^{-\kappa_1}\zeta_2^{-\kappa_2}w_0(\zeta)\frac{z_1}{\zeta_1 - z_1}\,\frac{\overline{z_2}}{\overline{\zeta_2 - z_2}}\,\frac{d\zeta_1}{\zeta_1}\,\frac{d\zeta_2}{\zeta_2}$$

$$= \frac{1}{(2\pi i)^2}\int_{\partial_0 \mathbb{D}^2} \zeta_1^{-\kappa_1-1}\zeta_2^{-\kappa_2}w_0(\zeta)\frac{z_1}{\zeta_1 - z_1}\,\frac{\overline{z_2}}{1 - \overline{z_2}\zeta_2}d\xi_1 d\eta_1 d\xi_2 d\eta_2$$

$$= \frac{z_1^{-\kappa_1}}{\pi}\int_{|\zeta_2|<1}\zeta^{-\kappa_2}2f_2(z_1,\zeta_2)\frac{\overline{z_2}}{1 - \overline{z_2}\zeta_2}d\xi_2 d\eta_2$$

$$+\frac{1}{\pi^2}\int_{\mathbb{D}^2}\zeta_1^{-\kappa_1-1}\zeta_2^{-\kappa_2}f_{2\overline{\zeta_1}}(\zeta)\frac{z_1}{\zeta_1 - z_1}\,\frac{\overline{z_2}}{1 - \overline{z_2}\zeta_2}d\xi_1 d\eta_1 d\xi_2 d\eta_2\,,$$

and by symmetry

$$\frac{1}{(2\pi i)^2}\int_{\partial_0 \mathbb{D}^2} \zeta_1^{-\kappa_1}\zeta_2^{-\kappa_2} w_0(\zeta)\frac{\overline{z_1}}{\overline{\zeta_1 - z_1}}\,\frac{z_2}{\zeta_2 - z_2}\,\frac{d\zeta_1}{\zeta_1}\,\frac{d\zeta_2}{\zeta_2}$$

$$= \frac{z_2^{-\kappa_2}}{\pi}\int_{|\zeta_1|<1} \zeta_1^{-\kappa_1} f_1(\zeta_1, z_2)\frac{\overline{z_1}}{1-\overline{z_1}\zeta_1} d\xi_1 d\eta_1$$

$$+\frac{1}{\pi^2}\int_{\mathbb{D}^2} \zeta_1^{-\kappa_1}\zeta_2^{-\kappa_2-1} f_{1\overline{\zeta_2}}(\zeta)\frac{\overline{z_1}}{1-\overline{z_1}\zeta_1}\,\frac{z_2}{\zeta_2 - z_2} d\xi_1 d\eta_1 d\xi_2 d\eta_2 \,.$$

Therefore

$$\mathrm{Re}\left\{\frac{2}{(2\pi i)^2}\int_{\partial_0 \mathbb{D}^2} \gamma(\zeta)\frac{z_1}{\zeta_1 - z_1}\,\frac{\overline{z_2}}{\overline{\zeta_2 - z_2}}\,\frac{d\zeta_1}{\zeta_1}\,\frac{d\zeta_2}{\zeta_2}\right\}$$

$$= \mathrm{Re}\left\{\frac{z_2^{-\kappa_2}}{\pi}\int_{|\zeta_1<1} \zeta_1^{-\kappa_1} f_1(\zeta_1, z_2)\frac{\overline{z_1}}{1-\overline{z_1}\zeta_1} d\xi_1 d\eta_1\right.$$

$$+\frac{z_1^{-\kappa_1}}{\pi}\int_{|\zeta_2|<1} \zeta_2^{-\kappa_2} f_2(z_1, \zeta_2)\frac{\overline{z_2}}{1-\overline{z_2}\zeta_2} d\xi_2 d\eta_2 \tag{4.32}$$

$$+\frac{1}{\pi^2}\int_{\mathbb{D}^2} \zeta_1^{-\kappa_1-1}\zeta_2^{-\kappa_2-1} f_{1\overline{\zeta_2}}(\zeta)$$

$$\left.\times\left[\frac{\overline{z_1}\zeta_1}{1-\overline{z_1}\zeta_1}\,\frac{z_2}{\zeta_2 - z_2} + \frac{z_1}{\zeta_1 - z_1}\,\frac{\overline{z_2}\zeta_2}{1-\overline{z_2}\zeta_2}\right] d\xi_1 d\eta_1 d\xi_2 d\eta_2\right\}.$$

Condition (4.13′) reads

$$\frac{2}{(2\pi i)^2}\int_{\partial_0 \mathbb{D}^2} \gamma(\zeta)\left[2(1 - z_1^{-\kappa_1}\overline{\zeta_1}^{-\kappa_1} z_2^{-\kappa_2}\overline{\zeta_2}^{-\kappa_2})\frac{1}{1-z_1\overline{\zeta_1}}\,\frac{1}{1-z_2\overline{\zeta_2}} - 1\right]\frac{d\zeta_1}{\zeta_1}\,\frac{d\zeta_2}{\zeta_2}$$

$$= \frac{1}{(2\pi i)^2}\int_{\partial_0 \mathbb{D}^2}\left\{\zeta_1^{-\kappa_1}\zeta_2^{-\kappa_2} w_0(\zeta) + \overline{\zeta_1}^{-\kappa_1}\overline{\zeta_2}^{-\kappa_2}\overline{w_0(\zeta)}\right\}$$

$$\times\left[2(1 - z_1^{-\kappa_1}\overline{\zeta_1}^{-\kappa_1} z_2^{-\kappa_2}\overline{\zeta_2}^{-\kappa_2})\frac{1}{1-z_1\overline{\zeta_1}}\,\frac{1}{1-z_2\overline{\zeta_2}} - 1\right]\frac{d\zeta_1}{\zeta_1}\,\frac{d\zeta_2}{\zeta_2}\,.$$

Here the first term on the right-hand side is

$$\frac{2}{(2\pi i)^2}\int_{\partial_0 \mathbb{D}^2}(\zeta_1^{-\kappa_1}\zeta^{-\kappa_2}-z_1^{-\kappa_1}z_2^{-\kappa_2})w_0(\zeta)\frac{d\zeta_1}{\zeta_1-z_1}\frac{d\zeta_2}{\zeta_2-z_2}$$

$$-\frac{1}{(2\pi i)^2}\int_{\partial_0 \mathbb{D}^2}\zeta_1^{-\kappa_1-1}\zeta_2^{-\kappa_2-1}w_0(\zeta)d\zeta_1 d\zeta_2$$

$$=\frac{2z_2^{-\kappa_2}}{\pi}\int_{|\zeta_1|<1}(\zeta_1^{-\kappa_1}-z_1^{-\kappa_1})f_1(\zeta_1,z_2)\frac{d\xi_1 d\eta_1}{\zeta_1-z_1}$$

$$+\frac{2z_1^{-\kappa_1}}{\pi}\int_{|\zeta_2|<1}(\zeta_2^{-\kappa_2}-z_2^{-\kappa_2})f_2(z_1,\zeta_2)\frac{d\xi_2 d\eta_2}{\zeta_2-z_2}$$

$$+\frac{2}{\pi^2}\int_{\mathbb{D}^2}(\zeta_1^{-\kappa_1}\zeta_2^{-\kappa_2}-z_1^{-\kappa_1}z_2^{-\kappa_2})f_{1\overline{\zeta_2}}(\zeta)\frac{d\xi_1 d\eta_1}{\zeta_1-z_1}\frac{d\xi_2 d\eta_2}{\zeta_2-z_2}$$

$$-\frac{1}{\pi^2}\int_{\mathbb{D}^2}\zeta_1^{-\kappa_1-1}\zeta_2^{-\kappa_2-1}f_{1\overline{\zeta_2}}(\zeta)d\xi_1 d\eta_1 d\xi_2 d\eta_2\,.$$

The second term is

$$\frac{1}{(2\pi i)^2}\int_{\partial_0 \mathbb{D}^2}\overline{\zeta_1}^{-\kappa_1-1}\overline{\zeta_2}^{-\kappa_2-1}\overline{w_0(\zeta)}$$

$$\times\left[2\left(1-z^{-\kappa_1}\overline{\zeta_1}^{-\kappa_1}z_2^{-\kappa_2}\overline{\zeta_2}^{-\kappa_2}\right)\frac{1}{1-z_1\overline{\zeta_1}}\frac{1}{1-z_2\overline{\zeta_2}}-1\right]d\overline{\zeta_1}d\overline{\zeta_2}$$

$$=\frac{1}{\pi^2}\int_{\mathbb{D}^2}\overline{\zeta_1}^{-\kappa_1-1}\overline{\zeta_2}^{-\kappa_2-1}\overline{f_{1\overline{\zeta_2}}(\zeta)}$$

$$\times\left[2(1-z^{-\kappa_1}\overline{\zeta_1}^{-\kappa_1}z_2^{-\kappa_2}\overline{\zeta_2}^{-\kappa_2})\frac{1}{1-z_1\overline{\zeta_1}}\frac{1}{1-z_2\overline{\zeta_2}}-1\right]d\xi_1 d\eta_1 d\xi_2 d\eta_2\,.$$

Thus condition (4.13′) in this case is

$$\frac{1}{(2\pi i)^2}\int_{\partial_0 \mathbb{D}^2}\gamma(\zeta)$$

$$\times\left[2(1-z^{-\kappa_1}\overline{\zeta_1}^{-\kappa_1}z_2^{-\kappa_2}\overline{\zeta_2}^{-\kappa_2})\frac{1}{\zeta_1-z_1}\frac{1}{\zeta_2-z_2}-\frac{1}{\zeta_1\zeta_2}\right]d\zeta_1 d\zeta_2$$

$$=\frac{z_2^{-\kappa_2}}{\pi}\int_{|\zeta_1|<1}(\zeta_1^{-\kappa_1}-z_1^{-\kappa_1})f_1(\zeta_1,z_2)\frac{d\xi_1 d\eta_1}{\zeta_1-z_1}$$

$$+\frac{z_1^{-\kappa_1}}{\pi}\int_{|\zeta_2|<1}(\zeta_2^{-\kappa_2}-z_2^{-\kappa_2})f_2(z_1,\zeta_2)\frac{d\xi_2 d\eta_2}{\zeta_2-z_2} \tag{4.33}$$

$$+\frac{1}{\pi^2}\int_{\mathbb{D}^2}\left\{\frac{(\zeta_1^{-\kappa_1}\zeta_2^{-\kappa_2}-z_1^{-\kappa_1}z_2^{-\kappa_2})f_{1\overline{\zeta_2}}(\zeta)}{(\zeta_1-z_1)(\zeta_2-z_2)}\right.$$

$$+\frac{(\overline{\zeta_1}^{-\kappa_1-1}\overline{\zeta_2}^{-\kappa_2-1}-z_1^{-\kappa_1}\overline{\zeta_1}^{-2\kappa_1-1}z_2^{-\kappa_2}\overline{\zeta_2}^{-2\kappa_2-1})\overline{f_{1\overline{\zeta_2}}(\zeta)}}{(1-z_1\overline{\zeta_1})(1-z_2\overline{\zeta_2})}$$

$$\left.-\frac{1}{2}\zeta_1^{-\kappa_1-1}\zeta_2^{-\kappa_2-1}f_{1\overline{\zeta_2}}(\zeta)-\frac{1}{2}\overline{\zeta_1}^{-\kappa_1-1}\overline{\zeta_2}^{-\kappa_2-1}\overline{f_{1\overline{\zeta_2}}(\zeta)}\right\}d\xi_1 d\eta_1 d\xi_2 d\eta_2\,.$$

For checking (4.31) to satisfy the boundary condition (4.11) equation (4.31) is multipied by $z_1^{-\kappa_1}z_2^{-\kappa_2}$ and (4.33) added to it. This gives

$$z_1^{-\kappa_1}z_2^{-\kappa_2}w(z)=\frac{1}{(2\pi i)^2}\int_{\partial_0\mathbb{D}^2}\gamma(\zeta)\left(2\frac{\zeta_1}{\zeta_1-z_1}\frac{\zeta_2}{\zeta_2-z_2}-1\right)\frac{d\zeta_1}{\zeta_1}\frac{d\zeta_2}{\zeta_2}$$

$$-\frac{z_2^{-\kappa_2}}{\pi}\int_{|\zeta_1|<1}\zeta_1^{-\kappa_1}f_1(\zeta_1,z_2)\frac{d\xi_1 d\eta_1}{\zeta_1-z_1}-\frac{z_1^{-\kappa_1}}{\pi}\int_{|\zeta_2|<1}\zeta_2^{-\kappa_2}f_2(z_1,\zeta_2)\frac{d\xi_2 d\eta_2}{\xi_2-z_2}$$

$$-\frac{1}{\pi^2}\int_{\mathbb{D}^2}\left\{\frac{\zeta_1^{-\kappa_1}\zeta_2^{-\kappa_2}f_{1\overline{\zeta_2}}(\zeta)}{(\zeta_1-z_1)(\zeta_2-z_2)}+\frac{\overline{\zeta_1}^{-\kappa_1-1}\overline{\zeta_2}^{-\kappa_2-1}\overline{f_{1\overline{\zeta_2}}(\zeta)}}{(1-z_1\overline{\zeta_1})(1-z_2\overline{\zeta_2})}\right.$$

$$\left.-\frac{1}{2}\zeta_1^{-\kappa_1-1}\zeta_2^{-\kappa_2-1}f_{1\overline{\zeta_2}}(\zeta)-\frac{1}{2}\overline{\zeta_1}^{-\kappa_1-1}\overline{\zeta_2}^{-\kappa_2-1}\overline{f_{1\overline{\zeta_2}}(\zeta)}\right\}d\xi_1 d\eta_1 d\xi_2 d\eta_2\,.$$

Therefore on $\partial_0\mathbb{D}^2$

$$\operatorname{Re}\{z_1^{-\kappa_1}z_2^{-\kappa_2}w(z)\}$$

$$=\frac{1}{(2\pi i)^2}\int_{\partial_0\mathbb{D}^2}\gamma(\zeta)\left(\frac{\zeta_1}{\zeta_1-z_1}\frac{\zeta_2}{\zeta_2-z_2}+\frac{\overline{\zeta_1}}{\overline{\zeta_1-z_1}}\frac{\overline{\zeta_2}}{\overline{\zeta_2-z_2}}-1\right)\frac{d\zeta_1}{\zeta_1}\frac{d\zeta_2}{\zeta_2}$$

$$-\operatorname{Re}\left\{\frac{z_2^{-\kappa_2}}{\pi}\int_{|\zeta_1|=1}\zeta_1^{-\kappa_1}f_1(\zeta_1,z_2)\frac{d\xi_1 d\eta_1}{\zeta_1-z_1}+\frac{z_1^{-\kappa_1}}{\pi}\int_{|\zeta_2|=1}\zeta_2^{-\kappa_2}f_2(z_1,\zeta_2)\frac{d\xi_2 d\eta_2}{\zeta_2-z_2}\right.$$

$$+\frac{1}{\pi^2}\int_{\mathbb{D}^2}\zeta_1^{-\kappa_1-1}\zeta_2^{-\kappa_2-1}f_{1\overline{\zeta_2}}(\zeta)$$

$$\left.\times\left[\frac{\zeta_1}{\zeta_1-z_1}\frac{\zeta_2}{\zeta_2-z_2}+\frac{z_1}{\zeta_1-z_1}\frac{z_2}{\zeta_2-z_2}-1\right]d\xi_1 d\eta_1 d\xi_2 d\eta_2\right\}\,.$$

Because

$$\frac{\zeta_1}{\zeta_1-z_1}\frac{\zeta_2}{\zeta_2-z_2}+\frac{z_1}{\zeta_1-z_1}\frac{z_2}{\zeta_2-z_2}-1=\frac{\zeta_1}{\zeta_1-z_1}\frac{z_2}{\zeta_2-z_2}+\frac{z_1}{\zeta_1-z_1}\frac{\zeta_2}{\zeta_2-z_2}$$

condition (4.32) shows on $\partial_0 \mathbb{D}^2$

$$\operatorname{Re}\{z_1^{-\kappa_1} z_2^{-\kappa_2} w(z)\}$$

$$= \frac{1}{(2\pi i)^2} \int_{\partial_0 \mathbb{D}^2} \gamma(\zeta) \left(\frac{\zeta_1}{\zeta_1 - z_1} + \frac{\overline{\zeta_1}}{\overline{\zeta_1 - z_1}} - 1 \right)$$

$$\times \left(\frac{\zeta_2}{\zeta_2 - z_2} + \frac{\overline{\zeta_2}}{\overline{\zeta_2 - z_2}} - 1 \right) \frac{d\zeta_1}{\zeta_1} \frac{d\zeta_2}{\zeta_2} = \gamma(z) \, .$$

That (4.31) satisfies system (4.21) can again be seen directly.

The theory of Riemann–Hilbert boundary value problems for inhomogeneous Cauchy–Riemann systems can be used to solve boundary value problems for inhomogeneous pluriharmonic and pluriholomorphic functions, i.e. for solutions to the systems

$$u_{z_k \overline{z_\ell}} = f_{k\ell} \; ; \; u_{\overline{z_k} \, \overline{z_\ell}} = f_{k\ell} \quad (1 \le k, \ell \le n) \, .$$

In [2, 3, 5] this was done for the Dirichlet problem in polydiscs.

References

[1] Begehr, H.: Complex analytic methods for partial differential equations. An introductory text. World Scientific, Singapore, 1994.

[2] Begehr, H.: Boundary value problems in $\mathbb{C}$ and $\mathbb{C}^n$. Analysis and mechanics in continuous media, ed. N.H. Anh et al. Publ. The HoChiMinh City Math. Sco. 3 (1995), 32–47; Acta Math. Vietnamica 22 (1997), no 2.

[3] Begehr, H.: Complex analytic methods for partial differential equations. ZAMM 76 (1996), S 2, 21–24.

[4] Begehr, H., Dzhuraev, A.: Schwarz problem for Cauchy–Riemann systems in several complex variables. Analysis and topology, ed. C. Andreian Cazacu, O. Lehto, Th.M. Rassias. World Scientific, Singapore, to appear.

[5] Begehr, H., Dzhuraev, A.: An introduction to several complex variables and partial differential equations. Addison Wesley Longman, Harlow, 1997.

[6] Begehr, H., Gilbert, R.P.: Transformations, transmutations, and kernel functions. I, II. Longman, Harlow, 1992, 1993.

[7] Begehr, H., Wen, G.–C.: Nonlinear elliptic boundary value problems and their applications. Addison Wesley Longman, Harlow, 1996.

[8] Efendiev, M.: Geometric properties of nonlinear mappings related to pseudodifferential operators and their topological degree. Habiliationsschrift, FU Berlin, 1997.

[9] Gakhov, F.D.: Boundary value problems. Pergamon Press, Oxford, 1966.

[10] Gilbert, R.P., Buchanan, J.: First order elliptic systems. A function theoretic approach. Acad. Press, New York, 1983.

[11] Govorov, N.V.: Riemann's boundary problem with infinite index. Introduction, appendix and editing by I.V.Ostrovsky. Birkhäuser, Basel, 1994.

[12] Kakichev, V.A.: The Hilbert problem for functions that are holomorphic in the bidisc, and some of its applications. Teor. Funkcii Funkcional. Anal., Prilozen Vyp 18 (1973), 3–18 (Russian).

[13] Kakichev, V.A., Zon, Le–Din: The integral and the problem of Schwarz for the bi–half–plane. Math. Anal., Work Collect., Rostov, 1978, 29–38.

[14] Muskhelishvili, N.: Singular integral equations. Nordhoff, Groningen, 1953.

[15] Vekua, I.N.: Generalized analytic functions. Pergamon Press, London, 1962.

[16] Wang, Li–Ping: Riemann–Hilbert boundary value problems for analytic functions in several bicylinder domains. J. Ocean Univ. Qingdao 21 (1991), 134–142 (Chinese).

[17] Wen, G.–C., Begehr, H.: Boundary value problems for elliptic equations and systems. Longman, Harlow, 1990.

[18] Wen, G.–C., Tain, M.–Y.: Some boundary value problems for several complex variables. J. Yantai Univ. Nat. Sci. Eng. 2 (1989), 1–14 (Chinese).

[19] Witte, J.: A generalized Pompeiu formula related to differential equations in $\mathbb{C}^n$ with higher order. Complex Variables, Theory Appl. 32 (1997), 7-27.

5 THE RIEMANN–HILBERT PROBLEM FOR NONLINEAR ELLIPTIC COMPLEX EQUATIONS OF FIRST ORDER IN MULTIPLY CONNECTED INFINITE DOMAINS

Guo–Chun Wen

Peking University
Beijing 100871
P.R. China

wengc@pku.pku.edu.cn

Abstract: There are many problems in mechanics and physics, the mathematical models of which are some boundary valye problems for nonlinear elliptic complex equations of first and second order in multiply connected infinite domains, see e.g. [3]. In this paper, we discuss the Riemann-Hilbert boundary value problem for nonlinear elliptic complex equations of first order in multiply connected domains including the infinite point.

5.1 FORMULATION OF THE RIEMANN-HILBERT PROBLEM FOR NONLINEAR COMPLEX EQUATIONS IN INFINITE DOMAINS

Let D be an $(N+1)$-connected domain including the infinite point with the boundary $\Gamma = \cup_{j=0}^{N}\Gamma_j \in C_\alpha^1(0<\alpha<1)$, without loss of generality, we may assume that D is a circular domain in $|z|>1$, whose boundary is $N+1$ circles $\Gamma_j = \{|z-z_j| = \gamma_j\}, j=1,\cdots,N$ and $\Gamma_0 = \Gamma_{N+1} = \{|z|=1\}$ and $\infty \in D$.

We consider the nonlinear uniformly elliptic complex equation

$$w_{\bar{z}} = Q_1 w_z + Q_2 \bar{w}_{\bar{z}} + A_1 w + A_2 \bar{w} + A_3,\ z \in D, \tag{5.1}$$

Suppose that the equation (5.1) satisfies the following conditions.

Condition C For any functions $w(z) \in C(\bar{D})$, $U(z), U_j(z) \in L_{p_0,2}(\bar{D})$ $(j =$

H.G.W. Begehr et al. (eds.), Partial Differential and Integral Equations, 85–95.

$1,2)$, $F(z,w,U)$ is measurable in $z\in D$ and continuous in $w\in\mathbb{C}$, and satisfy the conditions

$$|F(z,w,U_1)-F(z,w,U_2)|\le q_0|U_1-U_2|, \tag{5.2}$$

and

$$L_{p,2}(A_j,\bar D)\le k_0,\ j=0,1,2,\ L_{p,2}(A_3,\bar D)\le k_1, \tag{5.3}$$

in which $q_0, p_0\,(2<p_0<p), k_0, k_1$ are nonnegative constants. Besides, we can assume that $Q_j=0(j=1,2)$, $A_j=0(j=1,2,3)$, $z\notin D$.

Problem A The Riemann-Hilbert boundary value problem for (5.1) is to find a continuous solution $w(z)\in W^1_{p_0,2}(D)$ satisfying the boundary condition:

$$\operatorname{Re}[\overline{\lambda(z)}w(z)]=r(z),\ z\in\Gamma, \tag{5.4}$$

where $\lambda(z), r(z)$ satisfy

$$C_\alpha[\lambda(z),\Gamma]\le k_0,\ C_\alpha[r(z),\Gamma]\le k_2, \tag{5.5}$$

herein $\alpha\,(1/2<\alpha<1), k_0, k_2$ are nonnegative constants. Problem A with $A_3(z)=0$ in D, $r(z)=0$ on Γ is called Problem A_0. The index of Problem A and Problem A_0 is defined as

$$K=\frac{1}{2\pi}\Delta_\Gamma\arg\lambda(z). \tag{5.6}$$

Because Problem A with $K<N$ may not be solvable, similarly to [2], we modify the boundary conditions so that the problem becomes well-posed.

Problem B Find a continuous solution $w(z)\in W^1_{p_0,2}(D)$ satisfying the modified boundary conditions

$$\operatorname{Re}[\overline{\lambda(z)}w(z)]=r(z)+h(z), \tag{5.7}$$

$$h(z)=\begin{cases} 0,\ K\ge N, & \\ h_j,\ z\in\Gamma_j, j=1,\cdots,N-K & \\ 0,\ z\in\Gamma_j, j=N-K+1,\cdots,N+1 & \Big\}\ 0\le K<N, \\ h_j,\ z\in\Gamma_j, j=1,\cdots,N, & \\ h_0+\sum_{m=1}^{-K-1}\operatorname{Re}(h_m^+ + ih_m^-)z^m,\ z\in\Gamma_0 & \Big\}\ K<0, \end{cases} \tag{5.8}$$

where $h_j\ (j=0,1,\cdots,N)$ and $h_m^\pm\ (m=1,\cdots,-K-1, K<0)$ are appropriate real constants. Moreover, we can require that the solution $w(z)$ satisfies the point conditions

$$\operatorname{Im}[\overline{\lambda(a_j)}w(a_j)]=b_j,\ j\in J=\begin{cases}1,\cdots,2K-N+1,\ K\ge N,\\ N-K+1,\cdots,N+1,\ 0\le K<N,\\ \phi,\ K<0,\end{cases} \tag{5.9}$$

in which $a_j \in \Gamma_j (j=1,\cdots,N)$, $a_j \in \Gamma_0 (j=N+1,\cdots,2K-N+1, K\ge N)$ and $b_j (j\in J)$ are all real constants satisfying the conditions

$$|b_j| \le k_3,\ j\in J, \tag{5.10}$$

herein k_3 is a nonnegative constant. Problem B with $A_3(z)=0$ in D, $r(z)=0$ on Γ is called Problem B_0. In order to prove the uniqueness of solutions of Problem B, we need to add the following condition: For any $w_1(z), w_2(z) \in C(\bar{D})$, $U(z)\in L_{p_0,2}(\bar{D})$, there is

$$F(z,w_1,U) - F(z,w_2,U) = A_0(z)(w_1-w_2), \tag{5.11}$$

where $L_{p_0,2}(A_0,\bar{D}) < \infty$.

5.2 A PRIORI ESTIMATES OF SOLUTIONS OF THE RIEMANN-HILBERT PROBLEM FOR (5.1)

In oder to prove the solvability of Problem B for the complex equation (5.1), we need to give some estimates of solutions of Problem B for the complex equation (5.1). First of all, we give a representation theorem for Problem B.

Theorem 1 *Suppose that the complex equation* (5.1) *satisfies Condition C, and $w(z)$ is a solution of Problem B for* (5.1). *Then $w(z)$ is representable by*

$$w(z) = \Phi[\zeta(z)]e^{\phi(z)} + \psi(z), \tag{5.12}$$

where $\zeta(z)$ is a homeomorphism on $\bar{D}$, which quasiconformally maps D onto an $(N+1)$-connected circular domain G with boundary $L=\cup_{j=0}^{N} L_j$, where the $L_j=\zeta(\Gamma_j)\,(j=1,\cdots,N)$ are located in $|\zeta|<1$ bounded by $L_0=\zeta(\Gamma_0)=\{|\zeta|=1\}$, and $\zeta(\infty)=0$, $\Phi(\zeta)$ is an analytic function in G, $\psi(z),\phi(z),\zeta(z)$ and its inverse function $z(\zeta)$ satisfy the estimates

$$\tilde{C}_\beta[\tau,\bar{D}] \le k_4,\ \tau=\psi(z),\phi(z),\zeta(z),\ \tilde{C}_\beta[Z(\zeta),\bar{G}]\le k_4, \tag{5.13}$$

$$L_{p_0,2}[|\psi_{\bar z}|+|\psi_z|,\bar{D}] \le k_4,\ L_{p_0,2}[|\phi_{\bar z}|+|\phi_z|,\bar{D}]\le k_4, \tag{5.14}$$

$$L_{p_0,2}[|\zeta_{\bar z}|+|\zeta_z|,\bar{D}] \le k_5,\ L_{p_0,2}[|Z_{\bar\zeta}|+|Z_\zeta|,\bar{G}]\le k_5, \tag{5.15}$$

in which $Z(\zeta)=1/z(\zeta)$, $\tilde{C}_\beta[\tau,\bar{D}]=C_\beta[\tau,\bar{D}]+\max_{z\in\bar{D}}|\tau z^\beta|$, $\beta=1-2/p_0\le\alpha$, $p_0(2<p_0\le p)$, $k_j=k_j(q_0,p_0,k_0,k_1,D)(j=4,5)$ are nonnegative constants.

Proof Similarly to Theorem 2.4, Chapter 2 in [2], we substitute the solution $w(z)$ of Problem B into the coefficients of the equation (5.1) and consider the following system

$$\psi_{\bar z} = Q\psi_z + A_1\psi + A_2\bar\psi + A_3,\ Q=\begin{cases} Q_1+Q_2\overline{w_z}/w_z & \text{for } w_z\ne 0,\\ 0 & \text{for } w_z=0 \text{ or } z\notin D,\end{cases} \tag{5.16}$$

$$\phi_{\bar z} = Q\phi_z + A,\ A=\begin{cases} A_1+A_2\bar w/w & \text{for } w(z)\ne 0,\\ 0 & \text{for } w(z)=0 \text{ or } z\notin D,\end{cases} \tag{5.17}$$

$$W_{\bar{z}} = QW_z, \quad W(z) = \Phi[\zeta(z)]. \tag{5.18}$$

By using the continuity method and the principle of contracting mappings, we can find the solutions $\psi(z) = Tf = -\frac{1}{\pi}\iint_D[f(\zeta)/(\zeta - z)]d\sigma_\zeta$, $\phi = Tg$, $\zeta(z) = \Psi[\chi(z)]$, $\chi(z) = 1/z + Th$ of (5.16)-(5.18)(see Remark below), where $f(z), g(z), h(z) \in L_{p_0,2}(\bar{D})$, $2 < p_0 \le p$, $\chi(z)$ is a homeomorphism on $\bar{D}, \Psi(\chi)$ is a univalent analytic function, which conformally maps $E = \chi(D)$ onto an $(N+1)-$connected circular domain G, and $\Phi(\zeta)$ is an analytic function in G. We can verify that $\psi(z), \phi(z), \zeta(z)$ satisfy the estimates (5.13) and (5.14). It remains to prove that $\zeta(z)$ and its inverse function $z = z(\zeta)$ satisfy the estimate (5.15). In fact, we can find a homeomorphic solution of (15.18) in the form $\chi(z) = 1/z + Th$ such that $[\chi(z)]_z, [\chi]_{\bar{z}} \in L_{p_0,2}(\bar{D})$(see [1]). Next, we find a univalent analytic function $\zeta = \Psi(\chi)$, which maps $\chi(D)$ onto G, hence $\zeta = \zeta(z) = \Psi[\chi(z)]$. By the result on conformal mappings, applying the method of Lemma 2.1, Chapter 2 in [2], we can prove that (5.15) is true.

Secondly, we shall give a priori estimates of solutions for Problem B.

Theorem 2 *Under the same conditions as in Theorem* 1, *any solution* $w(z)$ *of Problem B for* (5.1) *satisfies*

$$\tilde{C}_\gamma[w(z), \bar{D}] \le M_1 = M_1(q_0, p_0, k, \alpha, K, D), \tag{5.19}$$

$$L_{p_0,2}[|w_{\bar{z}}| + |w_z|, \bar{D}] \le M_2 = M_2(q_0, p_0, k, \alpha, K, D), \tag{5.20}$$

where $k = k(k_0, k_1, k_2, k_3)$, $\gamma = \alpha\beta^2$, $2 < p_0 \le p$.

Proof On the basis of Theorem 1, the solution $w(z)$ of Problem B can be expressed as in (5.12), hence the boundary conditions (5.7)-(5.9) can be transformed into the boundary conditions of Problem $\tilde{B}$:

$$\mathrm{Re}[\overline{\Lambda(\zeta)}\Phi(\zeta)] = R(\zeta) + H(\zeta), \ \zeta \in L, \tag{5.21}$$

$$H(\zeta) = \begin{cases} 0, \zeta \in L, \ \text{if } K \ge N, \\ \left.\begin{array}{l} h_j, \zeta \in L_j, j = 1, \cdots, N-K, \\ 0, \zeta \in L_j, j = N-K+1, \cdots, N+1 \end{array}\right\} \text{if } 0 \le K < N, \\ \left.\begin{array}{l} h_j, \zeta \in L_j, j = 1, \cdots, N, \\ h_0 + \mathrm{Re}\sum\limits_{m=1}^{-K-1}(h_m^+ + ih_m^-)[z(\zeta)]^m, \zeta \in L_0 \end{array}\right\} \text{if } K < 0, \end{cases} \tag{5.22}$$

$$\mathrm{Im}[\overline{\Lambda(a_j')}\Phi(a_j')] = b_j', \ j \in J, \tag{5.23}$$

where

$$\overline{\Lambda(\zeta)} = \overline{\lambda[z(\zeta)]}e^{\phi[z(\zeta)]}, \ R(\zeta) = r[z(\zeta)] - \mathrm{Re}\{\overline{\lambda[z(\zeta)]}\psi[z(\zeta)]\},$$

$$a_j' = \zeta(a_j), \ b_j' = b_j - \mathrm{Im}[\overline{\lambda(a_j)}\psi(a_j)], \ j \in J.$$

By (5.13)-(5.15), it can be seen that $\Lambda(\zeta)$, $R(\zeta)$, b_j' $(j \in J)$ satisfy the conditions

$$C_{\alpha\beta}[\Lambda(\zeta), L] \le M_3, \ C_{\alpha\beta}[R(\zeta), L] \le M_3, \ |b_j'| \le M_3, \ j \in J, \tag{5.24}$$

where $M_3 = M_3(q_0, p_0, k, \alpha, K, D)$. If we can prove that the solution $\Phi(\zeta)$ of Problem $\tilde{B}$ satisfies the estimate

$$\tilde{C}_{\alpha\beta}[\Phi(\zeta), \bar{G}] \leq M_4, \ \tilde{C}_{\alpha\beta}[\Phi'(\zeta), \tilde{G}] \leq M_5, \tag{5.25}$$

in which $\tilde{G} = \zeta(\tilde{D})$, $\tilde{D}$ is any closed subset in D, $M_4 = M_4(q_0, p_0, k, \alpha, K, D)$, $M_5 = M_5(q_0, p_0, k, \alpha, K, \tilde{D})$, then from the representation (5.12) of the solution $w(z)$ of Problem B and the estimates (5.13)-(5.15), the estimates (5.19) and (5.20) can be derived.

It remains to prove that (5.25) holds. For this, we first verify the boundness of $\Phi(\zeta)$, i.e.

$$\hat{[\Phi, \bar{G}]} = C[\Phi(\zeta), \bar{G}] + \max_{\zeta \in \bar{G}} |\Phi(\zeta)\zeta| \leq M_6 = M_6(q_0, p_0, k, \alpha, K, D). \tag{5.26}$$

Suppose that (5.26) is not true. Then there exist sequences of functions $\{\Lambda_n(\zeta)\}$, $\{R_n(\zeta)\}$, $\{b'_{jn}\}$ satisfying the same conditions as $\Lambda(\zeta)$, $R(\zeta)$, b'_j, which uniformly converge to $\Lambda_0(\zeta)$, $R_0(\zeta)$, b'_{j0} $(j \in J)$ on L respectively. For the solution $\Phi_n(\zeta)$ of the boundary value problem(Problem B_n) corresponding to $\Lambda_n(\zeta), R_n(\zeta), b'_{jn}$ $(j \in J)$, we have $I_n = C[\Phi_n(\zeta), \bar{G}] \to \infty$ as $n \to \infty$. There is no harm in assuming that $I_n \geq 1$, $n = 1, 2, \cdots$. Obviously $\tilde{\Phi}_n(\zeta) = \Phi_n(\zeta)/I_n$ satisfies the boundary conditions

$$\text{Re}[\overline{\Lambda_n(\zeta)}\tilde{\Phi}_n(\zeta)] = [R_n(\zeta) + H(\zeta)]/I_n, \zeta \in L, \tag{5.27}$$

$$\text{Im}[\overline{\Lambda_n(a'_n)}\tilde{\Phi}_n(a'_n)] = b'_{jn}/I_n, \ j \in J. \tag{5.28}$$

Applying the Schwarz formula, the Cauchy formula and the method of symmetric extension (see Theorem 1.4, Chapter 1, [2]), the estimates

$$\begin{aligned} \hat{C}_{\alpha\beta}[\tilde{\Phi}_n(\zeta), \bar{G}] &= \hat{C}_{\alpha\beta}[\Phi_n(\zeta), \bar{G}] \\ &\quad + \max_{\zeta \in \bar{G}} |\Phi_n(\zeta)\zeta| \leq M_7, \ \hat{C}_{\alpha\beta}[\tilde{\Phi}'_n(\zeta), \tilde{G}] \leq M_8 \end{aligned} \tag{5.29}$$

can be obtained, where $M_7 = M_7(q_0, p_0, k, \alpha, K, D)$, $M_8 = M_8(q_0, p_0, k, \alpha, K, \tilde{D})$. Thus we can select a subsequence of $\{\tilde{\Phi}_n(\zeta)\}$, which uniformly converge to an analytic function $\tilde{\Phi}_0(\zeta)$ in $\bar{G}$, and $\tilde{\Phi}_0(\zeta)$ satisfies the homogeneous boundary conditions

$$\text{Re}[\overline{\Lambda_0(\zeta)}\tilde{\Phi}_0(\zeta)] = H(\zeta), \ \zeta \in L, \tag{5.30}$$

$$\text{Im}[\overline{\Lambda_0(a'_j)}\tilde{\Phi}_0(a'_j)] = 0, \ j \in J. \tag{5.31}$$

On the basis of the uniqueness theorem (see Theorem 4, we conclude that $\tilde{\Phi}_0(\zeta) = 0, \zeta \in \bar{G}$. However, from $\hat{C}[\tilde{\Phi}_n(\zeta), \bar{G}] = 1$, it follows that there exists a point $\zeta_* \in \bar{G}$, such that $|\tilde{\Phi}_0(\zeta_*)| > 0$. This contradiction proves that (5.26) holds. Afterwards using the method which leads from $\hat{C}[\tilde{\Phi}_n(\zeta), \bar{G}] = 1$ to (5.29), the estimate (5.25) can be derived.

Theorem 3 *Under the same conditions as in Theorem* 1, *any solution* $w(z)$ *of Problem B for* (5.1) *satisfies*

$$\tilde{C}_\gamma[w(z),\bar{D}] \le M_9k_*,\ L_{p_0,2}[|w_{\bar{z}}|+|w_z|,\tilde{D}] \le M_{10}k_*, \tag{5.32}$$

where γ, p_0, $\tilde{D}$ *are as stated in Theorem* 2, $k_* = k_1+k_2+k_3$, $M_9 = M_9(q_0,p_0, k_0,\alpha,K,D)$, $M_{10} = M_{10}(q_0,p_0,k_0,\alpha,K,\tilde{D})$.

Proof If $k_* = 0$, i.e. $k_1 = k_2 = k_3 = 0$, from Theorem 4, it follows that $w(z) = 0$, $z \in D$. If $k_* > 0$, it is easy to see that $W(z) = w(z)/k_*$ satisfies the complex equation and boundary conditions

$$W_{\bar{z}} - Q_1W_z - Q_2\overline{W_z} - A_1W - A_2\overline{W} = A_3/k_*, z \in D, \tag{5.33}$$

$$\mathrm{Re}[\overline{\lambda(z)}W(z)] = [r(z)+h(z)]/k_*, z \in \Gamma, \tag{5.34}$$

$$\mathrm{Im}[\overline{\lambda(a_j)}W(a_j)] = b_j/k_*, j \in J. \tag{5.35}$$

Noting that $L_{p,2}[A_3/k_*,\bar{D}] \le 1$, $C_\alpha[r(z)/k_*,\Gamma] \le 1$, $|b_j/k_*| \le 1, j \in J$ and according to the proof of Theorem 2, we have

$$\tilde{C}_\gamma[W(z),\bar{D}] \le M_9,\ L_{p_0,2}[|W_{\bar{z}}|+|W_z|,\tilde{D}] \le M_{10}. \tag{5.36}$$

From the above estimates, it immediately follows that (5.32) holds.

Remark Through the mapping $z = z(\zeta) = 1/\zeta$, the complex equation (5.1) can be reduced in the form

$$w_{\bar{\zeta}} = H(z,w,w_\zeta),\ H = \tilde{Q}_1w_\zeta+\tilde{Q}_2\bar{w}_{\bar{\zeta}}+\tilde{A}_1w+\tilde{A}_2\bar{w}+\tilde{A}_3,\ z \in G = \zeta(D), \tag{5.37}$$

in which

$$\tilde{Q}_1 = Q_1\zeta^2/\bar{\zeta}^2,\ \tilde{Q}_2 = Q_2,\ \tilde{A}_j = -A_j/\bar{\zeta}^2,\ j = 1,2,3,\ \zeta \in G,$$

and $\zeta = \zeta(z) = 1/z$. By Condition C, the above coefficients satisfy the conditions

$$|\tilde{Q}_1|+|\tilde{Q}_2| \le q_0,\ \zeta \in \tilde{D},\ L_p[\tilde{A}_j,\bar{G}] \le k_0,\ j = 1,2,\ L_p[\tilde{A}_3,\bar{G}] \le k_1. \tag{5.38}$$

If the function $w(z)$ is a solution of complex equation (5.1) with Condition C in D, then $w(z) = w[z(\zeta)] = w[1/\zeta]$ is a solution of the complex equation (5.37) in G. Noting that $w_z = -\zeta^2w_\zeta$, $w_{\bar{z}} = -\bar{\zeta}^2w_{\bar{\zeta}}$, we see that if $w(z) \in W^1_{p_0}(D)$, $2 < p_0 \le p$, then $w[z(\zeta)] \in W^1_{p_0,2}(G)$. The inverse result is also true. If $f(z) \in L_{p_0,2}(\bar{D})$, then

$$\begin{aligned} Tf &= -\frac{1}{\pi}\iint_D \frac{f(\zeta)}{\zeta - z}d\sigma_\zeta = -\frac{1}{\pi}\iint_G \frac{f(1/\zeta)}{\bar{\zeta}^2\zeta(1-\zeta z)}d\sigma_\zeta \\ &= S(0) - S(\frac{1}{z}),\ S(z) = -\frac{1}{\pi}\iint_G \frac{\tilde{f}(\zeta)}{\zeta - z}d\sigma_\zeta,\ \tilde{f}(\zeta) = f(\frac{1}{\zeta})/\bar{\zeta}^2, \end{aligned} \tag{5.39}$$

it shows that $\tilde{f}(z) \in L_{p_0}(\bar{G})$. Hence

$$\tilde{C}_\alpha[S(z),\bar{G}] \le ML_{p_0}[\tilde{f}(z),\bar{G}],\ \tilde{C}_\alpha[S(0)-S(\frac{1}{z}),\bar{G}] \le ML_{p_0}[\tilde{f}(z),\bar{G}], \quad (5.40)$$

in which $\alpha = 1 - 2/p_0$, $M = M(p_0)$. Thus by using the method of continuity and the principle of contracting mappings, we can prove that there exist the solutions $\psi(z) = Tf(z), \phi(z) = Tg(z) \in W^1_{p_0,2}(\bar{D})$ of (5.16), (5.17) respectively. Moreover, we can also find the solution $\chi(z) = 1/z + Th$ of the equation

$$W_{\bar{z}} = QW_z, \text{ or } h(z) = Q(z)[-\frac{1}{z^2} + \Pi h], \quad (5.41)$$

it is clear that $-Q(z)/z^2 \in L_{p,2}(\bar{D})$, and then $h(z) \in L_{p_0,2}(\bar{G})$. Due to the function $\chi(1/z) = z + S(0) - S(z) = z + S(0) - T[\tilde{h}]$ is a solution of

$$\tilde{h}(z) = \tilde{Q}(z)z^2/\bar{z}^2[1 + \Pi\tilde{h}] \text{ in } G,$$

where $\tilde{h}(\zeta) = h(1/\zeta)/\bar{\zeta}^2$, therefore $\chi(1/z) = z + S(0) - S(z) = z + S(0) - T[\tilde{h}]$ is a homeomorphism in G, obviously $\chi(z)$ is also a homeomorphism in D.

5.3 UNIQUENESS AND SOLVABILITY OF THE RIEMANN-HILBERT PROBLEM FOR (5.1)

Theorem 4 *If Condition C and* (5.11) *hold, then the solution of Problem B for* (5.1) *is unique.*

Proof Let $w_1(z), w_2(z)$ be two solutions of Problem B for (5.1). By Condition C and (5.11), we see that $w(z) = w_1(z) - w_2(z)$ satisfies the equation and boundary conditions

$$w_{\bar{z}} - \tilde{Q}w_z = \tilde{A}w,\ z \in D, \quad (5.42)$$

$$\mathrm{Re}[\overline{\lambda(z)}w(z)] = h(z),\ z \in \Gamma, \quad (5.43)$$

$$\mathrm{Im}[\overline{\lambda(a_j)}w(a_j)] = 0,\ j \in J, \quad (5.44)$$

where

$$\tilde{Q} = \begin{cases} [F(z,w_1,w_{1z}) - F(z,w_1,w_{2z})]/(w_1 - w_2)_z & \text{for } w_{1z} \ne w_{2z}, \\ 0 \text{ for } w_{1z} = w_{2z},\ z \in D, \end{cases}$$

$$\tilde{A} = \begin{cases} [F(z,w_1,w_{2z}) - F(z,w_2,w_{2z})]/(w_1 - w_2) & \text{for } w_1(z) \ne w_2(z), \\ 0 \text{ for } w_1(z) = w_2(z),\ z \in D, \end{cases}$$

and $|\tilde{Q}| \le q_0 < 1$, $z \in D$, $L_{p_0,2}(\tilde{A},\bar{D}) < \infty$. According to the representation (5.12), we have

$$w(z) = \Phi[\zeta(z)]e^{\phi(z)}, \quad (5.45)$$

where $\phi(z), \zeta(z), \Phi(\zeta)$ are as stated in Theorem 1. It can be seen that the analytic function $\Phi(z)$ satisfies the boundary conditions of Problem B_0 :

$$\mathrm{Re}[\overline{\Lambda(\zeta)}\Phi(\zeta)] = H(\zeta),\ \zeta \in L = \zeta(\Gamma), \quad (5.46)$$

$$\operatorname{Im}[\overline{\Lambda(a_j')}\Phi(a_j')] = 0,\ j \in J, \tag{5.47}$$

where $\Lambda(\zeta)$, $H(\zeta)$ $(\zeta \in L)$, a_j' $(j \in J)$ are as stated in (5.21)-(5.23). In accordance with Theorem 4.1, Chapter 2 in [2], we can see that when $K \geq 0$, there is $\Phi(\zeta) = 0, \zeta \in G = \zeta(D)$. Hence, $w(z) = \Phi[\zeta(z)]e^{\phi(z)} = 0$, i.e. $w_1(z) = w_2(z)$, $z \in D$.

Next, we discuss the case of $K < 0$. According to Theorem 4.12 of Chapter 4 in [1], the necessary and sufficient condition for solvability of Problem B_0 is that

$$\int_L [\overline{\Lambda(\zeta)}]^{-1}\Psi_n(\zeta)H(\zeta)\zeta'(s)ds = 0,\ n = 1,\cdots,N-2K-1, \tag{5.48}$$

where $\Psi_n(\zeta)(n = 1,\cdots,N-2K-1)$ is a complete system of linearly indepedent solutions for the boundary value problem B_0' for analytic functions with the boundary conditions

$$\operatorname{Re}\{[\overline{\Lambda(\zeta)}]^{-1}\Psi_n(\zeta)\zeta'(s)\} = 0,\ \zeta \in L,\ n = 1,\cdots,N-2K-1. \tag{5.49}$$

Taking into account the definition of $h(z)$ in (5.8), the conditions (5.48) can be rewritten as

$$\begin{aligned} &\sum_{j=0}^{N} h_j \int_{L_j} \operatorname{Im}[\overline{\Lambda(\zeta)}]^{-1}\Psi_n(\zeta)\zeta'(s)ds \\ &+\sum_{m=1}^{-K-1}\{h_m^+ \int_{L_0} \operatorname{Im}[\overline{\Lambda(\zeta)}]^{-1}\Psi_n(\zeta)\zeta'(s)\cos[m\arg z(\zeta)]ds \\ &-h_m^- \int_{L_0} \operatorname{Im}[\overline{\Lambda(\zeta)}]^{-1}\Psi_n(\zeta)\zeta'(s)\sin[m\arg z(\zeta)]ds\} = 0, \\ &\qquad n = 1,\cdots,N-2K-1. \end{aligned} \tag{5.50}$$

Suppose that $\sum_{j=0}^{N} | h_j | + \sum_{m=1}^{-K-1}[| h_m^+ | + | h_m^- |] \neq 0$, then the coefficients determinant of the algebraic system (5.50) certainly equals zero. Hence we can find real constants $c_1,\cdots,c_{N-2K-1}$, which are not all equal to zero, such that

$$\begin{aligned} &\int_{L_j} \operatorname{Im}[\overline{\Lambda(\zeta}]^{-1}\Psi(\zeta)\zeta'(s)ds = 0,\ j = 0,\cdots,N, \\ &\int_{L_0} \operatorname{Im}[\overline{\Lambda(\zeta)}]^{-1}\Psi(\zeta)\zeta'(s)\cos[m\arg z(\zeta)]ds = 0, \\ &\int_{L_0} \operatorname{Im}[\overline{\Lambda(\zeta)}]^{-1}\Psi(\zeta)\zeta'(s)\sin[m\arg z(\zeta)]ds = 0, \\ &\qquad m = 1,\cdots,-K-1, \end{aligned} \tag{5.51}$$

in which $\Psi(\zeta) = \sum_{j=1}^{N-2K-1} c_n\Psi_n(\zeta)$ is a solution of Problem B_0' and $\Psi(\zeta) \not\equiv 0$ in $G = \zeta(D)$. From (5.51) and Theorem 4.7 from Chapter 4 in [1], it can be derived that $\Psi(\zeta)$ has at least $2N - 2K - 1$ zero points on L. Noting that

$N-K-1=\frac{1}{2\pi}\Delta_\Gamma \arg[\Lambda(\zeta)]^{-1}\overline{\zeta'(s)}$ and according to the formula (4.19) from Chapter 4 in [1], we get the absurd inequality

$$2N-2K-1\le 2N_D+N_\Gamma=2N-2K-2.$$

This contradiction proves that

$$h_j=0,\ j=0,1,\cdots,N,\ \ h_m^{\pm}=0,\ m=1,\cdots,-K-1,$$

i.e. $H(\zeta)=0,\zeta\in L$, and so $\Phi(\zeta)=0,\zeta\in\bar{G}$. Consequently $w(z)=0$ in D, i.e. $w_1(z)=w_2(z),\ z\in D$.

Theorem 5 *Suppose that Condition C holds. Then Problem B for* (5.1) *is solvable.*

Proof Here we only discuss Problem B for (5.1). Let us introduce a closed, convex and bounded subset B_1 in the Banach space $B=L_{p_0,2}(\bar{D})\times L_{p_0,2}(\bar{D})\times L_{p_0,2}(\bar{D})$, whose elements are systems of functions $q=[Q(z),f(z),g(z)]$ with norms $\| q\|=L_{p_0,2}(Q,\bar{D})+L_{p_0,2}(f,\bar{D})+L_{p_0,2}(g,\bar{D})$, which satisfies the condition

$$|Q(z)|\le q_0<1\,(z\in D),\ L_{p_0,2}[f(z),\bar{D}]\le k_4,\ L_{p_0,2}[g(z),\bar{D}]\le k_4, \tag{5.52}$$

where q_0,k_4 are nonnegative constants as stated in (5.2), (5.14) and (5.15). Moreover introduce a closed and bounded subset B_2 in B, the elements of which are systems of functions $\omega=[f(z),g(z),h(z)]$ satisfying the condition

$$L_{p_0,2}[f(z),\bar{D}]\le k_4,\ L_{p_0,2}[g(z),\bar{D}]\le k_4,\ |h(z)|\le q_0|-\frac{1}{z^2}+\Pi h|, \tag{5.53}$$

where $\Pi h=-\frac{1}{\pi}\iint_D[h(\zeta)/(\zeta-z)^2]d\sigma_\zeta$. We mention that the space B_2 is not convex.

We arbitrarily select $q=[Q(z),f(z),g(z)]\in B_1$, and using the principle of contracting mappings, a unique solution $h(z)\in L_{p_0,2}(\bar{D})$ of the integral equation

$$h(z)=Q(z)[-\frac{1}{z^2}+\Pi h] \tag{5.54}$$

is found, which satisfies the third estimate in (5.52). Moreover, $\chi(z)=1/z+Th$ is a homeomorphism on $\bar{D}$. Now, we find a univalent analytic function $\zeta=\Phi(\chi)$, which maps $\chi(D)$ onto a circular domain G as stated in Theorem 1. Afterwards, we find an analytic function $\Phi(\zeta)$ in G satisfying the boundary conditions

$$\text{Re}[\overline{\Lambda(\zeta)}\Phi(\zeta)]=R(\zeta)+H(\zeta),\ \zeta\in L=\zeta(\Gamma), \tag{5.55}$$

$$\text{Im}[\overline{\Lambda(a_j')}\Phi(a_j')]=b_j',\ j\in J, \tag{5.56}$$

in which $\zeta(z)=\Psi[\chi(z)]$, $z(\zeta)$ is its inverse function, $\psi(z)=Tf$, $\phi(z)=Tg$, $\Lambda(\zeta)$, $R(\zeta)$, $H(\zeta)$, a_j', $b_j'(j\in J)$ are as stated in (5.21) – (5.23). In the following, we only consider the case of $K < 0$. It is sufficient to verify that the following conditions hold (see Chapter 2 in [2])

$$\int_L \overline{[\Lambda(\zeta)]^{-1}}\Psi_n(\zeta)[R(\zeta)+H(\zeta)]\zeta'(s)ds = 0,\ n = 1,\cdots,N-2K-1, \tag{5.57}$$

where $\Psi_n(\zeta)(n = 1,\cdots,N-2K-1)$ are as stated in (5.48), (5.49). It can be proved that the coefficients determinant of $h_j(j = 1,\cdots,N-2K-1)$ in (5.57) is not equal to zero. Hence from (5.57), we may find a unique system $h_j(j = 1,\cdots,N-2K-1)$. For the system $h_j(j = 1,\cdots,N-2K-1)$, the boundary value problem (5.55), (5.56) for analytic functions possesses a unique solution $\Phi(\zeta)$. Thus the function $w(z) = \Phi[\zeta(z)]e^{\phi(z)} + \psi(z)$ is determined.

Finally, we find out the solution $[f^*(z), g^*(z), h^*(z), Q^*(z)]$ of the system of integral equations

$$f^*(z) = F(z,w,\Pi f^*) - F(z,w,0) + A_1(z,w)Tf^* + A_2(z,w)\overline{Tf^*} + A_3(z,w), \tag{5.58}$$

$$Wg^*(z) = F(z,w,W\Pi g^* + \Pi f^*) - F(z,w,\Pi f^*) + A_1(z,w)W + A_2(z,w)\overline{W}, \tag{5.59}$$

$$\begin{aligned} S'(\chi)h^*(z)e^{\phi(z)} &= F[z,w,S'(\chi)(1+\Pi h^*)e^{\phi(z)} + W\Pi g^* + \Pi f^*] \\ &\quad - F(z,w,W\Pi g^* + \Pi f^*), \end{aligned} \tag{5.60}$$

$$Q^*(z) = h^*(z)/(-1/z^2 + \Pi h^*),\ S'(\chi) = [\Phi(\Psi(\chi))]_\chi, \tag{5.61}$$

and denote by $q^* = E(q)$ the mapping from $q = (Q, f, g)$ to $q^* = (Q^*, f^*, g^*)$. According to Theorem 2.1 from Chapter 4 in [2], we can prove that $q^* = E(q)$ continuously maps B_1 onto a compact subset in B_1. On the basis of the Schauder fixed-point theorem, there exists a system $q = (Q, f, g) \in B_1$, such that $q = E(q)$. Applying the above method, from $q = (Q, f, g)$, we can construct a function $w(z) = \Phi[\zeta(z)]e^{\phi(z)} + \psi(z)$, which is just a solution of Problem B for (5.1).

In a similar method as in the proof of Theorem 4.8, Chapter 2 in [2], from Theorem 5 the following result can be derived.

Theorem 6 *Under the same conditions as in Theorem 5, the following statements hold.*

(1) *If the index $K \geq N$, then Problem A for (5.1) is solvable.*

(2) *If $0 \leq K < N$, then the total number of solvability conditions for Problem A does not exceed $N - K$.*

(3) *If $K < 0$, then Problem A has $N - 2K - 1$ solvability conditions.*

References

[1] Vekua, I. N.: Generalized analytic functions. Pergamon, Oxford, 1962.

[2] Wen, Guo Chun, Begehr, H.: Boundary value problems for elliptic equations and systems. Longman Scientific and Technical, Harlow, 1990.

[3] Begehr, H., Wen, G.-C.: Nonlinear elliptic boundary value problems and their applications. Addison Wesley Longman, Harlow, 1996.

6 THE RIEMANN–HILBERT PROBLEM FOR FIRST ORDER COMPLEX EQUATIONS OF MIXED TYPE

Guo–Chun Wen

Peking University
Beijing 100871
P.R. China
wengc@pku.edu.cn

Abstract: In this paper we discuss the Riemann–Hilbert boundary value problem for first order linear complex equations of mixed(elliptic–hyperbolic) type in a simply connected domain. Firstly, we give a representation theorem and prove the uniqueness of solutions for the above boundary value problem, afterwards by using the method of successive iteration, the existence of solutions for the above problem is proved. In [1], A.V. Bitsadze discussed the Tricomi problem for some mixed equation of second order. By using the results in the present paper, we can derive the results from [1].

6.1 FORMULATION OF THE RIEMANN–HILBERT PROBLEM FOR FIRST ORDER COMPLEX EQUATIONS OF MIXED TYPE

Let D be a simply connected bounded domain in the complex plane $\mathbb{C}$ with the boundary $\partial D = \Gamma \cup L$, where $\Gamma(\subset \{y > 0\}) \in C^1_\alpha(0 < \alpha < 1)$ with the end points $z = 0$, 2 and $L = L_1 \cup L_2$, $L_1 = \{x = -y, 0 \le x \le 1\}$, $L_2 = \{x = y + 2, 1 \le x \le 2\}$, and denote $D_1 = D \cap \{y > 0\}$, $D_2 = D \cap \{y < 0\}$ and $z_1 = 1 - i$. Without loss of generality, we may assume that $\Gamma = \{|z - 1| = 1, y \ge 0\}$, otherwise through a conformal mapping, this requirement can be realized.

We discuss the first order linear system of mixed (elliptic–hyperbolic) type equations

$$\begin{cases} u_x - v_y = au + bv + f, \\ v_x + \operatorname{sgn} y\, u_y = cu + dv + g, \end{cases} \qquad z = x + iy \in D, \tag{6.1}$$

H.G.W. Begehr et al. (eds.), Partial Differential and Integral Equations, 97–105.

in which a, b, c, d, f, g are functions of $(x, y)(\in D)$. Its complex form is the following complex equation of first order

$$\left\{ \begin{array}{c} w_{\overline{z}} \\ w_{\overline{z^*}} \end{array} \right\} = F(z, w),\ F = A_1(z)w + A_2(z)\overline{w} + A_3(z) \text{ in } \left\{ \begin{array}{c} D_1 \\ D_2 \end{array} \right\}, \tag{6.2}$$

where

$$w = u + iv,\ z = x + iy,\ w_{\overline{z}} = \frac{1}{2}[w_x + iw_y],\ w_{\overline{z^*}} = \frac{1}{2}[w_x - i\overline{w}_y],$$

$$A_1 = \frac{a - ib + ic + d}{4},\ A_2 = \frac{a + ib + ic - d}{4},\ A_3 = \frac{f + ig}{2}.$$

Suppose that the complex equation (6.2) satisfies the following conditions (**Condition** C):

$A_j(z)\ (j = 1, 2, 3)$ are measurable in $z \in D_1$ and continuous on $\overline{D_2}$ on $D^* = \bar{D}\backslash\{0, 2\}$ and satisfy

$$L_p[A_j, \overline{D_1}] \le k_0,\ j = 1, 2,\ L_p[A_3, \overline{D_1}] \le k_1, \tag{6.3}$$

$$C[A_j, \overline{D_2}] \le k_0,\ j = 1, 2,\ C[A_3, \overline{D_2}] \le k_1 \text{ in } D_2, \tag{6.4}$$

where $p(> 2), k_0,\ k_1$ are nonnegative constants.

The Riemann–Hilbert boundary value problem for the complex equation (6.2) may be formulated as follows.

Problem M Find a continuous and bounded solution $w(z)$ of (6.2) on D^* satisfying the boundary conditions

$$\text{Re}[\overline{\lambda(z)}w(z)] = r(z),\ \ z \in \Gamma, \tag{6.5}$$

$$\text{Re}[\overline{\lambda(z)}w(z)] = r(z),\ \ z \in L_j(j = 1 \text{ or } 2),\ \text{Im}[\overline{\lambda(z_1)}w(z_1)] = b_1, \tag{6.6}$$

where $\lambda(z) = a(z) + ib(z), |\lambda(z)| = 1,\ z \in \Gamma \cup L_j(j = 1 \text{ or } 2),\ b_1$ is a real constant, and $\lambda(z),\ r(z),\ b_1$ satisfy the conditions

$$\begin{gathered} C_\alpha[\lambda(z), \Gamma] \le k_0,\ C_\alpha[r(z), \Gamma] \le k_2,\ |b_1| \le k_2, \\ C_\alpha[\lambda(z), L_j] \le k_0,\ C_\alpha[r(z), L_j] \le k_2,\ j = 1 \text{ or } 2, \end{gathered} \tag{6.7}$$

in which $\alpha\ (1/2 < \alpha < 1), k_0, k_2$ are nonnegative constants.

This boundary value problem for (6.2) with $A_3(z, w) = 0,\ z \in D, w \in \mathbb{C}$ and $r(z) = 0,\ z \in \Gamma \cup L_j(j = 1 \text{ or } 2),\ b_1 = 0$ will be called Problem M_0. The number

$$K = \frac{1}{2}(K_1 + K_2) \tag{6.8}$$

is called the index of Problem M and Problem M_0, where

$$K_j = \left[\frac{\phi_j}{\pi}\right] + J_j,\ J_j = 0 \text{ or } 1,\ e^{j\phi_j} = \frac{\lambda(t_j - 0)}{\lambda(t_j + 0)},\ \gamma_j = \frac{\phi_j}{\pi} - K_j,\ j = 1, 2, \tag{6.9}$$

in which $t_1 = 2$, $t_2 = 0$, $\lambda(t) = e^{-i\pi/4}$ on $(0, 2)$ and $\lambda(t_1 - 0) = \lambda(t_2 + 0) = \exp(-i\pi/4)$, or $\lambda(t) = e^{i\pi/4}$ on $(0, 2)$ and $\lambda(t_1 - 0) = \lambda(t_2 + 0) = \exp(i\pi/4)$. In this paper we only discuss the case of $K = -1/2$ on the boundary ∂D_1 of D_1. Let $e_1 = (1 + i)/2$, $e_2 = (1 - i)/2$, then we have

$$\begin{aligned} z &= x + iy = \mu e_1 + \nu e_2,\ \mu = x + y,\ \nu = x - y,\ x = (\mu + \nu)/2,\ y = (\mu - \nu)/2, \\ w &= u + jv = \xi e_1 + \eta e_2,\ \xi = u + v,\ \eta = u - v,\ u = (\xi + \eta)/2,\ v = (\xi - \eta)/2, \end{aligned} \tag{6.10}$$

and the linear hyperbolic complex equation (6.2) in D_2 can be rewritten in the form

$$\begin{aligned} \xi_\nu e_1 + \eta_\mu e_2 &= [A(z)\xi + B(z)\eta + E(z)]e_1 \\ &\quad +[C(z)\xi + D(z)\eta + F(z)]e_2 \text{ in } D_2, \end{aligned} \tag{6.11}$$

in which

$$A = (a - b + c + d)/4,\ B = (a + b + c - d)/4,\ C = (a - b - c - d)/4,$$

$$D = (a + b - c + d)/4,\ E = (f + g)/2,\ F = (f - g)/2.$$

The last two formulae of boundary condition (6.7) can be reduced to

$$\text{Re}[\overline{\lambda}(\xi e_1 + \eta e_2)] = r,\ \text{Im}[\overline{\lambda}(\xi e_1 + \eta e_2)]|_{z=z_1} = b_1, \tag{6.12}$$

where $\lambda = (a + b)e_1/2 + (a - b)e_2/2$. Moreover, the domain D is transformed into $Q = \{0 \le \mu \le 2, 0 \le \nu \le 2\}$, which is a rectangle and A, B, C, D, E, F are known functions of (μ, ν). There is no harm in assuming that $w(z_1) = 0$, otherwise through the transformatiom $W(z) = w(z) - [a(z_1) + ib(z_1)][r(z_1) + ib_1]$, this requirement can be realized.

6.2 REPRESENTATION AND UNIQUENESS OF SOLUTIONS OF THE RIEMANN–HILBERT PROBLEM FOR FIXED EQUATIONS

We first introduce a result from [2] as a lemma.

Lemma 1 *Suppose that the complex equation* (6.2) *satisfies Condition C. Then any solution of Problem M for* (6.2) *in* D_1 *with the boundary conditions* (6.5) *and*

$$\text{Re}[\overline{\lambda(x)}w(x)] = s(x),\ \lambda(x) = 1 - i \text{ or } 1 + i,\ x \in L_0,\ C_\alpha[s(x), L_0] \le k_3, \tag{6.13}$$

can be expressed as

$$w(z) = \Phi(z)e^{\phi(z)} + \psi(z),\ \ z \in \overline{D_1}, \tag{6.14}$$

where $\text{Im}[\phi(z)] = 0$, $z \in L_0 = (0, 2)$, *and* $\phi(z)$, $\psi(z)$ *satisfies the estimates*

$$C_\beta[\phi, \overline{D_1}] + L_{p_0}[\phi_{\bar z}, \overline{D_1}] \le M_1,\ C_\beta[\psi, \overline{D_1}] + L_p[\psi_{\bar z}, \overline{D_1}] \le M_1, \tag{6.15}$$

in which k_3, $\beta(0<\beta\le\alpha)$, $p_0(2<p_0\le 2)$, $M_1=M_1(p_0,\beta,k,D_1)$ *are nonnegative constants,* $k=(k_0,k_1,k_2,k_3)$, $\Phi(z)$ *is analytic in* D_1*, and* $w(z)$ *satisfies the estimate*

$$C_\beta[w(z)X(z),\overline{D_1}]\le M_2(k_1+k_2+k_3), \tag{6.16}$$

in which $X(z)=(z-t_1)^{2\gamma_1+\delta}(z-t_2)^{2\gamma_2+\delta}$*, herin* γ_1,γ_2 *are real constants as stated in* (6.9)*,* δ *is a sufficintly small positive constant, and* $M_2=M_2(p_0,\beta,k_0,D_1)$ *is a nonnegative constant.*

Theorem 1 *Suppose that the complex equation* (6.2) *satisfies Condition C. Then any solution of Problem M for* (6.2) *can be expressed as*

$$w(z)=w_0(z)+W(z), \tag{6.17}$$

where $w_0(z)$ *is a solution of Problem M for the equation*

$$\left\{\begin{array}{c} w_{\bar z}\\ w_{\bar z^*}\end{array}\right\}=0 \text{ in } \left\{\begin{array}{c} D_1\\ D_2\end{array}\right\}, \tag{6.18}$$

(see [5]*) and* $W(z)$ *possesses the form*

$$\begin{aligned}
&w(z)=w_0(z)+W(z)=\tilde\Phi(z)e^{\tilde\phi(z)}+\tilde\psi(z),\\
&\tilde\phi(z)=\tilde\phi_0(z)+Tg=\tilde\phi_0(z)-\frac{1}{\pi}\iint_{D_1}\frac{g(\zeta)}{\zeta-z}d\sigma_\zeta,\ \tilde\psi(z)=Tf \text{ in } D_1,\\
&\overline{W(z)}=\Phi(z)+\Psi(z),\ \Psi(z)=\int_2^\nu g_1(z)d\nu e_1+\int_0^\mu g_2(z)d\mu e_2,\ z\in D_2,
\end{aligned} \tag{6.19}$$

in which $\tilde\phi(z)=0$ *on* L_0, $e_1=(1+i)/2, e_2=(1-i)/2$, $\mu=x+y$, $\nu=x-y$, $\tilde\phi_0(z)$ *is an analytic function in* D_1*, and*

$$g(z)=\left\{\begin{array}{l} A_1+A_2\overline{w}/(w),\ w(z)\ne 0,\\ 0,\ w(z)=0,\end{array}\right.\quad f=A_1\tilde\phi+A_2\overline{\tilde\phi}+A_3 \text{ in } D_1, \tag{6.20}$$

$$g_1(z)=A\xi+B\eta+E,\ g_2(z)=C\xi+D\eta+F \text{ in } D_2,$$

where $\xi=\text{Re}w+\text{Im}w$, $\eta=\text{Re}w-\text{Im}w$*, and* $\tilde\phi(z)$, $\tilde\psi(z)$ *satisfy the estimates*

$$C_\beta[\tilde\phi(z),\overline{D_1}]+L_p[\tilde\phi_{\bar z},\overline{D_1}]\le M_1,\ C_\beta[\tilde\psi(z),\overline{D_1}]+L_p[\tilde\psi_{\bar z},\overline{D_1}]\le M_1, \tag{6.21}$$

where $M_1=M_1(p_0,\beta,k,D_1)$ *is a nonnegative constant,* $\tilde\Phi(z)$ *is analytic in* D_1 *and* $\Phi(z)$ *is a solution of the equation* (6.18) *in* D_2 *satisfying the boundary conditions*

$$\begin{aligned}
&\text{Re}[\overline{\lambda(z)}(e^{\tilde\phi(z)}\tilde\Phi(z)+\tilde\psi(z))]=r(z),\ z\in\Gamma,\\
&\text{Re}[\lambda(x)(\tilde\Phi(x)e^{\tilde\phi(x)}+\tilde\psi(x))]=s(x),\ x\in L_0,\\
&\text{Re}[\overline{\lambda(x)}\Phi(x)]=-\text{Re}[\overline{\lambda(x)}\Psi(x)],\ z\in L_0,\\
&\text{Re}[\overline{\lambda(z)}\Phi(z)]=-\text{Re}[\overline{\lambda(z)}\Psi(z)],\ z\in L_1 \text{ or } L_2,\\
&\text{Im}[\overline{\lambda(z_1)}\Phi(z_1)]=-\text{Im}[\overline{\lambda(z_1)}\Psi(z_1)].
\end{aligned} \tag{6.22}$$

Moreover, the solution $w_0(z)$ *of Problem* M *for* (6.18) *satisfies the estimate*

$$C_\beta[w_0(z)X(z),\overline{D}]\leq M_3(k_1+k_2) \tag{6.23}$$

(*see* [5]), *where* $X(z)$ *is as stated in* (6.16), $M_3=M_3(p_0,\beta,k_0,D)$ *is a nonnegative constant.*

Proof Let the solution $w(z)$ be substituted in the position of w in the complex equation (6.2) and (6.20), thus the functions $g_1(z)$, $g_2(z)$ and $\Psi(z)$ in $\overline{D_2}$ in (6.19),(6.20) can be determined. Moreover we can find the solution $\Phi(z)$ of (6.18) with the boundary conditions (6.22) and (6.13), where

$$s(x)=\begin{cases}\operatorname{Re}[\overline{\lambda(x/2)}r(x/2)]/[a(x/2)-b(x/2)] \text{ or}\\ \operatorname{Re}[\overline{\lambda(x/2+1)}r(x/2+1)]/[a(x/2+1)+b(x/2+1)],\end{cases} x\in L_0. \tag{6.24}$$

Thus

$$w(z)=w_0(z)+W(z)=\begin{cases}\tilde{\Phi}(z)e^{\tilde{\phi}(z)}+\tilde{\psi}(z) \text{ in } D_1,\\ w_0(z)+\overline{\Phi(z)}+\overline{\Psi(z)} \text{ in } D_2,\end{cases} \tag{6.25}$$

is a solution of Problem M for the complex equation

$$\begin{Bmatrix} w_{\overline{z}}\\ \bar{w}_{\overline{z^*}}\end{Bmatrix}=A_1w+A_2\bar{w}+A_3 \text{ in } \begin{Bmatrix} D_1\\ D_2\end{Bmatrix}, \tag{6.26}$$

which can be exprssed as in (6.17) and (6.19).

Theorem 2 *Suppose that the complex equation* (6.2) *satisfies Condition* C. *Then Problem* M *for* (6.2) *has at most a solution in* D.

Proof Let $w_1(z), w_2(z)$ be any two solutions of Problem M for (6.2). By Condition C, we see that $w(z)=w_1(z)-w_2(z)$ satisfies the homogeneous complex equation and boundary conditions

$$w_{\bar{z}}=\tilde{A}_1w+\tilde{A}_2\bar{w} \text{ in } D, \tag{6.27}$$

$$\operatorname{Re}[\overline{\lambda(z)}w(z)]=0,\ z\in L_1 \text{ or } L_2,\ \operatorname{Re}[\overline{\lambda(z_1)}w(z_1)]=0. \tag{6.28}$$

From Theorem 1, the solution $w(z)$ can be expressed in the form

$$w(z)=\begin{cases}\tilde{\Phi}(z)e^{\tilde{\phi}(z)},\ \tilde{\phi}(z)=\tilde{T}g \text{ in } D_1,\\ g(z)=\begin{cases}A_1+A_2\bar{w}/w,\ w(z)\neq 0,\ z\in D_1,\\ 0,\ w(z)=0,\ z\in D_1,\end{cases}\\ w_0(z)+\overline{\Phi(z)}+\overline{\Psi(z)},\\ \Psi(z)=\int_2^\nu [A\xi+B\eta]e_1d\nu+\int_0^\mu [C\xi+D\eta]e_2d\mu \text{ in } \overline{D_2},\end{cases} \tag{6.29}$$

where $\tilde{\Phi}(z)$ is analytic in D_1 and $\Phi(z)$ is a solution of (6.18) in D_2 satisfying the boundary condition (6.22), but $\psi(z) = 0$. Besides the function $\Phi(z)$ satisfies the boundary conditions

$$\begin{cases} \text{Re}[\overline{\lambda(x)}\Phi(x)] = s(x), \\ \text{Re}[\lambda(x)\Phi(x)] = \text{Re}[\lambda(x)(w(x) - w_0(z) - \overline{\Psi(x)})]e^{-\phi(x)}, \end{cases} \quad x \in L_0, \qquad (6.30)$$

where $s(x)$ is as stated in (6.24). Suppose $w(z) \not\equiv 0$ in the neighbourhood($\subset \bar{D}$) of two characteristic lines through the point $z_1 = 1 - i$. We may choose a positive number $R_0 < 1$, such that $8M_4M(2M_3 + 1)R_0 < 1$, where $M_4 = \max\{C[A, Q], C[B, Q], C[C, Q], C[D, Q]\}$, $M = 1 + 2k_0^2(1 + 2k_0^2)$, M_3 is a constant as stated in (6.23), and $m = C[w(z), \overline{Q_0}] > 0$, in which $Q_0 = \{0 \le \lambda \le R_0, 2 - R_0 \le \nu \le 2\}$. According to (6.29), we can derive the absurd inequality $m \le 8M_4M(2M_3 + 1)mR_0 < m$. It shows $w(z) = 0$, $z \in Q_0$. Moreover, we extend along the positive direction of μ and the negative direction of ν successively, and finally obtain $w(z) = 0$ for $z \in D_2$, and $\Psi(z) = 0$, $z \in \overline{D_2}$. Besides, noting that the analytic function $\tilde{\Phi}(z)$ in D_1 satisfies the boundary conditions

$$\text{Re}[\overline{\lambda(z)}e^{\phi(z)}\tilde{\Phi}(z)] = 0, \ z \in \Gamma, \ \text{Re}[\lambda(x)\tilde{\Phi}(x)] = 0, \ x \in L_0, \qquad (6.31)$$

and the index of the bounday value problem (6.31) is $K = -1/2$, hence $\tilde{\Phi}(z) = 0$ in D_1(see [3]), and then $w(z) = \tilde{\Phi}(z)e^{\tilde{\phi}(z)} = 0$ in D_1. Thus $w(z) = w_1(z) - w_2(z) = 0$ in D. This proves the uniqueness of solutions of Problem M for (6.2).

6.3 THE SOLVABILITY OF THE RIEMANN–HILBERT PROBLEM FOR FIRST ORDER COMPLEX EQUATIONS OF MIXED TYPE

Theorem 3 *Suppose that the mixed complex equation* (6.2) *satisfies Condition C. Then Problem M for* (6.2) *has a solution in D.*

Proof In order to find a solution $w(z)$ of Problem M in D, we can express $w(z)$ in the form (6.17)–(6.20). By the method in [2]–[5], we can find a solution $w(z)$ of the discontinuous Riemann-Hilbert boundary value problem(Problem M) for the complex equation (6.18) with the boundary conditions (6.5),(6.6), where the solution $w(z)$ satisfies the estimate (6.23), namely

$$C_\beta[w(z)X(z), \overline{D}] \le M_5 = M_5(p_0, \beta, k, D), \qquad (6.32)$$

where M_5 is a nonnegative constant and $k = (k_0, k_1, k_2)$. In the following, by using successive iteration, we shall find a solution of Problem M for the complex equation (6.2) in D_2. Firstly denoting the solution $w_0(z) = \xi_0 e_1 + \eta_0 e_2$ of Problem M for (6.18), and substituting w_0, ξ_0, η_0 into the positions of w, ξ, η in the right–hand side of (6.2), (6.20), similarly to (6.17)–(6.20), the function

$$\Psi_1(z) = \int_2^\nu [A\xi_0 + B\eta_0 + E]e_1 d\nu + \int_0^\mu [C\xi_0 + D\eta_0 + F]e_2 d\mu \text{ in } D_2 \qquad (6.33)$$

can be determined, where $\mu = x + y$, $\nu = x - y$. Afterwards we find a solution $\Phi_1(z)$ of (6.18) in D_2 satisfying the boundary conditions

$$\begin{aligned} \mathrm{Re}[\lambda(x)\Phi_1(x)] &= -\mathrm{Re}[\lambda(x)\Psi_1(x)],\ z \in L_0, \\ \mathrm{Re}[\overline{\lambda(z)}\Phi_1(z)] &= -\mathrm{Re}[\overline{\lambda(z)}\Psi_1(z)],\ z \in L_1 \text{ or } L_2, \\ \mathrm{Im}[\overline{\lambda(z_0)}\Phi_1(z_0)] &= -\mathrm{Im}[\overline{\lambda(z_0)}\Psi_1(z_0)], \end{aligned} \tag{6.34}$$

and

$$w_1(z) = w_0(z) + W_1(z) = w_0(z) + \overline{\Phi_1(z)} + \overline{\Psi_1(z)} \text{ in } \overline{D_2} \tag{6.35}$$

satisfies the estimate

$$C_\beta[w_1(z)X(z), \overline{D_2}] \le M_6 = M_6(p_0, \beta, k, D_2). \tag{6.36}$$

Moreover we substitute $w_1(z) = w_0(z) + W_1(z)$ and corresponding functions $\xi_1(z) = \mathrm{Re}w_1(z) + \mathrm{Im}w_1(z)$, $\eta_1(z) = \mathrm{Re}w_1(z) - \mathrm{Im}w_1(z)$ into the positions of $w(z)$, $\xi(z)$, $\eta(z)$ in (6.19),(6.20). Similarly to (6.33)–(6.35), we can find the corresponding functions $\Psi_2(z), \Phi_2(z)$ in $\overline{D_2}$, and $\overline{W_2(z)} = \Phi_2(z) + \Psi_2(z)$ in $\overline{D_2}$, and the function

$$w_2(z) = w_0(z) + W_2(z) = w_0(z) + \overline{\Phi_2(z)} + \overline{\Psi_2(z)} \text{ in } \overline{D_2} \tag{6.37}$$

satisfies a similar estimate of the form (6.36). Furthermore we substitue $w_1(z) = w_0(z) + W_1(z)$ and the corresponding functions $\xi_1(z) = \mathrm{Re}w_1(z) + \mathrm{Im}w_1(z)$, $\eta_1(z) = \mathrm{Re}w_1(z) - \mathrm{Im}w_1(z)$ into the positions of $w(z)$, $\xi(z)$, $\eta(z)$ in (6.19),(6.20). Similarly to (6.33)-(6.35), the corresponding functions $\Psi_2(z), \Phi_2(z)$ and $\overline{W_2(z)} = \Phi_2(z) + \Psi_2(z)$ in $\overline{D_2}$ can be found, and the function

$$w_2(z) = w_0(z) + W_2(z) = w_0(z) + \overline{\Phi_2(z)} + \overline{\Psi_2(z)} \text{ in } \overline{D_2} \tag{6.38}$$

satisfies an estimate of the form (6.36). Thus there is a sequence of functions $\{w_n(z)\}$,

$$\begin{aligned} w_n(z) &= w_0(z) + W_n(z) = w_0(z) + \overline{\Phi_n(z)} + \overline{\Psi_n(z)} \text{ in } D_2, \\ \Psi_n(z) &= \int_2^\nu [A\xi_{n-1} + B\eta_{n-1} + E]e_1 d\nu \\ &\quad + \int_0^\mu [C\xi_{n-1} + D\eta_{n-1} + F]e_2 d\mu \text{ in } \overline{D_2}, \end{aligned} \tag{6.39}$$

and then

$$\begin{aligned} &|[w_1(z) - w_0(z)]X(z)| \le |\Phi_1(z)X(z)| \\ &+\sqrt{2}|X(z)| \left[\left| \int_2^\nu [A\xi_0 + B\eta_0 + E]e_1 d\nu \right| + \left| \int_0^\mu [C\xi_0 + D\eta_0 + F]e_2 d\mu \right| \right] \\ &\le 4M_7 M(2M_3 + 1)(2m + 1)R' \text{ in } \overline{D_2}, \end{aligned} \tag{6.40}$$

where $M_7 = \max_{z\in\overline{D_2}}(|A|,|B|,|C|,|D|,|E|,|F|)$, $m = C[w_0(z)X(z),\overline{D_2}]$, $R' = 2$, $M = 1+2k_0^2(1+2k_0^2)$, M_3 is a constant as stated in (6.23). It is clear that $w_n(z) - w_{n-1}(z)$ satisfies

$$\overline{w_n(z)} - \overline{w_{n-1}(z)} = \Phi_n(z) - \Phi_{n-1}(z)$$

$$+\int_2^{\nu}[A(\xi_n-\xi_{n-1})+B(\eta_n-\eta_{n-1})]e_1 d\nu$$

$$+\int_0^{\mu}[C(\xi_n-\xi_{n-1})+D(\eta_n-\eta_{n-1})]e_2 d\mu \text{ in } \overline{D_2},$$

where $n = 1,2,\cdots$. From the above equality, it can be obtained that

$$|[w_n - w_{n-1}]X(z)| \leq [4M_7M(2M_3+1)(2m+1)]^n$$
$$\times\int_0^{R'}\frac{R'^{n-1}}{(n-1)!}dR' \leq \frac{[4M_7M(2M_3+1)(2m+1)R']^n}{n!} \text{ in } \overline{D_2}, \tag{6.41}$$

and then we can see that the sequence of functions $\{w_n(z)X(z)\}$, i.e.

$$w_n(z)X(z) = \{w_0(z)+[w_1(z)-w_0(z)]+\cdots+[w_n(z)-w_{n-1}(z)]\}X(z) \tag{6.42}$$

$(n = 1,2,\cdots)$ in $\overline{D_2}$ uniformly converges to a function $w_*(z)X(z)$, and $w_*(z)$ satisfies the equality

$$\overline{w_*(z)} = \overline{w_0(z)} + \Phi_*(z) + \Psi_*(z),$$
$$\Psi_*(z) = \int_2^{\nu}[A\xi_* + B\eta_* + E]e_1 d\nu + \int_0^{\mu}[C\xi_* + D\eta_* + F]e_2 d\mu \text{ in } \overline{D_2}, \tag{6.43}$$

where $\xi^* = \mathrm{Re}w^* + \mathrm{Im}w^*$, $\eta = \mathrm{Re}w^* - \mathrm{Im}w^*$, and $w_*(z)$ satisfies the estimate

$$C[w_*(z)X(z),\overline{D_2}] \leq M_8 = \exp(4M_7M(2M_3+1)(2m+1)R'). \tag{6.44}$$

Combining (6.32) and (6.44), it is obvious that the solution $w_*(z)$ of Problem M for (6.2) in $\bar{D}$ satisfies the estimate

$$C[w_*(z)X(z),\bar{D}] \leq M_9 = M_9(p_0,\beta,k,D), \tag{6.45}$$

where M_9 is a nonnegative constant.

From Theorem 2 and Theorem 3, we see that under Condition C, Problem M for the equation (6.2) has a unique solution $w(z)$, which can be found by using successive iteration, and $w(z)$ of Problem M satisfies the estimates

$$C[w(z)X(z),\bar{D}] \leq M_{10},\ C_\beta[w(z)X(z),\overline{D_1}] \leq M_{11}, \tag{6.46}$$

where

$$X(z) = \prod_{j=1}^{2}|z-t_j|^{\eta_j},\ \eta_j = 2\gamma_j+\delta,\ j=1,2, \tag{6.47}$$

in which $t_1 = 2$, $t_j = 0$, $\gamma_j (j = 1, 2)$ are as stated in (6.9), δ is a small positive number, and $\beta(0 < \beta < \delta)$, $M_j = M_j(p_0, \beta, k, D)$, $(j = 11, 12)$ are nonnegative constants, $k = (k_0, k_1, k_2)$. Moreover we can derive the following theorem.

Theorem 4 *Suppose that the equation* (6.2) *satisfies Condition* C. *Then any solution* $w(z)$ *of Problem* M *for* (6.2) *satisfies the estimates*

$$C[w(z)X(z), \bar{D}] \leq M_{13}(k_1 + k_2), \; C_\beta[w(z)X(z), \overline{D_1}] \leq M_{14}(k_1 + k_2), \qquad (6.48)$$

in which $M_j = M_j(p_0, \beta, k_0, D)(j = 13, 14)$ *are nonnegative constants.*

Proof When $k_1 + k_2 = 0$, from Theorem 2, it is easy to see that (6.47) holds. If $k_1 + k_2 > 0$, then it is seen that the function $W(z) = w(z)/(k_1 + k_2)$ is a solution of the homogeneous boundary value problem(Problem M_0). On the basis of the estimate (6.46), we can obtain the estimates

$$C[w(z)X(z), \bar{D}] \leq M_{13}, \; C_\beta[w(z)X(z), \overline{D_1}] \leq M_{14}, \qquad (6.49)$$

where $M_j = M_j(p_0, \beta, k_0, D)(j = 13, 14)$ are nonnegative constants. From (6.49), it follows the estimate (6.48).

From the estimates (6.46),(6.48), we can see the regularity of solutions of Problem M for (6.2).

Finally, we mention that if the index K is an arbitrary even integer or $2K$ is an arbitrary odd integer, the Riemann–Hilbert problem for (6.2) can be considered, but in general Problem M for $K \leq -1$ has $-2K - 1$ solvability conditions or their solutions for $K \geq 0$ include $2K + 1$ arbitrary real constants.

References

[1] Bitsadze, A.V.: Some classes of partial differential equations. Gordon and Breach, New York, etc., 1988.

[2] Wen, Guo-chun, Begehr, H.: Boundary value problems for elliptic equations and systems. Longman Scientific and Technical, Harlow, 1990.

[3] Wen, Guo-chun: Conformal mappings and boundary value problems. Amer. Math. Soc., Providence, RI, 1992.

[4] Begehr, H., Wen, Guo-chun: Nonlinear elliptic boundary value problems and their applications. Addison Wesley Longman, Harlow, 1996.

[5] Wen, Guo-chun: Some boundary value problems for linear mixed complex equations of first and second order. J. Yantai Univ., 1997, no.3, 157-164 (Chinese).

7 ON TOTALLY REAL NON–COMPACT MANIFOLDS GLOBALLY FOLIATED BY ANALYTIC DISCS

Elias Wegert †

Fakultät für Mathematik und Informatik
TU Bergakademie Freiberg
D–09596 Freiberg, Germany
wegert@mathe.tu-freiberg.de

Georgi Khimchiachvili

Institute of Mathematics
Georgian Academy of Sciences
M. Aleksidze str. 1
Tbilisi 380093, Georgia
and
Department of Theoretical Physics
University of Lodz
Pomorska str. 149/153
90–236 Lodz, Poland
khimiash@mvii.uni.lodz.pl

Ilya Spikovsky ‡

Department of Mathematics
The College of William and Mary
Williamsburg, VA 23187–8795, USA

†The first two authors were supported by Deutsche Forschungsgemeinschaft, grant We 1704/2-2
‡The third author was supported by NSF, grant DMS 9401848

H.G.W. Begehr et al. (eds.), Partial Differential and Integral Equations, 107–123.

Abstract: We prove the existence of analytic discs with boundaries in a non-compact target manifold $M \subset \mathbb{T} \times \mathbb{C}^2$ without exceptional points. It is shown that the manifolds M in the considered class are globally foliated by the solutions of related Riemann-Hilbert problems. The proof is based on a fixed-point equation for a compact operator and utilizes uniform norm estimates for the inverses of related Toeplitz operators.
Key words: Analytic disc, Riemann-Hilbert problem, Toeplitz operator

7.1 INTRODUCTION

An (embedded) *analytic disc* in $\mathbb{C}^n$ is the image of a continuous mapping w: $\overline{\mathbb{D}} \to \mathbb{C}^n$ which is holomorphic in the open complex unit disk $\mathbb{D}$.

Recent heightening of interest in analytic discs results from their interrelations with a variety of important and well investigated topics, such as polynomial hulls (Bedford [1], Forstnerič [8], [9], Slodkowski [18]), holomorphic motions (Slodkowski [18]), and H^∞-optimization (Helton and Howe [12], Helton and Marshall [13], Wegert [22]). Eliashberg's technique of filling with analytic discs [7] is a powerful tool for attacking hard questions in complex analysis, symplectic geometry (Gromov [11], McDuff and Salamon [16]), and quantum cohomology (McDuff and Salamon [16]).

In all these issues the central question pertains to the construction of analytic discs having boundaries in a prescribed smooth submanifold of a complex euclidean space. Though the problem goes back to Bishop's famous paper [3] in 1965 (while its prototype can actually be found already in Bernhard Riemann's thesis of 1851), it is far from being solved. Positive results were mainly obtained in special situations characterized by the presence of so-called exceptional points, where the tangent plane of the submanifold contains a complex line. As is well known, for compact submanifolds the existence of exceptional points may be established by topological methods [5], and this enables one to prove the local existence of analytic discs with boundaries in certain compact manifolds. However, it seems not to be known, how far these local discs do extend.

Another direction of research concerns the local structure and the stability of systems of analytic discs under perturbations (Bedford [1], Forstnerič [8], Globevnik [10]). The results were mainly obtained by linearization techniques.

Much less is known if exceptional points are absent and no a-priori assumption is made about the existence of analytic discs. This was the motivation for our research. The aim of this paper is to present *global* existence results for analytic discs with boundary in a non-compact target manifold without exceptional points.

The plan of the paper is as follows. Our setting, which is inspired by a geometric theory of non-linear Riemann-Hilbert problems, is described in Section 7.2.

Relevant material about Riemann-Hilbert problems and related Toeplitz operators is summarized in Section 7.3. We reformulate an existence result for solutions of linear Riemann-Hilbert problems in $\mathbb{C}^n$ in terms of analytic discs.

Section 7.4 contains the main result. Under certain explicitly given analytic conditions on the target manifold $M \subset \mathbb{T} \times \mathbb{C}^2$, we prove that M is globally foliated by a family of analytic discs with boundary in M.

We reduce the analytic disc problem to a fixed-point equation for a compact operator, which admits the application of standard tools of nonlinear analysis. In the simplest case considered here, the existence of a solution is established by Schauder's fixed point theorem.

Though situations which are geometrically more complicated require more refined techniques, such as degree theory, homotopy methods and a-priori estimates, we believe that in principle the proposed approach applies also to more involved problems. The crucial point is the understanding of the properties of certain associated Toeplitz operators and especially of their partial indices which control the local Fredholm theory of analytic discs (Globevnik [10]).

7.2 ANALYTIC DISKS AND RIEMANN-HILBERT PROBLEMS

Given a (real) submanifold M of $\mathbb{T} \times \mathbb{C}^n$, the *Riemann-Hilbert problem* consists in finding all continuous mappings w: $\overline{\mathbb{D}} \to \mathbb{C}^n$ which are holomorphic in the open disc $\mathbb{D}$, such that the graph of their boundary function lies in M,

$$(t, w(t)) \in M \quad \forall t \in \mathbb{T} := \partial\mathbb{D}. \tag{7.1}$$

We think of the *target manifold* M as of a manifold fibered over $\mathbb{T}$ and denote the fiber of M over $t \in \mathbb{T}$ by M_t.

Obviously, any solution of a Riemann-Hilbert problem with constant fibers $M_t \equiv M_1$ is an analytic disc with boundary in M_1. Conversely, if the boundary of an analytic disc w: $\overline{\mathbb{D}} \to \mathbb{C}^{n+1}$ lies in a target manifold $M \subset \mathbb{T} \times \mathbb{C}^n$, the first component of w is a Blaschke product. Once this component is fixed, a Riemann-Hilbert problem for the remaining n components of w emerges. The special choice $w_1(t) \equiv t$ leads to the Riemann-Hilbert problem (7.1).

An analytic form of writing the boundary condition of the Riemann-Hilbert problem is

$$F(t, w(t)) = 0 \quad \forall\, t \in \mathbb{T}, \tag{7.2}$$

where F: $\mathbb{T} \times \mathbb{C}^n \to \mathbb{R}^n$ is a defining function of the target manifold

$$M = \{(t, z) \in \mathbb{T} \times \mathbb{C}^n:\ F(t, z) = 0\}. \tag{7.3}$$

Those points $p = (t, z)$ of M at which the the tangent space TM_p contains a complex line are called *exceptional points* of M. If M is given by (7.3), a point $(t, z) \in M$ is exceptional if and only if

$$\det \partial_z F(t, z) = 0. \tag{7.4}$$

It is well known that exceptional points play a distinguished role for the solvability of (7.2). A manifold without exceptional points is said to be totally real. In what follows we study Riemann-Hilbert problems for non-compact totally real target manifolds.

7.3 LINEAR RIEMANN-HILBERT PROBLEMS

To begin with, we consider the case where F is a real (affine) linear function in $\mathbb{C}^n$ for each $t \in \mathbb{T}$. Since F is real-valued, we have

$$F(t,z) := A(t)\,z + \overline{A(t)}\,\overline{z} - f(t), \tag{7.5}$$

with $A\colon \mathbb{T} \to \mathbb{C}^{n\times n}$ and $f\colon \mathbb{T} \to \mathbb{R}^n$. In this case the boundary value problem (7.2) is the familiar (vector-valued) Riemann-Hilbert problem (see [19])

$$2\,\mathrm{Re}\,(A(t)w(t)) = f(t) \quad \forall t \in \mathbb{T}. \tag{7.6}$$

It should be mentioned that the next section relies on uniform norm estimates for the solutions of a family of such problems in appropriate function spaces. Despite the topic has been extensively investigated, the estimates we need do not seem to exist in the literature. So we derive them from scratch using certain algebraic manipulations which relate the problem (7.6) to an associated Toeplitz operator.

Let us start with some notations. By L^p, $1 \le p \le \infty$, we denote the Lebesgue spaces on $\mathbb{T}$. Since the meaning is clear from the context, we use the same notation for spaces of vector and matrix functions and do not indicate their dimension. The group of invertible matrices $A \in L^\infty$ with $A^{-1} \in L^\infty$ is referred to as GL^∞.

The Hardy spaces of holomorphic functions on $\mathbb{T}$ are denoted by H^p, $1 \le p \le \infty$. Alternatively, we use the notation $L^p_+ := H^p$, while L^p_- denotes the closed subspace of L^p spanned by $\{\ldots, t^{-2}, t^{-1}, 1\}$.

The orthogonal projection $P\colon L^2 \to H^2$ is (extends to) a bounded operator $L^p \to H^p$ for $p \in (1,\infty)$.

With a given matrix-function $A \in GL^\infty$ we associate the function $B := \overline{A}^{-1}A$ and study the following compressions of multiplication:

$$R_A\colon\ H^2 \to L^2_{\mathbb{R}},\ \ w \mapsto \mathrm{Re}\,(APw),$$

$$S_A\colon\ L^2_{\mathbb{R}} \to H^2,\ \ f \mapsto P\overline{A}^{-1}\mathrm{Re}\,f,$$
$$T_B\colon\ H^2 \to H^2,\ \ w \mapsto PBPw.$$

These (real linear) operators are bounded for each $p \in (1,\infty)$. The interrelations between the *Riemann-Hilbert operator* R_A, the *adjoint Riemann-Hilbert operator* S_A, and the *associated Toeplitz operator* T_B will be explained in a moment.

For the convenience of the reader we recall some basic facts from the theory of Toeplitz operators. The first result relates the Fredholm properties of T_B to the existence of a factorization of its symbol B into a holomorphic and an antiholomorphic factor.

By a *(generalized) Wiener-Hopf factorization* of the matrix-function $B \in L^\infty$ we understand a representation

$$B = B_- D B_+,$$

where B_-, B_+, and D are such that $B_-, B_-^{-1} \in L^2_-$, $B_+, B_+^{-1} \in L^2_+$, $D(t) = \operatorname{diag}(t^{\kappa_1}, \ldots, t^{\kappa_n})$ with $\kappa_j \in \mathbb{Z}$, and the mapping

$$w \mapsto B_+^{-1} P B_-^{-1} w \tag{7.7}$$

extends from the subspace of trigonometric polynomials to a bounded operator in L^2. The integers κ_j are said to be the *partial indices* of T_B.

Proposition 1 *Let $B \in L^\infty$.*

(i) *The Toeplitz operator T_B is Fredholm in H^2 if and only if B is in GL^∞ and admits a generalized Wiener-Hopf factorization.*

(ii) *If T_B is Fredholm, then*

$$\dim\ker T_B = -\sum_{\kappa_j<0} \kappa_j, \qquad \operatorname{codim}\operatorname{im} T_B = \sum_{\kappa_j>0} \kappa_j.$$

(iii) *If T_B is invertible, then T_B^{-1} is given by (the extension of) (7.7).*

Proof See [6], [4] Section 5.5, [15]. ∎

For our purposes we also need more concrete criteria. The first one concerns the Fredholm properties of Toeplitz operators with a continuous symbol.

Proposition 2 *Let $B \in GL^\infty$ be continuous.*

(i) *T_B is Fredholm in H^2 if and only if*

$$\det B(t) \neq 0 \ \ \forall\, t \in \mathbb{T}. \tag{7.8}$$

(ii) *If T_B is Fredholm, then*

$$\operatorname{ind} T_B = -\operatorname{wind}\det B.$$

(iii) *If T_B is Fredholm, and $B = B_- D B_+$ is a Wiener-Hopf factorization of B, then $B_-, B_-^{-1}, B_+, B_+^{-1} \in L^p$ for each $p \in (1, \infty)$.*

Proof See [4] 5.48–5.49; [15] Theorem 5.16. The third assertion follows from the fact that Toeplitz operators T_B with continuous symbol B are Fredholm in H^p simultaneously for all $p \in (1, \infty)$. ∎

In contrast to the scalar case, where T_B is invertible whenever $\operatorname{ind} T_B = 0$, invertibility of Toeplitz operators in higher dimensions is a more delicate question. Since we also need an estimate for the inverses of T_B, where B belongs to a certain *family* of matrix functions, we limit our considerations to a subclass of operators.

The matrix function $B \in L^\infty$ is said to be *sectorial* if there exist constant invertible matrices B_1, B_2 and a positive number m, such that

$$\operatorname{Re}(B_1 B(t) B_2\, z, z) \geq m\, \|z\|^2 \qquad \forall t \in \mathbb{T},\ z \in \mathbb{C}^n. \tag{7.9}$$

If B is sectorial, we denote by m_B the infimum of $(1/m)\,\|B_1^{-1}\|\,\|B_2^{-1}\|$ over all B_1, B_2, m for which (7.9) holds.

Lemma 1 *If $B \in GL^\infty$ is sectorial, then the Toeplitz operator T_B is invertible in H^2 and*

$$\|T_B^{-1}\| \le m_B.$$

Proof Since $T_{B_1BB_2}^{-1} = B_2^{-1}T_B^{-1}B_1^{-1}$, we can assume that (7.9) holds with $B_1 = B_2 = I$. Under the above assumptions, $B_{\mathbb{R}} := (1/2)\,(B+B^*)$ is positive definite and its inverse is bounded, and hence it admits a factorization $B_{\mathbb{R}} = B_+^* B_+$, where the matrix functions B_+ and B_+^{-1} belong to H^∞ (cf. [4] 5.13, or [15] Theorem 7.9). With $G := (B_+^*)^{-1}BB_+^{-1}$, the matrix B has the representation

$$B = B_+^* G B_+, \tag{7.10}$$

which, since B_+ and B_+^* are in L_+^∞ and L_-^∞, respectively, implies the factorization ([4], Proposition 2.14)

$$T_B = T_{B_+}^* T_G T_{B_+}.$$

Hence it is sufficient to prove that the inverses of T_G and T_{B_+} are uniformly bounded. For the Toeplitz operator T_G we have

$$T_G = P(B_+^*)^{-1}BB_+^{-1}P, \qquad T_G^* = P(B_+^*)^{-1}B^*B_+^{-1}P,$$

and consequently

$$(T_G)_{\mathbb{R}} := (1/2)(T_G + T_G^*) = P(B_+^*)^{-1}B_{\mathbb{R}}\,B_+^{-1}P = P.$$

Since P is the identity of H^2, we conclude that T_G is invertible and $\|T_G^{-1}\| \le 1$. Because B_+, $B_+^{-1} \in H^\infty$, we have $(T_{B_+})^{-1} = T_{B_+^{-1}}$, and therefore

$$\|T_{B_+}^{-1}\| = \|B_+^{-1}\|,$$

where the norm on the right-hand side denotes the norm of the operator of multiplication by B_+^{-1} in L^2. This norm is equal to $\sup_{t\in\mathbb{T}} 1/\eta(t)$, where

$$\eta(t) := \inf\left\{\left(B_+^*(t)B_+(t)\,z, z\right) : \|z\|=1\right\} = \inf\left\{\operatorname{Re}\left(B(t)\,z, z\right) : \|z\|=1\right\} \ge 1/m_B,$$

which completes the proof. ∎

Next, we investigate the relations between the operators R_A, S_A and T_B. In particular we show that, in an appropriate setting, the operators S_A are in fact adjoints of Riemann-Hilbert operators.

Since R_A and S_A are real linear but not complex linear, we decompose L^2 into its real and imaginary part,

$$L^2 \to L_{\mathbb{R}}^2 \oplus L_{\mathbb{R}}^2, \qquad f = a + ib \mapsto (a, b)$$

and introduce the real scalar product

$$\langle f, g\rangle := \langle \operatorname{Re} f, \operatorname{Re} g\rangle_{L^2_{\mathbb{R}}} + \langle \operatorname{Im} f, \operatorname{Im} g\rangle_{L^2_{\mathbb{R}}}.$$

With respect to this decomposition we may think of H^2 as the subspace

$$H^2 = \left\{(u, -Hu + c)\colon\ u \in L^2_{\mathbb{R}},\ c \in \mathbb{R}^n\right\},$$

where H denotes the (negative of) the conjugation operator

$$Hu(\mathrm{e}^{\mathrm{i}\tau}) := \frac{1}{2\pi}\int_0^{2\pi} u(\mathrm{e}^{\mathrm{i}\sigma}) \cot\frac{\sigma-\tau}{2}\,\mathrm{d}\sigma.$$

Note that

$$H(t^k) = \mathrm{i}\cdot \operatorname{sign} k \cdot t^k.$$

In the above setting we have $R_A^* \in \mathcal{L}(L^2_{\mathbb{R}}, H^2)$, $S_A^* \in \mathcal{L}(H^2, L^2_{\mathbb{R}})$, and $T_B^* \in \mathcal{L}(H^2)$.

If C denotes the operator of complex conjugation, $C\colon w \to \overline{w}$, then $P_0 := PCPC$ is the L^2 orthoprojection on the constants. Depending on the context, we consider $P_0 w$ either as a number or as a constant function. In order to shorten notations we abbreviate $A^{-\top} := (A^\top)^{-1}$.

Lemma 2 *Let $A \in GL^\infty$, $B := \overline{A}^{-1}A$. Then, in the above setting,*

(i) $2S_A R_A = T_B + P_0 C$,

(ii) $R_A^* = S_{A^{-\top}}$,

(iii) $T_B^* = T_{B^*}$.

Proof We suppress the subscripts of the operators.

1. We have

$$\begin{aligned} 2\,SR\,w &= P\overline{A}^{-1}(AP\,w + CAP\,w) \\ &= P\overline{A}^{-1}AP\,w + PCP\,w \qquad (7.11)\\ &= Tw + P_0 Cw. \end{aligned}$$

$$2\,SR\,w = P\overline{A}^{-1}(AP\,w + CAP\,w) = P\overline{A}^{-1}AP\,w + PCP\,w = Tw + P_0Cw.$$

2. Let $f \equiv a + \mathrm{i}b$ and $g \equiv c + \mathrm{i}d$. Since $2P = \mathrm{Id} - \mathrm{i}H + P_0$ and $\langle Hu, v\rangle_{L^2_{\mathbb{R}}} = -\langle u, Hv\rangle_{L^2_{\mathbb{R}}}$,

$$\begin{aligned} 2\langle Pf, g\rangle &= \langle(\mathrm{Id} - \mathrm{i}H + P_0)\,f, g\rangle \\ &= \langle(\mathrm{Id} + P_0)\,a + Hb, c\rangle + \langle(\mathrm{Id} + P_0)\,b - Ha, d\rangle \\ &= \langle a, c\rangle + \langle a, P_0 c\rangle - \langle b, Hc\rangle + \langle b, d\rangle + \langle b, P_0 d\rangle + \langle a, Hd\rangle \\ &= \langle a, c + P_0 c + Hd\rangle + \langle b, d + P_0 d - Hc\rangle \\ &= \langle f, (\mathrm{Id} - \mathrm{i}H + P_0)\,g\rangle \\ &= 2\langle f, Pg\rangle. \end{aligned}$$

A straightforward verification shows that the adjoint of multiplication by A is multiplication by A^* and $\mathrm{Re}^* = \mathrm{Re}$. Consequently,

$$R_A^* = (\mathrm{Re}\, AP)^* = P^* A^* \mathrm{Re}^* = PA^* \mathrm{Re} = S_{A^{-\top}},$$

$$T_B^* = (PBP)^* = P^* B^* P^* = PB^* P = T_{B^*}.$$

∎

We continue with studying the Fredholm properties of R and S.

Lemma 3 *Let $A \in GL^\infty$, $B := \overline{A}^{-1} A$. If T_B is invertible in H^2, then $S_A \in \mathcal{L}(L^2_{\mathbb{R}}, H^2)$ is injective and $R_A \in \mathcal{L}(H^2, L^2_{\mathbb{R}})$ is surjective.*

Proof 1. (R. Belch [2]) If $S_A x := P\overline{A}^{-1} x = 0$ for some $x \in L^2_{\mathbb{R}}$ we have $A^{-1}x \in H^2$ and

$$T_B A^{-1} x = P\overline{A}^{-1} A A^{-1} x = P\overline{A}^{-1} x = 0.$$

Since T_B is invertible, this implies that $x = 0$.

2. Since $\overline{(A^{-\top})}^{-1} A^{-\top} = B^*$, and $T_{B^*} = T_B^*$ is invertible simultaneously with T_B, the result of the first step shows that $R_A^* = S_{A^{-\top}}$ is injective, and hence the range $\mathrm{im}\, R_A$ is dense.

On the other hand, $S_A R_A$ is Fredholm, since $2\, S_A R_A = T_B + P_0 C$ and $P_0 C$ has finite rank. Consequently the image of S_A is closed and $\mathrm{codim}\,\mathrm{im} S_A$ is finite. Since S_A is injective, it is Fredholm, which, in conjunction with the Fredholmness of $S_A R_A$, implies that R_A is Fredholm. Hence its range is closed and coincides with the whole target. ∎

In order to study the kernel of R_A we introduce the matrix

$$\Lambda := P_0 T_B^{-1} I.$$

Here I is the identity matrix and T_B^{-1} acts on the columns of I. The operators

$$P_0^+ := \frac{1}{2}(\mathrm{Id} + \Lambda C), \qquad P_0^- := \frac{1}{2}(\mathrm{Id} - \Lambda C),$$

will be considered as real-linear operators in $\mathbb{C}^n \equiv \mathbb{R}^{2n}$.

Lemma 4 *Let $A \in GL^\infty$, $B := \overline{A}^{-1} A$. If the Toeplitz operator T_B is invertible in H^2 the operators P_0^+ and P_0^- are complementary (real) n-dimensional projections in $\mathbb{C}^n \equiv \mathbb{R}^{2n}$ with*

$$\ker P_0^- = \mathrm{i} \ker P_0^+, \qquad \mathbb{C}^n = \ker P_0^+ \oplus \mathrm{i} \ker P_0^+. \tag{7.12}$$

Proof 1. We show that $\overline{\Lambda}\Lambda = I$. Since T_B is invertible, its symbol B admits a Wiener-Hopf factorization $B = B_- B_+$. If this factorization is normalized by $P_0 B_- = I$, we have $T_B^{-1} I = B_+^{-1}$ (see Proposition 1) and hence $\Lambda = P_0 B_+^{-1}$.

Because $I = B\overline{B} = B_-B_+\overline{B}_-\overline{B}_+$ and $B_\pm^{\pm 1} \in L^2$, we have $B_-^{-1}\overline{B}_+^{-1} = B_+\overline{B}_- \in L^1$. Since P_0 is multiplicative on L^2_+ as well as on L^2_- and, on these subspaces, commutes with inversion,

$$\begin{aligned}\Lambda &= P_0\overline{B}_+^{-1} = P_0B_-^{-1} \cdot P_0\overline{B}_+^{-1} \\ &= P_0(B_-^{-1}\overline{B}_+^{-1}) = P_0(B_+\overline{B}_-) = P_0B_+ \cdot P_0\overline{B}_- = \overline{P_0B_+} = \overline{\Lambda}^{-1}.\end{aligned}$$

2. By virtue of $\overline{\Lambda}\Lambda = I$ we have $(\Lambda C)^2 = \mathrm{Id}$ and hence P_0^+ and P_0^- are complementary projections. Finally, for any $q \in \mathbb{C}^n$,

$$P_0^-(\mathrm{i}q) = \frac{1}{2}(\mathrm{Id} - \Lambda C)(\mathrm{i}q) = \frac{\mathrm{i}}{2}(\mathrm{Id} + \Lambda C)q = \mathrm{i}\,P_0^+q,$$

which proves (7.12). ∎

In order to determine the kernel of R_A we consider Riemann-Hilbert problems with side condition, $R_Aw = x$, $P_0w = q$. Here $x \in L^2_{\mathbb{R}}$ and $q \in \mathbb{C}^n$.

Lemma 5 *Let $A \in GL^\infty$, $B := \overline{A}^{-1}A$, $x \in L^2_{\mathbb{R}}$, $q \in \mathbb{C}^n$. If T_B is invertible in H^2, then the Riemann-Hilbert problem*

$$R_Aw = x, \qquad P_0w = q \tag{7.13}$$

is solvable in H^2 if and only if

$$P_0^+q = y, \tag{7.14}$$

where $y := P_0T_B^{-1}S_Ax$. If (7.14) *is satisfied, the unique solution of* (7.13) *is given by*

$$w = T_B^{-1}(2S_Ax - \overline{q}). \tag{7.15}$$

Proof (Following R. Belch [2]). We suppress the subscripts of the operators.

1. If w solves (7.13), then on account of Lemma 2 (i),

$$y := P_0T^{-1}Sx = P_0T^{-1}SRw = \frac{1}{2}P_0T^{-1}(T+P_0C)w = \frac{1}{2}(\mathrm{Id} + \Lambda C)P_0w = P_0^+q.$$

Moreover, $Tw = 2SRw - P_0Cw = 2Sx - \overline{q}$, which implies (7.15).

2. Conversely, if (7.14) is satisfied and w is given by (7.15), we have

$$P_0w = P_0T^{-1}(2Sx - \overline{q}) = 2y - \Lambda Cq = 2P_0^+q - \Lambda Cq = (\mathrm{Id} + \Lambda C)q - \Lambda Cq = q.$$

Using equations (7.14) and (7.15) again,

$$\begin{aligned}2SRw &= 2SRT^{-1}(2Sx - Cq) = (T + P_0C)T^{-1}(2Sx - Cq) \\ &= 2Sx + 2P_0CT^{-1}Sx - Cq - P_0CT^{-1}Cq \\ &= 2Sx + C\left(2P_0T^{-1}Sx - q - P_0T^{-1}Cq\right) \\ &= 2Sx + C\left(2P_0^+q - (\mathrm{Id} + \Lambda C)q\right) = 2Sx.\end{aligned}$$

Since S is injective, $Rw = x$ follows. ∎

Lemma 6 *Let $A \in GL^\infty$, $B := \overline{A}^{-1}A$. If T_B is invertible in H^2, then R_A and S_A are Fredholm with*

$$\dim\ker R_A = n, \quad \operatorname{codim}\operatorname{im} R_A = 0, \quad \dim\ker S_A = 0, \quad \operatorname{codim}\operatorname{im} S_A = n.$$

For each $z \in \mathbb{D}$ the set $\{w(z)\colon\ w \in \ker R_A\}$ is a totally real n-dimensional subspace of $\mathbb{C}^n$.

Proof According to Lemma 5, $R_A w = 0$ if and only if $w = -T_B^{-1}\overline{q}$ with $P_0^+ q = 0$; hence $\dim\ker R_A = \dim\ker P_0^+ = n$. Now $\operatorname{codim}\operatorname{im} S_A = n$ follows from $\operatorname{ind} S_A + \operatorname{ind} R_A = \operatorname{ind}(T_B + P_0 C) = 0$.

In order to prove the second assertion we can assume that $z = 0$. Since $P_0(\ker R_A) = \ker P_0^+$, Lemma 4 completes the proof. ∎

If the matrix A is continuously differentiable, we can 'lift' the result to the boundary of the circle.

Theorem 1 *If $A \in GL^\infty$ is continuously differentiable, $B := \overline{A}^{-1}A$, and T_B is invertible in H^2, then the linear-like target manifold*

$$M := \{(t,z) \in \mathbb{T} \times \mathbb{C}^n\colon\ \operatorname{Re} A(t)\, z = x(t)\}$$

is foliated by the graphs of the solutions to the Riemann-Hilbert problem $\operatorname{Re} Aw = x$.

Proof Since $\dim\ker R_A = n$ coincides with the dimension of the fibres M_t of the target manifold, it is sufficient to show that the graphs of two different solutions do not intersect. So we have to prove that $R_A w = 0$ and $w(t_0) = 0$, for some $t_0 \in \mathbb{T}$, is only possible if $w \equiv 0$.

Because A is continuously differentiable, the functions in the kernel of R_A belong to the Sobolev space W_2^1 on the circle. This, together with $w(t_0) = 0$, implies that the function $\tilde{w}$ defined by $\tilde{w}(t) := w(t)/(t - t_0)$ is in L^2. Applying $P\,(t-t_0)^{-1}\overline{A}^{-1}$ to $R_A w = 0$, we obtain

$$\begin{aligned} 0 &= P\,(t-t_0)^{-1}\overline{A}^{-1}\operatorname{Re}\left(A\,(t-t_0)\,\tilde{w}\right) \\ &= P\,\overline{A}^{-1}A\,\tilde{w} + P(t-t_0)^{-1}\overline{(t-1)}\,\overline{\tilde{w}} \\ &= T_B\,\tilde{w} - P\,\overline{t/t_0}\,\overline{\tilde{w}} = T_B\,\tilde{w}. \end{aligned}$$

Since T_B is invertible, $\tilde{w}$ and w vanish. ∎

Note that the assertion of Theorem 1 remains in force if A is merely Hölder continuous, but it does not extend to the case of continuous matrix functions.

7.4 NONLINEAR RIEMANN-HILBERT PROBLEMS

In this section we investigate nonlinear Riemann-Hilbert problems with target manifold $M \subset \mathbb{T} \times \mathbb{C}^2$. We assume that the boundary relation $F(., w) = 0$ can be solved with respect to one of the unknowns,

$$w_2 = G(., w_1), \tag{7.16}$$

with a given function G: $\mathbb{T} \times \mathbb{C} \to \mathbb{C}$. Since this problem is equivalent to the operator equation $(\mathrm{Id} - P)\, G(., Pw_1) = 0$, it might be thought of as a nonlinear Hankel operator. However, a closer look at its linear variant, where $G(., z) = g_0 + g_1 z + g_2 \overline{z}$, reveals that it is rather a mixture of a Toeplitz and a Hankel operator. The Toeplitz part is dominating in the so-called 'elliptic case' $|\partial_z G| < |\overline{\partial}_z G|$, while the Hankel part is dominating in the 'hyperbolic case' $|\overline{\partial}_z G| < |\partial_z G|$. The 'pure Toeplitz case' $\overline{\partial}_z G \equiv 0$ was studied in v. Wolfersdorf's paper [24].

It is not a big surprise that 'Toeplitz-like problems' have the better properties. For instance, if G is holomorphic in both variables, any holomorphic function w_1 yields a solution $w \equiv \big(w_1, G(., w_1)\big)$ of (7.16), which indicates a huge kernel of some 'Hankel-type problems'.

Another case of particular interest corresponds to real-valued G, pertaining to the 'parabolic case', since $|\partial_z G| = |\overline{\partial}_z G|$. In this situation w_2 must be real-valued and holomorphic and hence (7.16) is equivalent to the scalar Riemann-Hilbert problem (see [9], [17], [20], for instance)

$$G(t, w_1) = const. \tag{7.17}$$

Note that (7.16) can also be interpreted as a nonlinear transmission problem, if one of the unknown functions, w_1 say, is reflected about the circle $\mathbb{T}$. However, it should be mentioned that those transmission problems involve complex conjugation.

Throughout this section G is assumed to be continuously differentiable. The solutions are looked for in $H^\infty \cap W_r^1$, the space of holomorphic vector functions in $\mathbb{D}$ with boundary function in the Sobolev space $W_r^1(\mathbb{T})$. Under appropriate assumptions, the geometric structure of the solution set of (7.16) is similar to the scalar case (7.17).

Theorem 2 *Suppose that G: $\mathbb{T} \times \mathbb{C} \to \mathbb{C}$ is continuously differentiable and has uniformly bounded first derivatives.*

(i) *If there exists a positive constant δ such that*

$$|\overline{\partial}_z G(t, z)| - |\partial_z G(t, z)| \geq \delta > 0 \quad \forall (t, z) \in \mathbb{T} \times \mathbb{C} \tag{7.18}$$

and

$$\mathrm{Re}\, \overline{\partial}_z G(t, z) \geq \delta > 0 \quad \forall (t, z) \in \mathbb{T} \times \mathbb{C}, \tag{7.19}$$

then, for each $r \in (1, \infty)$, the target manifold of the Riemann-Hilbert problem (7.16) is covered by the solutions $w \in H^\infty \cap W_r^1$.

(ii) *If there exists a positive constant* δ *such that*

$$\operatorname{Re}\overline{\partial}_z G(t,z) - |\partial_z G(t,z)| \geq \delta > 0 \quad \forall (t,z) \in \mathbb{T} \times \mathbb{C}, \tag{7.20}$$

the covering is a foliation.

In other words, for each $q \in \mathbb{C}$ and each $t_0 \in \mathbb{T}$, the Riemann-Hilbert problem

$$w_2 = G(.,w_1), \qquad w_1(t_0) = q \tag{7.21}$$

has a solution $w = (w_1, w_2)$ belonging to $H^\infty \cap W_r^1$ for each $r \in (1,\infty)$. With the stronger assumption (7.20) this solution is unique.

The following constructions serve to transform the problem (7.21) into a fixed-point equation for a compact operator K. Obviously, it is sufficient to consider $t_0 = 1$. The idea is to differentiate the boundary relation along $\mathbb{T}$, which gives rise to a quasi-linear Riemann-Hilbert problem (v. Wolfersdorf [23]). The main ingredient of the operator K is a primitive of an appropriate solution of this auxiliary problem.

Fix $s \in (1,\infty)$. For a given scalar complex valued function $u \in W_s^1(\mathbb{T})$ we define

$$a(t) := \partial_z G(t,u(t)), \quad b(t) := \overline{\partial}_z G(t,u(t)), \quad f(t) := \mathrm{i}\, t\, \partial_t G(t,u(t)), \tag{7.22}$$

where $\mathrm{i}\, t\, \partial_t \equiv \partial_\tau$ denotes the derivative with respect to the polar angle τ of $t \equiv \mathrm{e}^{\mathrm{i}\tau}$. Note that a, b, and f are continuous.

Lemma 7 *Let* G *be subject to the assumptions of Theorem 2, fix* $r, s \in (1,\infty)$, *let* $u \in W_s^1$, *and let* a, b, *and* f *be given by* (7.22).

(i) *For each* $u \in W_s^1$ *the linear Riemann-Hilbert problem*

$$w_2 = a\, w_1 + b\, \overline{w_1} + f, \qquad w_1(0) = 0, \tag{7.23}$$

has a unique solution $w \equiv (w_1, w_2) \in H^r$.

(ii) *For each value of the constant* δ *in Theorem 2 there exists an* $r > 1$ *such that the* H^r*-norm of the solution* $w = (w_1, w_2)$ *to* (7.23) *is bounded by a constant not depending on the choice of* u.

Proof 1. With the definitions $\widetilde{w}(t) := (w_1(t)/t, w_2(t))$, $\widetilde{f} := -(\operatorname{Re} f, \operatorname{Im} f)$, and

$$A := \begin{bmatrix} \overline{b} + a & -1 \\ \mathrm{i}(\overline{b} - a) & \mathrm{i} \end{bmatrix} \begin{bmatrix} t & 0 \\ 0 & 1 \end{bmatrix} \tag{7.24}$$

the problem (7.23) is equivalent to

$$R\,\widetilde{w} := \operatorname{Re} A\, \widetilde{w} = \widetilde{f}. \tag{7.25}$$

Using the operator

$$Sx := P\,\overline{t}\,\overline{A}^{-1} \operatorname{Re} x, \tag{7.26}$$

a straightforward verification shows that $2\,SR = T := P\widetilde{B}P$, where

$$\widetilde{B} := \overline{t}\,\overline{A}^{-1}A = \frac{1}{b}\begin{bmatrix} 0 & -1 \\ -1 & 0 \end{bmatrix}\begin{bmatrix} |b|^2 - |a|^2 & \overline{at} \\ -at & 1 \end{bmatrix}. \tag{7.27}$$

In the next steps we prove that the Toeplitz operator T is invertible and that the adjoint Riemann-Hilbert operator S is injective, which implies that (7.25) has the unique solution

$$\widetilde{w} = 2T^{-1}S\widetilde{f}. \tag{7.28}$$

2. Remember that T depends on the choice of u in (7.22). In what follows we do not only prove that T is invertible in H^2 for all u (which would immediately follow from Corollary 9.6 of [15]), but also that the norms of the inverse $T^{-1} \in \mathcal{L}(H^{1/(r-1)}, H^r)$ is bounded by a constant not depending on the choice of u in (7.22), provided that $r > 1$ is sufficiently small. By virtue of the representation (7.27), we can replace T by $\widehat{T} := P\widehat{B}P$ with

$$\widehat{B} := \frac{1}{b}B, \qquad B := \begin{bmatrix} |b|^2 - |a|^2 & \overline{at} \\ -at & 1 \end{bmatrix}.$$

Since $\det \widehat{B} = \overline{b}$ and $\operatorname{Re} b > 0$, $\widehat{T}$ is Fredholm and $\operatorname{ind} \widehat{T} = 0$.

3. Because

$$\operatorname{Re}(Bz, z) = \left(|b|^2 - |a|^2\right)|z_1|^2 + |z_2|^2 \geq m\,\|z\|^2,$$

for some positive number $m = m(\delta)$, the Toeplitz operator T_B is invertible in H^2 and, according to Lemma 1, we have $\|T_B^{-1}\| \leq 1/m$ in H^2. Moreover, the (continuous) symbol B admits a generalized Wiener-Hopf factorization $B = B_-B_+$, where B_-, B_+, B_-^{-1}, $B_+^{-1} \in L^p$ for each $p < \infty$. Since $T_B^{-1} = B_+^{-1}PB_-^{-1}P$ and $B_+^{-1}PB_-^{-1} = B_+^{-1}PB_-^{-1}P$ on L^2, the multiplication operator

$$f \mapsto B_+^{-1}PB_-^{-1}f \tag{7.29}$$

is bounded on L^2 (uniformly with respect to $u \in W_s^1$).

4. Since $\operatorname{Re} b \geq \delta > 0$ and b is continuous, the scalar Toeplitz operator T_b is invertible on H^2 and b admits a Wiener-Hopf factorization $b = b_-b_+$ with $b_-, b_+ \in L^p$ for each $p < \infty$. If we normalize b_- by $P_0b_- = 1$, then $b_+^{-1} = T_b^{-1}1$ and hence $\|b_+^{-1}\|_2 \leq C$.

Repeating the same argument with b replaced by $1/b$, one proves that the norm of b_+ is bounded in L^2 (uniformly with respect to u), which implies that $\|b_-^{-1}\|_2 \leq C$.

In order to prove that the stronger estimates

$$\|b_+^{-1}\|_{2(1+\varepsilon)} \leq C(\delta), \qquad \|b_-^{-1}\|_{2(1+\varepsilon)} \leq C(\delta) \tag{7.30}$$

hold for some $\varepsilon > 0$, we remark that $\operatorname{Re} b^{1+\varepsilon}$ is bounded away from zero provided that ε is sufficiently small. It is easy to see that $b^{1+\varepsilon} = b_-^{1+\varepsilon}b_+^{1+\varepsilon}$ is the (normalized) Wiener-Hopf factorization of $b^{1+\varepsilon}$, and repeating what was said above the desired estimate (7.30) follows.

5. So far we have the factorization $\widehat{B} = b_- B_- B_+ b_+$ almost everywhere on $\mathbb{T}$. Using (7.29) and (7.30) we get that the operator

$$H^{1/(r-1)} \to H^r\colon\ w \mapsto b_+^{-1} B_+^{-1} P B_-^{-1} b_-^{-1} w \tag{7.31}$$

is bounded (uniformly with respect to u).

In order to prove that (7.31) is the inverse of $\widehat{T}$, we remark that $\widehat{T}w \equiv P b_- B_- B_+ b_+ w = f$ is equivalent to $B_+ b_+ w = P B_-^{-1} b_-^{-1} f$, (note that $B_+ b_+ w \in H^{2-\varepsilon}$ with $\varepsilon > 0$), which implies that $w = b_+^{-1} B_+^{-1} P B_-^{-1} b_-^{-1} f$ almost everywhere on $\mathbb{T}$.

6. In order to prove that S is injective, let $Sx \equiv P \bar{t} \overline{A}^{-1} x = 0$. Then $A^{-1} x \in \operatorname{im} P$ and we have

$$T A^{-1} x = P \bar{t} \overline{A}^{-1} A A^{-1} x = P \bar{t} \overline{A}^{-1} x = 0.$$

Since T is invertible, $A^{-1} x = 0$ and finally $x = 0$. ∎

We continue the construction of the fixed point equation. For any scalar complex valued function $u \in W_s^1$ we denote by $\widetilde{w} \equiv (\widetilde{w}_1, \widetilde{w}_2) \in H^r$ the solution of the associated Riemann-Hilbert problem

$$\widetilde{w}_2 = a\,\widetilde{w}_1 + b\,\overline{\widetilde{w}_1} + f, \qquad \widetilde{w}_1(0) = 0, \tag{7.32}$$

with a, b, and f from (7.22). By definition, the operator $K\colon W_s^1 \to W_r^1$ is given by

$$Ku(\mathrm{e}^{\mathrm{i}\tau}) := q + \int_0^\tau \widetilde{w}_1(\mathrm{e}^{\mathrm{i}\sigma})\,\mathrm{d}\sigma.$$

Lemma 8 (i) *The operator $K\colon W_s^1 \to W_r^1$ is compact for any $r, s \in (1, \infty)$.*

(ii) *The image of $K\colon W_r^1 \to W_r^1$ is bounded if $r > 1$ is sufficiently small.*

(iii) *The pair $(w_1, w_2) \in W_r^1$ is a solution of the Riemann-Hilbert problem (4.6) if and only if $Kw_1 = w_1$ and $w_2 = G(.,w_1)$.*

Proof 1. Because $\mathbb{T}$ is a one-dimensional manifold, the embedding $W_s^1(\mathbb{T}) \to C(\mathbb{T})$ is compact, and hence it suffices to prove that $K\colon C(\mathbb{T}) \to W_r^1(\mathbb{T})$ is continuous. Since the superposition operators

$$u \mapsto a := \partial_z G(.,u), \quad u \mapsto b := \overline{\partial}_z G(.,u), \quad u \mapsto f := \mathrm{i}t \cdot \partial_t G(.,u)$$

are continuous in $C(\mathbb{T})$, the associated Toeplitz operators $T := P\widetilde{B}P$ with $\widetilde{B}$ from (7.27) and the adjoint Riemann-Hilbert operators S from (7.26) depend continuously on u in $\mathcal{L}(H^r)$, and since all these operators are invertible, the solutions $\widetilde{w} \in H^r$ to the Riemann-Hilbert problems (7.23) also depend continuously on u (cf. (7.28)). Integrating these solutions along $\mathbb{T}$ proves the continuity of $K\colon\ C(\mathbb{T}) \to W_r^1(\mathbb{T})$.

2. If $r > 1$ is sufficiently small, according to Lemma 7, the solutions $\widetilde{w}$ of (7.23) are bounded in H^r uniformly with respect to the choice of $u \in W_r^1$, and hence the Ku are uniformly bounded in W_r^1.

3. Let $w = (w_1, w_2) \in H^\infty \cap W_r^1$ be a solution of $w_2 = G(., w_1)$. Differentiating this boundary relation with respect to the polar angle τ we obtain that $\widetilde{w} := \partial_\tau w \equiv \mathrm{i}t\partial_t w$ is the (unique) solution of the auxiliary Riemann-Hilbert problem (7.23). Consequently,

$$Kw_1(\mathrm{e}^{\mathrm{i}\tau}) = q + \int_0^\tau \widetilde{w}_1(\mathrm{e}^{\mathrm{i}\sigma})\ \mathrm{d}\sigma = w_1(1) + \int_0^\tau \partial_\tau w_1(\mathrm{e}^{\mathrm{i}\sigma})\,\mathrm{d}\sigma = w_1(\mathrm{e}^{\mathrm{i}\tau}).$$

4. Conversely, let $w_1 \in W_r^1$, $Kw_1 = w_1$, and $w_2 := G(., w_1)$. First of all $w_1(1) = q$ and $\partial_\tau w_1 = \widetilde{w}_1$. Since $\widetilde{w}_1$ is holomorphic, so is w_1 (look at the Laurent coefficients). Inserting $\widetilde{w}_1 = \partial_\tau w_1$ into (7.23) shows that

$$\widetilde{w}_2 = a\,\widetilde{w}_1 + b\,\overline{\widetilde{w}_1} + f = \frac{\mathrm{d}}{\mathrm{d}\tau} G(., w_1) = \partial_\tau w_2.$$

Consequently w_2 is holomorphic and $w := (w_1, w_2) \in H^\infty \cap W_r^1$ solves (7.16). ∎

By virtue of Lemma 8, the existence result (i) of Theorem 2 is a consequence of Schauder's fixed-point principle.

It remains to prove that the solution of (7.21) is unique under the assumption (7.20). Let $w^{(1)}$, $w^{(2)} \in H^\infty \cap W_1^r$ be two solutions of (7.21). Then the difference $\Delta w \equiv (\Delta w_1, \Delta w_2) := w^{(2)} - w^{(1)}$ solves the linear Riemann-Hilbert problem

$$\Delta w_2 = a\Delta w_1 + b\overline{\Delta w_1}, \tag{7.33}$$

where

$$a := \int_0^1 \partial_z G(., \lambda w_1^{(1)} + (1-\lambda) w_1^{(2)})\,\mathrm{d}\lambda, \quad b := \int_0^1 \overline{\partial}_z G(., \lambda w_1^{(1)} + (1-\lambda) w_1^{(2)})\,\mathrm{d}\lambda,$$

and $\Delta w(t_0) = 0$. The assumption (7.20) on $\partial_z G$ and $\overline{\partial}_z G$ ensures that

$$\operatorname{Re} b - |a| \geq \delta > 0 \quad \text{on} \quad \mathbb{T}. \tag{7.34}$$

The Toeplitz operator associated with the Riemann-Hilbert problem (7.33) has the symbol

$$B = \frac{1}{b}\begin{bmatrix} a & -1 \\ |a|^2 - |b|^2 & -\overline{a} \end{bmatrix},$$

which is sectorial by virtue of (7.34). Since T_B is invertible we infer from Lemma 5 that $\Delta w \equiv 0$, which completes the proof of Theorem 2. ∎

Remark 1. We conjecture that the solution is unique even without the strengthened assumption (7.20).

Remark 2. In order to meet the side condition $w_1(0) = 0$ the operator K can be modified to

$$Ku(e^{i\tau}) := \int_0^\tau \widetilde{w}_1(e^{i\sigma})\, d\sigma - \frac{1}{2\pi}\int_0^{2\pi}\int_0^\rho \widetilde{w}_1(e^{i\sigma})\, d\sigma\, d\rho.$$

If, additionally, w_1 is reflected in the unit circle, Theorem 2 can be reformulated as an existence result for solutions w_+, w_- of the nonlinear transmission problem

$$w_+ = G(\cdot, w_-),$$

where w_+ is holomorphic in $\mathbb{D}$ and w_- is holomorphic in $\mathbb{C}\backslash\overline{\mathbb{D}}$ with $w_-(\infty) = 0$.

Theorem 3 *If $G \in C^1$ has bounded derivatives, and*

$$\operatorname{Re} \partial_z G(t,z) - |\overline{\partial}_z G(t,z)| \geq \delta > 0 \quad \forall (t,z) \in \mathbb{T} \times \mathbb{C},$$

then the nonlinear transmission problem $w_+ = G(\cdot, w_-)$ has a unique solution $(w_-, w_+) \in W_2^1(\mathbb{T})$.

References

[1] Bedford, E.: Stability of the polynomial hull of $\mathbb{T}^2$. Ann. Scuola Norm. Sup. Pisa Cl. Sci. 8 (1982), 311–315.

[2] Belch, R.: Numerische Lösung linearer Riemann-Hilbert-Probleme. Technical report, TU Bergakademie Freiberg, 1995, 20p.

[3] Bishop, E.: Differentiable manifolds in complex Euclidean space. Duke Math. J. 32 (1965) 1–22.

[4] Böttcher, A., Silbermann, B.: Analysis of Toeplitz operators. Springer-Verlag, Berlin 1990.

[5] Chern, S.S., Spanier, E.: A theorem on orientable surfaces in four-dimensional space. Comment Math. Helv. 25 (1951), 205-209.

[6] Clancey, K., Gohberg, I.: Localization of singular integral operators. Math. Z. 169 (1979), 105–117.

[7] Eliashberg, J.: Filling by holomorphic discs and its applications. In: Geometry of low dimensional manifolds, London Math. Soc. Lecture Notes Ser. 152, Cambridge Univ. Press, Cambridge 1990, 45–67.

[8] Forstnerič, F.: Analytic discs with boundary in a maximal real submanifold of $\mathbb{C}^2$. Ann. Inst. Fourier 37 (1987), 1–44.

[9] Forstnerič, F.: Polynomial hulls of sets fibered over the circle. Indiana Univ. Math. J. 37 (1988) 4, 869–889.

[10] Globevnik, J.: Perturbing analytic discs attached to maximal real submanifolds of $\mathbf{C}^N$. Indag. Math. N.S. 7 (1996), 37-46.

[11] Gromov, M.: Pseudo-holomorphic curves in symplectic manifolds. Invent. Math. 82 (1985), 307-347.

[12] Helton, J.W., Howe, R.E.: A bang-bang theorem for optimization over spaces of analytic functions. J. Approx. Theory 47 (1986) 2, 101–121.

[13] Helton, J.W., Marshall, D.: Frequency domain design and analytic selections. Indiana Univ. Math. J. 39 (1990) 1, 157–184.

[14] Litvinchuk, G.S.: Boundary value problems and singular integral equations with shift. Nauka, Moscow, 1977 (Russian).

[15] Litvinchuk, G.S., Spitkovsky, I.: Factorization of measurable matrix functions. Akademie Verlag, Berlin, 1987.

[16] McDuff, D., Salamon, D.: J-holomorphic curves and quantum cohomology. Cambridge Univ. Press, Cambridge, 1994.

[17] Shnirelman, A.I.: The degree of quasi-linear like mapping and the nonlinear Hilbert problem. Math. Sb. 89 (1972) 3, 366–389.

[18] Slodkowski, Z.: Holomorphic motions and polynomial hulls. Proc. Amer. Math. Soc. 111 (1991) 2, 347–355.

[19] Vekua, N.P.: Systems of singular integral equations. Nordhoff, Groningen 1967.

[20] Wegert, E.: Nonlinear boundary value problems for holomorphic functions and singular integral equations. Akademie Verlag, Berlin, 1992.

[21] Wegert, E.: Topological methods for strongly nonliear Riemann-Hilbert problems for holomorphic functions. Math. Nachr. 134 (1987), 201–230.

[22] Wegert, E.: Boundary value problems and best approximation by holomorphic functions. J. Approx. Theory 16 (1990) 3, 322–334.

[23] v. Wolfersdorf, L.: A class of nonlinear Riemann-Hilbert problems for holomorphic functions. Math. Nachr. 116 (1984), 89–107.

[24] v. Wolfersdorf, L.: On the theory of the nonlinear Hilbert problem for holomorphic functions. In: H. Begehr, A. Jeffrey: Partial Differential Equations with Complex Analysis. Longman Scientific and Technical, Harlow, 1992, 134–149.

8 BOUNDARY BEHAVIOUR OF VECTOR–VALUED FUNCTIONS OF TWO CLASSES IN LOCALLY CONVEX SPACES

Chuan–Gan Hu †

Nankai University
Departement of Mathematics
Tianjin 300071
P.R. China

Ya-Hua Wang

Hubei National Colleg
Department of Mathematics
Enshi 445000
Hubei
P.R. China

Abstract: The isomorphism of a vector-valued Fourier transform is presented. The boundary values of vector-valued functions of two classes on the unit disk of a complete separated locally convex space exist almost everywhere on the unit circle in the senses of any seminorm and a nontangential limit.

Keywords: Complete separated locally convex space, Fourier transform
Mathematics Subject Classification: 30B30, 30G30, 46G20

8.1 INTRODUCTION

In [16], the Fourier transform has been studied with domain in the set $\mathbb{R}^0$ of all nonnegative numbers and range in the complex plane $\mathbb{C}$. In [8], the Fourier

†Project 19571040 supported by National Natural Science Foundation of China and the Construction Foundation of Major Subjects of the Educational Commission of Tianjin City.

H.G.W. Begehr et al. (eds.), Partial Differential and Integral Equations, 125–133.

transform and the convergence problem of boundary functions have been discussed with domain in the complex plane $\mathbb{C}$ and range in a Banach space. In [12], the Laplace transform of vector-valued functions with range in a locally convex space has been presented. In [4], the Fourier series of vector-valued functions with range in a Banach space has been researched. In [10], the bounded convergence of vector-valued analytic functions with range in a Banach space has also been deliberated. But many function spaces in distributed parameter systems are not a Banach space. For example, the set $C_0^\infty(\mathbb{R}^+)(\mathbb{R}^+$ is the set of all positive numbers) in [3] is not a Banach space, but it is a locally convex space. In order to extend the scope of ranges in H_∞-control theory, therefore we need inquire into the Fourier transform and the boundary behaviour of vector-valued functions in complete separated locally convex spaces.

8.2 PRELIMINARY CONCEPTS

Suppose that E is a complete separated locally convex space, that P is the sufficiently directed set $\{p_n\}_1^\infty$ of countable seminorms which generates the topology of E [13, 22], that E' is the dual space of E, and that $\mathbb{R}$ is the set of all real numbers.

Definition 1 A vector-valued function $f(t)$ on the interval $[\alpha, \beta]$ to E is of

(i) weak bounded variation in $[\alpha, \beta]$ if $\Phi((f))$ is of numerical bounded variation for every $\Phi \in E'$,

(ii) bounded variation if $\sup p\,[\sum_i (f(\beta_i) - f(\alpha_i))] < \infty$ over every choice of a finite number of non-overlapping intervals (α_i, β_i) in $[\alpha, \beta]$ for any $p \in P$,

(iii) strong bounded variation if $\sup \sum_i p[f(\alpha_i) - f(\alpha_{i-1})] < \infty$, where all possible finite partitions of $[\alpha, \beta]$ are allowed. The two suprema are known as the total and the strong total variation respectively.

Let $f(t)$ be a vector-valued analytic function in the unit disk $|t| < 1$ to E. Define for any $p \in P$

$$M_{\ell,p}(r, f) = \begin{cases} \left\{ \dfrac{1}{2\pi} \displaystyle\int_0^{2\pi} p^\ell[f(re^{i\theta})]d\theta \right\}^{1/\ell}, & 0 < l < \infty, \\ M(r, f) = \max\limits_{0 \le \theta \le 2\pi} p(f(re^{i\theta})), & l = 0. \end{cases}$$

Definition 2 *The class of all vector-valued functions $f(t)$ that are analytic in the unit disk satisfying $\lim_{r \to 1} M_{l,p}(r, f) < \infty$ for any $p \in P$ is denoted by VH_ℓ. The class of all bounded analytic vector-valued functions in the unit disk is denoted by VH_∞. The class of all vector-valued harmonic functions (see* [16]*) satisfying $M_{\ell,p}(r, u) < \infty$ in the unit disk is denoted by $Vh_{\ell,p}(0 < \ell \le \infty)$.*

8.3 ISOMORPHISM OF THE VECTOR-VALUED FOURIER TRANSFORM

Definition 3 *Let $\int_{-\infty}^{\infty} p(f(t))dt < \infty$ for any $p \in P$. If for any $p \in P, \lim_{t\to\infty}$ $\int_{-t}^{t} e^{-2\pi i s t} f(t)dt$ exists, we call this limit the Fourier transform of $f(t)$, denoted* *by $F(s)$, i.e.*

$$F(s)(\text{ or } \hat{f}(s)) = \int_{-\infty}^{\infty} e^{-2\pi i s t} f(t)dt,$$

where $F(s)$ and $\hat{f}(s)$ are vector-valued functions with domains in $\mathbb{R}$ and ranges in E, and s is a real number. The integral

$$(\hat{f})^{\vee}(t) = \int_{-\infty}^{\infty} F(s)e^{2\pi i s t}ds$$

is called the inverse Fourier transform of $F(s)$.

Definition 4 *Consider a vector-valued function $f(t)$ on a subinterval Q of $\mathbb{R}$ with range in E, which is square integrable with respect to t for any $p \in P$. The space of all such functions is denoted by $VL_2(Q)$. Subject to $\int_Q p^2(f(t))dt < \infty$ its 2-p norm is defined by $\|f\|_{2,p} = [\int_Q p^2(f(t))dt]^1/2$.*

Let $VC^{\infty}(\mathbb{R})$ be the class of all vector-valued functions $f(t)$ on $\mathbb{R}$ with range in E having continuous derivatives of all orders with respect to $t \in \mathbb{R}$ for any $p \in P$ with compact function, where "compact function" means that f has a compact support in the sense of the paranorm $|||\cdot||| = \sum_{j=1}^{\infty} 1/2^j \; p_j(\cdot)/[1+p_j(\cdot)]$.

$VC^{\infty}(\mathbb{R})$ is a locally convex space and is not a Banach space [2], [11], [17].

Let $VC_{\downarrow}^{\infty}(\mathbb{R})$ be the class of infinitely differentiable and rapidly decreasing functions on $\mathbb{R}$, where "infinitely differentiable"means that $f(t) \in VC^{\infty}(\mathbb{R})$, "rapidly decreasing" applies to $f, f', \cdots$, and means that $t^{\alpha}D^{\beta}f$ tends to 0 as $|t| \to \infty$ for any $p \in P$ and every nonnegative α and β, separately. Here D stands for differential $Df = f'$.

To define the Fourier transform for vector-valued functions f of a class in $VL_2(\mathbb{R})$, it is not enough merely to write down the formal integral for $\hat{f}$ since the integrand is not necessarily summable for any $p \in P$. If a vector-valued function $f \in VC_{\downarrow}^{\infty}(\mathbb{R})$, then by partial integration

$$(f')^{\wedge} = 2\pi i s\hat{f}.$$

Proposition 1 (a) *The Fourier transform maps $VC_{\downarrow}^{\infty}(\mathbb{R})$ onto itself.*

(b) $(\hat{f})^{\vee} = f$.

(c) $\|f\|_{2,p} = \|\hat{f}\|_{2,p}$.

Proof (a) Pick a function $f(t) \in VC_{\downarrow}^{\infty}(\mathbb{R})$, then by partial integration

$$\Phi[(f')^{\wedge}] = 2\pi i s\Phi(\hat{f}),$$

here $\Phi \in E'$. Because E is separated, we have

$$(f')^{\wedge} = 2\pi i s\hat{f}.$$

Also, by the rapid decrease of f, we obtain $(-2\pi i t f)^{\wedge} = (\hat{f})'$, and so, by induction,

$$(2\pi i s)^{\alpha}D^{\beta}\hat{f} = [D^{\alpha}(-2\pi i t)^{\beta}f]^{\wedge},$$

for any nonnegative integer α and β. Therefore $\hat{f}$ also belongs to $VC_{\downarrow}^{\infty}(\mathbb{R})$.

(b) Similar to classical methods we have that $\sum_{n=-\infty}^{\infty} e^{2\pi int/T}T^{-1}\hat{f}(n/T)$ (a Riemann sum) converges to $\int_{-\infty}^{\infty}\hat{f}(s)e^{2\pi ist}ds$, as $T\to\infty$ in the sense of $|||\cdot|||$. So the proof is complete.

(c)Let

$$f_c(t)=\begin{cases} f(t), & |t|\le c,\\ \tilde{\theta}(\text{zero element}), & |t|>c.\end{cases}$$

Then $\int_{-\infty}^{\infty}p^2(f(t)-f_c(t))dt\to 0(c\to\infty)$. Let $F_c(s)=\int_{-\infty}^{\infty}e^{-2\pi ist}\ f_c(s)dt$. Because $f_c(t)\in VL_1(\mathbb{R})$, $\int_{-\infty}^{\infty}e^{-2\pi ist}f_c(s)dt$ exists.

Setting $\delta>0$ it follows that

$$\begin{aligned} I &= \int_{-\infty}^{\infty} e^{-\frac{1}{2}\delta^2s^2}p^2(F_c(s))ds\\ &= \int_{-\infty}^{\infty}p(f_c(t))dt\int_{-\infty}^{\infty}p(f_c(\gamma))d\gamma\int_{-\infty}^{\infty}exp\left\{-\frac{1}{2}\delta^2s^2+2\pi is(\gamma-t)\right\}ds.\end{aligned}$$

Consequently

$$\int_{-\infty}^{\infty}\exp\left\{-\frac{1}{2}\delta^2s^2+2\pi is(\gamma-t)\right\}ds=\frac{\sqrt{2\pi}}{\delta}e^{-2\pi^2(\gamma-t)^2/\delta^2},$$

so

$$\begin{aligned} I &= \frac{\sqrt{2\pi}}{\delta}\int_{-\infty}^{\infty}p(f_c(t))dt\int_{-\infty}^{\infty}p(f_c(\gamma))e^{-2\pi^2(\gamma-t)^2/\delta^2}d\gamma\\ &\le \left\{\int_{-\infty}^{\infty}\int_{-\infty}^{\infty}p^2(f_c(t))e^{-2\pi^2(\gamma-t)^2/\delta^2}dtd\gamma\right\}^{1/2}\\ &\quad\times\left\{\int_{-\infty}^{\infty}\int_{-\infty}^{\infty}p^2(f_c(\gamma))e^{-2\pi^2(\gamma-t)^2/\delta^2}dtd\gamma\right\}^{1/2}\frac{\sqrt{2\pi}}{\delta}\\ &= \int_{-\infty}^{\infty}p^2(f_c(t))dt.\end{aligned}$$

Hence $F_c(t)\in VL_2(\mathbb{R})$. And

$$I=\frac{\sqrt{2\pi}}{\delta}\int_{-\infty}^{\infty}e^{-2\pi^2\gamma^2/\delta^2}\psi(\gamma)d\gamma, \tag{8.1}$$

where $\psi(\gamma)=\int_{-\infty}^{\infty}p(f_c(t))p(f_c(t+\gamma))dt$. Because $\psi(\gamma)\le\int_{-\infty}^{\infty}p^2(f(t))dt$, $\psi(\gamma)$ is bounded for any $p\in P$. Clearly

$$\begin{aligned}\psi(\gamma_1)-\psi(\gamma_2) &\le \left\{\int_{-\infty}^{\infty}p^2(f_c(t))dt\right\}^{1/2}\\ &\quad\times\left\{\int_{-\infty}^{\infty}p^2[f_c(t+\gamma_1)-f_c(t+\gamma_2)]dt\right\}^{1/2},\end{aligned}$$

thus $\psi(\gamma)$ is continuous for any $p \in P$. It follows that

$$\lim_{\delta \to 0} \frac{\sqrt{2\pi}}{\delta} \int_{-\infty}^{\infty} e^{-2\pi^2\gamma^2/\delta^2} \psi(\gamma) d\gamma = \psi(0) = \int_{-\infty}^{\infty} p^2(f_c(t)) dt.$$

As $\delta \to 0$, by (8.1) we have

$$\int_{-\infty}^{\infty} p^2(F_c(t)) dt = \int_{-\infty}^{\infty} p^2(f_c(t)) dt. \tag{8.2}$$

Take $c \to \infty$, by (8.2) we have the required result.

Q.E.D.

Theorem 1 *The Fourier transform is an isomorphism from $VL_2(\mathbb{R})$ onto $VL_2(\mathbb{R})$.*

This proof is similar to the classical method.

8.4 BOUNDARY PROPERTIES OF VECTOR-VALUED FUNCTIONS

Definition 5 *Let D denote a bounded domain, ∂D its boundary, and $\overline{D}$ the closure of D, and let $g(t)$ be a vector-valued continuous function defined in D. Suppose that for any subdomain B with $\overline{B} \subset D$, a vector-valued function $U(t)$ is harmonic in B and continuous on $\overline{B}$ with $p(g(t)) \leq p(U(t))(t \in \partial B$, for any $p \in P)$ implying $p(g(t)) \leq p(U(t))$ in B. Then any vector-valued function that assumes the same properties as $g(t)$ described above is called a subharmonic function in D, where ranges of g and U are in E.*

Using similar methods as in the classical case we obtain the following conclusions.

Theorem 2 *A necessary and sufficient condition that a continuous vector-valued function $g(t)$ be vector-valued subharmonic in D is that for each $t_0 \in D$ there exists a $\rho_0 > 0$ such that the disk $|t - t_0| < \rho_0$ is in D and*

$$p(g(t_0)) \leq \frac{1}{2\pi} \int_0^{2\pi} p(g(t_0 + \rho e^{i\theta})) d\theta \tag{8.3}$$

for every $\rho < \rho_0, g(t) \in E$ and $p \in P$.

Let $\nu(r, \theta) = (1 - r^2)/(1 - 2r\cos\theta + r^2)$ be the Poisson kernel. Then the following vector-valued Poisson-Stieltjes integral

$$u(t) = u(re^{i\theta}) = \frac{1}{2\pi} \int_0^{2\pi} \nu(r, \theta - z) d\mu(z) \tag{8.4}$$

holds, $t = re^{i\theta}, 0 \leq r < 1$. It is used to indicate the relationship between the vector-valued harmonic function and its boundary values within the unit disk. Here $\mu(z)$ is of bounded variation on $[0, 2\pi]$ and u and μ are vector-valued functions with domain in $\mathbb{C}$ and ranges in the semireflexive E [20]. In particular, if

$$u(t) = u(re^{i\theta}) = \frac{1}{2\pi}\int_0^{2\pi} \nu(r,\theta - z)\mu(e^{iz})dz,$$

then $u(t)$ is called the Poisson integral of μ.

Definition 6 *Let $u(t)$ be vector-valued harmonic in the closed disk $|t| \le r$ with range in E, then*

$$u(a) = \frac{1}{2\pi}\int_{|t|=r} \frac{r^2 - |a|^2}{|t-a|^2} u(t)d\theta = \frac{1}{2\pi}\int_{|t|=r} \mathrm{Re}\frac{t+a}{t-a}u(t)d\theta$$

is called the vector-valued Poisson formula.

Proposition 2 *Let $u(t)$ be vector-valued harmonic in the closed disk $|t| \le r, u(t) \in E, t = \rho e^{i\theta}, 0 \le \theta \le 2\pi, r$ and ρ positive. Then*

$$\frac{\rho - r}{\rho + r} p(u(0)) \le p(u(t)) \le \frac{\rho + r}{\rho - r} p(u(0)) \tag{8.5}$$

for any fixed $p \in P$.

Theorem 3 *Consider a sequence of vector-valued functions $u_n(t)$, each defined and harmonic in a certain region Ω_n, let Ω be a region such that every point in Ω has a neighborhood contained in all but a finite number of the Ω_n, and assume, moreover, that in this neighborhood $p(u_n(t)) \le p(u_{n+1}(t))$ for any $p \in P$ as soon as n is sufficiently large. Then there are only two possibilities. Either $\lim_{n\to\infty} p(u_n(t)) = +\infty$ for every $t \in \Omega_n$ and some $p \in P$ on every compact subset of Ω or $u_n(t)$ tends to a vector-valued harmonic limit function $u(t)$ in the sense of seminorms in Ω, uniformly on compact sets, where the ranges of $u_n(t)$ and $u(t)$ are in E.*

Definition 7 *If $\lim_{t\to 0}(\mu(\theta_0+t)-\mu(\theta_0-t))/2t$ exists in the sense of seminorms, denoted by $D\mu(\theta_0)$, i.e. $D\mu(\theta_0) = \lim_{t\to 0}(\mu(\theta_0+t)-\mu(\theta_0-t))/2t$ for any $p \in P$, then $D\mu(\theta_0)$ is defined as the symmetric derivative at a point $\theta = \theta_0$, where μ is continuously differentiable in the sense of seminorms.*

Theorem 4 *Let $u(t)$ be a Poisson-Stieltjes integral of form (8.4), where μ is of bounded variation. If the symmetric derivative $D\mu(\theta_0)$ exists at a point θ_0, then the radial limit $\lim_{r\to 1} u(re^{i\theta_0})$ exists in the sense of seminorms and has the value $Du(\theta_0)$.*

If $f(t)$ is a vector-valued analytic function [4] in a domain D, then $f(t)$ is vector-valued subharmonic in D. This means that in each disk contained in $D, f(t)$ is dominated by a vector-valued harmonic function.

Definition 8 *A vector-valued function $g(t)$ is said to have a vector -valued harmonic majorant in D if there is a vector-valued function $u(t)$ harmonic in D such that $p(g(t)) \le p(u(t))$ for all t in D and any $p \in P$.*

If g is strongly continuous and has a harmonic majorant, it is obviously subharmonic, but the converse is false.

Definition 9 *Let $D = \{t = x + iy : y > 0\}$. The space $V\overline{H}_\ell$ $(0 < \ell < \infty)$ consists of vector-valued analytic functions in D such that $p^\ell(f(x+iy))$ is integrable for each $y > 0$ and any $p \in P$ and*

$$M_{l,p}^0(y,t) = \left(\int_{-\infty}^{\infty} p^\ell(f(x+iy))dx\right)^{1/\ell}$$

is bounded.

Similarly to Definition 8 and 9 we can define concepts of a space VL_l, a vector-valued Jordan curve and a nontangential limit.

By Harnack's principle (see [1], p.236) and the bounded convergence theorem we have the following result.

Theorem 5 *If $0 < \ell < \infty$ and $f \in V\overline{H}_\ell$, then $f \in V\overline{H}_\ell(D)$.*

Definition 10 *A vector-valued function f analytic in D is said to be of class $VN_\ell(D)$ if there exists a sequence of rectifiable Jordan curves $J_1, J_2, \cdots$ in D, tending to the boundary of D in the sense that J_n eventually surrounds each compact subdomain of D such that $\int_{J_n} p^\ell(f(t))|dt| \le M_p < \infty$ for any $p \in P$.*

Suppose now that D is a domain bounded by a rectifiable Jordan curve J. Let $\varphi(w)$ map $|w| < 1$ conformally onto D, and let $\psi(t)$ be the inverse mapping. Since J is rectifiable, it makes sense to speak of a tangential direction almost everywhere.

Theorem 6 *Each vector-valued function f of class $VN_\ell(D)$ or $VH_\ell(D)$ has a nontangential limit almost everywhere on J, which cannot vanish on a set of positive measure unless $p(f(t)) = 0$ for any $p \in P$. Furthermore,*

$$\int_J p^\ell(f(t))|dt| < \infty, \quad \text{if } f \in VN_\ell(D)$$

and

$$\int_J p^\ell(f(t))|\psi^1(t)||dt| < \infty \text{ iff } f \in VH_\ell(D).$$

Proof If $f \in VN_\ell(D)$, then $F(w) = f(\varphi(w))[\varphi'(w)]^{1/\ell} \in VH_\ell$. From Section 3.5 in [5] we derive that φ preserves sets of measure zero on the boundary, and it preserves angles at almost every boundary point. But $F(w)$ has a nontangential limit, and so does $[\varphi'(w)]^{1/\ell}$, since $\varphi' \in H_1$. Hence $f(\varphi(w))$, and therefore $f(t)$, has a nontangential limit almost everywhere. If $\Phi(f(t)) = 0$ for any $\Phi \in E'$ on a boundary set of positive measure, the same is true for $F(w)$. The separation property of E implies $F(w) = \tilde{\theta}$, and $p(f(t)) = 0$ for any $p \in P$. The integrability of $p^\ell(f(t))$ follows from the fact that $p^\ell(F(w))$ is integrable over $|w| = 1$. The discussion is similar for $f \in VH_\ell(D)$.

Corollary 1 *If $f \in V\overline{H}_\ell$ and if $f(t) \neq \tilde{\theta}$ for any $t \in D$ (here $\tilde{\theta}$ stands for the zero element in $V\overline{H}_\ell$), then the boundary function $f(x) = \lim_{y\to 0} f(x+iy)$ exists* *almost everywhere, for any* $p \in P, f \in VL_\ell$ *and*

$$\int_{-\infty}^{\infty} \frac{\log p(f(x))}{1+x^2} dx > -\infty. \tag{8.6}$$

Proof As in the proof of Theorem 6, the existence (more generally) of a nontangential limit $f(x) = \lim_{t\to x} f(t)$ for any $p \in P$, follows from the fact that $F(w) = f(\varphi(w))$ is in VH_ℓ. Fatou's lemma shows $p^\ell(f(x))$ is integrable over $(-\infty, \infty)$. Finally (8.6) follows from Theorem 2.2 in [5] and that

$$\int_{|w|=1} \log p(F(w))|dw| > -\infty.$$

It is also true that $f(x+iy)$ tends to $f(x)$ for any $p \in P$ in the VL_ℓ mean. Before showing this, it is convenient to prove a Poisson integral representation for $V\overline{H}_1$.

References

[1] Ahlfors, L.V.: Complex analysis. 2nd ed., McGraw-Hill, New York, 1966.

[2] Bourbaki, N.: Topological vector spaces. Springer-Verlag, Berlin, Heidelberg, New York, London, Paris, Tokyo, 1987.

[3] Curtain, R.F., Pritchard, A.J.: Infinite dimensional linear systems theory. Springer-Verlag, Berlin, Heidelberg, New York, 1978.

[4] Davies, B.: Integral transforms and their applications. Springer-Verlag, New York, Berlin, Heidelberg, Tokyo, 1978.

[5] Duren, P.L.: Theory of H^p spaces. Academic Press, New York, San Francisco, London, 1970.

[6] Dym, H., McKean, H.P.: Fourier series and integrals. Academic Press, New York, San Francisco, London, 1972.

[7] Edwards, R.E.: Fourier series: a modern introduction. V.1., 2nd ed. Springer-Verlag, New York, Heidelberg, Berlin, 1979.

[8] Francis, B.A.: A course in H_∞ control theory. Springer-Verlag, Berlin, Heidelberg, New York, London, Paris, Tokyo, 1978.

[9] Hille, E., Phillips, R.S.: Functional analysis and semi-groups. Amer. Math. Soc. Coll. Publ., Providence, R.I., 1957.

[10] Hu, C.G.: Boundary properties of a class of vector-valued analytic functions in the unit disc. Acta Sci. Nat. Univ. Nankaiensis 2(1984), 109-116.

[11] Hu, C.G.: Linear systems of a class in a locally convex space. Chinese Sci. Bull. 14(1993), 1154-1157.

[12] Hu, C.G.: The linear systems of a class in locally convex spaces. Complex Analysis and Its Applications. Pitman Research Notes in Math. Ser. 305, Longman, Harlow, 1994.

[13] Hu, C.G., Beg, I.: Cauchy-Stieltjes integral in locally convex spaces. Complex Variables Theory Appl. 15 (1990), 233-239.

[14] Hu, C.G., Wang, Z.M.: Bounded variation functions in locally convex spaces. Acta Sci. Nat. Univ. Nankaiensis 4(1991), 81-83.

[15] Hu, C.G., Yang, C.C.: Vector-valued functions and their applications. Kluwer Acad. Publ., Dordrecht, Boston, London, 1992.

[16] Lai, X.J.: Some summable $(C, 1)$ problems of vector-valued trigonometric series. Acta Sci. Nat. Univ. Nankaiensis 2(1984), 117-125.

[17] Daley, R.E.A.C., Wiener, N.: Fourier transforms in complex domain. Amer. Math. Soc. Coll. Publ. 19, New York, 1934.

[18] Rudin, W.: Real and complex analysis. McGraw-Hill, New York, 1974.

[19] Rudin, W.: Functional analysis. McGraw-Hill, New York, 1991.

[20] Schaeffer, H.H.: Topological vector spaces. Springer-Verlag, New York, 1971.

[21] Stein, E.M., Weiss, C.: Introduction to Fourier analysis on Euclidean spaces. Princeton Univ. Press, Princeton, New Jersey, 1984.

[22] Wilansky, A.: Modern methods in topologicol vector spaces. McGraw-Hill, New York, 1978.

[23] Zadeh, L.A., Desoer, C.A.: Linear systems theory. McGraw-Hill, New York, 1963.

9 ON SINGULAR INTEGRALS OF THE TIME-HARMONIC MAXWELL EQUATIONS THEORY

Reynaldo Rocha-Chávez

Departamento de Matemáticas, ESFM-IPN, Mexico City, MEXICO.
rrocha@esfm.ipn.mx

Michael Shapiro

Departamento de Matemáticas, ESFM-IPN, Mexico City, MEXICO.
shapiro@esfm.ipn.mx

Abstract: We establish an analogue of the well-known Poincaré-Bertrand formula for the theory of time-harmonic electromagnetic vector fields, as well as for some versions of quaternionic analysis.

9.1 INTRODUCTION

Let Γ be a closed Liapunov curve in the complex plane; let f be a Hölder function on $\Gamma \times \Gamma$. Then everywhere on Γ the following equality is true:

$$\frac{1}{\pi i} \int_{\Gamma_\tau} \frac{d\tau}{\tau - t} \frac{1}{\pi i} \int_{\Gamma_{\tau_1}} \frac{f(\tau, \tau_1)\, d\tau_1}{\tau_1 - \tau}$$

$$= f(t, t) + \frac{1}{\pi i} \int_{\Gamma_{\tau_1}} d\tau_1 \frac{1}{\pi i} \int_{\Gamma_\tau} \frac{f(\tau, \tau_1)\, d\tau}{(\tau - t)(\tau_1 - \tau)}, \tag{9.1}$$

which is usually called the Poincaré-Bertrand (P-B) formula. All the integrals in (9.1) do not converge, in general, as improper integrals and they are understood in the sense of the Cauchy principal value. Being formally an assertion

H.G.W. Begehr et al. (eds.), Partial Differential and Integral Equations, 135–153.

from the theory of singular integrals, (9.1) is, in a certain sense, a bridge between that theory and one-dimensional complex analysis (o.c.a.). Indeed, all the integrals in (9.1) contain the (singular) Cauchy kernel, and it would be trivial to explain its importance for o.c.a. Furthermore, it follows immediately from (9.1) that the singular Cauchy integral

$$S[f](t) := \frac{1}{\pi i} \int_{\Gamma} \frac{f(\tau)\, d\tau}{\tau - t},$$

generates an involution on $\mathbb{C}^{0,\mu}(\Gamma;\mathbb{C})$, $0 < \mu < 1$, and hence the operator $P := \frac{1}{2}(I + S)$, I the identity operator, is a projector. More exactly P is the projector of the space of Hölder functions onto the traces of functions holomorphic inside Γ.

Thus any intention to generalize (9.1) should, in our opinion, take into consideration that relation and look for an appropriate analogue of the Cauchy kernel.

Below we present a number of generalizations of (9.1). In Section 9.1 the reader can find the Poincaré-Bertrand formula for time-harmonic electromagnetic field theory (i.e., theory of solutions of the time-harmonic Maxwell equations). Why we consider it to be a good analogue of (9.1), is explained in Section 9.4 (where a proof of it is given as well). The proof is based on contents of Sections 9.2 and 9.3. In Section 9.2 we present the P-B formula for α-hyperholomorphic quaternionic function theory developed in [3] and our formula contains, as specific cases, some of its predecessors, compare with [1] and [6] for instance.

In the paper [4] a notion of the quaternionic Maxwell operator was introduced and studied. On one hand, its null-set contains all null-solutions of the time-harmonic Maxwell equations. On the other, the null-set is in one-to-one correspondence with pairs consisting of an α-hyperholomorphic and an α-"antihyperholomorphic" function. We obtain, using Section 9.2 and this correspondence, the P-B formula for the quaternionic Maxwell operator.

In all situations described above we introduce the corresponding "Cauchy singular operator" and prove its involutiveness.

Acknowledgments *This work was done with partial support of COFAA-IPN and DEPI-IPN (México). The authors are participants of the joint German-Mexican project "Quaternionic Analysis and Partial Differential Equations" supported by DFG (Germany), CONACyT and IPN (México) and this participation was useful in realizing some important aspects of the problem under consideration.*

9.2 THE POINCARÉ-BERTRAND FORMULA FOR THE TIME-HARMONIC ELECTROMAGNETIC FIELDS THEORY

9.2.1 Let Ω denote a domain in $\mathbb{R}^3$, let $\Gamma = \partial\Omega$ be its boundary, and let $\vec{E}$ and $\vec{H}$ be, respectively, the electrical and magnetic components of an electromagnetic field in Ω. It is assumed that Ω is filled up with a homogeneous medium and that there are no currents and charges in Ω. If the electromagnetic field $\left(\vec{E}, \vec{H}\right)$ is time-harmonic (= monochromatic) then it satisfies the following Maxwell equations:

$$rot\ \vec{H} = \sigma\vec{E}, \qquad rot\ \vec{E} = i\omega\mu\vec{H}, \tag{9.2}$$

$$div\ \vec{H} = 0, \qquad div\ \vec{E} = 0. \tag{9.3}$$

Here $\sigma := \sigma^* - i\omega\varepsilon$ is the complex electrical conductivity; ε is the dielectric constant; μ is a magnetic permeability; σ^* is the medium electrical conductivity being inverse to its electrical resistivity: $\sigma^* = \frac{1}{\rho}$. Mathematically, $\vec{E}$ and $\vec{H}$ are complex-valued vector fields:

$$\vec{E},\ \vec{H} : \Omega \subset \mathbb{R}^3 \to \mathbb{C}^3,$$

and they satisfy the homogeneous Helmholtz equation

$$\Delta\vec{E} + \lambda\vec{E} = 0, \tag{9.4}$$

$$\Delta\vec{H} + \lambda\vec{H} = 0, \tag{9.5}$$

where $\lambda := i\omega\mu\sigma^* + \omega^2\mu\varepsilon = i\omega\mu\sigma \in \mathbb{C}$ and its square root $\alpha := \sqrt{\lambda}$ ($Re\ \alpha > 0$) is called a medium wave number.

For any complex-valued vectors $\vec{f}$ and $\vec{g}$ by definition

$$\left\langle \vec{f},\ \vec{g} \right\rangle := \sum_{k=1}^{3} f_k g_k \in \mathbb{C} \tag{9.6}$$

is the scalar (= interior) product, and

$$\vec{f} \times \vec{g} := \begin{vmatrix} \vec{i}_1 & \vec{i}_2 & \vec{i}_3 \\ f_1 & f_2 & f_3 \\ g_1 & g_2 & g_3 \end{vmatrix} \in \mathbb{C}^3 \tag{9.7}$$

is the vector (= cross) product. The equality (9.6) determines a $\mathbb{C}$-valued bilinear form over $\mathbb{C}$; the equality (9.7) determines a $\mathbb{C}^3$-valued bilinear form over $\mathbb{C}$.

9.2.2 Remarks 1. It is easy to see that the equalities (9.3) can be deduced directly from (9.2), and thus there is no need, at least formally, to include them into the Maxwell system. But in many physical sources, books and articles, the conditions (9.3) are written down explicitly, and we follow here this tradition. Besides, there are some substantial mathematical reasons for doing this (see below).

2. It is known (see [2], e.g.) that for any solution $\left(\vec{E},\ \vec{H}\right)$ of the system (9.2)–(9.3) its components are orthogonal with respect to the bilinear form (9.6):

$$\left\langle \vec{E},\ \vec{H} \right\rangle = 0. \tag{9.8}$$

9.2.3 Set

$$\mathcal{M} := \begin{pmatrix} \sigma & -rot \\ rot & -i\omega\mu \end{pmatrix}, \tag{9.9}$$

then the system (9.2) becomes

$$\mathcal{M}\left[\begin{pmatrix} \vec{E} \\ \vec{H} \end{pmatrix}\right] = 0. \tag{9.10}$$

The operator $\mathcal{M}$ acts naturally on the space $C^1(\Omega;\ \mathbb{C}^3 \times \mathbb{C}^3)$. Taking into account (9.3) let us introduce, for $k \in \mathbb{N} \cup \{0\}$, its subspaces

$$\hat{C}^k := \hat{C}^k(\Omega;\ \mathbb{C}^3 \times \mathbb{C}^3) := \left\{\begin{pmatrix} \vec{f} \\ \vec{g} \end{pmatrix} \in C^k(\Omega;\ \mathbb{C}^3 \times \mathbb{C}^3) \mid div\ \vec{f} = div\ \vec{g} = 0\right\}.$$

will be termed "the time-harmonic Maxwell operator". It is essential to note that $\hat{\mathcal{M}}$ maps a solenoidal (= divergenceless) vector field into a solenoidal one reducing of course, the smoothness; if $\begin{pmatrix} \vec{f} \\ \vec{g} \end{pmatrix} \in \hat{C}^2$, $\begin{pmatrix} \vec{u} \\ \vec{v} \end{pmatrix} := \hat{\mathcal{M}}\left[\begin{pmatrix} \vec{f} \\ \vec{g} \end{pmatrix}\right]$ then $\begin{pmatrix} \vec{u} \\ \vec{v} \end{pmatrix} \in \hat{C}^1$.

9.2.4 Let $\theta_\lambda^\pm$ denote the fundamental solution of the Helmholtz operator:

$$\theta_\lambda^\pm(x) = -\frac{1}{4\pi}\frac{1}{|x|}e^{\pm i\alpha|x|}, \qquad x \in \mathbb{R}^3 \setminus \{0\},$$

where $\alpha^2 = \lambda$, $Re\ \alpha > 0$. In particular, for $\alpha = 0$ we obtain $\theta_0 := \theta_0^\pm$, the fundamental solution of the Laplace operator. For calculations the following correlation which can be easily obtained is useful:

$$\theta_\lambda^\pm(x) \underset{|x|\to 0}{=} \theta_0(x) \mp i\alpha\frac{1}{4\pi} + o(|x|).$$

Below $\theta_\lambda := \theta_\lambda^-$ participates only.

Introduce the following notation:

$$\mathcal{U}(\xi - \zeta) := \begin{pmatrix} grad\ \theta_\lambda(\xi - \zeta) \times \vec{n}_\zeta & i\omega\mu\theta_\lambda(\xi - \zeta)\vec{n}_\zeta \\ \sigma\theta_\lambda(\xi - \zeta)\vec{n}_\zeta & grad\ \theta_\lambda(\xi - \zeta) \times \vec{n}_\zeta \end{pmatrix}$$

where $\vec{n}_\zeta$ is the outward pointing normal unit vector to Γ at $\zeta \in \Gamma$.

Denote by $\mathcal{K}_{Mxw}$ the following "ordered pair of a scalar field and a matrix-valued field", that is:

$$\mathcal{K}_{Mxw}(\xi,\ \zeta) := (\langle grad\ \theta_\lambda(\xi - \zeta),\ \vec{n}_\zeta\rangle,\ -\mathcal{U}(\xi - \zeta)) \tag{9.11}$$

for any fixed $\xi,\ \zeta \in \mathbb{R}^3$, with $\xi \neq \zeta$. For it and any $\vec{g} := \begin{pmatrix} \vec{e} \\ \vec{h} \end{pmatrix} : \Gamma \to \mathbb{C}^3 \times \mathbb{C}^3$ we define a new operation "$\star$":

$$\mathcal{K}_{Mxw}(\xi,\ \zeta) \star \vec{g}(\zeta) := \langle grad\ \theta_\lambda(\xi - \zeta),\ \vec{n}_\zeta\rangle\vec{g}(\zeta) - \mathcal{U}(\xi - \zeta) \times \vec{g}(\zeta) \tag{9.12}$$

where

$$\mathcal{U}(\xi-\zeta)\times\vec{g}(\zeta):=\begin{pmatrix}(grad\,\theta_\lambda(\xi-\zeta)\times\vec{n}_\zeta)\times\vec{e}+i\omega\mu\theta_\lambda(\xi-\zeta)\vec{n}_\zeta\times\vec{h}\\ \sigma\theta_\lambda(\xi-\zeta)\vec{n}_\zeta\times\vec{e}+(grad\,\theta_\lambda(\xi-\zeta)\times\vec{n}_\zeta)\times\vec{h}\end{pmatrix}.$$

$\mathcal{K}_{Mxw}$ will be called "The time-harmonic Maxwell-Cauchy kernel". The reasons for introducing the term, as well as those for introducing the operation "$\star$", the reader can find in Sections 9.4 and 9.5 below.

To obtain the main theorems written in a compact form we need more additional notations. Let $V=(\vec{v}_{pq})_{1\le p,\ q\le 2}$ be a matrix with vector entries, $W=(w_{pq})_{1\le p\ q\le 2}$ be a complex matrix , and let "$*$" stand for the usual matrix multiplication. Then

$$\mathcal{K}_{Mxw}(\xi,\zeta)\circ W\ :=(\langle grad\,\theta_\lambda(\xi-\zeta),\ \vec{n}_\zeta\rangle W,\ -\mathcal{U}(\xi-\zeta)*W);$$

$$V\diamond\begin{pmatrix}\vec{e}\\ \vec{h}\end{pmatrix}\ :=\begin{pmatrix}\langle\vec{v}_{11},\ \vec{e}\rangle+\left\langle\vec{v}_{12},\ \vec{h}\right\rangle\\ \langle\vec{v}_{21},\ \vec{e}\rangle+\left\langle\vec{v}_{22},\ \vec{h}\right\rangle\end{pmatrix}.$$

9.2.5 **Theorem** *(Poincaré-Bertrand formula for time-harmonic electromagnetic theory). Let* Γ *be a Liapunov surface;* $\vec{e},\ \vec{h}\in C^{0,\mu}(\Gamma\times\Gamma;\ \mathbb{C}^3)$ *the Hölder space of* $\mathbb{C}^3$*-valued vector-fields. Then the following equalities hold everywhere on* Γ*:*

$$\int\limits_{\Gamma_{\tau_1}}\int\limits_{\Gamma_\tau}\mathcal{K}_{Mxw}(t,\tau)\star\left(\mathcal{K}_{Mxw}(\tau,\tau_1)\star\begin{pmatrix}\vec{e}(\tau_1,\tau)\\ \vec{h}(\tau_1,\tau)\end{pmatrix}\right)ds_\tau ds_{\tau_1}$$

$$-\int\limits_{\Gamma_{\tau_1}}\int\limits_{\Gamma_\tau}\mathcal{U}(t-\tau)*\left(\mathcal{U}(\tau-\tau_1)\diamond\begin{pmatrix}\vec{e}(\tau_1,\tau)\\ \vec{h}(\tau_1,\tau)\end{pmatrix}\right)ds_\tau ds_{\tau_1}+\frac{1}{4}\begin{pmatrix}\vec{e}(t,t)\\ \vec{h}(t,t)\end{pmatrix}$$

$$=\int\limits_{\Gamma_\tau}\int\limits_{\Gamma_{\tau_1}}\mathcal{K}_{Mxw}(t,\tau)\star\left(\mathcal{K}_{Mxw}(\tau,\tau_1)\star\begin{pmatrix}\vec{e}(\tau_1,\tau)\\ \vec{h}(\tau_1,\tau)\end{pmatrix}\right)ds_{\tau_1}ds_\tau$$

$$-\int\limits_{\Gamma_\tau}\int\limits_{\Gamma_{\tau_1}}\mathcal{U}(t-\tau)*\left(\mathcal{U}(\tau-\tau_1)\diamond\begin{pmatrix}\vec{e}(\tau_1,\tau)\\ \vec{h}(\tau_1,\tau)\end{pmatrix}\right)ds_{\tau_1}ds_\tau;\qquad(9.13)$$

$$\int\limits_{\Gamma_{\tau_1}}\int\limits_{\Gamma_\tau}\mathcal{U}(t-\tau)\diamond\left(\mathcal{K}_{Mxw}(\tau,\tau_1)\star\begin{pmatrix}\vec{e}(\tau_1,\tau)\\ \vec{h}(\tau_1,\tau)\end{pmatrix}\right)ds_\tau ds_{\tau_1}$$

$$+\int\limits_{\Gamma_{\tau_1}}\int\limits_{\Gamma_\tau}\langle grad\,\theta_\lambda(t-\tau),\ \vec{n}_\tau\rangle\left(\mathcal{U}(\tau-\tau_1)\diamond\begin{pmatrix}\vec{e}(\tau_1,\tau)\\ \vec{h}(\tau_1,\tau)\end{pmatrix}\right)ds_\tau ds_{\tau_1}$$

$$=\int\limits_{\Gamma_\tau}\int\limits_{\Gamma_{\tau_1}}\mathcal{U}(t-\tau)\diamond\left(\mathcal{K}_{Mxw}(\tau,\tau_1)\star\begin{pmatrix}\vec{e}(\tau_1,\tau)\\ \vec{h}(\tau_1,\tau)\end{pmatrix}\right)ds_{\tau_1}ds_\tau$$

$$+\int\limits_{\Gamma_\tau}\int\limits_{\Gamma_{\tau_1}} \langle grad\, \theta_\lambda (t-\tau),\ \vec{n}_\tau\rangle \left(\mathcal{U}(\tau-\tau_1)\diamond\begin{pmatrix}\vec{e}(\tau_1,\ \tau)\\ \vec{h}(\tau_1,\ \tau)\end{pmatrix}\right) ds_{\tau_1} ds_\tau. \quad (9.14)$$

The proof and comments can be found in Section 9.4.

9.2.6 Let Γ be as above; $\vec{e},\ \vec{h} \in C^{0,\mu}(\Gamma\times\Gamma;\mathbb{C}^3)$. Then the following singular integral exists in the sense of the Cauchy principal value:

$$\int\limits_{\Gamma_\tau} \mathcal{K}_{Mxw}(t,\ \tau)\star\begin{pmatrix}\vec{e}(\tau)\\ \vec{h}(\tau)\end{pmatrix} ds_\tau. \quad (9.15)$$

But the corresponding operator looses basic properties of its predecessor from one-dimensional complex analysis. That is why we introduce the condition

$$\int\limits_{\Gamma} \mathcal{U}(t-\tau)\diamond\begin{pmatrix}\vec{e}(\tau)\\ \vec{h}(\tau)\end{pmatrix} ds_\tau = 0 \quad \text{for any} \quad t\in\Gamma \quad (9.16)$$

(compare with (4.5) below) and we set

$$C^{0,\mu}_{Mxw} := C^{0,\mu}_{Mxw}(\Gamma) := \left\{\begin{pmatrix}\vec{e}\\ \vec{h}\end{pmatrix} \in C^{0,\mu}(\Gamma;\mathbb{C}^3\times\mathbb{C}^3) \mid \ (16)\ \text{holds}\right\}.$$

Let now $\begin{pmatrix}\vec{e}\\ \vec{h}\end{pmatrix} \in C^{0,\mu}_{Mxw}(\Gamma)$; we set

$$S_{Mxw}\left[\begin{pmatrix}\vec{e}\\ \vec{h}\end{pmatrix}\right](t) := 2\int\limits_{\Gamma_\tau} \mathcal{K}_{Mxw}(t,\ \tau)\star\begin{pmatrix}\vec{e}(\tau)\\ \vec{h}(\tau)\end{pmatrix} ds_\tau.$$

The operator S_{Mxw} will be referred to as "the electromagnetic (time-harmonic) Cauchy singular operator".

9.2.7 **Theorem** *(Involutiveness of the electromagnetic Cauchy singular operator). For* $0<\mu<1$, *the operator* S_{Mxw} *acts invariantly on* $C^{0,\mu}_{Mxw}$ *and is an involution:*

$$S^2_{Mxw} = I.$$

Again the proof and comments can be found in Section 9.4.

9.3 HYPERHOLOMORPHIC FUNCTION THEORY

9.3.1 Let $\mathbb{H}$ be the set of real quaternions. This means that elements of $\approx$ are of the form $a = \sum_{k=0}^{3} a_k i_k$, where $\{a_k \mid k\in\mathbb{N}_3^0 := \mathbb{N}_3\cup\{0\};\ \mathbb{N}_3 := \{1,\ 2,\ 3\}\}\subset\mathbb{R}$; i_0 is the unit; $i_1,\ i_2,\ i_3$ are called the imaginary units, and they define arithmetic rules in $\mathbb{H}$: by definition $i_k^2 = -i_0$, $k\in\mathbb{N}_3$; $i_1 i_2 = -i_2 i_1 = i_3$, $i_2 i_3 = -i_3 i_2 = i_1$, $i_3 i_1 = -i_1 i_3 = i_2$.

Natural operations of addition and multiplication in $\mathbb{H}$ turn it into a non-commutative field (= a skew-field). There is a main involution in $\mathbb{H}$. It is called the quaternionic conjugation, and it plays an exceptionally significant role. This involution is defined in the following way:

$$\bar{i}_0 := i_0, \; \bar{i}_k = -i_k, \; k \in \mathbb{N}_3,$$

and it extends onto $\mathbb{H}$ by $\mathbb{R}$-linearity, that is, for $a \in \mathbb{H}$

$$\bar{a} := \sum_{k=0}^{3} \overline{a_k i_k} = \sum_{k=0}^{3} a_k \bar{i}_k = a_0 - \sum_{k=1}^{3} a_k i_k =: Z(a). \tag{9.17}$$

Note that

$$\bar{a}a = a\bar{a} = \sum_{k=0}^{3} a_k^2 = |a|_{\mathbb{R}^4}^2 =: |a|_{\mathbb{H}}^2 . \tag{9.18}$$

Therefore for $a \in \mathbb{H} \setminus \{0\}$ a quaternion $a^{-1} := \frac{1}{a\bar{a}}\bar{a} = \frac{1}{|a|^2}\bar{a}$ is an inverse to a. We mention also a very important property of the quaternionic conjugation: for any $a, \; b \in \mathbb{H}$

$$\overline{ab} = \bar{b}\bar{a} \tag{9.19}$$

or

$$Z(ab) = Z(b)\, Z(a) . \tag{9.20}$$

9.3.2 For our purposes we need a notion of a complex quaternion. The set of complex quaternions $\mathbb{H}(\mathbb{C})$ consists of the elements $a = \sum_{k=1}^{3} a_k i_k$ where $\{i_k\}$ is as above, $\{a_k \mid k \in \mathbb{N}_3^0\} \subset \mathbb{C}$, a set of usual complex numbers with the imaginary unit i. By definition

$$i \cdot i_k = i_k \cdot i, \; k \in \mathbb{N}_3^0.$$

Arithmetic rules are defined in $\mathbb{H}(\mathbb{C})$ just as in $\mathbb{H}$. It is obvious that $\mathbb{H}$ is a real subalgebra in $\mathbb{H}(\mathbb{C})$. Each element $a \in \mathbb{H}(\mathbb{C})$ can be represented in the form

$$a = a^1 + i \cdot a^2 = a^1 + a^2 \cdot i, \tag{9.21}$$

where $a^1, \; a^2 \in \mathbb{H}$. Quaternionic conjugation acts on quaternionic units only, not on i, in (9.17).

9.3.3 Denoting for $a \in \mathbb{H}$ or $a \in \mathbb{H}(\mathbb{C})$, $a_0 =: Sc(a)$; $\vec{a} := \sum_{k=1}^{3} a_k i_k =: Vec(a)$ we can write $a = a_0 + \vec{a}$. a_0 is called the scalar part of the quaternion a, $\vec{a}$ is called the vector part, or the purely imaginary part. Using notations (9.6), (9.7) we have for any complex quaternions a and b:

$$\begin{aligned} ab &= (a_0 + \vec{a})\left(b_0 + \vec{b}\right) = \\ &= a_0 b_0 - \left\langle \vec{a}, \; \vec{b} \right\rangle + a_0 \vec{b} + b_0 \vec{a} + \vec{a} \times \vec{b}. \end{aligned} \tag{9.22}$$

The most impressive form this equality takes if $a_0 = b_0 = 0$:

$$\vec{a}\vec{b} = -\left\langle \vec{a}, \; \vec{b} \right\rangle + \vec{a} \times \vec{b}. \tag{9.23}$$

Equality (9.23) contains three kinds of multiplication of 3-dimensional vectors, and it can be made use of to imply and to explain all the rules of the scalar and vector products.

9.3.4 We shall consider $\mathbb{H}(\mathbb{C})$-valued functions defined in Ω:

$$f : \Omega \to \mathbb{H}(\mathbb{C}) .$$

Notations $C^p(\Omega; \mathbb{H}(\mathbb{C}))$, $p \in \mathbb{N} \cup \{0\}$, have the usual component-wise meaning.

Let $\psi := (\psi^1, \psi^2, \psi^3)$, $\psi^1, \psi^2, \psi^3 \in \mathbb{H}$ and $Sc(\psi^1) = Sc(\psi^2) = Sc(\psi^3) = 0$. Thus ψ is a set of 3-dimensional vectors. Assume additionally that ψ is an orthonormalized system. Sometimes we shall call ψ a structural set. On a set $C^1(\Omega; \mathbb{H}(\mathbb{C}))$ define the operators ${}^{\psi}D$ and D^{ψ} by the equalities

$${}^{\psi}D[f] := \sum_{k=1}^{3} \psi^k \frac{\partial f}{\partial x_k} =: \sum_{k=1}^{3} \psi^k \partial_k [f], \tag{9.24}$$

$$D^{\psi}[f] := \sum_{k=1}^{3} \frac{\partial f}{\partial x_k} \psi^k =: \sum_{k=1}^{3} \partial_k [f] \psi^k. \tag{9.25}$$

Let Δ be the 3-dimensional Laplace operator: $\Delta = \sum_{k=1}^{3} \partial_k^2$. Then on $C^2(\Omega;$ $\mathbb{H}(\mathbb{C})$ the following equalities are true:

$$\Delta = {}^{\psi}D^{\overline{\psi}}D = -{}^{\psi}D^{\psi}D = D^{\psi}D^{\overline{\psi}} = -D^{\psi}D^{\psi}, \tag{9.26}$$

where $\overline{\psi} := \left(\overline{\psi}^1, \overline{\psi}^2, \overline{\psi}^3\right)$. Let $\lambda \in \mathbb{C} \setminus \{0\}$, and let α be its complex square root: $\alpha \in \mathbb{C}$, $\alpha^2 = \lambda$, $Re\ \alpha > 0$. The 3-dimensional Helmholtz operator

$$\Delta_\lambda := \Delta + \lambda \tag{9.27}$$

acts on $C^2(\Omega; \mathbb{H}(\mathbb{C}))$. For an arbitrary fixed set ψ and for the stated above α the operators

$${}^{\psi}D_\alpha := \alpha + {}^{\psi}D, \qquad {}^{\overline{\psi}}D_\alpha := \alpha - {}^{\psi}D, \tag{9.28}$$

$${}_{\alpha}D^{\psi} := \alpha + D^{\psi}, \qquad {}_{\alpha}D^{\overline{\psi}} := \alpha - D^{\psi}, \tag{9.29}$$

are called the left (and right respectively) mutually conjugate Cauchy-Riemann operators. From (9.26) it follows that the following equalities are true:

$$\Delta_\lambda = {}^{\psi}D_\alpha {}^{\overline{\psi}}D_\alpha = {}^{\overline{\psi}}D_\alpha {}^{\psi}D_\alpha = {}_{\alpha}D^{\psi}\, {}_{\alpha}D^{\overline{\psi}} = {}_{\alpha}D^{\overline{\psi}}\, {}_{\alpha}D^{\psi}. \tag{9.30}$$

They give the factorization of the Helmholtz operator with a complex parameter λ. Thus a family (depending on the parameters ψ and α) of the operators (9.28) and (9.29), factorizing the Helmholtz operator, is obtained.

The sets

$${}^{\psi}\mathfrak{M}_\alpha(\Omega; \mathbb{H}(\mathbb{C})) := \ker {}^{\psi}D_\alpha, {}_{\alpha}\mathfrak{M}^{\psi}(\Omega; \mathbb{H}(\mathbb{C})) := \ker {}_{\alpha}D^{\psi} \tag{9.31}$$

will be referred to as left-(ψ, α)-hyperholomorphic (correspondingly right) in Ω, $\mathbb{H}(\mathbb{C})$-valued functions shortening this name if misunderstanding can not arise. The sets (9.31) will be called the classes of (ψ, α)-hyperholomorphy briefly.

The equalities (9.30) mean that (ψ, α)-hyperholomorphic functions of the class C^2 are metaharmonic in Ω, that is belong to ker Δ_λ.

9.3.5 The investigation of the time-harmonic Maxwell operator requires to consider, as a rule, the case $\psi = \psi_{st} := \{i_1, i_2, i_3\}$, for which we will use notations

$$^{\psi_{st}}D_\alpha =: {}^{st}D_\alpha =: D_\alpha, \qquad {}_\alpha D^{\psi_{st}} =: {}_\alpha D^{st} =: {}_\alpha D,$$

$$^{\overline{\psi}_{st}}D_\alpha =: {}^{\overline{st}}D_\alpha =: \overline{D}_\alpha, \qquad {}_\alpha D^{\overline{\psi}_{st}} =: {}_\alpha D^{\overline{st}} =: {}_\alpha \overline{D}. \tag{9.32}$$

For this special case the vector representation of a quaternion (see Subsection 9.3.3) gives rise to the following representation of the operators ${}^{st}D$ and ${}^{st}D_\alpha$: for any $f \in C^1(\Omega; \mathbb{H}(\mathbb{C}))$

$$\begin{aligned} D[f] &= \sum_{k=1}^{3} i_k \frac{\partial f}{\partial x_k} = \sum_{k=1}^{3} i_k \frac{\partial}{\partial x_k}\left(f_0 + \vec{f}\right) \\ &= \sum_{k=1}^{3} i_k \frac{\partial f_0}{\partial x_k} - \sum_{k=1}^{3}\left\langle i_k, \frac{\partial \vec{f}}{\partial x_k}\right\rangle + \sum_{k=1}^{3} i_k \times \frac{\partial \vec{f}}{\partial x_k} \\ &= grad\, f_0 - div\, \vec{f} + rot\, \vec{f}. \end{aligned} \tag{9.33}$$

Hence

$${}^{st}D_\alpha\left[f_0 + \vec{f}\right] = -div\, \vec{f} + \alpha_0 f_0 + grad\, f_0 + rot\, \vec{f} + f_0\vec{\alpha} + \alpha_0\vec{f}, \tag{9.34}$$

where $\alpha_0 = \alpha \in \mathbb{C}$.

9.3.6 The fundamental solution (= the Cauchy kernel for the corresponding theory) of the operator ${}^{\psi}D_\alpha$ is given by the formula

$$\begin{aligned} \mathcal{K}_{\psi,\alpha}(x) &:= {}_\alpha D^{\overline{\psi}}[\theta_\lambda](x) = {}^{\overline{\psi}}D_\alpha[\theta_\lambda](x) \\ &= \frac{1}{|x|^2}\left(\alpha|x|^2 + (i\alpha|x| + 1)\sum_{k=1}^{3}\psi^k x_k\right)\theta_\lambda(x). \end{aligned} \tag{9.35}$$

Let $\sigma_{\psi,x} := \sum_{k=1}^{3}(-1)^{k-1}\psi^k d\hat{x}^k$, where $d\hat{x}^k$ denotes, as usual, the differential form $dx^1 \wedge dx^2 \wedge dx^3$ with the factor dx^k omitted. Note that if $d\Gamma$ is the element of the surface area in $\mathbb{R}^3$ then $|\sigma_{\psi,x}| = d\Gamma$. Moreover, if Γ is a smooth surface then $\sigma_{\psi,\tau} = \sum_{k=1}^{3}\psi^k n_k(\tau)\, d\Gamma$ where $\vec{n} = (n_1, n_2, n_3)$ is the unit vector of the outward normal on Γ at the point $\tau \in \Gamma$. Also we will denote $\vec{n}_\tau := \vec{n}(\tau)$ for any $\tau \in \Gamma$.

9.3.7 **Theorem** *(Poincaré-Bertrand formula for (ψ,α)-hyperholomorphic function theory with $\alpha \in \mathbb{C}$). Let Γ be a Liapunov surface in $\mathbb{R}^3$without boundary, and let $f \in C^{0,\mu}(\Gamma \times \Gamma; \mathbb{H}(\mathbb{C}))$, then the following equality holds everywhere on Γ*

$$\int_{\Gamma_{\tau_1}}\int_{\Gamma_\tau} \mathcal{K}_{\psi,\alpha}(t-\tau)\,\sigma_{\psi,\tau}\mathcal{K}_{\psi,\alpha}(\tau-\tau_1)\,\sigma_{\psi,\tau_1} f(\tau_1,\tau) + \frac{1}{4} f(t,t)$$

$$= \int_{\Gamma_\tau}\int_{\Gamma_{\tau_1}} \mathcal{K}_{\psi,\alpha}(t-\tau)\,\sigma_{\psi,\tau}\mathcal{K}_{\psi,\alpha}(\tau-\tau_1)\,\sigma_{\psi,\tau_1} f(\tau_1,\tau). \tag{9.36}$$

Here $C^{0,\mu}(\Gamma \times \Gamma; \mathbb{H}(\mathbb{C}))$, $0 < \mu \leq 1$, denotes a space of $\mathbb{H}(\mathbb{C})$-valued Hölder functions of $(\tau_1,\tau) \in \Gamma \times \Gamma$.

Proof We shall follow basically the line of the case $\alpha = 0$ in [6] where the reader can find some details omitted here. We mention that a comprehensive justification was done in [7]. First of all one has:

$$-\int_{\Gamma_\tau} \mathcal{K}_{\psi,\alpha}(t-\tau)\,\sigma_{\psi,\tau} = \frac{1}{2}, \quad \text{for any} \quad t \in \Gamma. \tag{9.37}$$

Indeed given $t \in \Gamma$, let $\mathbb{B}(t,\ \varepsilon) := \{x \in \mathbb{R}^3 \mid |x-t| < \varepsilon\}$, $\Lambda_\varepsilon := \partial(\Omega^+ \backslash \mathbb{B}(t,\ \varepsilon))$, $\gamma_\varepsilon := \overline{\mathbb{B}(t,\ \varepsilon)} \cap \Lambda_\varepsilon$, $\Omega_\varepsilon^+ := \Omega \backslash \overline{\mathbb{B}(t,\ \varepsilon)}$. Then

$$\begin{aligned} -\int_{\Gamma_\tau} \mathcal{K}_{\psi,\alpha}(t-\tau)\,\sigma_{\psi,\tau} &= \lim_{\varepsilon\to 0}\left[-\int_{\Lambda_\varepsilon} \mathcal{K}_{\psi,\alpha}(t-\tau)\,\sigma_{\psi,\tau}\right] \\ &\quad + \lim_{\varepsilon\to 0}\int_{\gamma_\varepsilon} \mathcal{K}_{\psi,\alpha}(t-\tau)\,\sigma_{\psi,\tau} \\ &= \lim_{\varepsilon\to 0}\int_{\gamma_\varepsilon} \mathcal{K}_{\psi,\alpha}(t-\tau)\,\sigma_{\psi,\tau} = \frac{1}{2}. \end{aligned}$$

Secondly, the following equality holds for any t, τ_1 on Γ with $t \neq \tau_1$:

$$\int_{\Gamma_\tau} \mathcal{K}_{\psi,\alpha}(t-\tau)\,\sigma_{\psi,\tau}\mathcal{K}_{\psi,\alpha}(\tau-\tau_1) = 0. \tag{9.38}$$

Indeed, with the same notations we have:

$$\begin{aligned} &\int_{\Gamma_\tau} \mathcal{K}_{\psi,\alpha}(t-\tau)\,\sigma_{\psi,\tau}\mathcal{K}_{\psi,\alpha}(\tau-\tau_1) \\ &= \lim_{\varepsilon\to 0}\left[-\int_{\Lambda_\varepsilon} \mathcal{K}_{\psi,\alpha}(t-\tau)\,\sigma_{\psi,\tau}\left(-\mathcal{K}_{\psi,\alpha}(\tau-\tau_1)\right)\right] \\ &\quad - \lim_{\varepsilon\to 0}\left[-\int_{\gamma_\varepsilon} \mathcal{K}_{\psi,\alpha}(t-\tau)\,\sigma_{\psi,\tau}\left(-\mathcal{K}_{\psi,\alpha}(\tau-\tau_1)\right)\right]. \end{aligned}$$

Let $\beta_{\tau_1}(\tau) := -\mathcal{K}_{\psi,\alpha}(\tau-\tau_1)$. Since $\beta_{\tau_1} \in {}^{\psi}\mathfrak{M}_{\alpha}(\mathbb{R}^3\setminus\{\tau_1\};\ \mathbb{H}(\mathbb{C}))$ it follows that

$$-\int\limits_{\Lambda_\varepsilon} \mathcal{K}_{\psi,\alpha}(t-\tau)\,\sigma_{\psi,\tau}\beta_{\tau_1}(\tau)$$

$$= -\frac{1}{2}\beta_{\tau_1}(t) + \lim_{\substack{s\to t\\ s\in\Omega_\varepsilon^+}}\left[-\int\limits_{\Lambda_\varepsilon} \mathcal{K}_{\psi,\alpha}(s-\tau)\,\sigma_{\psi,\tau}\beta_{\tau_1}(\tau)\right]$$

$$= -\frac{1}{2}\beta_{\tau_1}(t) + \lim_{\substack{s\to t\\ s\in\Omega_\varepsilon^+}}\beta_{\tau_1}(s) = -\frac{1}{2}\mathcal{K}_{\psi,\alpha}(t-\tau_1).$$

In analogy to the above arguments it is proved that

$$\lim_{\varepsilon\to 0}\int\limits_{\gamma_\varepsilon} \mathcal{K}_{\psi,\alpha}(t-\tau)\,\sigma_{\psi,\tau}\mathcal{K}_{\psi,\alpha}(\tau-\tau_1) = -\frac{1}{2}\mathcal{K}_{\psi,\alpha}(t-\tau_1)$$

which leads to (9.38).

Consider now the following:

$$\int\limits_{\Gamma_\tau} \mathcal{K}_{\psi,\alpha}(t-\tau)\,\sigma_{\psi,\tau}\int\limits_{\Gamma_{\tau_1}} \mathcal{K}_{\psi,\alpha}(\tau-\tau_1)\,\sigma_{\psi,\tau_1}f(\tau_1,\tau)$$

$$= \int\limits_{\Gamma_\tau} \mathcal{K}_{\psi,\alpha}(t-\tau)\,\sigma_{\psi,\tau}\int\limits_{\Gamma_{\tau_1}} \mathcal{K}_{\psi,\alpha}(\tau-\tau_1)\,\sigma_{\psi,\tau_1}\left[f(\tau_1,\tau)-f(\tau,\tau)\right]$$

$$+\int\limits_{\Gamma_\tau} \mathcal{K}_{\psi,\alpha}(t-\tau)\,\sigma_{\psi,\tau}\int\limits_{\Gamma_{\tau_1}} \mathcal{K}_{\psi,\alpha}(\tau-\tau_1)\,\sigma_{\psi,\tau_1}\left[f(\tau,\tau)-f(t,t)\right]$$

$$+\int\limits_{\Gamma_\tau} \mathcal{K}_{\psi,\alpha}(t-\tau)\,\sigma_{\psi,\tau}\int\limits_{\Gamma_{\tau_1}} \mathcal{K}_{\psi,\alpha}(\tau-\tau_1)\,\sigma_{\psi,\tau_1}f(t,t).$$

On the right-hand side the integrals of the first and the second terms exist as improper integrals (because of f being a Hölder function), hence one can change the order of integration. The last term, by (9.37), is equal to $\frac{1}{4}f(t,t)$, and we have that

$$\int\limits_{\Gamma_\tau} \mathcal{K}_{\psi,\alpha}(t-\tau)\,\sigma_{\psi,\tau}\int\limits_{\Gamma_{\tau_1}} \mathcal{K}_{\psi,\alpha}(\tau-\tau_1)\,\sigma_{\psi,\tau_1}f(\tau_1,\tau)$$

$$
\begin{aligned}
&= \int\limits_{\Gamma_{\tau_1}}\int\limits_{\Gamma_\tau} \mathcal{K}_{\psi,\alpha}(t-\tau)\,\sigma_{\psi,\tau}\mathcal{K}_{\psi,\alpha}(\tau-\tau_1)\,\sigma_{\psi,\tau_1}\left[f(\tau_1,\tau)-f(\tau,\tau)\right] \\
&\quad + \int\limits_{\Gamma_{\tau_1}}\int\limits_{\Gamma_\tau} \mathcal{K}_{\psi,\alpha}(t-\tau)\,\sigma_{\psi,\tau}\mathcal{K}_{\psi,\alpha}(\tau-\tau_1)\,\sigma_{\psi,\tau_1}\left[f(\tau,\tau)-f(t,\ t)\right] \\
&\quad + \frac{1}{4}f(t,t) \\
&= \int\limits_{\Gamma_{\tau_1}}\int\limits_{\Gamma_\tau} \mathcal{K}_{\psi,\alpha}(t-\tau)\,\sigma_{\psi,\tau}\mathcal{K}_{\psi,\alpha}(\tau-\tau_1)\,\sigma_{\psi,\tau_1}f(\tau_1,\tau) \\
&\quad - \int\limits_{\Gamma_{\tau_1}}\left[\int\limits_{\Gamma_\tau} \mathcal{K}_{\psi,\alpha}(t-\tau)\,\sigma_{\psi,\tau}\mathcal{K}_{\psi,\alpha}(\tau-\tau_1)\right]\sigma_{\psi,\tau_1}f(t,\ t)+\frac{1}{4}f(t,t) \\
&= \int\limits_{\Gamma_{\tau_1}}\int\limits_{\Gamma_\tau} \mathcal{K}_{\psi,\alpha}(t-\tau)\,\sigma_{\psi,\tau}\mathcal{K}_{\psi,\alpha}(\tau-\tau_1)\,\sigma_{\psi,\tau_1}f(\tau_1,\tau)+\frac{1}{4}f(t,t)\,.
\end{aligned}
$$

Note that (9.38) was used here.

The Theorem is completely proved.

9.3.8 Let $f \in C^{0,\mu}(\Gamma\times\Gamma;\mathbb{H}(\mathbb{C}))$, $0<\mu<1$; set

$$
{}^{\psi}S_\alpha[f](t) := -2\int\limits_{\Gamma} \mathcal{K}_{\psi,\alpha}(t-\tau)\,\sigma_{\psi,\tau}f(\tau) \tag{9.39}
$$

for any $t\in\Gamma$. The corresponding operator ${}^{\psi}S_\alpha$ will be referred to as the Cauchy singular operator of the (ψ,α)-hyperholomorphic function theory. It can be proved that ${}^{\psi}S_\alpha$ acts invariantly on $C^{0,\mu}$ for $0<\mu<1$.

9.3.9 **Corollary** *(Involutiveness of the Cauchy singular operator of the (ψ,α)-hyperholomorphic function theory). For $0<\mu<1$, the operator ${}^{\psi}S_\alpha$ is an involution on $C^{0,\mu}(\Gamma;\mathbb{H}(\mathbb{C}))$:*

$$
{}^{\psi}S_\alpha{}^2 = I,
$$

I the identity operator on $C^{0,\mu}$.

Proof Indeed let $f(\tau_1,\ \tau)=f(\tau_1)$ do not depend on $\tau\in\Gamma$. Then by (9.36) and (9.38):

$$
\begin{aligned}
{}^{\psi}S_\alpha{}^2[f](t) &= 2\int\limits_{\Gamma_\tau} \mathcal{K}_{\psi,\alpha}(t-\tau)\,\sigma_{\psi,\tau}\left[2\int\limits_{\Gamma_{\tau_1}} \mathcal{K}_{\psi,\alpha}(\tau-\tau_1)\,\sigma_{\psi,\tau_1}f(\tau_1)\right] \\
&= 2\int\limits_{\Gamma_{\tau_1}}\left[2\int\limits_{\Gamma_\tau} \mathcal{K}_{\psi,\alpha}(t-\tau)\,\sigma_{\psi,\tau}\mathcal{K}_{\psi,\alpha}(\tau-\tau_1)\right]\sigma_{\psi,\tau_1}f(\tau_1) \\
&\quad + f(t,t) = f(t)\,.
\end{aligned}
$$

9.3.10 This Corollary was proved in [3] [Subsection 5.5] with completely different techniques. It implies immediately that the operators ${}^{\psi}P_{\alpha} := \frac{1}{2}\left(I + {}^{\psi}S_{\alpha}\right)$, ${}^{\psi}Q_{\alpha} := \frac{1}{2}\left(I - {}^{\psi}S_{\alpha}\right)$ are projections onto the traces of functions hyperholomorphic in $\Omega =: \Omega^{+}$ and $\Omega^{-} := \mathbb{R}^{3}\backslash\overline{\Omega}^{+}$ respectively. This gives, next, the necessary and sufficient conditions of $(\psi,\ \alpha)$-hyperholomorphic extension from the boundary of a given Hölder function (see [3], Subsections 5.5 and 5.9).

9.4 THE QUATERNIONIC MAXWELL OPERATOR

9.4.1 In [5], [4] (see also [3]), there was established an intimate relation between time-harmonic electromagnetic fields and hyperholomorphic functions. We start this Section with reproducing some details of that relation.

Equality (9.33) means, in particular, that on solenoidal vector fields D acts as the operator rot: if $f_0 = 0$, $div\,\vec{f} = 0$ then

$$D\left[\vec{f}\right] = rot\,\vec{f}. \tag{9.40}$$

Thus the following matrix operator (compare with (9.9)) arises quite naturally:

$$\mathcal{N} := \begin{pmatrix} \sigma & -D \\ D & -i\omega\mu \end{pmatrix}. \tag{9.41}$$

We shall consider it on the set $C^{1}\left(\Omega;\ Mat_{2\times 2}\left(\mathbb{H}(\mathbb{C})\right)\right)$, $Mat_{2\times 2}\left(\mathbb{H}(\mathbb{C})\right)$ being a set of 2×2 matrices with entries from $\mathbb{H}(\mathbb{C})$. Denote by $Mat_{2\times 1}$ its subset consisting of matrices of the form $\begin{pmatrix} a & 0 \\ b & 0 \end{pmatrix}$ which are identified naturally with columns $\begin{pmatrix} a \\ b \end{pmatrix}$. Abusing a little we shall not distinguish between $\begin{pmatrix} a & 0 \\ b & 0 \end{pmatrix}$ and $\begin{pmatrix} a \\ b \end{pmatrix}$.

It is important to observe that in [4] the operator $\mathcal{N}$ was considered just on $C^{1}\left(\Omega;\ Mat_{2\times 1}\right)$ but we shall not change the notation. Hence for us

$$\mathcal{N} : C^{1}\left(\Omega;\ Mat_{2\times 2}\left(\mathbb{H}(\mathbb{C})\right)\right) \to C^{0}\left(\Omega;\ Mat_{2\times 2}\left(\mathbb{H}(\mathbb{C})\right)\right).$$

Its restriction $\tilde{\mathcal{N}} := \mathcal{N}\Big|_{C^{1}\left(\Omega;\ Mat_{2\times 1}(\mathbb{C}^{3})\right)}$ onto, in fact, pairs of $\mathbb{C}^{3}$-valued functions has a form

$$\tilde{\mathcal{N}} = \begin{pmatrix} \sigma & div - rot \\ -div + rot & -i\omega\mu \end{pmatrix}$$

and hence does not coincide with $\mathcal{M}$ (moreover, $\tilde{\mathcal{N}}$ maps such pairs onto pairs of $\mathbb{H}(\mathbb{C})$-valued functions). But the restriction $\mathcal{N}\Big|_{\hat{C}^{1}}$ gives just $\hat{\mathcal{M}}$:

$$\hat{\mathcal{M}} = \mathcal{N}\Big|_{\hat{C}^{1}}.$$

By this reason $\mathcal{N}$ was termed "the quaternionic Maxwell operator" in [4] where the notation $\mathfrak{N}(\Omega) := \ker \mathcal{N}$ was introduced as well. We shall use also the notations

$$\tilde{\tilde{\mathcal{N}}} := \mathcal{N}\Big|_{C^1(\Omega;\ Mat_{2\times 1}(\mathbb{H}(\mathbb{C})))},\ \tilde{\mathfrak{N}}(\Omega) := \ker \tilde{\mathcal{N}},\ \tilde{\tilde{\mathfrak{N}}}(\Omega) := \ker \tilde{\tilde{\mathcal{N}}}.$$

Set

$$A_1 := \begin{pmatrix} \alpha & -\sigma \\ -\alpha & -\sigma \end{pmatrix},\ B_1 := \frac{1}{2}\begin{pmatrix} \sigma^{-1} & -\sigma^{-1} \\ \alpha^{-1} & \alpha^{-1} \end{pmatrix} \tag{9.42}$$

Then

$$A_1 * \mathcal{N} * B_1 = \begin{pmatrix} \overline{D}_\alpha & 0 \\ 0 & D_\alpha \end{pmatrix}. \tag{9.43}$$

Analogously for

$$A_2 := \begin{pmatrix} -\alpha & -\sigma \\ \alpha & -\sigma \end{pmatrix},\ B_2 := \frac{1}{2}\begin{pmatrix} -\sigma^{-1} & \sigma^{-1} \\ \alpha^{-1} & \alpha^{-1} \end{pmatrix} \tag{9.44}$$

one has

$$A_2 * \mathcal{N} * B_2 = \begin{pmatrix} D_\alpha & 0 \\ 0 & \overline{D}_\alpha \end{pmatrix}, \tag{9.45}$$

(all the matrices A_1, B_1, A_2, B_2 are invertible). This means, in particular, that

$$\ker \mathcal{N} = B_2\left(\ker \begin{pmatrix} D_\alpha & 0 \\ 0 & \overline{D}_\alpha \end{pmatrix}\right) = B_2\begin{pmatrix} \mathfrak{M}_\alpha & \mathfrak{M}_\alpha \\ \overline{\mathfrak{M}}_\alpha & \overline{\mathfrak{M}}_\alpha \end{pmatrix} \tag{9.46}$$

where it is clear what the new notation means. Hence there exists a one-to-one correspondence between elements of $\mathcal{N}$ and matrices of the form $\begin{pmatrix} f_{11} & f_{12} \\ f_{21} & f_{22} \end{pmatrix}$ with f_{11}, f_{12} being α-hyperholomorphic functions and f_{21}, f_{22} being α-"anti-hyperholomorphic":

$$\mathcal{N} = \left\{\begin{pmatrix} \frac{1}{2\sigma}(f_{21} - f_{11}) & \frac{1}{2\sigma}(f_{22} - f_{12}) \\ \frac{1}{2\alpha}(f_{21} + f_{11}) & \frac{1}{2\alpha}(f_{22} + f_{12}) \end{pmatrix}\right\}.$$

The "quaternionic Maxwell-Cauchy kernel", i.e. the fundamental solution of $\mathcal{N}$ is given by

$$\mathcal{K}_{\mathcal{N},\alpha} := B_1 * A_2 * \mathcal{N} * B_2 * A_1 * \begin{pmatrix} \theta_\lambda & 0 \\ 0 & \theta_\lambda \end{pmatrix}, \tag{9.47}$$

that is, $\mathcal{N}[\mathcal{K}_{\mathcal{N},\alpha}] = \delta\begin{pmatrix} 1 & 0 \\ 0 & 1 \end{pmatrix}$.

There exists a direct connection between $\mathcal{K}_{\mathcal{N},\alpha}$ and the corresponding hyperholomorphic Cauchy kernels $\mathcal{K}_\alpha$ and $\overline{\mathcal{K}}_\alpha$:

$$\begin{aligned} \mathcal{K}_{\mathcal{N},\alpha} &= \frac{1}{2}\begin{pmatrix} \sigma^{-1}\alpha\left(\overline{\mathcal{K}}_\alpha + \mathcal{K}_\alpha\right) & -\left(\overline{\mathcal{K}}_\alpha - \mathcal{K}_\alpha\right) \\ \overline{\mathcal{K}}_\alpha - \mathcal{K}_\alpha & -\alpha^{-1}\sigma\left(\overline{\mathcal{K}}_\alpha + \mathcal{K}_\alpha\right) \end{pmatrix} \\ &= \frac{1}{2}\begin{pmatrix} \sigma^{-1} & \sigma^{-1} \\ \alpha^{-1} & -\alpha^{-1} \end{pmatrix} * \begin{pmatrix} \alpha\overline{\mathcal{K}}_\alpha & -\sigma\overline{\mathcal{K}}_\alpha \\ \alpha\mathcal{K}_\alpha & \sigma\mathcal{K}_\alpha \end{pmatrix} \end{aligned}, \tag{9.48}$$

where $\overline{\mathcal{K}}_\alpha$ is the Cauchy kernel for $\overline{D}_\alpha$.

Let

$$C := \frac{1}{2}\begin{pmatrix} \sigma^{-1} & \sigma^{-1} \\ \alpha^{-1} & -\alpha^{-1} \end{pmatrix}.$$

One more notation is necessary:

$$\tilde{\sigma} := \begin{pmatrix} 0 & \overline{\sigma}_{st} \\ \sigma_{st} & 0 \end{pmatrix}$$

where σ_{st} is defined in Subsection 9.3.6.

9.4.2 **Theorem** *(Poincaré-Bertrand formula for the quaternionic Maxwell-Cauchy kernel). Let Γ be a Liapunov surface in $\mathbb{R}^3$ without boundary. Let $f \in C^{0,\mu}(\Gamma \times \Gamma; Mat_{2\times 2}(\mathbb{H}(\mathbb{C})))$ with $0 < \mu \leq 1$. The following equality holds everywhere on Γ:*

$$\int_{\Gamma_{\tau_1}}\int_{\Gamma_\tau} \mathcal{K}_{\mathcal{N},\alpha}(t-\tau) * \tilde{\sigma}_\tau * \mathcal{K}_{\mathcal{N},\alpha}(\tau-\tau_1) * \tilde{\sigma}_{\tau_1} * f(\tau,\tau_1) + \frac{1}{4}f(t,t)$$

$$= \int_{\Gamma_\tau}\int_{\Gamma_{\tau_1}} \mathcal{K}_{\mathcal{N},\alpha}(t-\tau) * \tilde{\sigma}_\tau * \mathcal{K}_{\mathcal{N},\alpha}(\tau-\tau_1) * \tilde{\sigma}_{\tau_1} * f(\tau,\tau_1). \qquad (9.49)$$

Proof We have:

$$\mathcal{K}_{\mathcal{N},\alpha}(t-\tau) * \tilde{\sigma}_\tau * \mathcal{K}_{\mathcal{N},\alpha}(\tau-\tau_1) * \tilde{\sigma}_{\tau_1}$$

$$= C * \begin{pmatrix} \alpha\overline{\mathcal{K}}_\alpha(t-\tau) & -\sigma\overline{\mathcal{K}}_\alpha(t-\tau) \\ \alpha\mathcal{K}_\alpha(t-\tau) & \sigma\mathcal{K}_\alpha(t-\tau) \end{pmatrix} * \begin{pmatrix} 0 & \overline{\sigma}_{\tau,st} \\ \sigma_{\tau,st} & 0 \end{pmatrix}$$

$$* \frac{1}{2}\begin{pmatrix} \sigma^{-1} & \sigma^{-1} \\ \alpha^{-1} & -\alpha^{-1} \end{pmatrix} * \begin{pmatrix} \alpha\overline{\mathcal{K}}_\alpha(\tau-\tau_1) & -\sigma\overline{\mathcal{K}}_\alpha(\tau-\tau_1) \\ \alpha\mathcal{K}_\alpha(\tau-\tau_1) & \sigma\mathcal{K}_\alpha(\tau-\tau_1) \end{pmatrix} * \begin{pmatrix} 0 & \overline{\sigma}_{\tau_1,st} \\ \sigma_{\tau_1,st} & 0 \end{pmatrix}$$

$$= C * \begin{pmatrix} \sigma\overline{\mathcal{K}}_\alpha(t-\tau)\overline{\sigma}_\tau\overline{\mathcal{K}}_\alpha(\tau-\tau_1)\overline{\sigma}_{\tau_1} & \alpha\overline{\mathcal{K}}_\alpha(t-\tau)\overline{\sigma}_\tau\overline{\mathcal{K}}_\alpha(\tau-\tau_1)\overline{\sigma}_{\tau_1} \\ \sigma\mathcal{K}_\alpha(t-\tau)\sigma_\tau\mathcal{K}_\alpha(\tau-\tau_1)\sigma_{\tau_1} & -\alpha\mathcal{K}_\alpha(t-\tau)\sigma_\tau\mathcal{K}_\alpha(\tau-\tau_1)\sigma_{\tau_1} \end{pmatrix}.$$

This means that, using (9.36), we have

$$\int_{\Gamma_\tau}\int_{\Gamma_{\tau_1}} \mathcal{K}_{\mathcal{N},\alpha}(t-\tau) * \tilde{\sigma}_\tau * \mathcal{K}_{\mathcal{N},\alpha}(\tau-\tau_1) * tilde\sigma_{\tau_1} * f(\tau,\ \tau_1)$$

$$= C * \int_{\Gamma_\tau}\int_{\Gamma_{\tau_1}} \begin{pmatrix} \sigma\overline{\Theta}_{t,\ \tau,\ \tau_1}f_{11} + \alpha\overline{\Theta}_{t,\ \tau,\ \tau_1}f_{21} & \sigma\overline{\Theta}_{t,\ \tau,\ \tau_1}f_{12} + \alpha\overline{\Theta}_{t,\ \tau,\ \tau_1}f_{22} \\ \sigma\Theta_{t,\ \tau,\ \tau_1}f_{11} - \alpha\Theta_{t,\ \tau,\ \tau_1}f_{21} & \sigma\Theta_{t,\ \tau,\ \tau_1}f_{12} - \alpha\Theta_{t,\ \tau,\ \tau_1}f_{22} \end{pmatrix}$$

$$= C * \left(\int_{\Gamma_{\tau_1}}\int_{\Gamma_\tau} \begin{pmatrix} \sigma\overline{\Theta}_{t,\ \tau,\ \tau_1}f_{11} + \alpha\overline{\Theta}_{t,\ \tau,\ \tau_1}f_{21} & \sigma\overline{\Theta}_{t,\ \tau,\ \tau_1}f_{12} + \alpha\overline{\Theta}_{t,\ \tau,\ \tau_1}f_{22} \\ \sigma\Theta_{t,\ \tau,\ \tau_1}f_{11} - \alpha\Theta_{t,\ \tau,\ \tau_1}f_{21} & \sigma\Theta_{t,\ \tau,\ \tau_1}f_{12} - \alpha\Theta_{t,\ \tau,\ \tau_1}f_{22} \end{pmatrix} \right.$$

$$+\begin{pmatrix} \sigma f_{11}+\alpha f_{21} & \sigma f_{12}+\alpha f_{22} \\ \sigma f_{11}-\alpha f_{21} & \sigma f_{12}-\alpha f_{22} \end{pmatrix}\Bigg)\Bigg)$$

$$=\int_{\Gamma_{\tau_1}}\int_{\Gamma_\tau} \mathcal{K}_{\mathcal{N},\alpha}(t-\tau)*\tilde{\sigma}_\tau*\mathcal{K}_{\mathcal{N},\alpha}(\tau-\tau_1)*\tilde{\sigma}_{\tau_1}*f(\tau,\tau_1)+\frac{1}{4}f(t,t),$$

where

$$\Theta_{t,\ \tau,\ \tau_1} \quad : \quad =\mathcal{K}_\alpha(t-\tau)\,\sigma_\tau\mathcal{K}_\alpha(\tau-\tau_1)\,\sigma_{\tau_1},$$
$$\overline{\Theta}_{t,\ \tau,\ \tau_1} \quad : \quad =\overline{\mathcal{K}}_\alpha(t-\tau)\,\overline{\sigma}_\tau\overline{\mathcal{K}}_\alpha(\tau-\tau_1)\,\overline{\sigma}_{\tau_1}.$$

9.4.3 Let $g\in C^{0,\mu}(\Gamma;\ Mat_{2\times2}(\mathbb{H}(\mathbb{C})))$, $0<\mu<1$; set

$$S_{\mathcal{N},\alpha}[g](t):=-2\int_\Gamma \mathcal{K}_{\mathcal{N},\alpha}(t-\tau)*\tilde{\sigma}_\tau*g(\tau) \tag{9.50}$$

for any $t\in\Gamma$. The corresponding operator $S_{\mathcal{N},\alpha}$ will be referred to as "the quaternionic Maxwell-Cauchy singular operator". From (9.48) it follows that

$$S_{\mathcal{N},\alpha}=\frac{1}{2}\begin{pmatrix} \sigma^{-1} & \sigma^{-1} \\ \alpha^{-1} & -\alpha^{-1} \end{pmatrix}*\begin{pmatrix} \sigma\overline{S}_\alpha & \alpha\overline{S}_\alpha \\ \sigma S_\alpha & -\alpha S_\alpha \end{pmatrix}, \tag{9.51}$$

where $\overline{S}_\alpha$ has the kernel $\overline{\mathcal{K}}_\alpha$. Together with Subsection 9.3.8 this means that $S_{\mathcal{N},\alpha}$ acts invariantly on $C^{0,\mu}(\Gamma;\ Mat_{2\times2}(\mathbb{H}(\mathbb{C})))$.

9.4.4 **Theorem** *(Involutiveness of the quaternionic Maxwell-Cauchy singular operator). For $0<\mu<1$, the operator $S_{\mathcal{N},\alpha}$ is an involution on $C^{0,\mu}(\Gamma;$ $Mat_{2\times2}(\mathbb{H}(\mathbb{C}))$:*

$$S^2_{\mathcal{N},\alpha}=I. \tag{9.52}$$

Proof It is clear from (9.51) that we need realizing some computations only.

$$S^2_{\mathcal{N},\alpha}=\frac{1}{4}\begin{pmatrix} \overline{S}_\alpha+S_\alpha & \frac{\alpha}{\sigma}(\overline{S}_\alpha-S_\alpha) \\ \frac{\sigma}{\alpha}(\overline{S}_\alpha-S_\alpha) & \overline{S}_\alpha+S_\alpha \end{pmatrix}^2$$

which gives (9.52) after taking Corollary 9.3.9 into account.

9.5 THE POINCARÉ-BERTRAND FORMULA FOR THE MAXWELL EQUATIONS

9.5.1 The aim of this section is to prove all facts from Section 1 using Theorem 9.4.2 and Theorem 9.4.4. In a sense this means that one has "to restrict" formulas (9.49) and (9.52) onto certain subsets of $C^{0,\mu}(\Gamma\times\Gamma;\ Mat_{2\times2}(\mathbb{H}(\mathbb{C})))$. For doing this, we shall use some specific "vectorial operations" with quaternionic matrices.

Let $Q,\ R \in Mat_{2\times 2}(\mathbb{H}(\mathbb{C}))$, denote $Q_0 := Sc(Q) := (Sc(q_{m,\ n}))_{1\le m,n\le 2}$, $\vec{Q} := Vec(Q) := (\vec{q}_{m,\ n})_{1\le m,n\le 2}$. Set

$$\left\langle \vec{Q},\ \vec{R}\right\rangle_{Mat} := \begin{pmatrix} \langle \vec{q}_{11},\ \vec{r}_{11}\rangle + \langle \vec{q}_{12},\ \vec{r}_{21}\rangle & \langle \vec{q}_{11},\ \vec{r}_{12}\rangle + \langle \vec{q}_{12},\ \vec{r}_{22}\rangle \\ \langle \vec{q}_{21},\ \vec{r}_{11}\rangle + \langle \vec{q}_{22},\ \vec{r}_{21}\rangle & \langle \vec{q}_{21},\ \vec{r}_{12}\rangle + \langle \vec{q}_{22},\ \vec{r}_{22}\rangle \end{pmatrix}, \tag{9.53}$$

$$\vec{Q} \underset{Mat}{\times} \vec{R} := \begin{pmatrix} \vec{q}_{11}\times\vec{r}_{11} + \vec{q}_{12}\times\vec{r}_{21} & \vec{q}_{11}\times\vec{r}_{12} + \vec{q}_{12}\times\vec{r}_{22} \\ \vec{q}_{21}\times\vec{r}_{11} + \vec{q}_{22}\times\vec{r}_{21} & \vec{q}_{21}\times\vec{r}_{12} + \vec{q}_{22}\times\vec{r}_{22} \end{pmatrix}. \tag{9.54}$$

This gives immediately the following analogue for the equality (9.22):

$$Q * R = Q_0 * R_0 - \left\langle \vec{Q},\ \vec{R}\right\rangle_{Mat} + \vec{Q} \underset{Mat}{\times} \vec{R} + Q_0 * \vec{R} + \vec{Q} * R_0. \tag{9.55}$$

Thus we have "matricial scalars" and "matricial vectors" as well as "matricial vector operations", all of them considered as "pieces" of "matricial quaternions". These objects possess a majority of usual properties of quaternions and vectors with some obvious differences; for instance: the multiplication of a "matricial vector" by a "matricial scalar" is not commutative in general; analogously the "matricial scalar product" is not commutative.

9.5.2 Let now $\mathcal{K}^0_{Mxw}(\xi,\zeta)$ denote the quaternionic matrix generated by $\mathcal{K}_{Mxw}(\xi,\zeta)$ (see (9.11)): $\mathcal{K}^0_{Mxw}(\xi,\zeta) := \langle grad\ \theta_\lambda(\xi-\zeta),\ \vec{n}_\zeta\rangle \begin{pmatrix} 1 & 0 \\ 0 & 1 \end{pmatrix} - \mathcal{U}(\xi-\zeta)$. One can verify directly that $\mathcal{K}_{\mathcal{N},\alpha}(\xi-\zeta) * \tilde{\sigma}_\zeta = \mathcal{K}^0_{Mxw}(\xi,\ \zeta)\, d\Gamma_\zeta$. Furthermore if $\vec{f} \in Mat_{2\times 1}(\mathbb{C}^3)$ then

$$\begin{aligned} \mathcal{K}_{\mathcal{N},\alpha}(\xi-\zeta) * \tilde{\sigma}_\zeta * \vec{f}(\zeta) &= \mathcal{K}^0_{Mxw} * \vec{f}(\zeta)\, d\Gamma_\zeta \\ &= \left(\left\langle \mathcal{U}(\xi-\zeta),\ \vec{f}(\zeta)\right\rangle_{Mat} + \mathcal{K}_{Mxw}(\xi,\ \zeta) \star \vec{f}(\zeta)\right) d\Gamma_\zeta, \end{aligned} \tag{9.56}$$

see (9.12) for definition of "$\star$".

If, in addition, $\vec{f}$ is a boundary value of a function from $\ker \mathcal{N}$ then, by the Cauchy integral formula, $\vec{f}(x) = K_{\mathcal{N},\alpha}\left[\vec{f}\right](x)$ for any $x \in \Omega$ which is equivalent to the system

$$0 = \int_\Gamma \left\langle \mathcal{U}(x-\tau),\ \vec{f}(\tau)\right\rangle_{Mat} d\Gamma_\tau, \tag{9.57}$$

$$\vec{f}(x) = \int_\Gamma \mathcal{K}_{Mxw}(x,\ \tau) \star \vec{f}(\tau)\, d\Gamma_\tau, \tag{9.58}$$

for any $x \in \Omega$. The condition (9.57) will prove to be important not only for traces of functions from $\ker \mathcal{N}$.

9.5.3 Applying (9.56) we get subsequently for $\vec{f} \in C^{0,\mu}(\Gamma\times\Gamma;\ Mat_{2\times 1}(\mathbb{C}^3))$:

$$\mathcal{K}_{\mathcal{N},\alpha}(t-\tau) * \tilde{\sigma}_\tau * \mathcal{K}_{\mathcal{N},\alpha}(\tau-\tau_1) * \tilde{\sigma}_{\tau_1} * f(\tau,\tau_1)$$

$$= \Big\{ \mathcal{K}_{\mathcal{N},\alpha}(t-\tau) * \tilde{\sigma}_\tau * \Big\langle \mathcal{U}(\tau-\tau_1),\ \vec{f}(\tau,\tau_1) \Big\rangle_{Mat}$$

$$+ \mathcal{K}_{\mathcal{N},\alpha}(t-\tau) * \tilde{\sigma}_\tau * \Big(\mathcal{K}_{Mxw}(\tau,\ \tau_1) \star \vec{f}(\tau,\tau_1) \Big) \Big\} d\Gamma_\tau d\Gamma_{\tau_1}$$

$$= \Big\{ \mathcal{K}^0_{Mxw}(t,\tau) * \Big\langle \mathcal{U}(\tau-\tau_1),\ \vec{f}(\tau,\tau_1) \Big\rangle_{Mat}$$

$$+ \mathcal{K}^0_{Mxw}(t,\ \tau) * \Big(\mathcal{K}_{Mxw}(\tau,\ \tau_1) \star \vec{f}(\tau,\ \tau_1) \Big) \Big\} d\Gamma_\tau d\Gamma_{\tau_1}$$

$$= \Big\{ \mathcal{K}^0_{Mxw}(t,\ \tau) * \Big\langle \mathcal{U}(\tau-\tau_1),\ \vec{f}(\tau,\ \tau_1) \Big\rangle_{Mat}$$

$$+ \Big\langle \mathcal{U}(\tau-\tau_1), \mathcal{K}_{Mxw}(\tau,\ \tau_1) \star\ \vec{f}(\tau,\ \tau_1) \Big\rangle$$

$$+ \mathcal{K}_{Mxw}(t,\ \tau) \star \Big(\mathcal{K}_{Mxw}(\tau,\tau_1) \star\ \vec{f}(\tau,\tau_1) \Big) \Big\} d\Gamma_\tau d\Gamma_{\tau_1}.$$

Substituting this into (9.49) and separating scalar and vector components we arrive at (9.13) and (9.14).

9.5.4 To prove Theorem 9.2.7 let us compare the operators $S_{\mathcal{N},\alpha}$ and S_{Mxw}. First of all, S_{Mxw} is, obviously, a restriction of $S_{\mathcal{N},\alpha}$ onto $C^{0,\mu}_{Mxw}$. Besides, the condition (9.16) is of the form $\Big\langle Vec(S_{\mathcal{N},\alpha}),\ \vec{f} \Big\rangle_{Mat}$, and one can apply reasonings from Subsection 3.8 of article [8] which leads to the invariance of S_{Mxw} on $C^{0,\mu}_{Mxw}$. It suffices now to apply Theorem 9.4.4 from above.

References

[1] Gürlebeck, K., Sprössig, W.: Quaternionic analysis and elliptic boundary value problems. Birkhäuser, Basel, 1990.

[2] Ilinski, A., Kravtsov, V., Sveshnikov, A.: Mathematical models of electrodynamics, Visshaya Shkola, Moscow, 1991 (Russian).

[3] Kravchenko, V., Shapiro, M.: Integral representations for spatial models of mathematical physics. Pitman Research Notes in Mathematics Series, 351, Longman, Harlow, 1996.

[4] Kravchenko, V., Shapiro, M.: Quaternionic time-harmonic Maxwell operator. J. Phys. A 28 (1995), 5017-5031.

[5] Kravchenko, V., Shapiro, M.: Hemholtz operator with a quaternionic wave number and associated function theory. Deformations of Mathematical Structures. Hurwitz-type structures and applications to surface physics. Ed. J. Lawrynowicz. Kluwer Academic Publishers, Dodrecht, 1993, 101-128.

[6] Mitelman, I., Shapiro, M.: Formulae of changing of integration order and of inversion for some multidimensional singular integrals and hypercomplex analysis. J. Natural Geometry 5 (1994), 11-27.

[7] Rocha-Chávez, R.: Fundamentación de algunos hechos del análisis cuaterniónico. ESFM-IPN, Mexico, 1997 (Spanish).

[8] Shapiro, M.: Structure of the quaternionic modules and some properties of the involutive operators. J. Natural Geometry 1 (1992), 9-37.

10 SYSTEMS OF FIRST ORDER PARTIAL DIFFERENTIAL EQUATIONS – A HYPERCOMPLEX APPROACH

Heinrich Begehr

Freie Universität Berlin
Fachbereich Mathematik und Informatik
I. Mathematisches Institut
Arnimallee 3
D-14195 Berlin
Germany
begehr@math.fu-berlin.de

Abstract: Hypercomplex analysis is useful for treating elliptic systems in plane domains. A modified hypercomplex Pompeiu operator is introduced leading to a singular hypercomplex integral operator. It serves to solve the Schwarz problem for the inhomogeneous hypercomplex Cauchy–Riemann equation. A hypercomplex approach is also used to solve some boundary value problem for a linear first order system in two complex variables.

Elliptic systems of first order partial differential equations in the plane can be treated in a natural way by utilizing a hyperanalytic approach. The advantage is that the main part of the system appears as a Cauchy–Riemann operator in a proper hypercomplex variable rather than in the complex variable z. This operator leads to the hyperanalytic functions introduced in [9]. If some lower order terms are included in the system one is lead to generalized analytic functions treated at first by Gilbert and Hile [14, 12], see [5, 6, 11, 13]. In the case of systems in several complex variables the hypercomplex approach can be used too, see [2, 10]. Systems in two complex variables are considered having analytic coefficients. They are reducible to a system of ordinary differential equations with respect to one of the variables and hence are equivalent to a system of Volterra integral equation. Solving this iteratively gives the solutions

H.G.W. Begehr et al. (eds.), Partial Differential and Integral Equations, 155–175.

to the system. Even weakly coupled systems can be treated and well–posed boundary value problems are available.

10.1 FIRST ORDER SYSTEMS IN THE PLANE

A first order system

$$Au_x + Bu_y = f(x, y, u) \tag{10.1}$$

in two real variables x, y with given real $n \times n$ matrices A, B, a given real n–dimensional vector–valued function f and the vector–valued unknown u is called elliptic if

$$\det A \neq 0 \ , \ \det[\lambda A + B] \neq 0 \ \text{ for all } \ \lambda \in I\!R \ .$$

It therefore can always be written in the form

$$u_x + Au_y = f(x, y, u) \tag{10.2}$$

where

$$\det[A + \lambda E] = 0 \ \text{ for all } \ \lambda \in I\!R \ . \tag{10.3}$$

Let B be a regular $n \times n$ matrix such that

$$B^{-1}AB = \Lambda = \begin{bmatrix} \Lambda_1 & & \\ & \ddots & \\ & & \Lambda_k \end{bmatrix} , \ \Lambda_\kappa = \begin{bmatrix} \lambda_\kappa & & 0 \\ 1 & \ddots & \\ 0 & 1 & \lambda_\kappa \end{bmatrix} , \ 1 \leq \kappa \leq k \ ,$$

is the Jordan normal form of A with $\lambda_\kappa \in \mathbb{C}$, and Λ_κ being an $n_\kappa \times n_\kappa$ matrix, $1 \leq \kappa \leq k, \sum_{\kappa=1}^{k} n_\kappa = n$. Introducing the complex vector $w = B^{-1}u$ gives

$$w_x + \Lambda w_y = f(x, y, w) \tag{10.4}$$

with an altered right–hand side f. Denoting $\lambda_\kappa = \mu_\kappa + i\nu_\kappa, w_\kappa = u_\kappa + iv_\kappa = (w_{\kappa 0}, \ldots, w_{\kappa n_\kappa - 1}), f_\kappa = g_\kappa + ih_\kappa = (f_{\kappa 0}, \ldots, f_{\kappa n_\kappa - 1}), w = (w_1, \ldots, w_k), f = (f_1, \ldots, f_k)$ the subsystem

$$w_{\kappa x} + \Lambda_\kappa w_{\kappa y} = f_\kappa \ , \tag{10.5}$$

in real form written component wise is

$$u_{\kappa 0x} + \mu_\kappa u_{\kappa 0y} - \nu_\kappa v_{\kappa 0y} = g_{\kappa 0}, v_{\kappa 0x} + \mu_\kappa v_{\kappa 0y} + \nu_\kappa u_{\kappa 0y} = h_{\kappa 0} \ ,$$

$$u_{\kappa \tau x} + \mu_\kappa u_{\kappa \tau y} - \nu_\kappa v_{\kappa \tau y} - u_{\kappa \tau - 1y} = g_{\kappa \tau},$$

$$v_{\kappa \tau x} + \mu_\kappa v_{\kappa \tau y} + \nu_\kappa u_{\kappa \tau y} - v_{\kappa \tau - 1y} = h_{\kappa \tau}, 1 \leq \tau \leq n_\kappa - 1 \ .$$

Introducing for fixed $\kappa, 1 \leq \kappa \leq k$, the new independent variables

$$\xi_\kappa := x, \eta_\kappa = -\frac{\mu_\kappa}{\nu_\kappa} x + \frac{1}{\nu_\kappa} y$$

such that

$$\frac{\partial}{\partial x} = \frac{\partial}{\partial \xi_\kappa} - \frac{\mu_\kappa}{\nu_\kappa}\frac{\partial}{\partial \eta_\kappa} , \quad \frac{\partial}{\partial y} = \frac{1}{\nu_\kappa}\frac{\partial}{\partial \eta_\kappa} , \quad \frac{\partial}{\partial x} + \mu_\kappa \frac{\partial}{\partial y} = \frac{\partial}{\partial \xi_\kappa}$$

leads to the system

$$u_{\kappa 0 \xi_\kappa} - v_{\kappa 0 \eta_\kappa} = g_{\kappa 0} , \quad v_{\kappa 0 \xi_\kappa} + u_{\kappa 0 \eta_\kappa} = h_{\kappa 0} ,$$

$$u_{\kappa \tau \xi_\kappa} - v_{\kappa \tau \eta_\kappa} - \frac{1}{\nu_\kappa} u_{\kappa \tau - 1 \eta_\kappa} = g_{\kappa \tau} ,$$

$$v_{\kappa \tau \xi_\kappa} + u_{\kappa \tau \eta_\kappa} - \frac{1}{\nu_\kappa} v_{\kappa \tau - 1 \eta_\kappa} = h_{\kappa \tau} , 1 \le \tau \le n_k - 1 .$$

In complex form this is with $z_\kappa = \xi_\kappa + i\eta_\kappa, 2\partial/\partial z_\kappa = \partial/\partial \xi_\kappa - i\partial/\partial \eta_\kappa, 2\partial/\partial \overline{z_\kappa} = \partial/\partial \xi_\kappa + i\partial/\partial \eta_\kappa$,

$$w_{\kappa 0 \overline{z_\kappa}} = \frac{1}{2} f_{\kappa 0} , \quad w_{\kappa \tau \overline{z_\kappa}} - \frac{i}{2\nu_\kappa} \left(w_{\kappa \tau - 1 z_\kappa} - w_{\kappa \tau - 1 \overline{z_\kappa}} \right) = \frac{1}{2} f_{\kappa \tau} , 1 \le \tau \le n_\kappa - 1 .$$

Returning to vector notation

$$w_{\kappa \overline{z_\kappa}} + \frac{i}{2\nu_\kappa} e_\kappa \left(w_{\kappa z_\kappa} - w_{\kappa \overline{z_\kappa}} \right) = \frac{1}{2} f_\kappa , \quad e_\kappa := \begin{bmatrix} 0 & & 0 \\ 1 & \ddots & \\ & \ddots & \ddots \\ 0 & 1 & 0 \end{bmatrix} , \tag{10.6}$$

where e_κ is a nilpotent $n_\kappa \times n_\kappa$ matrix satisfying $e_\kappa^{n_\kappa} = 0$. It is convenient to identify the vectors w_κ, f_κ with the respective hypercomplex numbers

$$w_\kappa = \sum_{\tau=0}^{n_\kappa - 1} w_{\kappa \tau} e^\tau , \quad f_\kappa = \sum_{\tau=0}^{n_\kappa - 1} f_{\kappa \tau} e^\tau .$$

Then (10.6) is a hypercomplex differential equation in the complex variables $\overline{z_\kappa}, z_\kappa$. In case f_κ depends only on w_κ and not on the other subvectors $w_\rho, 1 \le \rho \le k, \rho \ne \kappa$, of w, system (10.5) is a decomposed system of k subsystems which can be treated separately.

Let now f_κ depend only on $\overline{z_\kappa}, z_\kappa$. Then (10.6) is just a linear hypercomplex partial differential equation which will now be considered dropping the index κ. More generally we consider the equation

$$w_{\overline{z}} + q_1 w_z + q_2 w_{\overline{z}} = f \tag{10.7}$$

with given hypercomplex quantities

$$q_k = \sum_{\tau=1}^{n-1} q_{k\tau} e^\tau , k = 1, 2, f = \sum_{\tau=0}^{n-1} f_\tau e^\tau$$

and unknown hypercomplex w. Here $q_k, k = 1, 2$, is nilpotent. As (10.7) can be written as

$$w_{\overline{z}} + q w_z = f_0 , \quad q = \frac{q_1}{1 + q_2} , \quad f_0 = \frac{f}{1 + q_2}$$

we can assume $q_2 \equiv 0$. Following Douglis [9], see [11, 12, 14, 15] a generating solution for the differential operator

$$D := \frac{\partial}{\partial \overline{z}} + q_1 \frac{\partial}{\partial z} + q_2 \frac{\partial}{\partial \overline{z}} \tag{10.8}$$

is constructed having the form

$$t(z) = z + \tau(z)\ ,\ \tau(z) = \sum_{k=1}^{n-1} t_k(z) e^k \tag{10.9}$$

and satisfying $Dt = 0$. As the components of this hypercomplex equation are

$$\frac{\partial z}{\partial \overline{z}} = 0\ ,\ \frac{\partial t_k}{\partial \overline{z}} + q_{1k} + \sum_{\tau=1}^{k-1} \left[q_{1k-\tau} \frac{\partial t_\tau}{\partial z} + q_{2k-\tau} \frac{\partial t_\tau}{\partial \overline{z}} \right] = 0 \tag{10.10}$$

the t_k are determined inductively. A particular solution to the complex inhomogeneous Cauchy-Riemann equation $\partial t_k / \partial \overline{z} = f_k$ is given by the Pompeiu operator

$$Tf_k(z) := -\frac{1}{\pi} \int_G f_k(\zeta) \frac{d\xi d\eta}{\zeta - z} \tag{10.11}$$

provided $f_k \in L_p(\overline{G}), 1 < p$, in case the domain G under consideration is bounded and $f_k \in L_{p,2}(\overline{G})$ if G is unbounded, see [1, 5, 7, 17, 18]. The general solution is $\varphi + Tf_k$ with analytic function φ. Choosing $\varphi = 0$ then

$$t_k = -Tf_k\ ,\ f_k = q_{1k} + \sum_{\tau=1}^{k-1} \left[q_{1k-\tau} \frac{dt_\tau}{\partial z} + q_{2k-\tau} \frac{dt_\tau}{\partial \overline{z}} \right] \tag{10.12}$$

is determined for $1 \leq k \leq n-1$. In order to guarantee $f_k \in L_p(\overline{G})$ e.g. one has to assume $q_{k\tau} \in L_p(\overline{G})$ and use that

$$\frac{\partial T f(z)}{\partial z} = \Pi f(z) = -\frac{1}{\pi} \int_D f(\zeta) \frac{d\xi d\eta}{(\zeta - z)^2}$$

where Π is a Calderon–Zygmund operator mapping $L_p(\overline{G})$ into itself, see [17].

It is convenient for differentiable hypercomplex functions to introduce $(t, \overline{t})$ as new variables instead of $(z, \overline{z})$. The Jacobian of this transformation is

$$J := \begin{vmatrix} t_z & t_{\overline{z}} \\ \overline{t_{\overline{z}}} & \overline{t_{\overline{z}}} \end{vmatrix} = t_z \overline{t_z} - t_{\overline{z}} \overline{t_{\overline{z}}} = 1 + \tau_z + \overline{\tau_z} + \tau_z \overline{\tau_z} - \tau_{\overline{z}} \overline{\tau_{\overline{z}}} \tag{10.13}$$

$$= t_z \overline{t_z} \left(1 - \frac{q_1}{1 + q_2} \frac{\overline{q_1}}{1 + \overline{q_2}} \right) \neq 0\,.$$

For the respective differential operators one has

$$\frac{\partial}{\partial z} = t_z \frac{\partial}{\partial t} + \overline{t_{\overline{z}}} \frac{\partial}{\partial \overline{t}}\ ,\ \frac{\partial}{\partial \overline{z}} = t_{\overline{z}} \frac{\partial}{\partial t} + \overline{t_z} \frac{\partial}{\partial \overline{t}} \tag{10.14}$$

such that

$$\frac{\partial}{\partial t} = \frac{1}{J}\left(\overline{t_z}\frac{\partial}{\partial z} - \overline{t_{\overline{z}}}\frac{\partial}{\partial \overline{z}}\right), \ \frac{\partial}{\partial \overline{t}} = \frac{1}{J}\left(t_z\frac{\partial}{\partial \overline{z}} - t_{\overline{z}}\frac{\partial}{\partial z}\right). \tag{10.15}$$

From here $\partial t/\partial t = 1, \partial t/\partial \overline{t} = 0, \partial \overline{t}/\partial t = 0, \partial \overline{t}/\partial \overline{t} = 1$ follows which mean that $(t, \overline{t})$ can be considered as independent variables as $(z, \overline{z})$ can too. Moreover, as $q_1 t_z = t_z q_1$

$$\begin{aligned}\frac{\partial w}{\partial \overline{t}} = \frac{1}{J}(t_z w_{\overline{z}} - t_{\overline{z}} w_z) &= \frac{t_z}{J(1+q_2)}[(1+q_2)w_{\overline{z}} + q_1 w_z] \\ &= \frac{t_z}{J(1+q_2)}Dw = \frac{Dw}{D\overline{t}}.\end{aligned} \tag{10.16}$$

Hence, $Dw = 0$ is equivalent to $\partial w/\partial \overline{t} = 0$ so that w is an analytic function in $t = t(z) = t(z, \overline{z})$. And $Dw = f$ is equivalent to $\partial w/d\overline{t} = \hat{f}$ with $\hat{f} := t_z f/[J(1+q_2)]$.

As Douglis [9], see also [16], has pointed out the hyperanalytic functions, i.e. the solutions to $Dw = 0$ are given in the form $\varphi(t(z))$ with arbitrary complex–valued analytic function φ. Locally they are representable as power series in $t(z) - t(z_0)$,

$$\varphi(t(z)) = \sum_{k=0}^{\infty} a_k (t(z) - t(z_0))^k = \sum_{k=0}^{n-1} \frac{1}{k!}\varphi^{(k)}(z)(\tau(z) - \tau(z_0))^k. \tag{10.17}$$

In order to find the general solution to the inhomogeneous equation $Dw = f$ a particular solution has to be found. In analogy to the Pompeiu operator define

$$\widehat{T}f(z) := -\frac{1}{\pi}\int_G \frac{t_\zeta(\zeta)f(\zeta)}{1+q_2(\zeta)}\,\frac{d\xi d\eta}{t(\zeta) - t(z)}, \tag{10.18}$$

where according to (10.16) one has $t_\zeta(\zeta)D\overline{t(\zeta)} = J(\zeta)(1+q_2(\zeta))$.

The operator $\widehat{T}^*$ is attained from the following Cauchy–Pompeiu representation.

Lemma 1 *Let q_1, q_2 be continuously differentiable in the bounded smooth domain $G \subset \mathbb{C}$ and $w \in C(\overline{G}) \cup C^1/G)$. Then*

$$\begin{aligned}\int_{\partial D} w(z)dt(z) \quad &+ \quad \int_G \frac{\partial w(z)}{\partial \overline{t(z)}}dt(z)d\overline{t(z)} = 0, \\ \int_{\partial D} w(z)d\overline{t(z)} \quad &- \quad \int_G \frac{\partial w(z)}{dt(z)}dt(z)d\overline{t(z)} = 0,\end{aligned} \tag{10.19}$$

where $dt(z)d\overline{t(z)} = J(z)dzd\overline{z}$.

Proof For the last formula observe from the complex Gauß formula, see e.g. [1],

$$\int_{\partial G} w(z)d\overline{t(z)} = \int_{\partial G} w(z)(\overline{t_{\bar z}(z)}dz + \overline{t_z(z)}d\bar z)$$

$$= -\int_G [w_{\bar z}(z)\overline{t_z(z)} - w_z(z)\overline{t(z)}]dzd\bar z = \int_G J(z)\frac{\partial w(z)}{\partial t(z)}dzd\bar z.$$

The first formula follows in the same way.

Applying (19) for $G_\varepsilon := G\backslash\{(\zeta - z) \leq \varepsilon\}$ gives

$$\int_{\partial G_\epsilon} w(\zeta)\frac{dt(\zeta)}{t(\zeta)-t(z)} + \int_{G_\epsilon} \frac{\partial w(\zeta)}{\partial \overline{t(\zeta)}}\frac{dt(\zeta)d\overline{t(\zeta)}}{t(\zeta)-t(z)} = 0.$$

Letting ε tend to zero leads on the basis of, see [9],

$$\frac{dt(\zeta)}{t(\zeta)-t(z)} = \frac{d\zeta}{\zeta - z} + d\left(\sum_{\tau=1}^{n-1}\frac{(-1)^{k-1}}{k}\left(\frac{\tau(\zeta)-\tau(z)}{\zeta - z}\right)^k\right)$$

to the following representation.

Cauchy Pompeiu formula Let $G \subset \mathbb{C}$ be a bounded smooth domain and $w \in C(\overline{G}) \cup C^1(G)$. Then

$$w(z) = \frac{1}{2\pi i}\int_{\partial G} w(\zeta)\frac{dt(\zeta)}{t(\zeta)-t(z)} + \frac{1}{2\pi i}\int_G \frac{\partial w(\zeta)}{\partial \overline{t(\zeta)}}\frac{dt(\zeta)d\overline{t(\zeta)}}{t(\zeta)-t(z)},$$

$$w(z) = -\frac{1}{2\pi i}\int_{\partial G} w(\zeta)\frac{d\overline{t(\zeta)}}{\overline{t(\zeta)-t(z)}} + \frac{1}{2\pi i}\int_G \frac{\partial w(\zeta)}{\partial t(\zeta)}\frac{dt(\zeta)d\overline{t(\zeta)}}{\overline{t(\zeta)-t(z)}}. \tag{10.20}$$

The operator

$$\underset{\sim}{T} f(z) := \frac{1}{2\pi i}\int_G f(\zeta)\frac{dt(\zeta)d\overline{t(\zeta)}}{t(\zeta)-t(z)} = -\frac{1}{\pi}\int_G J(\zeta)f(\zeta)\frac{d\xi d\eta}{t(\zeta)-t(z)} \tag{10.21}$$

satisfies $\partial \underset{\sim}{T} f(z)/\partial\overline{t(z)} = f$, as can be proved as in [15]. Thus the operator $\widehat{T}$ from (10.18) because of (10.16) solves $Dw = f$. For simplicity in the following $q_2 = 0$ is assumed and the notation $q := q_1$ used.

Theorem 1 *For $G \in C^{1+\alpha}$ the operator $\underset{\sim}{T}$ is compact from $C^\alpha(\overline{D})$ into $C^{1+\alpha}(\overline{D})$. Moreover,*

$$\partial \underset{\sim}{T} f/\partial\bar t = f\ ,\ \partial \underset{\sim}{T} f/\partial t = \underset{\sim}{\Pi} f,\ \underset{\sim}{\Pi} f(z) := \frac{1}{2\pi i}\int_G f(\zeta)\frac{dt(\zeta)d\overline{t(\zeta)}}{(t(\zeta)-t(z))^2}, \tag{10.22}$$

$\underset{\sim}{\Pi}$ *being a singular operator from* $C^\alpha(\overline{G})$ *into* $C^\alpha(\overline{G})$ *to be interpreted as a Cauchy principle value integral operator.*

Proof 1. For $f \in C^\alpha(\overline{G})$ and $G_\varepsilon := G\backslash\{\zeta : |\zeta - z| \leq \varepsilon\}$

$$\begin{aligned}
\underset{\sim}{\Pi} f(z) &= \lim_{\varepsilon\to 0} \frac{1}{2\pi i} \int_{G_\varepsilon} f(\zeta) \frac{dt(\zeta)d\overline{t(\zeta)}}{(t(\zeta)-t(z))^2} \\
&= \lim_{\varepsilon\to 0} \frac{1}{2\pi i} \int_{G_\varepsilon} \left[f(\zeta)J(\zeta)\left(\frac{\zeta - z}{t(\zeta)-t(z)}\right)^2 - \frac{f(z)J(z)}{(t_z(z))^2}\right] \frac{d\zeta d\overline{\zeta}}{(\zeta - z)^2} \\
&\quad + \frac{f(z)J(z)}{t_z^2(z)} \lim_{\varepsilon\to 0} \frac{1}{2\pi i} \int_{G_\varepsilon} \frac{d\zeta d\overline{\zeta}}{(\zeta - z)^2},
\end{aligned}$$

$$\frac{1}{2\pi i}\int_{G_\varepsilon} \frac{d\zeta d\overline{\zeta}}{(\zeta - z)^2} = -\frac{1}{2\pi i}\int_{\partial G_\varepsilon} \frac{d\overline{\zeta}}{\zeta - z} = -\frac{1}{2\pi i}\int_{\partial G} \frac{\overline{\zeta} d\zeta}{(\zeta - z)^2} = -\phi_0'(z),$$

$$\phi_0(z) := \frac{1}{2\pi i}\int_{\partial G} \frac{\overline{\zeta} d\zeta}{\zeta - z}.$$

Thus

$$\begin{aligned}
\underset{\sim}{\Pi} f(z) &= \frac{1}{2\pi i} \int_{G} \left[f(\zeta)J(\zeta)\left(\frac{\zeta - z}{t(\zeta)-t(z)}\right)^2 - \frac{f(z)J(z)}{t_z^2(z)}\right] \frac{d\zeta d\overline{\zeta}}{(\zeta - z)^2} \\
&\quad - \frac{f(z)J(z)}{t_z^2(z)} \phi_0'(z).
\end{aligned}$$

Because by the assumption on q and f the function

$$\begin{cases} f(\zeta)J(\zeta)\left(\dfrac{\zeta - z}{t(\zeta)-t(z)}\right)^2, & \zeta \neq z, \\ \dfrac{f(z)J(z)}{t_z^2(z)}, & \zeta = z, \end{cases}$$

is Hölder continuous in $\overline{D}$ so that the last integral exists in the improper sense. Hence, $\underset{\sim}{\Pi} f$ exists as a Cauchy principal value integral.

2. Next $\underset{\sim}{\Pi} f$ is shown to belong to $C^\alpha(\overline{G})$ for $f \in C^\alpha(\overline{G})$.

$$\begin{aligned}\underset{\sim}{\Pi} f(z_2) - \underset{\sim}{\Pi} f(z_1) &= \frac{t(z_2)-t(z_1)}{2\pi i} \int_G \frac{(f(\zeta)-f(z_2))dt(\zeta)\overline{dt(\zeta)}}{(t(\zeta)-t(z_1))(t(\zeta)-t(z_2))^2} \\ &+ \frac{t(z_2)-t(z_1)}{2\pi i} \int_G \frac{(f(\zeta)-f(z_1))dt(\zeta)\overline{dt(\zeta)}}{(t(\zeta)-t(z_1))^2(t(\zeta)-t(z_2))} \\ &+ \frac{t(z_2)-t(z_1)}{2\pi i} \int_G \frac{f(z_2)dt(\zeta)\overline{dt(\zeta)}}{(t(\zeta)-t(z_1))(t(\zeta)-t(z_2))^2} \\ &+ \frac{t(z_2)-t(z_1)}{2\pi i} \int_G \frac{f(z_1)dt(\zeta)\overline{dt(\zeta)}}{(t(\zeta)-t(z_1))^2(t(\zeta)-t(z_2))},\end{aligned}$$

$$\begin{aligned}&\int_G \frac{(t(z_2)-t(z_1))dt(\zeta)\overline{dt(\zeta)}}{(t(\zeta)-t(z_1))(t(\zeta)-t(z_2))^2} \\ &= \int_G \left\{\frac{1}{(t(\zeta)-t(z_2))^2} - \frac{1}{t(z_2)-t(z_1)}\left(\frac{1}{t(\zeta)-t(z_2)} - \frac{1}{t(\zeta)-t(z_1)}\right)\right\} dt(\zeta)\overline{dt(\zeta)}.\end{aligned}$$

With the hyperanalytic function

$$\phi(z) := \frac{1}{2\pi i}\int_{\partial G} \frac{\overline{t(\zeta)}dt(\zeta)}{t(\zeta)-t(z)} = \overline{t(z)} - \frac{1}{2\pi i}\int_G \frac{dt(\zeta)\overline{dt(\zeta)}}{t(\zeta)-t(z)},$$

we have

$$\begin{aligned}-\frac{1}{2\pi i}\int_{G_\epsilon} \frac{dt(\zeta)\overline{dt(\zeta)}}{(t(\zeta)-t(z))^2} &= \frac{1}{2\pi i}\int_{\partial G_\epsilon} \frac{\overline{dt(\zeta)}}{t(\zeta)-t(z)} \\ &= \frac{\partial}{\partial t(z)} \frac{1}{2\pi i}\int_{\partial G_\epsilon} \frac{\overline{t(\zeta)}dt(\zeta)}{t(\zeta)-t(z)} = \frac{d\phi(z)}{dt(z)}\end{aligned}$$

because

$$\begin{aligned}&\lim_{\varepsilon\to 0} \frac{1}{2\pi i}\int_{|\zeta-z|=\varepsilon} \frac{\overline{t(\zeta)}dt(\zeta)}{t(\zeta)-t(z)} \\ &= \lim_{\varepsilon\to 0} \frac{1}{2\pi i}\int_{|\zeta-z|=\varepsilon} \overline{t(\zeta)}t_\zeta(\zeta)\frac{(\zeta-z)}{t(\zeta)-t(z)}\frac{d\zeta - q(\zeta)d\overline{\zeta}}{\zeta-z} = \overline{t(z)}.\end{aligned}$$

Hence,

$$\frac{t(z_2)-t(z_1)}{2\pi i}\int_G \frac{dt(\zeta)\overline{dt(\zeta)}}{(t(\zeta)-t(z_1))(t(\zeta)-t(z_2))^2}$$

$$= -\frac{d\phi(z_2)}{dt(z_2)} + \frac{1}{t(z_2) - t(z_1)} \left[\phi(z_2) - \phi(z_1) - \overline{t(z_2)} + \overline{t(z_1)}\right].$$

By the assumptions

$$\frac{d\phi(z)}{dt(z)} = \frac{1}{2\pi i} \int\limits_{\partial G} \frac{d\overline{t(\zeta)}}{t(\zeta) - t(z)} = \frac{1}{2\pi i} \int\limits_{\partial G} \frac{\overline{t_\zeta(\zeta)(\zeta'(s)} - \overline{q(\zeta)\zeta'(s))}}{\zeta'(s)(t(\zeta) - t(z))/(\zeta - z)} \frac{d\zeta}{\zeta - z}$$

is Hölder continuous. Moreover, see [1], p. 87,

$$\left| \frac{1}{2\pi i} \int\limits_G \frac{(f(\zeta) - f(z_2))dt(\zeta)d\overline{t(\zeta)}}{(t(\zeta) - t(z_1))(t(\zeta) - t(z_2))^2} \right|$$
$$\leq \frac{H_\alpha(f)}{2\pi i} \int\limits_G \frac{|J(\zeta)|d\zeta d\bar{\zeta}}{\left|\frac{t(\zeta) - t(z_1)}{\zeta - z_1}\right| \left|\frac{t(\zeta) - t(z_2)}{\zeta - z_2}\right|^2 |\zeta - z_1||\zeta - z_2|^{2-\alpha}}$$
$$\leq H_\alpha(f) M(\alpha, q)|z_1 - z_2|^{\alpha - 1}.$$

Therefore

$$\left| \underset{\sim}{\Pi} f(z_2) - \underset{\sim}{\Pi} f(z_1) \right| \leq H_\alpha(f) M(\alpha, q)|z_1 - z_2|^\alpha$$
$$+ |f(z_2) - f(z_1)| \left| \frac{d\phi(z_2)}{dt(z_2)} - \frac{\phi(z_2) - \phi(z_1)}{t(z_2) - \bar{t}(z_1)} + \frac{\overline{t(z_2) - t(z_1)}}{t(z_2) - t(z_1)} \right|$$
$$+ |f(z_1)| \left| \frac{d\phi(z_2)}{d\bar{t}(z_2)} - \frac{d\phi(z_1)}{dt(z_1)} \right|$$

which shows the Hölder continuity of $\underset{\sim}{\Pi} f$.

3. For the derivatives of $\underset{\sim}{T} f$ observe

$$\frac{\underset{\sim}{T} f(z) - \underset{\sim}{T} f(z_0)}{t(z) - t(z_0)} - \underset{\sim}{\Pi} f(z_0) = \frac{1}{2\pi i} \frac{1}{t(z) - t(z_0)}$$
$$\times \int\limits_G \left[\frac{1}{t(\zeta) - t(z)} - \frac{1}{t(\zeta) - t(z_0)} - \frac{t(z) - t(z_0)}{(t(\zeta) - t(z_0))^2} \right] f(\zeta) dt(\zeta) d\overline{t(\zeta)}$$
$$= \frac{t(z) - t(z_0)}{2\pi i} \int\limits_G \frac{f(\zeta) - f(z_0)}{(t(\zeta) - t(z_0))^2 (t(\zeta) - t(z))} dt(\zeta) d\overline{t(\zeta)}$$
$$+ f(z_0) \left[\frac{d\phi(z_0)}{dt(z_0)} - \frac{\phi(z) - \phi(z_0)}{t(z) - t(z_0)} + \frac{\overline{t(z) - t(z_0)}}{t(z) - t(z_0)} \right]$$

so that

$$\left|\frac{\underset{\sim}{T} f(z) - \underset{\sim}{T} f(z_0)}{t(z)-t(z_0)} - \underset{\sim}{\Pi} f(z_0) - \frac{\overline{t(z)-t(z_0)}}{t(z)-t(z_0)} f(z_0)\right|$$

$$\leq H_\alpha(f) M(\alpha, q)|z-z_0|^\alpha + |f(z_0)| \left|\frac{\phi(z)-\phi(z_0)}{t(z)-t(z_0)} - \frac{d\phi(z_0)}{dt(z_0)}\right|.$$

Let $z = z_0 + \varepsilon e^{i\varphi}$ and ε tend to zero then for $\varphi = 0$ one gets for $z = z_0$

$$\frac{\partial \underset{\sim}{T} f(z)}{\partial x} \Big/ \frac{\partial t(z)}{\partial x} - \underset{\sim}{\Pi} f(z) - f(z) \frac{\partial \overline{t(z)}}{\partial x} \Big/ \frac{\partial t(z)}{\partial x} = 0$$

i.e.

$$\frac{\partial \underset{\sim}{T} f(z)}{\partial t(z)} + \frac{\partial \underset{\sim}{T} f(z)}{\partial \overline{t(z)}} \frac{\partial \overline{t(z)}}{\partial x} \Big/ \frac{\partial t(z)}{\partial x} - \underset{\sim}{\Pi} f(z) - f(z) \frac{\partial \overline{t(z)}}{\partial x} \Big/ \frac{\partial t(z)}{\partial x} = 0$$

and for $\varphi = \pi/2$

$$\frac{\partial \underset{\sim}{T} f(z)}{\partial y} \Big/ \frac{\partial t(z)}{\partial y} - \underset{\sim}{\Pi} f(z) - f(z) \frac{\partial \overline{t(z)}}{\partial y} \Big/ \frac{\partial t(z)}{\partial y} = 0$$

i.e.

$$\frac{\partial \underset{\sim}{T} f(z)}{\partial t(z)} + \frac{\partial \underset{\sim}{T} f(z)}{\partial \overline{t(z)}} \frac{\partial \overline{t(z)}}{\partial y} \Big/ \frac{\partial t(z)}{\partial y} - \underset{\sim}{\Pi} f(z) - f(z) \frac{\partial \overline{t(z)}}{\partial y} \Big/ \frac{\partial t(z)}{\partial y} = 0.$$

Subtracting both equations gives

$$\left[\frac{\partial \underset{\sim}{T} f(z)}{\partial \overline{t(z)}} - f(z)\right] \left[\frac{\partial \overline{t(z)}}{\partial x} \Big/ \frac{\partial t(z)}{\partial x} - \frac{\partial \overline{t(z)}}{\partial y} \Big/ \frac{\partial t(z)}{\partial y}\right] = 0.$$

As

$$\frac{\partial \overline{t(z)}}{\partial x} \frac{\partial t(z)}{\partial y} - \frac{\partial t(z)}{\partial x} \frac{\partial \overline{t(z)}}{\partial y} = 2i\, J,$$

$$\frac{\partial t(z)}{\partial x} \frac{\partial t(z)}{\partial y} = i \left[\left(\frac{\partial t(z))}{\partial z}\right)^2 - \left(\frac{\partial t(z)}{\partial \bar{z}}\right)^2\right] = i \left(\frac{\partial t(z)}{\partial z}\right)^2 (1-q^2) \neq 0$$

we have

$$\frac{\partial \underset{\sim}{T} f(z)}{\partial \overline{t(z)}} = f\,, \quad \frac{\partial \underset{\sim}{T} f(z)}{\partial t(z)} = \underset{\sim}{\Pi} f.$$

4. For the compactness assume more generally $f \in L_p(\overline{G}), 2 < p$. Then

$$\underset{\sim}{T} f(z) = \frac{1}{2\pi i} \int_G J(\zeta) f(\zeta) \frac{\zeta - z}{t(\zeta) - t(z)} \frac{d\zeta d\overline{\zeta}}{\zeta - z},$$

$$\underset{\sim}{T} f(z_2) - \underset{\sim}{T} f(z_1) = \frac{1}{2\pi i} \int_G f(\zeta) \frac{t(z_2) - t(z_1)}{(t(\zeta) - t(z_1))(t(\zeta) - t(z_2))} dt(\zeta) d\overline{t(\zeta)}.$$

Because of

$$M^{-1} \leq \left| \frac{t(z_2) - t(z_1)}{z_2 - z_1} \right| \leq M \,, \quad z_1, z_2 \in \overline{G}, \ |J(z)| \leq M, \ z \in \overline{G},$$

for proper constant $M \geq 1$, see [9],

$$\begin{aligned} |\underset{\sim}{T} f(z)| &\leq M(p,q,G) \|f\|_{L_p(\overline{G})}, \\ \left| \underset{\sim}{T} f(z_2) - \underset{\sim}{T} f(z_1) \right| &\leq \frac{M_0}{2\pi i} \int_G |f(\zeta)| \frac{|z_2 - z_1|}{|\zeta - z_1||\zeta - z_2|} d\zeta d\overline{\zeta} \\ &\leq M(p,q) \|f\|_{L_p(\overline{G})} |z_1 - z_2|^{(p-2)/p}, \ z_1, z_2 \in \overline{G}, \end{aligned}$$

see [17]. Hence,

$$\begin{aligned} C_{1+\alpha}(Tf; \overline{G}) &= C_0(Tf; \overline{G}) + C_\alpha((Tf)_z; \overline{G}) + C_\alpha((Tf)_{\overline{z}}); \overline{G}) \\ &\leq M(\alpha, q, G) C_\alpha(f; \overline{G}). \end{aligned}$$

As $\underset{\sim}{T} f$ turns out to be a particular solution to the hypercomplex inhomogeneous Cauchy–Riemann equation

$$w_{\overline{t}} = f \tag{10.23}$$

the general solution to this equation is

$$w = \varphi + \underset{\sim}{T} f \tag{10.24}$$

with an arbitrary hyperanalytic function. φ can be determined by proper boundary valule conditions on w. In [3, 4] this was done with the Riemann–Hilbert boundary condition for solutions to more general first order systems namely for generalized hyperanalytic functions, see also [6, 13]. Exemplarily here only the Schwarz problem

$$\operatorname{Re} w = \gamma \quad \text{on} \quad \partial G \tag{10.25}$$

will be considered for the simple equation (10.23). Moreover, G is assumed to be the unit disc $\boldsymbol{D}$. This means for the hyperanalytic φ

$$\operatorname{Re} \varphi = \gamma - \operatorname{Re} \underset{\sim}{T} f \ \text{ on } \partial \boldsymbol{D}. \tag{10.26}$$

In [8] the solution to this problem is shown to be

$$\varphi = (S + 2\, \underset{\sim}{T}_q C')(\gamma - \operatorname{Re}\, \underset{\sim}{T} f) + ic \tag{10.27}$$

where c is an arbitrary hyperreal constant, $c \in \mathcal{A}_\nu$, S is the Schwarz operator

$$S\gamma(z) := \frac{1}{2\pi i} \int_{\partial \mathbb{D}} \gamma(\zeta) \frac{\zeta + z}{\zeta - z} \frac{d\zeta}{\zeta}, \tag{10.28}$$

C' is the derivative of the Cauchy operator

$$C'\gamma(z) := \frac{1}{2\pi i} \int_{\partial \mathbb{D}} \gamma(\zeta) \frac{d\zeta}{(\zeta - z)^2} \tag{10.29}$$

and $\underset{\sim}{T}_q$ is the hypercomplex integral operator

$$\underset{\sim}{T}_q := \sum_{k=1}^{n-1} e^k \sum_{\rho=1}^{k} (-1)^\rho \sum_{\nu_1 + \ldots + \nu_\rho = k} \widetilde{T}_{q_{\nu_1}} \widetilde{\Pi}_{q_{\nu_2}} \ldots \widetilde{\Pi}_{q_{\nu_\rho}}. \tag{10.30}$$

Here the last sum has to be taken for all $\nu_\sigma, 1 \leq \nu_\sigma \leq k, 1 \leq \sigma \leq \rho$ with $\nu_1 + \ldots + \nu_\rho = k$ and

$$\widetilde{T}_q f(z) := \frac{1}{2\pi i} \int_{\mathbb{D}} \left(\frac{q(\zeta) f(\zeta)}{\zeta - z} + \frac{z\overline{q(\zeta) f(\zeta)}}{1 - z\overline{\zeta}} \right) d\zeta d\overline{\zeta} \tag{10.31}$$

and

$$\widetilde{\Pi}_q f(z) := \frac{\partial}{\partial z} \widetilde{T}_q f(z) = \frac{1}{2\pi i} \int_{\mathbb{D}} \left(\frac{q(\zeta) f(\zeta)}{(\zeta - z)^2} + \frac{\overline{q(\zeta) f(\zeta)}}{(1 - z\overline{\zeta})^2} \right) d\zeta d\overline{\zeta}. \tag{10.32}$$

Instead of proving (10.27) it will be verified to solve problem (10.26). To this end observe

$$\begin{aligned} \frac{\partial}{\partial \overline{z}} \underset{\sim}{T}_q &= \sum_{k=1}^{n-1} e^k \sum_{\rho=1}^{k} (-1)^\rho \sum_{\nu_1 + \ldots + \nu_\rho = k} q_{\nu_1} \widetilde{\Pi}_{q_{\nu_2}} \ldots \widetilde{\Pi}_{q_{\nu_\rho}}, \\ \frac{\partial}{\partial z} \underset{\sim}{T}_q &= \underset{\sim}{\Pi}_q := \sum_{k=1}^{n-1} e^k \sum_{\rho=1}^{k} (-1)^\rho \sum_{\nu_1 + \ldots + \nu_\rho = k} \widetilde{\Pi}_{q_{\nu_1}} \ldots \widetilde{\Pi}_{q_{\nu_\rho}}, \end{aligned}$$

from which

$$\frac{\partial}{\partial \overline{z}} \underset{\sim}{T}_q C' \tilde{\gamma} = -q(C'\tilde{\gamma} + \underset{\sim}{\Pi}_q C'\tilde{\gamma}) = -q \frac{\partial}{\partial z} (C + \underset{\sim}{T}_q C') \tilde{\gamma}$$

i.e.

$$D(C + \underset{\sim}{T}_q C')\tilde{\gamma} = 0$$

for any $\tilde{\gamma} \in C(\partial \boldsymbol{D}; \mathcal{A})$. Thus the hypercomplex operator $C + \underset{\sim}{T}_q C'$ maps $C(\partial \boldsymbol{D}; \mathcal{A})$, into the set of in $\boldsymbol{D}$ hyperanalytic functions. This is true for $S + 2\underset{\sim}{T}_q C'$ too because

$$S\tilde{\gamma} = 2C\tilde{\gamma} + \frac{1}{2\pi i}\int_{\partial \boldsymbol{D}} \tilde{\gamma}(\zeta)\frac{d\zeta}{\zeta}, \frac{\partial}{\partial z}S\tilde{\gamma} = 2C'\tilde{\gamma}$$

so that $(S + 2\underset{\sim}{T}_q C')\gamma$ differs from $(C + \underset{\sim}{T}_q C')\gamma$ only by some constant term.

That $\operatorname{Re}(S + 2\underset{\sim}{T}_q C')\tilde{\gamma}$ coinsides with $\tilde{\gamma}$ on $\partial \boldsymbol{D}$ for hyperreal $\tilde{\gamma}$ follows from the respective property of the Schwarz operators and from

$$\operatorname{Re} \underset{\sim}{T}_q f = 0 \text{ on } \partial \boldsymbol{D}.$$

Thus the solution to (10.23), (10.25) for $G = \boldsymbol{D}$ is

$$w = (S + 2\underset{\sim}{T}_q C')(\gamma - \operatorname{Re} \underset{\sim}{T} f) + \underset{\sim}{T} f + ic. \tag{10.33}$$

Theorem 2 *Let* $f \in L_p(\overline{\boldsymbol{D}}; \mathcal{A}), 2 < p$, *and* $\gamma \in C(\partial \boldsymbol{D}; \mathcal{A}_r)$. *Then the problem* (10.23), (10.25) *for the unit disc* $\boldsymbol{D}$ *is solvable. The solution is with some arbitrary hyperreal constant* $c, c \in \mathcal{A}_r$, *given by* (10.33) *where the integral operators involved are given by* (10.21), (10.28)-(10.30).

In [8] this result is used to solve boundary value problems for some second order systems.

10.2 FIRST ORDER SYSTEMS IN $\mathbb{C}^2$ WITH ANALYTIC COEFFICIENTS

Similarly as system (10.1) in the two real variables x and y can be reduced to system (10.4) this can be done in case of complex variables. The first order system

$$w_{z_2} + Aw_{z_1} = f(z_1, z_2, w) \tag{10.34}$$

with an $N \times N$ matrix A with analytic entries in z_1 and $z_2, (z_1, z_2) \in G_0 \subset \mathbb{C}^2$, an N–dimensional vector–function f with analytic components and the unknown N–dimensional vector w can be reduced to the form

$$w_{z_2} + \Lambda w_{z_1} = \tilde{f}(z_1, z_2, w) \tag{10.35}$$

where Λ is the Jacobian normal form of A. This will be possible if the eigenvalues of A have constant multiplicities throughout G_0, see [1],[2],[10]. In the following the particular case is investigated where Λ is just one block matrix

$$\Lambda = \begin{bmatrix} \lambda & & 0 \\ 1 & \ddots & \\ & \ddots & \ddots \\ 0 & 1 & \lambda \end{bmatrix},$$

$\lambda = \lambda(z_1, z_2)$ is an analytic function in G_0 and $\tilde{f}$ is independent of w. This case is essential for treating the general system (10.35). In component form (10.35) then is

$$w_{0z_2} + \lambda w_{0z_1} = \tilde{f}_0, \ w_{\nu z_2} + \lambda w_{\nu z_1} + w_{\nu-1 z_1} = \tilde{f}_\nu, \ 1 \le \nu \le n-1. \quad (10.35')$$

Assuming the differential form $dz_1 - \lambda dz_2$ to have an integrating factor $\mu = \mu(z_1, z_2)$ different from zero and analytic in D_0 then

$$d\zeta_1 := \mu dz_1 - \mu\lambda dz_2$$

is a total differential of a function $\zeta_1 = \zeta_1(z_1, z_2)$. Choosing $\zeta_2 = \zeta_2(z_1, z_2) = z_2$ gives a mapping from G_0 into $\mathbb{C}^2$ by the analytic vector-valued function (ζ_1, ζ_2) the Jacobian of which is

$$\begin{vmatrix} \zeta_{1z_1} & \zeta_{1z_2} \\ \zeta_{2z_1} & \zeta_{2z_2} \end{vmatrix} = \begin{vmatrix} \mu & -\mu\lambda \\ 0 & 1 \end{vmatrix} = \mu \neq 0.$$

This means locally there is an inverse mapping $z_1 = z_1(\zeta_1, \zeta_2), z_2 = \zeta_2$ the Jacobian of which is

$$\begin{vmatrix} z_{1\zeta_1} & z_{1\zeta_2} \\ z_{2\zeta_1} & z_{2\zeta_2} \end{vmatrix} = \begin{vmatrix} 1/\mu & \lambda \\ 0 & 1 \end{vmatrix} = \frac{1}{\mu}.$$

Furtheron the mapping (ζ_1, ζ_2) is assumed to be one-to-one on G_0 mapping G_0 onto a domain $G \subset \mathbb{C}^2$. Then for $w(z_1, z_2)$ analytic in G_0

$$\omega(\zeta_1, \zeta_2) := w(z_1(\zeta_1, \zeta_2), z_2(\zeta_1, \zeta_2))$$

is an analytic function in G satisfying

$$\omega_{\zeta_1} = \frac{1}{\mu} w_{z_1}, \ \omega_{\zeta_2} = \lambda w_{z_1} + w_{z_2}.$$

Hence, (10.35′) is equivalent to

$$\omega_{0\zeta_2} = f_0, \ \omega_{\nu\zeta_2} + \mu\omega_{\nu-1\zeta_1} = f_\nu, \ 1 \le \nu \le n-1. \quad (10.36)$$

Integration leads to

$$\begin{aligned} \omega_0(\zeta_1, \zeta_2) &= \varphi_0(\zeta_1) + \int_0^{\zeta_2} f_0(\zeta_1, t)dt, \\ & \qquad\qquad (10.37) \\ \omega_\nu(\zeta_1, \zeta_2) &= \varphi_\nu(\zeta_1) + \int_0^{\zeta_2} \{f_\nu(\zeta_1, t) - \mu(\zeta_1, t)\omega_{\nu-1\zeta_1}(\zeta_1, t)\} dt, 1 \le \nu \le n-1. \end{aligned}$$

In vector form (10.36) is just

$$\omega_{\zeta_2} + \begin{bmatrix} 0 & & 0 \\ \mu & \ddots & \\ 0 & \mu & 0 \end{bmatrix} w_{\zeta_1} = f. \tag{10.36'}$$

With the nilpotent matrix

$$e = \begin{bmatrix} 0 & & 0 \\ 1 & \ddots & \\ 0 & 1 & 0 \end{bmatrix}, e^n = 0,$$

and the hypercomplex representation of the vectors ω and f,

$$\omega = \sum_{k=0}^{n-1} \omega_k e^k, \ f = \sum_{k=0}^{n-1} f_k e^k$$

the hypercomplex form of (10.36′) is

$$\sum_{k=0}^{n-1} \omega_{k\zeta_2} e^k + \sum_{k=1}^{n-1} \mu \omega_{k-1\zeta_1} e^k = \sum_{k=0}^{n-1} f_k e^k,$$

or with the hypercomplex differential operator

$$D := \frac{\partial}{\partial \zeta_2} + \mu e \frac{\partial}{\partial \zeta_1} \tag{10.38}$$

just

$$D\omega = f. \tag{10.39}$$

A generating solution

$$t = \sum_{k=0}^{n-1} \tau_k e^k$$

for D with $\tau_0 = \tau_0(\zeta_1, \zeta_2) = \zeta_1$ and $Dt = 0$ can be found via (10.38). On can just choose

$$\tau_k = \tau_k(\zeta_1, \zeta_2) = (-1)^k \mu_k(\zeta_1, \zeta_2), \ \ 1 \le k \le n-1, \tag{10.40}$$

with

$$\mu_0(\zeta_1, \zeta_2) = \zeta_1, \mu_k(\zeta_1, \zeta_2) = \int_0^{\zeta_2} \mu(\zeta_1, s) \mu_{k-1\zeta_1}(\zeta_1, s) ds, \ \ 1 \le k \le n-1.$$

Introducing the hypercomplex variables (t_1, t_2), where

$$t_1(\zeta_1, \zeta_2) := t(\zeta_1, \zeta_2), \ t_2(\zeta_1, \zeta_2) = \zeta_2$$

equation (10.39) can be further simplified. The Jacobian of this transformation is

$$J := \begin{vmatrix} t_{1\zeta_1} & t_{1\zeta_2} \\ t_{2\zeta_1} & t_{2\zeta_2} \end{vmatrix} = t_{1\zeta_1} = 1 + \sum_{k=1}^{n-1} \tau_{k\zeta_1} e^k \neq 0.$$

From

$$\frac{\partial}{\partial \zeta_1} = J \frac{\partial}{\partial t_1}, \quad \frac{\partial}{\partial \zeta_2} = -\mu e J \frac{\partial}{\partial t_1} + \frac{\partial}{\partial t_2}$$

it follows

$$\frac{\partial}{\partial t_2} = D, \quad \frac{\partial}{\partial t_1} = \frac{1}{J} \frac{\partial}{\partial \zeta_1}. \tag{10.41}$$

Besides $Dt_1 = 0$ one has $Dt_2 = 1$. The first equation is $\partial t_1 / \partial t_2 = 0$. As also $J \partial t_2 / \partial t_1 = \partial \zeta_2 / \partial \zeta_1 = 0, t_1, t_2$ can be considered as independent hypercomplex variables. With (10.41) equation (10.39) becomes

$$\frac{\partial}{\partial t_2} \omega(\zeta_1(t_1, t_2), t_2) = f(\zeta_1(t_1, t_2), t_2). \tag{10.42}$$

A particular solution to this equation is

$$\omega_0(t_1, t_2) := \int_0^{t_2} f(\zeta_1(t_1, t), t) dt.$$

The general equation to $D\varphi = 0$ is with arbitrary analytic hypercomplex function φ given by $\varphi(t_1(\zeta_1, \zeta_2))$. Thus the general solution to (10.42) is

$$\omega(\zeta_1, \zeta_2) = \varphi(t_1(\zeta_1, \zeta_2)) + \omega_0(t_1(\zeta_1, \zeta_2), \zeta_2).$$

Theorem 3 *The general solution to system* (10.39) *with analytic f in G is*

$$\omega(\zeta_1, \zeta_2) = \varphi(t_1(\zeta_1, \zeta_2)) + \int_0^{\zeta_2} f(\zeta_1(t_1(\zeta_1, \zeta_2), t), t) dt \tag{10.43}$$

where φ is an arbitrary analytic hypercomplex function provided (t_1, t_2) is a one-to-one mapping on G.

Verification Differentiation gives

$$\begin{aligned} \omega_{\zeta_1} = \varphi'(t_1) t_{1\zeta_1} &+ \int_0^{\zeta_2} f_{\zeta_1}(\zeta_1(t_1(\zeta_1, \zeta_2), t), t) \frac{\partial \zeta_1(t_1(\zeta_1, \zeta_2), t)}{\partial t_1} \frac{\partial t_1(\zeta_1, \zeta_2)}{\partial \zeta_1} dt \\ \omega_{\zeta_2} = \varphi'(t_1) t_{1\zeta_2} &+ \int_0^{\zeta_2} f_{\zeta_1}(\zeta_1(t_1(\zeta_1, \zeta_2), t), t) \frac{\partial \zeta_1(t_1(\zeta_1, \zeta_2), t)}{\partial t_1} \frac{dt_1(\zeta_1, \zeta_2)}{\partial \zeta_2} dt \\ &+ f(\zeta_1(t_1(\zeta_1, \zeta_2), \zeta_2), \zeta_2) \end{aligned}$$

so that because of $Dt_1 = 0$ one has

$$\begin{aligned} D\omega = \omega_{\zeta_2} + \mu e \omega_{\zeta_1} &= \varphi'(t_1(\zeta_1,\zeta_2))Dt_1(\zeta_1,\zeta_2) \\ &+ \int_0^{\zeta_2} f_{\zeta_1}(\zeta_1(t_1(\zeta_1,\zeta_2),t),t)\frac{\partial\zeta_1(t_1(\zeta_1,\zeta_2),t)}{\partial t_1}Dt_1(\zeta_1,\zeta_2)dt_2 \\ &+ f(\zeta_1(t_1(\zeta_1,\zeta_2),\zeta_2),\zeta_2) = f(\zeta_1,\zeta_2). \end{aligned}$$

Remark φ can be determined by boundary conditions on the boundary of the domain

$$G_1 := G \cap \{(z_1,z_2) \in \mathbb{C} : z_2 = 0\} \subset \mathbb{C} .$$

For example the Riemann–Hilbert condition

$$\operatorname{Re}\{\overline{\chi(\zeta_1)}\omega(\zeta_1,0)\} = g(\zeta) \text{ on } \partial G_1,$$

with hypercomplex nonvanishing χ on ∂G_1 imply

$$\operatorname{Re}\{\overline{\chi(\zeta_1)}(\varphi(\zeta_1)\} = g(\zeta) \text{ on } \partial G_1, \tag{10.44}$$

as $t_1(\zeta_1,0) = \zeta_1$, see (10.40). (10.44) is just the Riemann–Hilbert boundary condition for the analytic hypercomplex function φ of one complex variable. In [1] semilinear weakly coupled systems (10.35) are treated via the before explained method. The system is reduced to some system of Volterra integral equations.

As an application of this treatment of first order systems in two complex variables with analytic coefficients some higher order equations are investigated.

Example

$$u_{z_2z_2} - u_{z_1} = 0 \text{ in } |z_1|^2 + |z_2|^2 < 1. \tag{10.45}$$

Introducing

$$w_0 := u_{z_2},\ w_1 = u_{z_1},\ w = (w_0, w_1) \tag{10.46}$$

equations (10.45) becomes

$$w_{z_2} - Aw_{z_1} - Bw = 0,\ A := \begin{pmatrix} 0 & 0 \\ 1 & 0 \end{pmatrix},\ B := \begin{pmatrix} 0 & 1 \\ 0 & 0 \end{pmatrix}, \tag{10.47}$$

which is of the form (10.36′). Here with $e = A$

$$D = \frac{\partial}{\partial z_2} - e\frac{\partial}{\partial z_1},\ t_1(z_1,z_2) = z_1 + ez_2,\ z_1 = t_1 - et_2,\ t_2 = z_2$$

so that (10.43) gives for the solution w the integral equation

$$w(z_1,z_2) = \varphi(z_1 + ez_2) + \int_0^{z_2} (w_1, 0)(z_1 + ez_2 - et, t)dt \tag{10.48}$$

where $\varphi = (\varphi_0, \varphi_1)$ is an analytic vector. Replacing the complex variable in the power series development of φ by the matrix

$$z_1 + ez_2 = \begin{bmatrix} z_1 & 0 \\ z_2 & z_1 \end{bmatrix}$$

shows

$$\varphi(z_1 + ez_2) = (\varphi_0(z_1), \varphi_0'(z_1)z_2 + \varphi_1(z_1))$$

and, similarly, for the analytic vector $(w_1, 0)$ in (z_1, z_2)

$$(w_1, 0)(z_1 + ez_2, t) = (w_1(z_1, t), w_{1z_1}(z_1, t)z_2).$$

Thus (10.48) is the system

$$\begin{aligned} w_0(z_1, z_2) &= \varphi_0(z_1) + \int_0^{z_2} w_1(z_1, t)dt, \\ w_1(z_1, z_2) &= \varphi_0'(z_1)z_2 + \varphi_1(z_1) + \int_0^{z_2} w_{1z_1}(z_1, t)(z_2 - t)dt. \end{aligned} \tag{10.49}$$

The last equation can be solved iteratively

$$\begin{aligned} w_{10}(z_1, z_2) &= \varphi_0'(z_1)z_2 + \varphi_1(z_1), \\ w_{1k}(z_1, z_2) &= \varphi_0'(z_1)z_2 + \varphi_1(z_1) + \int_0^{z_2} w_{1k-1z_1}(z_1, t)(z_2 - t)dt, \ k \in \mathbb{N}. \end{aligned}$$

Then

$$w_{1k}(z_1, z_2) = \sum_{\nu=1}^{k+1} \varphi_0^{(\nu)}(z_1) \frac{z_2^{2\nu-1}}{(2\nu - 1)!} + \sum_{\nu=0}^{k} \varphi_1^{(\nu)}(z_1) \frac{z_2^{2\nu}}{(2\nu)!}$$

follows inductively, so that

$$\begin{aligned} w_1(z_1, z_2) &= \sum_{\nu=1}^{\infty} \varphi_0^{(\nu)}(z_1) \frac{z_2^{2\nu-1}}{(2\nu - 1)!} + \sum_{\nu=0}^{\infty} \varphi_1^{(\nu)}(z_1) \frac{z^{2\nu}}{(2\nu)!}, \\ w_0(z_1, z_2) &= \sum_{\nu=0}^{\infty} \varphi_0^{(\nu)}(z_1) \frac{z_2^{2\nu}}{(2\nu)!} + \sum_{\nu=0}^{\infty} \varphi_1^{(\nu)}(z_1) \frac{z_2^{2\nu+1}}{(2\nu + 1)!}. \end{aligned} \tag{10.50}$$

Obviously, $w_{1z_2} = w_{0z_1}$, so that system (10.50) is integrable giving

$$\begin{aligned} u(z_1, z_2) &= \int_0^{z_1} w_1(t,0)dt + \int_0^{z_2} w_0(z_1,t)dt + c \\ &= \sum_{\nu=0}^{\infty} \varphi_0^{(\nu)}(z_1)\frac{z_2^{2\nu+1}}{(2\nu+1)!} + \sum_{\nu=0}^{\infty} \varphi_1^{(\nu)}(z_1)\frac{z_2^{2\nu+2}}{(\nu+2)!} \\ &\quad + \int_0^{z_1} \varphi_1(t)dt + c. \end{aligned} \tag{10.51}$$

From here

$$u(z_1, 0) = \int_0^{z_1} \varphi_1(t)dt + c, \; u_{z_2}(z_1, 0) = \varphi_0(z_1), \tag{10.52}$$

follows.

Theorem 4 *The solution to* (10.45) *satisfying*

$$\operatorname{Re} u(z_1, 0) = \gamma_1 \ , \ \operatorname{Re} u_{z_2}(z_1, 0) = \gamma_0 \ \text{ on } \ |z_1| = 1 \tag{10.53}$$

is with arbitrary real constant c_0 and arbitrary complex constant c_1

$$\begin{aligned} u(z_1, z_2) &= \sum_{\nu=0}^{\infty} \frac{\nu! z_2^{2\nu}}{(2\nu)!\pi i} \int_{|\zeta|=1} \left\{ \frac{z_2}{2\nu+1}\gamma_0(\zeta) + \gamma_1(\zeta) \right\} \frac{d\zeta}{(\zeta - z_1)^{\nu+1}} \\ &\quad - \frac{1}{2\pi i} \int_{|\zeta|=1} (z_2\gamma_0(\zeta) + \gamma_1(\zeta)\} \frac{d\zeta}{\zeta} + ic_0 z_2 + c_1 \end{aligned} \tag{10.54}$$

Proof From (10.47), (10.53) $\varphi_0(z) = S\gamma_0 + ic_0$ with the Schwarz operator S and arbitrary real constant c_0 is seen. Moreover, with $z_1 = e^{it}$

$$\operatorname{Re} \frac{d}{dt} u(e^{it}, 0) = \operatorname{Re}\{iz_1 u_{z_1}(z_1, 0)\} = \operatorname{Re}\{iz_1, \varphi_1(z_1)\} = \frac{d}{dt}\gamma_1(e^{it})$$

is seen. As

$$\int_0^{2\pi} \frac{\partial}{\partial t}\gamma_1(e^{it})dt = 0 \tag{10.55}$$

φ_1 is uniquely given by this boundary condition, see e.g.[1], namely $iz_1\varphi_1 = S(\partial\gamma_1(e^{it})/\partial t)$ giving with (10.55)

$$\varphi_1(z_1) = -\frac{1}{\pi} \int_0^{2\pi} \frac{d}{dt}\gamma_1(e^{it}) \frac{de^{it}}{(e^{it} - z_1)e^{it}} = \frac{1}{\pi i} \int_{|\zeta|=1} \gamma_1(\zeta) \frac{d\zeta}{(\zeta - z_1)^2}.$$

Together with (10.51) this gives

$$\begin{aligned} u(z_1, z_2) &= \sum_{\nu=1}^{\infty} \frac{\nu! z^{2\nu+1}}{(2\nu+1)!\pi i} \int\limits_{|\zeta|=1} \gamma_0(\zeta) \frac{d\zeta}{(\zeta - z_1)^{\nu+1}} \\ &+ \sum_{\nu=1}^{\infty} \frac{\nu! z^{2\nu}}{(2\nu)!\pi i} \int\limits_{|\zeta|=1} \gamma_1(\zeta) \frac{d\zeta}{(\zeta - z_1)^{\nu+1}} \\ &+ \frac{z_2}{\pi i} \int\limits_{|\zeta|=1} \gamma_0(\zeta) \frac{d\zeta}{\zeta - z} - \frac{z_2}{2\pi i} \int\limits_{|\zeta|=1} \gamma_0(\zeta) \frac{d\zeta}{\zeta} + i c_0 z_2 \\ &+ \frac{1}{\pi i} \int\limits_{|\zeta|=1} \gamma_1(\zeta) \frac{d\zeta}{\zeta - z_1} - \frac{1}{\pi i} \int\limits_{|\zeta|=1} \gamma_1(\zeta) \frac{d\zeta}{\zeta} + c \end{aligned}$$

which is (10.54).

In [1], [10] instead of (10.47) the equation

$$\frac{\partial^m n}{\partial z_2^m} - \frac{\partial u}{\partial z_1} = 0 \text{ in } |z_1|^2 + |z_2|^2 < 1$$

is treated for $3 \leq m$. System (10.47) then becomes a weakly coupled system with an $(m-2) \times (m-2)$ and a 2×2 matrix. The related system of Volterra integral equations can be solved iteratively.

Remark Equation (10.45) of course can be solved in a different way through the integral representation formula

$$u(z_1, z_2) = \varphi_0(z_1) + \varphi_1(z_1) z_2 + \int\limits_0^{z_2} (z_2 - t) u_{z_1}(z_1, t) dt$$

where φ_0 and φ_1 are arbitrary analytic functions.

References

[1] Begehr, H.: Complex analytic methods for partial differential equations. An introductory text. World Scientific, Singapore, 1994.

[2] Begehr, H., Dzhuraev, A.: An introduction to several complex variables and partial differential equations. Addison Wesley Longman, Harlow, 1997.

[3] Begehr, H., Gilbert, R.P.: Randwertaufgaben ganzzahliger Charakteristik für verallgemeinerte hyperanalytische Funktionen. Appl. Anal. 6 (1977), 189-205.

[4] Begehr, H., Gilbert, R.P.: Boundary value problems associated with first order elliptic systems in the plane. Contemporary Math. 11 (1982), 13–48.

[5] Begehr, H., Gilbert, R.P.: Pseudohyperanalytic functions. Complex Variables, Theory Appl. 9 (1988), 343–357.

[6] Begehr, H., Gilbert, R.P.: Transformations, transmutations, and kernel functions; I, II. Longman, Harlow, 1992, 1993.

[7] Begehr, H., Wen, G.C.: Nonlinear elliptic boundary value problems and their applications. Addison Wesley Longman, Harlow, 1996.

[8] Begehr, H., Wen, G.C.: Some second order systems in the complex plane. Preprint, FU Berlin, 1997.

[9] Douglis, A.: A function–theoretic approach to elliptic systems of equations in two variables. Comm. Pure Appl. Math. 6 (1953), 259–289.

[10] Dzhuraev, A., Begehr, H.: On a boundary value problem for a first order holomorphic system in $\mathbb{C}^2$. Ross. Akad. Nauk Doklady 339 (1994), 297–300; Russ. Acad. Sci. Dokl. Math. 50 (1995), 418–422.

[11] Gilbert, R.P.: Constructive methods for elliptic equations. Lecture Notes in Math. 365, Springer–Verlag, Berlin, 1974.

[12] Gilbert, R.P., Hile, G.N.: Generalized hypercomplex function theory. Trans. Amer. Math. Soc. 195 (1974), 1–29.

[13] Gilbert, R.P., Buchanan, J.L.: First order elliptic systems. A function theoretic approach. Acad. Press, New York, 1983.

[14] Hile, G.N.: Hypercomplex function theory applied to partial differential equations. Ph.D. thesis, Indiana University, Bloomington, Indiana, 1972.

[15] Huang, S.X.: On properties of some operators in Douglis albegra and their applications to pde. J. Part. Diff. Eq. 1 (1988), 21–30.

[16] Kühn, E.: Über die Funktionentheorie und das Ähnlichkeitsprinzip einer Klasse elliptischer Differentialgleichungen in der Ebene. Dissertation, Univ. Dortmund, Dortmund, 1974.

[17] Vekua, I.N.: Generalized analytic functions. Pergamon Press, Oxford, 1962.

[18] Wen, G.C., Begehr, H.: Boundary value problems for elliptic equations and systems. Longman, Harlow, 1990.

11 TO ELLIPTIC THEORY FOR DOMAINS WITH PIECEWISE SMOOTH BOUNDARY IN THE PLANE

Alexandre P. Soldatov

Novgorod State University
Department of Mathematics and Informatic
B.St. Peterburgskaya 41
173003 Novgorod
Russian Federation
sold@lan.novsu.ac.ru

Abstract: A new approach to elliptic boundary value problems for domains with piecewise smooth boundary in the plane is developed with the help of (in Douglis sense) hyperanalytic functions. With respect to the elliptic system with constant and only leading coefficients these functions play the same role as the usual analytic functions do for the Laplace equation. Some applications to the mixed problem for the Lamé system of plane elasticity theory are given.

11.1 FUNCTIONS WHICH ARE ANALYTIC IN THE DOUGLIS SENSE

A general solution of elliptic system in the plane with constant and only leading coefficients can be represented with the help of analytic functions by Bitsadze's formula [1]. This approach is classic in the study of elliptic boundary value problems. Nevertheless this representation contains the derivatives of analytic functions. As a consequence there are some difficulties in the case of domains with piecewise smooth boundary. The situation becomes more simple when we use the so-called Douglis analytic functions.

Let the matrix $J \in C^{s\times s}$ be non-singular and all its eigenvalues ν lie in the upper half-plane $\mathrm{Im}\nu > 0$. The complex s - vector function

$$\phi(x) = (\phi_1(x), \dots, \phi_s(x)), \quad x = (x_1, x_2) \in R^2$$

H.G.W. Begehr et al. (eds.), Partial Differential and Integral Equations, 177–186.

is called J - analytic function (or function which is analytic in the Douglis sense) if it satisfies the system

$$\frac{\partial \phi}{\partial x_2} - J\frac{\partial \phi}{\partial x_1} = 0. \tag{11.1}$$

There is a rather complete analogue of the analytic function theory for this equation. For the first time the equation (11.1) was studied by A.Douglis [2] under the assumption that the matrix J is triangular and its entries J_{kr} depend only on $k-r$. Later on the theory has been developed by D.Pascali [3], B.Bojarski [4], R.Gilbert [5, 6] G.Hile [7] and others.

Let us connect the $s \times s$ matrix $[x] = x_1 \cdot 1 + x_2 \cdot J$, $\det[x] \neq 0$ as $x \neq 0$, with the point $x \in R^2$. Then every solution $\phi \in C^1(D)$ is infinitely differentiable and it might be expanded into a generalized power series in a neighborhood of every point $x_0 \in D$:

$$\phi(x) = \sum_{k=0}^{\infty} [x - x_0]^k \phi^{(k)}(x_0) \, .$$

Here and later on $\phi^{(k)}$ denotes the partial derivative with respect to x_1 of order k. Under this condition the Cauchy theorem

$$\int_{\partial D_0} [dy]\phi(y) = 0, \quad [dy] = 1 \cdot dy_1 + J dy_2,$$

and the Cauchy formula

$$\frac{1}{2\pi i}\int_{\partial D_0} [dy][y-x]^{-1}\phi(y) = \phi(x), \quad x \in D_0$$

are valid. There is a formula, that establishes a one-to-one correspondence between J - analytical functions ϕ and analytical vector-functions ϕ_0. In the case of $\sigma(J) = \{\nu\}$ this formula has a very simple form:

$$\phi(x) = \sum_{k=0}^{s-1} \frac{x_2^k}{k!}(J - \nu)^k \phi_0^{(k)}(x_1 + \nu x_2). \tag{11.2}$$

In the domain D we consider the elliptic system

$$\sum_{k+r=n} a_{k,r} \frac{\partial^n u}{\partial x_1^k \partial x_2^r} = 0 \tag{11.3}$$

of order n with the constant coefficients $a_{k,r} \in R^{l \times l}$. The ellipticity condition is

$$\det\left(\sum_{k+r=n} \xi_1^k \xi_2^r a_{k,r}\right) \neq 0, \; \xi \neq 0.$$

In particular the number $nl = 2s$ is even, and the equation $\det P(\nu) = 0$, determined by the characteristic polinomial

$$P(z) = \sum_{k+r=n} a_{k,r} z^r,$$

has exactly s roots in the upper half-plane. By σ we denote the set of these roots. Consider the matrix equation

$$\sum_{k+r=n} a_{k,r} b J^r = 0 \tag{11.4}$$

with the unknown rectangular $l \times s$ - matrix $b \in C^{l\times s}$ and the square $s \times s$ - matrix $J \in C^{s\times s}$. This equation has a solution satisfying the condition:

$$\sigma(J) = \sigma, \quad \det \begin{pmatrix} b & \overline{b} \\ bJ & \overline{bJ} \\ \vdots & \vdots \\ bJ^{n-1} & \overline{bJ^{n-1}} \end{pmatrix} \neq 0. \tag{11.5}$$

Moreover, the matrix J could be chosen as a block diagonal matrix, constructed of Jordan cells. In addition columns $x_0, x_1, \ldots$ of the matrix b, that correspond to the ν - Jordan block, form a chain of eigenvectors and adjoint vectors for the bundle $P(z)$:

$$\sum_{r=0}^{k} \frac{P^{(r)}(\nu)}{r!} x_{k-r} = 0, \quad k = 0, 1 \ldots. \tag{11.6}$$

If the s - vector-function ϕ satisfies (11.1), then it follows from (11.4), that the l - vector functions $b\phi$ and $\overline{b\phi}$ are solutions of (11.3). The inverse preposition is also valid.

Theorem 1 (A.Soldatov [8], R.Yeh [9]). *Let the domain D be simply connected. Then any solution $u \in C^n(D)$ of the system* (11.3) *may be represented in the form*

$$u(x) = \mathrm{Re} b\phi(x), \tag{11.7}$$

where the function ϕ is determined uniquely by the condition

$$\mathrm{Re} bJ^r \phi(x_0) = 0, \quad r = 1, \ldots, n-1,$$

in the fixed point $x_0 \in D$.

Similarly to the case of analytical functions with the help of the representation (11.7), an elliptical boundary value problem

$$\sum_{k+r\leq n-1} \alpha_{k,r}(y) \frac{\partial^{k+r} u}{\partial x_1^k \partial x_2^r}(y) = f(y), \quad y \in \partial D,$$

where $\alpha_{k,r}(y)$ are real $s \times l$ - matrix functions, can be reduced to the Riemann-Hilbert – type problem

$$\mathrm{Re} \sum_{i=0}^{n-1} G_i(y) \phi^{(i)}(y) = f(y), \quad y \in \partial D,$$

with $s \times s-$ matrix coeffitients

$$G_i(y) = \sum_{k+r=i} \alpha_{k,\tau}(y) bJ^{\tau}, \quad y \in \partial D.$$

If the coefficients $a_{k,\tau}$ are complex then the roots of the characteristic equation $\det P(z) = 0$ are to be distributed in the upper and lower half planes arbitrarily. Let $s_1(s_2)$ be the amount of these roots $\nu \in \sigma_1$ $(\nu \in \sigma_2)$ in the upper (lower) half-plane. In this case the relation (11.3) and its complex conjugate are replaced by the relations

$$\sum_{k+r=n} a_{k,r} b_1 J_1^r = 0, \quad \sum_{k+r=n} a_{k,r} \overline{b_2 J_2^r} = 0$$

for matrices $b_i \in C^{l \times l}$, $J_i \in C^{s_i \times s_i}$.

Respectively, the condition (11.5) transforms to

$$\sigma(J_i) = \sigma_i, \quad \det \begin{pmatrix} b_1 & \overline{b_2} \\ b_1 J_1 & \overline{b_2 J_2} \\ \vdots & \vdots \\ b_1 J_1^{n-1} & \overline{b_2 J_2^{n-1}} \end{pmatrix} \neq 0,$$

and the representation (11.7) has the form

$$u = b_1 \phi_1 + \overline{b_2 \phi_2}. \tag{11.8}$$

Consider the case $l = 1$ of a single equation. In this case the unique Jordan cell and the chain (11.6) correspond to every eigenvalue $\nu \in \sigma_i$. For example an equation

$$\left(\frac{\partial}{\partial x_2} - i \frac{\partial}{\partial x_1} \right)^n u = 0,$$

that defines polyanalytic functions [10] would have the representation

$$u = \phi_1,$$

where ϕ_1 is the first component of the J - analytical function ϕ with the Jordan cell J, corresponding to $\nu = i$.

We can also receive a similar representation for systems, that are elliptic in the Douglis–Nirenberg sense (N.Zhura [11]). For example a general solution of the linearized Stokes system

$$\frac{\partial^2 u_1}{\partial x^2} + \frac{\partial^2 u_1}{\partial y^2} - \frac{\partial p}{\partial x} = 0,$$

$$\frac{\partial^2 u_2}{\partial x^2} + \frac{\partial^2 u_2}{\partial y^2} - \frac{\partial p}{\partial y} = 0,$$

$$\frac{\partial u_1}{\partial x} + \frac{\partial u_2}{\partial y} = 0,$$

where $u = (u_1, u_2)$ is the velocity vector and p - is the pressure, has the form [12]

$$u = \mathrm{Re}\begin{pmatrix} 1 & 0 \\ i & -1 \end{pmatrix}\phi, \quad p = \mathrm{Re}\,(2i\phi_2').$$

Here the function $\Phi = (\Phi_1, \Phi_2)$ is analytic in the Douglis sense with the matrix

$$J = \begin{pmatrix} i & 1 \\ 0 & i \end{pmatrix}.$$

As above this representation admits to study general boudary value problems for such systems. (N.Zhura [13]).

11.2 THE RIEMANN-HILBERT PROBLEM

On the Riemann sphere consider the simply connected domain D, bounded by the piecewise smooth contour Γ. The Riemann-Hilbert problem is to find a J - analytical function ϕ in the domain D under the boundary condition

$$\mathrm{Re} G\phi\Big|_{\Gamma} = g. \tag{11.9}$$

Without loss of generality we can assume that the matrix J is triangular. We suppose that the matrix-function $G(y)$ is piecewise continuous on Γ, and $\det G(y) \neq 0$ for all $y \in \Gamma$, including limit values at discontinuity points. By $\xi(y) = (\xi_1(y), \xi_2(y))$ denote a piecewise continuous unit tangent vector on Γ, such that it's direction corresponds to the orientation of Γ, which leaves D to the left-hand side. Let F be a finite set of points on Γ, containing all discontinuity points of G and ξ. If the boundary of Γ is a non compact set then the point ∞ is added to F. We suppose that the contour Γ has no cusps, i.e. $\xi(\tau+0) \neq -\xi(\tau-0), \quad \tau \in \Gamma$. Therefore, eigenvalues of the matrix

$$Q_\tau = [\xi(\tau+0)]^{-1}[\xi(\tau-0)]$$

do not lie on the negative semiaxis, so we can introduce degrees $(Q_\tau)^\zeta$, $\zeta \in C$, as functions of Q_τ u^ζ, $|\arg u| < \pi$. Let us consider block $2s \times 2s$ - matrices

$$X(\zeta,\tau) = \begin{pmatrix} \overline{G(\tau-0)} & e^{\pi i\zeta} G(\tau-0) Q_\tau^\zeta \\ e^{\pi i\zeta}\overline{G(\tau+0)}\,\overline{(Q_\tau)}^{-\zeta} & G(\tau+0) \end{pmatrix}, \tag{11.10}$$

where Q_τ should be replaced by Q_τ^{-1} in the case of $\tau = \infty$.

Let X_0 correspond to $G(\tau \pm 0) = 1$. If the number $\varepsilon > 0$ is rather small (small enough), then the function $\det(XX_0^{-1})(\zeta,\tau)$ is non-vanishing in the strip $-\varepsilon < \mathrm{Re}\zeta < 0$ and the continuous branch of it's argument has limit values $\arg\det(XX_0^{-1})(\pm i\infty, \tau)$ as $\mathrm{Im}\zeta \to \pm\infty$. Similarly, let $\arg\det G(y)$ mean countinuous argument's branches on connected components of $\Gamma \setminus F$. Then

$$\kappa_0 = \sum_{\tau\in\Gamma_1} \left\{ \frac{1}{\pi}\arg\det G(\tau+0) - \frac{1}{\pi}\arg\det G(\tau-0) \right\}$$

$$-\sum_{\tau\in F}\left\{\frac{1}{2\pi}\text{arg det}\left(XX_0^{-1}\right)(+i\infty,\tau)-\frac{1}{2\pi}\text{arg det}\left(XX_0^{-1}\right)(-i\infty,\tau)\right\}$$

is an integer and it does not depend on the choice of argument's branches.

Let us describe classes, where this problem is considered. First let the closure $\overline{D}$ be a compact set in the plane, i.e. $\infty\notin\overline{D}$. By $H(\overline{D})$ denote the class of functions, that satisfy the Hölder condition on $\overline{D}$. Let $\overset{\circ}{H}(\overline{D};F)$ consist of all functions $\phi\in H(\overline{D})$, vanishing on F. Further we define the class $\overset{\bullet}{H}(\overline{D};F)$ by the condition $\prod_{\tau\in F}|x-\tau|^{\varepsilon}\phi(x)\in\overset{\circ}{H}$ for any $\varepsilon>0$. In the same way with respect to each connected component of $\Gamma\setminus F$ we define classes $\overset{\circ}{H}(\Gamma;F)$, $H(\Gamma,F)$ and $\overset{\bullet}{H}(\Gamma,F)$ on the boundary of Γ.

If the set $\overline{D}$ is unbounded on the plane, then we choose a point $x^*\notin\overline{D}$ and consider the linear-fractional transformation

$$x_1+ix_2\to\frac{1}{x_1-x_1^*+i(x_2-x_2^*)}.$$

By definition the image of Γ is a piecewise smooth contour under this transformation. The H-classes naturally transfer to the considered case with the help of this transformation.

In what follows we assume the smoothness conditions

$$G,\xi\in H(\Gamma;F)$$

to hold for the functions G and the curve Γ.

Theorem 2 (a) *The homogeneous problem* (11.9) *has a finite number p of linearly independent solutions in the class $\overset{\bullet}{H}$. The non-homogeneous problem is solvable if and only if the function g satisfies to a finite number q of the linearly independent orthogonality conditions*

$$\int_{\Gamma}g(y)f_i(y)ds_y=0,\quad j=1,\ldots,q,$$

and the problem's index $p-q=\kappa_0+s$.

(b) *If* $\det X(\zeta)\neq 0$ *as* $\text{Re}\zeta=0$, $\zeta\neq 0$ *and* $X^{-1}(\zeta)=O(1)$ *as* $\zeta\to 0$, *then any solution $\phi\in\overset{\bullet}{H}$ of the problem with the right hand side $g\in\overset{\circ}{H}$ belongs to $H(\overline{D})$. If* $\det X(\zeta)\neq 0$ *as* $\text{Re}\zeta\neq 0$ *this statement is also valid for $g\in H(\Gamma;F)$.*

The statement (a) of the theorem also remains in force in the class $\overset{\circ}{H}$, if we put κ_0+s-s_0 for the index κ_0+s, here s_0 is the number of the function $\det X(\zeta)$ zeros on the line $\text{Re}\zeta=0$, including multiplicities.

Let us note that Theorem 2 for domains with compact boundary is proved in [3]. This theorem is based on the following analogue representation of Vekua-

Muskhelishvili for analytic functions by a Cauchy type integral with real density.

Theorem 3 (a) *Suppose $\infty \notin \overline{D}$, then any J - analytical function $\phi \in \overset{\bullet}{H}$ can be uniquely represented in the form*

$$\phi(x) = \frac{1}{2\pi i}\int_{\Gamma}[dy][y-x]^{-1}\varphi(y) + i\eta, \tag{11.11}$$

where φ is the real vector-function from $\overset{\bullet}{H}(\Gamma; F)$, and $\eta \in R^s$.

(b) *Suppose $\infty \in D$, then formula (11.11) takes place, where $\eta \in C^s$ and*

$$\int_{\Gamma}\varphi(y)d\rho_y = 0.$$

(c) *Suppose $\infty \in \Gamma$ and the function $\phi \in \overset{\circ}{H}$ in the neighborhood of $\tau = \infty$. Then* (11.11) *holds, where $\eta \in C^s$ and the real function $\varphi \in \overset{\bullet}{H}$ belongs to $\overset{\circ}{H}$ in a neighborhood of $\tau = \infty$.*

The Riemann-Hilbert problem (11.9) reduces to the equivalent system of singular integral equations with the help of the representation (11.11) and the analogue of Sokhotsky-Plemelj formulas. Besides the Cauchy-kernel these equations contain noncompact integral operators with kernels, that are approximately homogeneous of degree -1 with respect to the distance to the points $\tau \in F$. The theory of these equations was constructed by A.Soldatov [15, 16], R.Duduchava [17], and Theorem 2 is based on the results of this theory.

11.3 A MIXED PROBLEM OF PLANE ELASTICITY THEORY

To illustrate the function theoretical approach, shown above , we consider the Lamé system

$$\mu\Delta u + (\lambda+\mu)\text{grad}\,(\text{div}\,u) = 0 \tag{11.12}$$

for the displacement vector $u = (u_1, u_2)$.

For this system we have the representation (11.7) with the matrices

$$b = \begin{pmatrix} 1 & 0 \\ i & -\kappa \end{pmatrix}, \quad J = \begin{pmatrix} i & 1 \\ 0 & i \end{pmatrix}, \tag{11.13}$$

where we denote $\kappa = (\lambda + 3\mu)/(\lambda+\mu)$.

We note that substituting the representation

$$\phi(x) = \phi_0(x_1 + ix_2) + \begin{pmatrix} 0 & 1 \\ 0 & 0 \end{pmatrix}\phi_0'(x_1 + ix_2),$$

by (11.2), we arrive at the known Kolosov-Muskhelishvili formulas.

The mixed problem in the domain D is to define a vector u on the part Γ_I of the boundary Γ and the normal component

$$(\sigma n)_1 = \sigma_1 n_1 + \sigma_3 n_2, \quad (\sigma n)_2 = \sigma_3 n_1 + \sigma_2 n_2,$$

of the stress tensor σ on the other part Γ_{II}. As it is known the stress tensor σ, is determined by the elements

$$\sigma_i = \lambda\left(\frac{\partial u_1}{\partial x_1} + \frac{\partial u_2}{\partial x_2}\right) + 2\mu\frac{\partial u_i}{\partial x_i},\ i = 1, 2;\quad \sigma_3 = \mu\left(\frac{\partial u_1}{\partial x_2} + \frac{\partial u_2}{\partial x_1}\right),$$

where $n = (\xi_2, -\xi_1)$ is the external unit normal vector on Γ.

It is clear that in notation (11.7)

$$\sigma n = \mu \mathrm{Re} c\frac{\partial \phi}{\partial \xi},\quad c = \begin{pmatrix} -2i & \kappa - 1 \\ 2 & i(\kappa + 1) \end{pmatrix}. \tag{11.14}$$

As a result, the mixed problem reduces to the following Riemann-Hilbert problem

$$\mathrm{Re} b\phi\Big|_{\Gamma_I} = f_I,\quad \mathrm{Re} c\phi\Big|_{\Gamma_{II}} = f_{II} + \eta. \tag{11.15}$$

Here η is an arbitrary vector, which is constant on connected components of Γ_{II}. This vector has to be defined together with ϕ. According to Theorem 1 the solution Φ is subject to the additional condition

$$\mathrm{Re} bJ\Phi(x_0) = 0 \tag{11.16}$$

at a fixed point x_0.

As an example we consider the quadrant $D = \{x \mid x_1 > 0, x_2 > 0\}$ with the sides Γ_I and Γ_{II} along the axis x_1 and x_2, respectively. For this domain in notation (11.10) we have $Q_0 = Q_\infty^{-1} = -J$ and

$$X(\zeta, 0) = \begin{pmatrix} \bar{c} & e^{\pi i\zeta} c(-J)^{\zeta} \\ e^{\pi i \zeta}\bar{b}(-\bar{J})^{-\zeta} & b \end{pmatrix}.$$

From (1.13) and (11.14) direct calculations show that

$$\det X(\zeta, 0) = \det X(\zeta, \infty) =$$

$$= 16\kappa e^{\pi i\zeta}\left(\sin^2\frac{\pi\zeta}{2} + \frac{\zeta^2}{\kappa} - \frac{(\kappa+1)^2}{4\kappa}\right),$$

$$\det X_0(\zeta, 0) = \det X_0(\zeta, \infty) = -2e^{\pi i\zeta}\cos\pi\zeta.$$

Hence the index of the Riemann-Hilbert problem (11.9) is equal to zero in the considered case. Therefore, the index of the problem (11.15), (11.16), which is equivalent to the initial mixed problem for the Lamé system, is zero too. We shall mention that function theoretical approaches for the problems ([19, 18]) of plane elasticity theory has been studied by R.Gilbert, W.Wendland [6], R.Gilbert, W.Lin [20], Hua Loo Keng [21], H.Begehr, W.Lin [22], R.Duduchava [23], A.Soldatov, N.Zhura [25], A.Soldatov, S.Mitin [26] and others.

References

[1] Bitsadze, A.V.: Boundary value problems for second order elliptic equations. Nauka, Moskow, 1966.

[2] Douglis, A.: A function theoretic approach to elliptic systems of equations in two variables. Comm. Pure Appl. Math. 6 (1953), 259–289.

[3] Pascali, D.: Vecteurs analytiques généralisés. Rev. Roumaine Math. Pure Appl. 10 (1965), 779–808.

[4] Bojarski, B.: Theory of generalized analytic vectors. Ann. Polon. Math. 17 (1966), 281–320 (Russian).

[5] Gilbert, R.P., Hile G.N.: Degenerate elliptic systems whose coefficient matric has a group inverse. Complex Variables, Theory Appl. 1 (1982), 61–87.

[6] Gilbert, R.P., Wendland W.L.: Analytic, generalized hyperanalytic function theory and an application to elasticity. Proc. Roy. Soc. Edinburgh 73A (1975), 317–371.

[7] Hile, G.N.: Elliptic systems in the plane with first order terms and constant coefficients. Comm. Partial Diff. Eq. 3 (1978), 949–977.

[8] Soldatov, A.P.: High order elliptic systems. Diff. Eq. 25 (1989), 109–115.

[9] Yeh, R.Z.: Hyperholomorphic functions and higher order partial differential equations in the plane. Pacif. J. Math. 142 (1990), 379–399.

[10] Balk, M.B.: Polyanalytic functions. Akademie Verlag, Berlin, 1991.

[11] Zhura, N.A.: A general boundary value problem for systems ellipitic in the Douglis–Nirenberg sense in domains with piecewise smooth boundary. Russ. Acad. Sci. Dokl. Math. 47 (1993), 421–426.

[12] Zhura, N.A.: A nonlocal boundary value problem for the steady–state linearised two–dimensional system of Navier–Stokes equations. Soviet Math. Dokl. 41 (1990), 264–268.

[13] Zhura, N.A.: Bitsadze–Smarskij type boundary value problems for Douglis–Nirenberg elliptic systems. Diff. Eq. 28 (1992), 79–88.

[14] Soldatov, A.P.: A general boundary value problem in the theory of functions. Soviet Math. Dokl. 37 (1988), 486–489.

[15] Soldatov, A.P.: On the Noether theory of operators. One–dimensional singular integral operators of general form. Diff. Eq. 14 (1978), 498–508.

[16] Soldatov, A.P.: Singular integral operators and boundary value problems of function theory. Visshaya Shkola, Moscow, 1991 (Russian).

[17] Duduchava, R.V.: Integral equations of convolution type with discontinuous presymbols, singular integral equations with fixed singularities and their applications to some problems of mechanics. Trudy Tbil. Mat. Inst. AN GSSR **60** (1979), 1–135 (Russian).

[18] Muskhelishvili, N.I.: Some basic problems of the mathematical theory of elasticity. Noordhoff, Groningen, 1953.

[19] Parton, V.Z., Perlin, P.I.: Integral equations of elasticity. Nauka, Moscow, 1977 (Russian).

[20] Gilbert, R.P., Lin, W.: Function theoretic solutions to problems of orthogonal elasticity. J. Elasticity 15 (1985), 143–154.

[21] Hua Loo Keng, Lin, W., Wu Ci-Quian: Second order systems of partial differential equations in the plane. Research Notes in Mathematics 128. Pitman, London, 1985.

[22] Begehr, H., Lin W.: A mixed-contact boundary problem in orthotropic elasticity. Partial diff. equations with real analysis. Longman, Harlow, 1992, 219–239.

[23] Duduchava, R.V.: General singular integral equations and basic problems of planar elasticity theory. Trudy Tbil. Mat. Inst. AN GSSR 82 (1986), 45–89 (Russian).

[24] Soldatov, A.P.: A mixed problem of the plane theory of elasticity in regions with a piecewise-smooth boundary. Diff. Eq. 23 (1987), 2–27.

[25] Soldatov, A.P., Zhura, N.A.: A mixed contact problem from plane elasticity theory in domains with piecewise-smooth boundaries. Diff. Eq. 24 (1988), 42–50.

[26] Soldatov, A.P., Mitin S.P.: On the solvability of mixed-contact problems in two-dimensional elasticity theory. Diff. Eq. 29 (1993), 760–763.

12 BOUNDARY VALUE PROBLEMS FOR A SYSTEM OF FIRST ORDER EQUATIONS OF A PRINCIPAL TYPE IN THE PLANE

N. Zhura

People's Friendship University
Ordzhonicidze 3
117923 Moscow
Russian Federation

Abstract: The paper deals with boundary value problems for a system of equations of principal type with constant real coefficients in some domains in the plane. For the regular solutions of these systems a representation is obtained by means of solutions of a (canonical) elliptic and a (canonical) hyperbolic system. A formula for the index of these problems are given.

12.1

In the (x_1, x_2) - plane $\mathbb{R}^2$ we consider a system of linear partial differential equations

$$\partial u/\partial x_2 = a\partial u/\partial x_1, \tag{12.1}$$

where $u(x_1, x_2)$ is an unknown l - vector-valued function, $a \in \mathbb{R}^{l\times l}$ is a constant matrix. The system (12.1) is said to be of composite type (principal one [1], [2]) if it's characteristic equation

$$\det(a - \nu) = 0 \tag{12.2}$$

has $s_1 \geq 1$ complex roots with positive imaginary parts and $s_2 \geq 1$ real roots, $2s_1 + s_2 = l$. Let $b_1 \in \mathbb{C}^{s_1\times l}$, $b_2 \in \mathbb{R}^{s_2\times l}$ be constant matrices such that the

H.G.W. Begehr et al. (eds.), Partial Differential and Integral Equations, 187–191.

nonsingular matrix $b = (b_1|\overline{b_1}|b_2)$ reduces a to the Jordan normal form

$$b^{-1}ab = \mathrm{diag}\,(J_1, \overline{J_1}, J_2), \tag{12.3}$$

where $J_1 \in \mathbb{C}^{s_1 \times s_1}$ is a triangular matrix whose eigenvalues all have a positive imaginary part and $J_2 \in \mathbb{R}^{s_2 \times s_2}$ is a triangular matrix with real eigenvalues.

Theorem 1 [4] *Any regular solution u of the system* (12.1) *can be represented in the form*

$$u = 2\mathrm{Re}\, b_1\Phi + b_2\Psi, \tag{12.4}$$

where Φ is a regular solution of the canonical elliptic system

$$\partial\Phi/\partial x_2 = J_1 \partial\Phi/\partial x_1, \tag{12.5}$$

and Ψ is a regular solution of the canonical hyperbolic system

$$\partial\Psi/\partial x_2 = J_2 \partial\Psi/\partial x_1. \tag{12.6}$$

Solutions of the system (12.5) are said to be J_1 – analytical functions. Moreover, there exists an l-vector-valued function $\psi = (\psi_k)^1_{k=2s_1+1}$, $\psi_k \equiv \psi_k(x_1 + \nu_k x_2)$, such that

$$\Psi(x_1, x_2) = (\exp(x_2 J_{2,0} \partial/\partial x_1))\psi, \;\; J_2 = \nu + J_{2,0}. \tag{12.7}$$

12.2

We consider a boundary value problem

$$Cu = f \text{ on } \Gamma, \tag{12.8}$$

for the system (12.1) where C is an $(s_1 + s_2) \times l$ matrix-valued function, and f is an $(s_1 + s_2)$ vector-valued function on $\Gamma = \partial D$.

Let $H_\mu(\overline{D})$ be the space of functions satisfying a Hölder condition on the closure $\overline{D}$ of the domain D with exponent $0 < \mu \leq 1$ (and bounded if D is infinite). The space $H^k_\mu(\overline{D})$ consists of the functions with partial derivatives in $H^{k-1}_\mu(\overline{D})$, $k \geq 1$, $(H^0_\mu \equiv H_\mu)$. The space $H^k_\mu(\overline{D})$ is a Banach space with respect to the corresponding norm.

This also applies to the classes H^k_μ on Γ. We consider the problem (12.1),(12.8) in the class $H^1_\mu(\overline{D})$ of solutions (12.1) such that the functions Φ and Ψ belong to the class $H^1_\mu(\overline{D})$ in the representation (12.4). Let us put $\overline{s_2} = s_2$ for definiteness, so there exists only one block in J_2. Then we say that the vector-valued function Ψ defined by (12.7) belongs to the class H^1_μ if the components of ψ belong to the class $H^{l-2s_1+1-k}_\mu$, $k = 1, \ldots, s_2$, as functions of one variable. In the general case this definition is regarded with respect to each Jordan block of J_2. In what follows it is assumed that

$$\gamma'(t) \in H^k_{\mu+\varepsilon}, \quad t \in \overline{\mathbb{R}}, \; k = \overline{s_2}, \varepsilon > 0, 0 < \mu < 1, \tag{12.9}$$

and

$$\delta'_\nu(t) \neq 0, \quad t \in \overline{\mathbb{R}}, \tag{12.10}$$

for any real root ν of equation (12.2), where $\gamma: \mathbb{R} \to \Gamma$ is the parametrization of Γ, $\delta_\nu(t) = \text{Re}\gamma + \nu\text{Im}\gamma$, and $\overline{s_2}$ is the degree of the real root ν in the minimal polynom of the matrix a. It is assumed also, that

$$C, f \in H^k_\mu(\Gamma), \; k = \overline{s_2}, \tag{12.11}$$

and

$$|(b^{-1}u)_1| \leq \text{const}(|x_1| + |x_2|)^{-1} \tag{12.12}$$

as $|x_1| + |x_2|$ tends to infinity, where by $(b^{-1}u)_1$ we denote the first s_1 elements of the vector $b^{-1}u$.

Let us put

$$C = \begin{pmatrix} C_1 \\ C_2 \end{pmatrix}, \quad f = \begin{pmatrix} f_1 \\ f_2 \end{pmatrix}. \tag{12.13}$$

where C_k is an $(s_k \times l)$ matrix and f_k is an s_k vector, $k = 1, 2$.

Whithout loss of generality we can assume that

$$\det C_2 b_2 \neq 0 \text{ on } \Gamma. \tag{12.14}$$

Let us consider

$$A = C_1(1 - b_2(C_2 b_2)^{-1} C_2) b_1. \tag{12.15}$$

We say that (12.1),(12.8) is a normal type problem if

$$\det A \neq 0 \quad \text{on } \overline{\Gamma}. \tag{12.16}$$

Theorem 2 *Suppose that the conditions* (12.9), (12.10) *for the contour* Γ *and condition* (12.11) *for* C, f *are fullfilled. Then the problem* (12.1), (12.8) *is Noetherian in* $H^1_\mu(\overline{D})$ *if and only if the normality condition* (12.16) *is satisfied, and its index is*

$$\kappa = -\frac{1}{\pi} \arg \det A \Big|_{\overline{\Gamma}} + s_1. \tag{12.17}$$

12.3

Suppose now that the boundary ∂D of the infinite domain D consists of two smooth curves Γ_1 and Γ_2 that connect two corner points $\tau_1 = (0, 0)$ and $t_2 = (1, 0)$. We consider the following boundary value problem

$$C_j u = f_j \text{ on } \Gamma_j, \quad j = 1, 2, \tag{12.18}$$

for the system (12.1), where C_j is an $(s_1 + s_2) \times l$ matrix and f_j is an $(s_1 + s_2)$ vector.

We consider the problem (12.1),(12.18) in the class $H^1_{\mu,\lambda}(D)$ of solutions of (12.1) such that the functions Φ and Ψ belong to $H^1_{\mu,\lambda}(D_0)$, $D_0 = D^1 \cup D^2$, in the representation (12.4).

The solution u of the problem (12.1), (12.18) may be defined only in the domains D^1 and D^2 such that $D^1 \cup D^2 = D \cap (\cap_j D_j)$, where D_j is a strip with the boundary lines

$$x_1 + \nu_j x_2 = \tau_k, \quad k = 1,2, \; j = 2s_1 + 1, \ldots, l \,.$$

We say that the J_1 - analytic function Φ belongs to $H_{\mu,\lambda}(D)$ with the weight order $\lambda = (\lambda_0, l_1)$, if

$$\Phi(z) = |z|^{l_0 - \mu} |z-1|^{\lambda_1 - \mu} \Phi_*(z),$$

where $\Phi_*(z) \in H_\mu(\overline{D})$ and $\Phi_*(\tau_j) = 0, j = 1,2$. The classes $H^k_{\mu,\lambda}(D)$ of differentiable functions are introduced by induction. The estimate (12.12) is assumed to be satisfied. By θ_j denote the angle of the sector corresponding to the corner τ_j. Evidently, $\theta_j = \arg q_k - \arg q_r$, $0 < \arg q < 2\pi$, $0 < \theta_j < 2\pi$, $P_j = \overline{k,r}$, $j = 1,2$, (more presicely, $P_1 = \overline{1,3}$, $P_2 = \overline{4,2}$), where the rotation from the vector q_k to q_r about τ_j within the domain is clock-wise and the vector q is identified with the complex numbers $q_{2i-1} = \gamma_i'(0)$, $q_{2i} = -\gamma_i'$ (12.1).

Let us define the functions of matrices

$$m_j(\zeta) = (\text{Re } q_k + (\text{Im } q_k) J_1)^\zeta (\text{Re } q_r + (\text{Im } q_r) J_1)^{-\zeta}, \; \overline{k,r} = P_j,$$

and let be

$$\begin{aligned} x_0(\zeta) &= \left(e^{2\pi i \zeta} - 1\right)^{-1} \left(A_1(0) w_1(\zeta) + \overline{A_1(0)} v_1(\zeta) w_1(\zeta)\right), \\ x_1(\zeta) &= \left(e^{2\pi i \zeta} - 1\right)^{-1} \left(\overline{A_2(1)} v_2(\zeta) w_2(\zeta) + A_2(1) w_2(\zeta)\right), \end{aligned} \tag{12.19}$$

where

$$v_j = \left(\begin{array}{cc} \dfrac{0}{m_j^{-1}(\overline{\zeta})} & \begin{array}{c} m_j(\zeta) \\ 0 \end{array} \end{array} \right), \quad w_j(\zeta) = e^{2\pi i \zeta} \overline{v_j(\overline{\zeta})} - 1, \; j = 1,2,$$

$$A_j = C_{j,1}(1 - b_2 (C_{j,2} b_2)^{-1} C_{j,2}) b_1, \; j = 1,2.$$

Theorem 3 *Suppose that the conditiones* (12.9), (12.10) *for the curves* Γ_1 *and* Γ_2 *including the corners* τ_1, τ_2 *are fullfilled. Let* C, f *belong to* $H^k_{\mu,\lambda}$, $k = \overline{s_2}$, *and the condition* (12.12) *holds. Suppose that the normality condition*

$$\det A_j(t) \neq 0, \quad 0 \le t \le 1, \; j = 1,2\,, \tag{12.20}$$

is satisfied. Then the problem (12.1), (12.18) *is Noetherian in* $H^1_{\mu,\lambda}(D)$ *if and only if*

$$\det x_k(\zeta) \neq 0, \quad \text{Re}\zeta = \lambda_k, \; k = 0,1, \tag{12.21}$$

and its index is

$$\kappa = -\frac{1}{\pi} \arg \det (A_1(t) A_2^{-1}(t)) \Big|_0^1 - \frac{1}{2\pi} \sum_{k=0}^{1} \arg \det x_k(\zeta) \Big|_{\zeta = \lambda_k - i\infty}^{\lambda_k + i\infty} - s_1. \tag{12.22}$$

12.4

We now consider the problem

$$C_j u = f_j \text{ on } \Gamma_j,\ j = 1, 2, \tag{12.23}$$

in finite domains D, whose boundary ∂D consists of two curves Γ_1 and Γ_2 with the corners τ_1 and τ_2. We assume that the matrix $C_1(C_2)$ of order $(s_1 + s_2) \times l\ (s_1 \times l)$, and the vector $f_1(f_2)$ of order $s_1 + s_2\ (s_1)$ are prescribed on $\Gamma_1(\Gamma_2)$ and f_j, C_j are belonging to $H^k_{\mu,\lambda}$, $k = \overline{s_2}$, $j = 1, 2$. Here the curves Γ_j satisfy conditions (12.9), (12.10).

Theorem 4 *The assertion of Theorem* 3 *remains in force also for the problem* (12.1), (12.23), *provided only that A_j mean the matrices $A_1 = C_{1,1}(1 - b_2(C_{1,2}b_2)^{-1}\ C_{1,2})b_1, A_2 = C_2 b_1$, and the last term $-s_1$ in the formula* (12.22) *must be replaced by s_1.*

12.5

We also have studied the questions of asymptotics of the solutions near the corner points and the smoothness of the solutions up to the boundary. We generalized this approach for systems of higher order and for a class of admissible finite domains with piecewise smooth boundary. If the order of C_1 in the last problem is not equal to $s_1 + s_2$ then we investigated this problem only for the case $\overline{s_2} = 1$.

Our study is carried out in the framework of the function-theoretic methods [3]. The scheme of this method is as follows. First of all we express a general regular solution in terms of regular solutions Φ and Ψ and use an anologue of a theorem of Vekua on integral representations of Φ and some notions about Ψ which arises from (12.7). By substituting that into the corresponding boundary conditions we reduce the problem to system of integral equations on the boundary of the domain. For another approach see [5].

References

[1] Hadamard, J.: Propriétés d'une équation linéaire aux dérivées partielle du quatrième ordre. Tohoky Math. J. 37 (1933), 133-150.

[2] Hörmander, L.: Linear partial differential operators. Springer–Verlag, Berlin, 1963.

[3] Soldatov, A.P.: Singular integral operators and boundary value problems of function theory. Visshaya Shkola, Moscow, 1991 (Russian).

[4] Zhura, N.A.: A general boundary value problem for systems elliptic in the Douglis–Nirenberg sense in domains with piecewise smooth boundary. Russ. Acad. Sci. Dokl. Math. 47 (1993), 421–426.

[5] Eskin, G.I.: Boundary value problems for equations with constant coefficients on the plane. Mat. Sbornik (N.S.) 59 (101) (1962), Suppl., 67-124 (Russian).

13 THE STABILITY OF THE RIEMANN BOUNDARY VALUE PROBLEM FOR VECTOR–VALUED M–ANALYTIC FUNCTIONS

Chuan–Gan Hu †

Nankai University
Department of Mathematics
Tianjin 300071
P.R. China

Dan–Ping Zhang

Beijing Institute of Machinery Industry
Research Section of Mathematics
Beijing 100085
P.R. China

Abstract: In this paper, certain boundary value problems are solved in a Banach space which is defined by vector–valued regular solutions of the system $f_x + Mf_y = 0$ of elliptic partial differential equations, where M is a constant $m \times m$ matrix without real eigenvalues and f is an $m \times q$ matrix. The properties of the integral of Cauchy type for vector-valued M-analytic functions are discussed. Next, the representation of the solutions of the Riemann boundary value problem and its pertubed problem for vector-valued M-analytic functions are presented.

†Project 19571040 supported by the National Natural Science Foundation of China and the Construction Foundation of Major Subject of Education Commission of Tianjin City.

H.G.W. Begehr et al. (eds.), Partial Differential and Integral Equations, 193–205.

Key words: M-analytic functions, Riemann boundary value problem, perturbation, Banach space, stability.
AMS Classification: 30E20, 30E25, 30D35, 46G20, 45N05.

13.1 INTRODUCTION

G.N.Hill introduced the concept of M-analytic functions when he studied a system and showed many properties from complex analysis of one variable, for example, Cauchy's integral formula, Taylor's expansion, Laurent's expansion and Morera's theorem etc[1]. On the above foundations, W.Lin and D.Dai gave some properties of the Cauchy type integral on M-analytic functions and their application [8]. In this paper, we shall extend some conclusions in [8] to the vector-valued case with range in a Banach space and research further the Riemann boundary value problem of vector-valued M-analytic functions and its perturbed problem with range in a Banach space.

13.2 CAUCHY TYPE INTEGRAL

Let $D = \{w : |w| < 1\}$ and $\bar{D} = \{w : |w| \leq 1\}$ and let $U = \{f = [a_{ij}(w)]_{m\times q}\}$ be the set of all $m \times q$ M-analytic functions in D and continuous on $\overline{D}$.

Definition 1 *The norm $|||f|||$ of f is defined by* $|||f||| = \sup_{|w|\leq 1} \left(\sum_{i,j} |a_{ij}(w)|^2\right)^{1/2}$.

It is easy to verify that U endowed with this norm is a Banach space.

Suppose that Ω^+ is a bounded simply connected domain in the complex plane $\mathbb{C}$ and its boundary is a simply closed smooth curve L, that $\Omega^- = \mathbb{C}\backslash\overline{\Omega}^+$ where $\overline{\Omega}^+$ is the closure of Ω^+, i.e. Ω^- is the exterior of L, and that the set $H_\alpha(L)$ of all $m \times q$ matrix-valued functions on L is defined by

$$\{\varphi : |||\varphi|||_\alpha = \max_{s\in L} \|\varphi(s)\| + \sup_{\substack{s_1,s_2\in L \\ s_1\neq s_2}} \frac{\|\varphi(s_1) - \varphi(s_2)\|}{\|s_1 - s_2\|^\alpha} < \infty\},$$

where $0 < \alpha < 1$ and $\|\varphi(s)\| = \left[\sum_{i,j} |\varphi_{ij}(s)|^2\right]^{1/2}$ $(\varphi(s) = [\varphi_{ij}(s)]_{m\times q})$.

Note, since there is an one-to-one correspondence between a point $z = x + iy$ in $\mathbb{C}$ and $xI - yM^{-1}$ where M is a constant $m \times m$ matrix without real eigenvalue and I is the unit matrix, we use them in the sense of an one-to-one correspondence afterwards, i.e. $(x, y) \equiv z \equiv xI - yM^{-1}$.

Definition 2 *Suppose that $\varphi(s, w)$ is an $m \times q$ matrix-valued bounded function and integrable with respect to s over L, that $\varphi(s, w)$ (rewrite $\varphi_s(w)$) $\in U$ for any fixed $s \in L$ and that the integral*

$$\Phi(z, w) = Q^{-1} \int_L (s - z)^{-1} ds\, \varphi(s, w) \tag{13.1}$$

exists as $z \notin L$, then the integral of the right-hand side in (13.1) *is called the integral of the Cauchy type with density $\varphi(s,w)$, where Q is a proper constant matrix.* Clearly, $(s-z)^{-1}$ is an M-analytic function of z as $s \neq z$ (see [1]). So the function $\Phi(z,w)$(rewrite $\Phi_w(z)$) in (13.1) is a piecewise M-analytic function for any fixed $w \in D$, i.e. $\Phi_w(z)$ is M-analytic in Ω^+ and Ω^- for any fixed $w \in D$, respectively, and $\Phi_w(\infty) = 0$.

Definition 3 *Suppose that $t_0 \in L$, that $l(t_0,\varepsilon)(\subset L)$ is an arc, whose endpoints t' and t'' satisfy the condition $|t_0 - t'| = |t_0 - t''| = \varepsilon$ with center t_0, and that the following limit*

$$\lim_{\varepsilon \to 0} Q^{-1} \int_{L - l(t_0,\varepsilon)} (s-t)^{-1} ds\, \varphi(s,w) \tag{13.2}$$

exists, then this limit (13.2) *is called the Cauchy principal value of the integral of $\varphi(s,w)$ over L or the Cauchy principal value for short, denoted by $\Phi(t_0,w) = Q^{-1} \int_L (s-t_0)^{-1} ds\, \varphi(s,w)$.*

Theorem 1 *If $\varphi(s,w)(=\varphi_s(w)) \in U$ for any fixed $s \in L$ and $\varphi(s,w)(= \varphi_w(s)) \in H_\alpha(L)$ for any fixed $w \in D$, then there are the following results:*

(i) *For any fixed $z \notin L$, the integral of Cauchy type in* (13.1) *$\Phi_z(w)(= \Phi(z,w)) \in U$.*

(ii) *The Cauchy's principal value* (13.2) *exists and*

$$\begin{aligned} Q^{-1} \int_L (s-t)^{-1} ds\, \varphi(s,w) &= Q^{-1} \int_L (s-t)^{-1} ds\, [\varphi(s,w) - \varphi(t,w)] \\ &\quad + Q^{-1} \textstyle\int_L (s-t)^{-1} ds\, \varphi(t,w) \end{aligned} \tag{13.3}$$

as $t \in L$ and $w \in D$.

Assume that L is a continuous curve in $\mathbb{C}$, that for any z in $\mathbb{C}$, $d(z,L)$ is the distance between z and L, and that z_L in L is the closest point to z, i.e.$d(z,L) = d(z,z_L)$, where $z_L \in L$. It follows that $|t - z_L| \le |t-z| + |z - z_L| \le 2|t-z|$.

Lemma 1 *If $\varphi_w(s) \in H_\alpha(L)$ for any fixed $w \in D$ and $\varphi_s(w) \in U$ for any fixed $s \in L$ and if $l(\subset L)$ is a subarc with arc length $v(l)$, then*

$$\left\| \int_l (s-z)^{-1} ds\, [\varphi(s,w) - \varphi(z_l,w)] \right\| \le c_1 [v(l)]^\alpha \tag{13.4}$$

for any $z \in \mathbb{C}$ and

$$\left\| \int_l (s-z)^{-2} ds\, [\varphi(s,w) - \varphi(z_l,w)] \right\| \le c_2 |z - z_l|^{\alpha - 1} \tag{13.5}$$

for any $z \notin l$, here c_1 and c_2 are constants.

$\Phi^+(t_0,w)$and $\Phi^-(t_0,w)$ denote the limit values of a function $\Phi(z,w)$ as z approaches to $t_0 \in L$ from the left-hand and the right-hand neighborhoods of the curve L respectively. Then there are the following relationship between $\Phi^+(t_0,w)$ and $\Phi^-(t_0,w)$.

Theorem 2 *If $\varphi(s,w)$ satisfies conditions in Theorem 1, then the limit value of the integral of the Cauchy type in* (13.1) *exists and there are the following Plemelj's formulas*

$$\Phi^+(t_0,w) = \lim_{\substack{z\in\Omega^+ \\ z\to t_0}} \Phi(z,w) = \frac{1}{2}\varphi(t_0,w) + Q^{-1}\int_L (s-t_0)^{-1} ds\, \varphi(s,w) \quad (13.6)$$

and

$$\Phi^-(t_0,w) = \lim_{\substack{z\in\Omega^- \\ z\to t_0}} \Phi(z,w) = -\frac{1}{2}\varphi(t_0,w) + Q^{-1}\int_L (s-t_0)^{-1} ds\, \varphi(s,w). \quad (13.7)$$

Lemma 2 (Extended Residue Theorem)*Suppose that L is a piecewise smooth curve, the interior of which is a domain J, with a positive direction, that $\Phi(z)$ is an $m\times q$ matrix-valued and M-analytic function in J except for singularities $z_1,\cdots,z_M$, that $\Phi(z)$ has poles $t_1,\cdots,t_n$ in L with orders $n_1,\cdots,n_N$ respectively, and that*

$$\frac{d^{n-1}}{dt^{n-1}}\left[\prod_{j=1}^{N}(t-t_j)^{n_j}\Phi(t)\right] \in H_\alpha(L),$$

where $n = \max_j n_j$, then

$$Q^{-1}\int_L dt\,\Phi(t) = \sum_{k=1}^{M}\operatorname{Res}\Phi(z_k) + \sum_{j=1}^{N}\frac{1}{2}\operatorname{Res}\Phi(t_j).$$

Proof We give the proof in the following three steps.

Step 1. Assume that $\Phi(z)$ is M-analytic in J and has a pole t_0 in L with order n.

We rewrite $\Phi(z) = (z-t_0)^{-n}f(z)$ with $f(t_0)\neq 0$ for any $z\in\overline{J}$, where $f(z)$ is M-analytic in J and continuous on $\overline{J}$ which is the closure of J. By assumptions we know $d^{n-1}/dt^{n-1}[f(t)]\in H_\alpha(L)$ and

$$\int_L dt\,\Phi(t) = \int_L (t-t_0)^{-n}dt\, f(t) = \frac{1}{(n-1)!}\int_L (t-t_0)^{-1}dt\, f^{(n-1)}(t).$$

Since $f^{(n-1)}(z)$ is M-analytic in J and continuous of $\overline{J}$, Cauchy's formulas [2, 7] yield

$$Q^{-1}\int_L (t-z)^{-1}dt\, f^{(n-1)}(t) = f^{(n-1)}(z).$$

Due to $f^{(n-1)}(t)\in H_\alpha(L)$, then from the Plemelj formulas we derive

$$f^{(n-1)}(t_0) = \frac{1}{2}f^{(n-1)}(t_0) + Q^{-1}\int_L (t-t_0)^{-1}dt\, f^{(n-1)}(t).$$

So

$$Q^{-1}\int_L (t-t_0)^{-1}dt\, f^{(n-1)}(t) = \frac{1}{2} f^{(n-1)}(t_0).$$

It follows that

$$Q^{-1}\int_L dt\, \Phi(t) = \frac{Q^{-1}}{(n-1)!}\int_L (t-t_0)^{-1}dt\, f^{(n-1)}(t)$$
$$= \frac{1}{2}\mathrm{Res}\,\Phi(t)|_{t=t_0} = \frac{1}{2}\mathrm{Res}\,\Phi(t_0).$$

Step 2. Assume that $\Phi(z)$ is M-analytic in J and has poles $t_1,\cdots,t_N$ in L with orders $n_1,\cdots,n_N$ respectively. Setting $f(t) = \Pi_{j=1}^{N}(t-t_j)^{n_j}\Phi(t)$ and using the assumptions we know $f^{(n-1)}(t) \in H_\alpha(L)$ as $t \in L$. For each t_j we make an arc $l_j \subset L$ and a circular arc γ_j in J with center t_j and radius r_j small enough such that $l_j \bigcup \gamma_j$ is a closed curve and $t_k \notin J_j$ as $k \neq j$ where $J_j = l_j \bigcup \gamma_j$. Because $\Phi(z)$ is M-analytic, the Cauchy theorem implies

$$Q^{-1}\int_L dt\, \Phi(t) = \sum_{j=1}^{N} Q^{-1}\int_{J_j} dt\, \Phi(t).$$

Consequently

$$Q^{-1}\int_L dt\, \Phi(t) = \sum_{j=1}^{N} \frac{1}{2}\mathrm{Res}\,\Phi(t_j).$$

Step 3. If $\Phi(z)$ has singularities $z_1,\cdots,z_M$ in J and poles $t_1,\cdots,t_N$ in L with orders $n_1,\cdots,n_N$ respectively. We make mutually disjoint circles C_k : $|z-z_k| = \rho_k$ (for $k=1,\cdots,M$) in each neighborhood of z_k in J for sufficiently small ρ_k and $l_j \bigcup \gamma_j$ according to the method of Step 2 for $j=1,\cdots,N$. From the properties of $\Phi(z)$ and the Cauchy Theorem we derive

$$Q^{-1}\int_L dt\, \Phi(t) = \sum_{k=1}^{M} \mathrm{Res}\,\Phi(z_k) + \sum_{j=1}^{N} \frac{1}{2}\mathrm{Res}\,\Phi(t_j).$$

This completes the proof of the lemma.

13.3 THE PROBLEM OF DETERMING AN M-ANALYTIC FUNCTION FROM A JUMP

Suppose that MCF denotes the set of all $m \times q$ matrix-valued continuous functions with ranges in U, that L is a simple smooth closed curve in the complex plane $\mathbb{C}$ with positive direction as a jump curve, that the interior of L is the inner domain, denoted by Ω^+ with $0 \in \Omega^+$and the exterior of L is the external domain, denoted by Ω^- and that $\varphi(t,w) \in U$ for any fixed t in L and $\varphi(t,w) \in H_\alpha(L)$ for any fixed $w \in D$ as $(t,w) \in L \times D$.

We find a piecewise M-analytic function $\Phi(z,w) \in U$ for any fixed $z \in \mathbb{C}$ with finite order at $z=\infty$ and a smooth jump curve L such that

$$\Phi^+(t,w) - \Phi^-(t,w) = \varphi(t,w) \quad \text{for} \quad (t,w) \in L \times D, \tag{13.8}$$

where $\Phi^{\pm}(t,w)$ are the limit values on L of $\Phi(z,w)$ from the inner domain Ω^+ and the external domain Ω^- relative to L respectively.

Now we solve problem (13.8) as follows. Since $\Phi(z,w)$ is M-analytic in Ω^+ for any fixed $w \in D$ with the limit value $\Phi^+(t,w)$ on L, from the Cauchy Theorem we derive

$$\Phi(z,w) = Q^{-1}\int_L (s-z)^{-1}ds\,\Phi^+(s,w) \text{ for } z \in \Omega^+, \tag{13.9}$$

$$0 = Q^{-1}\int_L (s-z)^{-1}ds\,\Phi^+(s,w) \text{ for } z \in \Omega^-. \tag{13.10}$$

Similarly, we have

$$p(z,w) = Q^{-1}\int_L (s-z)^{-1}ds\,\Phi^-(s,w) \text{ for } z \in \Omega^+ \tag{13.11}$$

$$-\Phi(z,w) + p(z,w) = Q^{-1}\int_L (s-z)^{-1}ds\,\Phi^-(s,w) \text{ for } z \in \Omega^-, \tag{13.12}$$

where $p(z,w) = \sum_{n=0}^{k} z^n A_n(w)$ (here $A_n(w) \in U$ for any positive integer n) is a polynomial matrix and the principal part of $\Phi(z,w)$ at $z = \infty$.

From formulas (13.9)-(13.12) we derive

$$\begin{cases} \Phi(z,w)-p(z,w) = Q^{-1}\int_L (s-z)^{-1}ds\,[\Phi^+(s,w)-\Phi^-(s,w)], \; z \in \Omega^+ \\ \Phi(z,w)-p(z,w) = Q^{-1}\int_L (s-z)^{-1}ds\,[\Phi^+(s,w)-\Phi^-(s,w)], \; z \in \Omega^-, \end{cases}$$

that is

$$\Phi(z,w) - p(z,w) = Q^{-1}\int_L (s-z)^{-1}ds\,\left[\Phi^+(s,w) - \Phi^-(s,w)\right] \text{ for } z \notin L.$$

Combining the above we can get the general solution of problem (13.8)

$$\Phi(z,w) = Q^{-1}\int_l (s-z)^{-1}ds\,\varphi(s,w) + p(z,w) \text{ for } z \notin L.$$

13.4 THE RIEMANN BOUNDARY VALUE PROBLEM

We find a piecewise M-analytic function $\Phi(z,w)$, with finite order at $z=\infty$ in MCF and jump curve L, which satisfies the condition

$$\Phi^+(t,w) = G(t,w)\Phi^-(t,w) + g(t,w), t \in L \tag{13.13}$$

and $\Phi(z,w) \in U$ for any fixed $z \in \mathbb{C}$, where $g(t,w) \in H_\alpha(L)$ for any fixed $w \in D, g(t,w) \in U$ for any fixed t in L, $G(t,w)$ is an operator-valued function from $L \times D$ to $\mathcal{L}(U,U)$ (the set of all bounded linear operators from U to U), $G^{-1}(t,w)$ exists; $MG(t,w) = G(t,w)M$ and $\Phi^{\pm}(t,w) \in H_\alpha(L)$ for any fixed $w \in D$.

We are going to reduce problem (13.13) to a singular integral equation with Cauchy kernel.

A new unknown $m \times q$ bounded matrix-valued function $\Phi(t, w)$, is introduced with $\varphi(t, w) \in U$ for any fixed $t \in L$ through

$$\Phi(z,w) = \begin{cases} Q^{-1} \int_L (s-z)^{-1} ds\,[G(s,w)\varphi(s,w) + g(s,w)], & \text{for } z \in \Omega^+, \\ -Q^{-1} \int_L (s-z)^{-1} ds\,\varphi(s,w) + p(z,w), & \text{for } z \in \Omega^-, \end{cases} \tag{13.14}$$

where $p(z, w)$ is a polynomial matrix and the principal part of $\Phi(z, w)$ at $z = \infty$. The solution of problem (13.13) can be written certainly the form of (13.14).

In (13.14), let $z \to t$ for z in Ω^+ and substitute it into (13.13), we have

$$\begin{aligned} & Q^{-1} \int_L (s-t)^{-1} ds\,[G(s,w)\varphi(s,w) + g(s,w)] \\ = \; & G(t,w) \left[-Q^{-1} \int_L (s-t)^{-1} ds\,\varphi(s,w) + p(t,w) \right] + g(t,w). \end{aligned}$$

Using the assumptions, we know that $G^{-1}(t, w)$ exists and commutes with M, thereby,

$$\begin{aligned} & G^{-1}(t,w) Q^{-1} \int_L (s-t)^{-1} ds\,[G(s,w)\varphi(s,w) + g(s,w)] \\ & + Q^{-1} \int_L (s-t)^{-1} ds\,\varphi(s,w) \\ = \; & p(t,w) + G^{-1}(t,w) g(t,w). \end{aligned}$$

It follows that

$$\begin{aligned} & Q^{-1} \int_L (s-t)^{-1} ds\, G^{-1}(t,w) G(s,w)\varphi(s,w) \\ & + Q^{-1} \int_L (s-t)^{-1} ds\,\varphi(s,w) \\ = \; & p(t,w) + f(t,w), \end{aligned}$$

where $f(t, w) = Q^{-1} \int_L (s-t)^{-1} ds\, G^{-1}(t, w)[2g(t, w) - g(s, w)]$. By Theorem 2.4 in [8], we see that $f(t, w) \in H_\alpha(L)$ for any fixed $w \in D$, and by Theorem 1 we have $f(t, w) \in U$ for any fixed $t \in L$. Consequently, $\varphi(s, w)$ satisfies the following equation

$$K\varphi = p(t,w) + f(t,w) \quad \text{for } t \in L, \tag{13.15}$$

where K is a singular integral operator which maps $H_\alpha(L)$ into itself. Clearly if equation (13.15) for some $p(t, w)$ of polynomial matrices has a solution $\varphi(t, w)$, then substituting $\varphi(t, w)$ into (13.14) we obtain a solution of (13.13). Therefore the non-homogeneous Riemann boundary value problem (13.13) is equivalent to equation (13.15).

The integral operator H of the Cauchy principal value is denoted by

$$H\omega = 2Q^{-1}\int_L (t-t_0)^{-1}dt\,\omega(t,w).$$

Theorem 2.5 in [8] implies $H^2 = I$. From (13.15) we derive

$$HK\varphi = Hp + F(t_0, w)$$

where $F(t_0,w) = 2Q^{-1}\int_L (t-t_0)^{-1}dt\,f(t,w), F(t,w) \in U$ for any fixed $t \in L$, $F(t,w) \in H_\alpha(L)$ for any fixed $w \in D$, and $K_1(t,s,w) = [G^{-1}(t,w)G(s,w) - I]\varphi(s,w)$.

It is easily checked that $K_1(t,s,w) \in H_\alpha(L \times L)$ for any fixed $w \in D$. Theorem 3.1 in [8] and the relation $G^{-1}(t_0,w)G(t_0,w) - I = 0$ yield

$$HK\varphi = Hp + F(t_0, w). \tag{13.16}$$

Lemma 2 implies

$$Hp = 2Q^{-1}\int_L (t-t_0)^{-1}dt\,p(t,w) = p(t_0,w)$$

and

$$Q^{-1}\int_L (t-t_0)^{-1}(s-t)^{-1}dt\,K_1(t,s,w) = (s-t_0)^{-1}K^*(s,t_0,w)\varphi(s,w),$$

where

$$\begin{aligned} K^*(s,z,w) &= Q^{-1}\left\{\int_L (t-z)^{-1}dt\,[G^{-1}(t,w)G(s,w) - I]\right. \\ &\quad \left. - \int_L (t-s)^{-1}dt\,[G^{-1}(t,w)G(s,w) - I]\right\} \text{ for } z \in \Omega^+. \end{aligned}$$

It follows that equation (13.16) becomes

$$HK\varphi = p(t_0, w) + F(t_0, w). \tag{13.17}$$

Thus boundary value problem (13.13) is equivalent to equation (13.17).

Now we turn to solving equation (13.17). Let

$$\Phi_0(z,w) = Q^{-1}\int_L (s-z)^{-1}ds\,K^*(s,z,w)\varphi(s,w), \quad z \in \Omega^+. \tag{13.18}$$

Then, by Lemma 4.4 in [8], we get

$$\Phi_0^+(t,w) = Q^{-1}\int_L (s-t_0)^{-1}ds\,K^*(s,t_0,w)\varphi(s,w) \text{ for } t_0 \in L. \tag{13.19}$$

Substituting (13.19) into (13.17), we come to

$$\varphi(t_0,w) = p(t_0,w) + F(t_0,w) - 2\Phi_0^+(t_0,w). \tag{13.20}$$

Substituting (13.20) into (13.18) yields

$$\begin{aligned}\Phi_0(z,w) &= Q^{-1}\int_L (s-z)^{-1}ds\,K^*(s,z,w)p(s,w)\\ &\quad +Q^{-1}\int_L (s-z)^{-1}ds\,K^*(s,z,w)F(s,w)\\ &\quad -2\mathrm{Res}[(s-z)^{-1}K^*(s,z,w)\Phi_0^+(s,w)]_{s=z}.\end{aligned}$$

It follows that

$$\Phi_0(t_0,w) = Q^{-1}\int_L (s-t_0)^{-1}ds\,K^*(s,t_0,w)(p(s,w)+F(s,w)) \tag{13.21}$$

as $z \to t_0 \in L$, and substituting (13.21) into (13.20) yields

$$\begin{aligned}\varphi(t_0,w) &= p(t_0,w)+F(t_0,w)\\ &\quad -2Q^{-1}\int_L (s-t_0)^{-1}ds\,K^*(s,t_0,w)(p(s,w)+F(s,w)).\end{aligned} \tag{13.22}$$

Since $p(t_0,w)$ and $F(t_0,w) \in U$ for any fixed t_0 in L, we obtain $K^*(s,t_0,w)$ $(p(t_0,w)+F(t_0,w)) \in U$ for any fixed $t_0, s \in L$. So $\varphi(t_0,w) \in U$ for any fixed $t_0 \in L$.

Combining the above we get the following conclusion.

Theorem 3 *In the Riemann boundary value problem* (13.13), *if we only demand that $\Phi(z,w)$ has a pole with finite order at $z=\infty$, then problem* (13.13) *is solvable and the general solution of.* (13.13) *is*

$$\Phi(z,w) = \begin{cases} Q^{-1}\int_L (s-z)^{-1}ds\,[G(s,w)\varphi(s,w)+g(s,w)], & z \in \Omega^+,\\ -Q^{-1}\int_L (s-z)^{-1}ds\,\varphi(s,w)+p(z,w), & z \in \Omega^-.\end{cases}$$

13.5 THE PERTURBED RIEMANN BOUNDARY VALUE PROBLEM

Definition 4 *Suppose that $\Phi_n^\pm, \Phi^\pm, g, g_n \in H_\alpha(L)$ and that $G(t)$ and $\Delta_n(t)$ are operator-valued functions of t from U to U in $\mathcal{L}(U,U)$, then the boundary value problem*

$$\Phi_n^+(t) - \Phi_n^-(t) = [G(t)+\Delta_n(t)]\Phi_n^-(t) + g_n(t), t \in L \tag{13.23}$$

is called the perturbed boundary value problem

$$\Phi^+(t) - \Phi^-(t) = G(t)\Phi^-(t) + g(t), \quad t \in L, \tag{13.24}$$

where $|||g_n - g|||_\alpha \to 0$, $||\Delta_n|| \to 0$ as $n \to \infty$, and $||\cdot||$ is the norm of an operator.

It is also interesting to ask: Is there a piecewise M-analytic function, in MCF which satisfies (13.23) on L as a smooth jump curve, with finite order at $z=\infty$?

We shall solve this problem and research the relation of solutions between (13.23) and (13.24). Firstly let

$$\Phi_{n,k}(z,w) = Q^{-1}\int_L (s-z)^{-1}ds\,\big[(G(s,w) + \Delta_n(s,w))\Phi^-_{n,k-1}(s,w) + g_n(s,w)\big]$$

for $k = 0, 1, 2, \cdots$. Let for any fixed $s \in L$

$$\Phi^-_{n,0}(s,w) = 0,\; g_n(s,w) \in U. \tag{13.25}$$

It follows that

$$\begin{aligned} &\Phi_{n,k+1}(z,w) - \Phi_{n,k}(z,w) \\ =\;& Q^{-1}\int_L (s-z)^{-1}ds\,[G(s,w) + \Delta_n(s,w)][\Phi^-_{n,k}(s,w) - \Phi^-_{n,k-1}(s,w)]. \end{aligned}$$

The Plemelj formulas imply

$$\|\Phi^-_{n,k+1}(t,w) - \Phi^-_{n,k}(t,w)\| \le \beta\|G + \Delta_n\|\; |||\Phi^{-1}_{n,k} - \Phi^-_{n,k-1}|||_\alpha$$

where β is a constant. Thus

$$|||\Phi^-_{n,k+1} - \Phi^-_{n,k}|||_\alpha \le [\beta\|G + \Delta_n\|]^k\; |||\Phi^-_{n,1}|||_\alpha. \tag{13.26}$$

Formula (13.25) yields

$$\Phi^-_{n,1}(t,w) = -\frac{1}{2}g_n(t,w) + Q^{-1}\int_L (s-t)^{-1}ds\, g_n(s,w).$$

Since $\{g_n\}$ is bounded in $H_\alpha(L)$, $|||\Phi^-_{n,1}|||_\alpha \le \beta_0$, where β_0 is a constant.

If $\beta\|G + \Delta_n\| < 1$ as $n > N$, then

$$|||\Phi^-_{n,k} - \Phi^-_n|||_\alpha \to 0 \quad \text{as} \quad k \to \infty. \tag{13.27}$$

Similarly, we have

$$|||\Phi^+_{n,k} - \Phi^+_n|||_\alpha \to 0 \quad \text{as} \quad k \to \infty. \tag{13.28}$$

The Plemelj formulas and (13.25) yield

$$\begin{aligned} &\Phi^\pm_n(t,w) = \pm\frac{1}{2}\{[G(t,w) + \Delta_n(t,w)]\Phi^-_n(t,w) + g_n(t,w)\} \\ &+Q^{-1}\int_L (s-t)^{-1}ds\,\{[G(s,w) + \Delta_n(s,w)]\Phi^-_n(s,w) + g_n(s,w)\}. \end{aligned} \tag{13.29}$$

Consequently,

$$\Phi_n(z,w) = Q^{-1}\int_L (s-z)^{-1}ds\Big\{[G(s,w) + \Delta_n(s,w)]\Phi^-_n(s,w) + g_n(s,w)\Big\}$$

satisfies boundary value problem (13.23). Obviously $\Phi_n(z,w)$ is piecewise M-analytic with respect to z and with $\Phi(\infty)=0$ and $\Phi_n(z,w)\in U$ for any fixed $z\in\Omega^{\pm}$ and n.

Combining the above we obtain the following statement.

Proposition 1 *If* $\beta\|G+\Delta_n\|<1$, *then the perturbed problem* (13.23) *of problem* (13.24) *has a unique solution vanishing at* $z=\infty$.

Next we shall give the relationship between the solution of the vector-valued singular integral equation

$$a_n(t,w)\varphi_n(t,w)+2Q^{-1}b_n(t,w)\int_L(s-t)^{-1}ds\,\varphi_n(s,w)=f_n(t,w) \tag{13.30}$$

and the solution of the perturbed problem (13.23), where a_n, b_n are operator-valued functions on $L\times D$ to $\mathcal{L}(U,U)$ and $\varphi_n, f_n\in H_\alpha(L)$.

Proposition 2 *If*

$$(a_n+b_n)^{-1} \tag{13.31}$$

exists, then the solution of the integral equation (13.30) *is equivalent to the solution of* (13.23) *vanishing at* $z=\infty$.

Proof We introduce a piecewise holomorphic function of z,

$$\Phi_n(z,w)=Q^{-1}\int_L(s-z)^{-1}ds\,\varphi_n(s,w)\quad\text{for } z\notin L. \tag{13.32}$$

The Plemelj formulas imply

$$\begin{cases}\Phi_n^+(t,w)-\Phi_n^-(t,w)=\varphi_n(t,w)\\ \Phi_n^+(t,w)+\Phi_n^-(t,w)=2Q^{-1}\int_L(s-t)^{-1}ds\,\varphi_n(s,w).\end{cases} \tag{13.33}$$

Substitution of the formulas in (13.33) into (13.30) yields

$$\begin{aligned}&[a_n(t,w)+b_n(t,w)]\Phi_n^+(t,w)-[a_n(t,w)+b_n(t,w)]\Phi_n^-(t,w)\\ =\;&f_n(t,w)-2b_n(t,w)\Phi_n^-(t,w).\end{aligned}$$

Let $g_n(t,w)=[a_n(t,w)+b_n(t,w)]^{-1}f_n(t,w)$ and

$$\Delta_n(t,w)=-G(t,w)-2b_n(t,w)[a_n(t,w)+b_n(t,w)]^{-1}.$$

Then integral equation (13.30) becomes

$$\Phi_n^+(t,w)-\Phi_n^-(t,w)=[G(t,w)+\Delta_n(t,w)]\Phi_n^-(t,w)+g_n(t,w). \tag{13.34}$$

Formula (13.32) implies $\Phi_n(\infty,w)=0$.

From the above statement we derive that if equation (13.30) has a solution $\varphi_n(t,w)\in H_\alpha(L)$ for any fixed $w\in D$, then the piecewise M-analytic function $\Phi_n(z,w)$ by definition in (13.32) is a solution of the Riemann boundary value problem (13.34) in MCF.

Conversely, if problem (13.34) has a solution $\Phi_n(z,w)$ in MCF with $\Phi_n(\infty, w) = 0$, then the first formula in (13.33) and $\Phi_n(\infty, w) = 0$ yield that formula (13.32) holds. It follows that the second formula in (13.33) holds as well. Substitution of the formulas in (13.33) into (13.34) yields that $\varphi_n(t,w)$ satisfies equation (13.30).

Suppose that operators K_n and K are defined by the following formulas

$$K_n\varphi \equiv a_n(t,w)\varphi(t,w) + 2Q^{-1}b_n(t,w)\int_L (s-t)^{-1}ds\,\varphi(s,w)$$

and

$$K\varphi \equiv a(t,w)\varphi(t,w) + 2Q^{-1}b(t,w)\int_L (s-t)^{-1}ds\,\varphi(s,w)$$

respectively, here a_n, a, b_n, b are operator-valued functions and $\varphi \in H_\alpha(L)$.

Clearly, the operators K_n and K are bounded linear operators on $H_\alpha(L)$. For any small enough $r > 0$, let $A_r = \{(a_n, b_n) : \|K_n\| \geq r\}$, where $\|\cdot\|$ is the norm of an operator.

Proposition 3 *If*

$$\|a_n - a\| \to 0, \ \|b_n - b\| \to 0 \text{ and } |||f_n - f|||_\alpha \to 0 \tag{13.35}$$

as $n \to \infty$ and if there exist r and N such that

$$(a_n, b_n) \in A_r \tag{13.36}$$

for $n > N$, then the following conclusions hold:

(i) *Integral equation* (13.30) *has a unique solution φ_n.*

(ii) *The integral equation*

$$a(t,w)\varphi(t,w) + 2Q^{-1}b(t,w)\int_L (s-t)^{-1}ds\,\varphi(s,w) = f(t,w)$$

has a unique solution.

(iii) $|||\varphi_n - \varphi|||_\alpha \to 0$ *for $n \to \infty$.*

Proposition 4 *If conditions* (13.31), (13.35) *and* (13.36) *hold and if* $\|2b_n(a_n + b_n)^{-1} + G\| \to 0$ *for $n \to \infty$, then*

$$\begin{aligned}\lim_{n\to\infty} \Phi_n(z) = \Phi(z) &= \lim_{n\to\infty} Q^{-1}\int_L (s-t)^{-1}ds\,\varphi_n(s,w) \\ &= Q^{-1}\textstyle\int_L (s-t)^{-1}ds\,\varphi(s,w)\end{aligned} \tag{13.37}$$

is a solution of problem (13.24).

References

[1] Hill, G.N.: Elliptic systems in the plane with first order terms and constant coefficients. Comm. Part. Diff. Eq. 3(1978), 947–977.

[2] Hille E., Phillips, R.S.: Functional Analysis and Semi-Groups. Amer. Math. Soc. Colloq. Publ., Amer. Math. Soc, Providence, RI, 1957.

[3] Hu, C.G.: The disturbance of vector-valued doubly-periodic Riemann boundary value problems. J. Math. Anal. Appl. 131(1988), 373–391.

[4] Hu, C.G.: Vector-valued boundary value problems of a class, Proceedings on International Conference on Integral Equations and Boundary Value Problems, ed. G.–C. Wen, Z. Zhao. World Scientific, Singapore, New Jersey, London, Hong Kong, 1991, 57-64.

[5] Hu, C.G.: Elliptic functions and the stability of integral equations of a class in a locally convex space. Complex Variables Theory Appl. 24(1994), 209–217.

[6] Hu, C.G.: Linear systems of a class in locally convex spaces, Complex analysis and its applications, ed. C.C. Yang et. al. Pitman Research Notes in Mathematics Series 305, Longman Scientific & Technical, Harlow, 1994, 90–199.

[7] Hu, C.G., Yang,C.C.: Vector-valued functions and their applications. Kluwer Acad. Publ., Dordrecht, Boston, London, 1992.

[8] Lin W., Dai,D.: The Cauchy type integral for M-analytic function and its applications. Acta Math. Sci. 2(1990), 149–166.

[9] Lu, J.K.: Boundary value problems for analytic functions. World Scientific, Singapore, New Jersey, London, Hong Kong, 1993.

[10] Muskhelishvili,N.I.: Singular integral equations. Noordhoff, Groningen, 1953.

14 A CLASS OF INVERSE HILBERT BOUNDARY VALUE PROBLEMS

Shi Qiang Liu

Ninxia University
Yinchuan 750021
P.R. China
jydu@whu.edu.cn

Abstract: In this paper, we give a mathematical formulation and solution of a class of inverse Hilbert boundary value problems (H^{-1}) and obtain theorems of the solvability.

14.1 FORMULATION OF THE INVERSE HILBERT PROBLEM

Let L be a closed smooth contour oriented anticlockwise in positive direction. S^+ is the inner region surrounded by L, and S^- is the external region in the complex plane.

Problem H^{-1}: Find a pair of functions $(\Phi^+(z), \psi(t))$, where $\Phi^+(z) = u(z) + iv(z)$ is an analytic function in S^+, continuous on $S^+ \cup L$, and $\psi(t) \in H(L)$ satisfying the Hilbert boundary value problem

$$\begin{aligned} \operatorname{Re}\{[a_1(t) + ib_1(t)]\Phi^+(t)\} &= c_1(t)\psi(t), \\ \operatorname{Re}\{[a_2(t) + ib_2(t)]\Phi^+(t)\} &= c_2(t)\psi(t), \end{aligned} \quad t \in L, \tag{14.1}$$

where $a_j(t)$, $b_j(t)$, $c_j(t)(j = 1, 2)$ are all known real functions on L, and assume $c_1(t) \neq c_2(t)$ on L.

14.2 SOLUTIONS OF THE INVERSE PROBLEM

Similar to [2], after a conformal mapping, we just need to discuss the problem on $S^+ := \{|z| < 1\}$, $L := \{|t| = 1\}$.

H.G.W. Begehr et al. (eds.), Partial Differential and Integral Equations, 207–210.

Introducing the symmetrical extension of $\Phi(z)$ on S^-(exterior of $|z| \leq 1$), (see [1], [2]), namely

$$\Phi_*(z) = \overline{\Phi(\frac{1}{\bar{z}})} = \overline{\Phi}(\frac{1}{z}) \tag{14.2}$$

we define

$$\Omega(z) = \begin{cases} \Phi(z), & \text{when } z \in S^+, \\ \Phi_*(z), & \text{when } z \in S^-. \end{cases} \tag{14.3}$$

From [1] we know that $\Omega(z)$ is a sectionally analytic function with a jump on L satisfying the following boundary condition

$$\Omega^-(t) = \overline{\Omega^+(t)}, \quad t \in L, \tag{14.4}$$

and

$$\Omega^*(t) = \Omega^+(t), \quad t \in L. \tag{14.5}$$

So, the boundary condition (1) can be rewritten as

$$\begin{aligned} \Omega^+(t) &= G_1(t)\Omega^-(t) + g_1(t)\psi(t), \\ \Omega^+(t) &= G_2(t)\Omega^-(t) + g_2(t)\psi(t), \end{aligned} \quad t \in L, \tag{14.6}$$

where

$$G_j(t) = -\frac{g_j(t) - ib_j(t)}{a_j(t) + ib_j(t)}, \; g_j(t) = \frac{2c_j(t)}{a_j(t) + ib_j(t)}.$$

Because $\Phi(0)$ is a bounded quantity, it is easy to see that $\Omega(\infty)$ is bounded, too. From (14.6) we get

$$g(t)\Omega^+(t) = G(t)\Omega^-(t), \quad t \in L, \tag{14.7}$$

in which

$$g(t) = \begin{vmatrix} 1 & g_1(t) \\ 1 & g_2(t) \end{vmatrix}, \; G(t) = \begin{vmatrix} G_1(t) & g_1(t) \\ G_2(t) & g_2(t) \end{vmatrix}, \; t \in L.$$

14.2.1 The normal case: $g(t) \neq 0$, $G(t) \neq 0$

From (14.7) we get the boundary condition of the homogeneous Riemann problem

$$\Omega^+(t) = \tilde{G}(t)\Omega^-(t), \quad t \in L, \tag{14.8}$$

where

$$\tilde{G}(t) = G(t)/g(t).$$

A. When $\kappa = 0$, by using the solutions of the classical homogeneous Riemann boundary value problem, see [1], we obtain

$$\Omega(z) = CX(z), \tag{14.9}$$

where C is an arbitrary constant and

$$X(z) = \exp\left\{\frac{1}{2\pi i}\int_L \frac{\log \tilde{G}(t)}{t-z}dt\right\}, \quad z \notin L, \tag{14.10}$$

$$\kappa = \frac{1}{2\pi}[\arg \tilde{G}(t)]_L = \mathrm{Ind}_L G(t) - \mathrm{Ind}_L g(t). \tag{14.11}$$

So, the solution of the inverse problem in this case can be obtained as

$$(\Phi(z), \Psi(t)) = (CX(z), CD_1(t)X^-(t)), \tag{14.12}$$

in which

$$D_1(t) = \frac{1}{g(t)}\begin{vmatrix} G_1(t) & 1 \\ G_2(t) & 1 \end{vmatrix}.$$

B. When $\kappa > 0$, we obtain

$$(\Phi(z), \Psi(t)) = (Q_\kappa(z), P_\kappa(t)D_1(t)X^-(t)), \tag{14.13}$$

where

$$Q_\kappa(z) = C_0 z^\kappa + C_1 z^{\kappa-1} + \cdots + \overline{C_1}z + \overline{C_0},$$

$$P_\kappa(t) = C_0 t^\kappa + C_1 t^{\kappa-1} + \cdots + \overline{C_{\kappa-1}}z + \overline{C_\kappa}.$$

Finally we have the following result for the normal case.

Theorem 1 *When the index $\kappa \geq 0$, the inverse problem (Problem H^{-1}) has $\kappa + 1$ linearly indenpendent solutions. When $\kappa < 0$, Problem H^{-1} only has the trivial solution. 14.2.2 The non-normal case*

Similarly to [5], we have

$$g(t) = \prod_{j=1}^{m}(t-\alpha_j)^{\lambda_j} g_0(t), \quad g_0(t)(\neq 0) \in, \quad \lambda_j > 0,$$

$$G(t) = \prod_{k=1}^{n}(t-\beta_k)^{\mu_k} G_0(t), \quad G_0(t)(\neq 0) \in H, \quad \mu_k > 0,$$

where α_j $(j = 1, \cdots, m)$, β_k $(k = 1, \cdots, n)$ are different from zero points of $g(t)$ and $G(t)$ respectively, $\alpha_j \neq \beta_k$, λ_j, μ_k are positive integers, m, n are natural numbers, and

$$\lambda = \sum_{j=1}^{m}\lambda_j, \quad \mu = \sum_{k=1}^{n}\mu_k.$$

Denote

$$\Pi_1(t) = \prod_{k=1}^{n}(t-\beta_k)^{\mu_k}, \quad \Pi_2(t) = \prod_{j=1}^{m}(t-\alpha_j)^{\lambda_j}.$$

Similarly to [2], [5] we get for $z \in S^+$

$$(\Phi(z), \Psi(t)) = \begin{cases} [X(z)(\Pi_1(z)P_{\kappa-\mu}(z) + \Pi_1^*(z)P_{\kappa-\mu}^*/2), \\ \qquad D_2(t)P_{\kappa-\mu}(t)], \\ [X(z)(\Pi_2(z)P_{\kappa-\mu}(z) + \Pi_2^*(z)P_{\kappa-\mu}^*/2), \\ \qquad D_2(t)X^-(t)P_{\kappa-\mu}(t)], \end{cases} \tag{14.14}$$

in which

$$\Pi_1^*(z) = \prod_{k=1}^{n}(1-\overline{\beta_k}z)^{\mu_k}, \ \Pi_2^*(z) = \prod_{j=1}^{m}(1-\overline{\alpha_j}z)^{\lambda_j},$$

$$P_{\kappa-\mu}^*(z) = \overline{C_0} + \overline{C_1}z + \cdots + \overline{C_{\kappa-\mu}}z^{\kappa-\mu}, \ D_2(t) = \frac{G_1(t) - G_2(t)}{g_0(t)}.$$

Theorem 2 *When the index $\kappa \geq \mu$, then there exists the solution of inverse problem (Problem H^{-1}) in the non-normal case, which can be expressed as* (14.14). *When $\kappa < \mu$, Problem H^{-1} only has the trival solution.*

References

[1] Muskhelishvili, N.I.: Singular integral equations. Noordhoff, Groningen, 1953.

[2] Lu, C.K.: Boundary value problem for analytic functions. Shanghai Sci. Tech. Press, Shanghai, 1977.

[3] Liu Shiqiang: On the inverse Riemann boundary value problem on a closed contour. J. Ningxia Univ. 17 (1996), No.1, 1-2 (Chinese).

[4] Li Xing: The inverse Riemann boundary value jump problem. Proc. Int. Conf. Computation Engineering Science, Tech. Publ., Atlanta, Ga, USA, 1992, 360-361.

[5] Li Xing: The inverse problem of the periodic homogeneous Riemann–Hilbert problem. Inverse Problem, Tech. Publ. Atlanta, Ga, USA, 1993, 61-67.

15 CAUCHY FORMULA, INTEGRAL OF CAUCHY TYPE AND HILBERT PROBLEM FOR BIANALYTIC FUNCTIONS

Zhao Zhen †

Beijing Normal University
Department of Mathematics
Beijing 100875
P.R. China

zhaozhen@bnu.edu.cn

Abstract: In this paper the Cauchy formula, the integral of Cauchy type and the Hilbert problem for bianalytic functions are kinvestigated.

15.1 INTRODUCTION

Due to their wonderful properties, analytic functions possess a lot of applications in mechanics and mathematical physics. For example, when we study a physical field in the plane without source and curl, the theory of analytic functions display its strength. But if we discuss a physical field with sources or curls, then this important tool of analytic functions will be of no use.

In [1] we have suggested the concept of bianalytic functions and have considered some of their properties. Furthermore some boundary value problems (i.e. Dirichlet's problem, Dirichlet's problem in a changed form, Riemann-Hilbert's problem, Schwarz's problem, Problem A, etc.) are also considered, see [1], [2]–[4]. In this paper we will investigate the Cauchy formula and the integral of Cauchy type for bianalytic functions and prove some useful theorems for them. Finially, we also will investigate the Hilbert problem for bianalytic functions, and prove the theorems of its solvability.

†This work is supported by the National Natural Science Foundation of China (19571010).

H.G.W. Begehr et al. (eds.), Partial Differential and Integral Equations, 211–218.

The complex form of the $C-R$ equations is

$$\frac{\partial W}{\partial \overline{z}} = 0, \tag{15.1}$$

where $\partial/\partial\overline{z} = 1/2(\partial/\partial x + i\partial/\partial y)$, $W = u + iv$, $z = x + iy$.

In addition we also define the derivatives: $\partial/\partial z = 1/2(\partial/\partial x - i\partial/\partial y)$ and $\Delta \equiv \partial^2/\partial\overline{z}\partial z = \partial^2/\partial z\partial\overline{z}$.

Definition 1 *Suppose G is a region in the complex plane and define in G a complex function $W(z)$, for which the second order derivative $\partial^2 W/\partial\overline{z}^2$ exists. Then we will call W a bianalytic function, if $W(z)$ satisfies the partial differential equation*

$$\frac{\partial^2 W}{\partial \overline{z}^2} = 0, \quad z \in G. \tag{15.2}$$

We denote the set of all bianalytic functions by $\mathcal{D}_2(G)$.

We know, see [1], that a bianalytic function can be represented as

$$W(z) = \overline{z}\varphi_1(z) + \varphi_2(z), \tag{15.3}$$

where $\varphi_1(z)$, $\varphi_2(z)$ are analytic functions in G. It should be noted that in (15.3) instead of $\overline{z}$ one can use $\overline{(z-a)}$, where a is a certain exterior point of the domain G. For simplicity we will call $\varphi_1(z)$ and $\varphi_2(z)$ the analytic factor and analytic addition of the bianalytic function W, respectively. Later sometimes the analytic addition $\varphi_2(z)$ will be neglected (i.e. $\varphi_2(z) \equiv 0$). In general we denote all functions in the form $w^*(z) = \overline{z}\varphi(z)$, where $\varphi(z)$ is an arbitray analytic function, by $\mathcal{D}_2^*(G) \subset \mathcal{D}_2(G)$.

When we use the methods of function theory to solve problems of elasticity in the plane, the stress function $U(x, y)$ which is biharmonic, plays a very important role. We can successfully represent the stress function $U(x, y)$ as the real part of a certain bianalytic function. i.e. $U(x, y) = Re[W(z)]$, where $W(z)$ is bianalytic. This means that investigation of bianalytic functions is of important practical interest.

15.2 CAUCHY FORMULA FOR BIANALYTIC FUNCTIONS

It is well-known that the Cauchy formula plays a very important role in the theory of analytic functions. By this formula an analytic function is given as soon as its boundary values are known. In the theory of bianalytic functions we also can obtain an analogue formula – the Cauchy formula for bianalytic functions. The following theorem holds.

Theorem 1 *Suppose G is a bounded region in the plane, L is its boundary and $W(z)$ is a bianalytic function in G, the derivative $(\partial W/\partial\overline{z})$ of which is continuous on $G \cup L$. Then we have the Cauchy formula*

$$W(z) = \frac{1}{2\pi i}\int_L \frac{W(t)}{t-z}\,dt - \frac{1}{2\pi i}\int_L \frac{\overline{t-z}}{t-z}\frac{\partial W}{\partial \overline{t}}\,dt, \quad z \in G. \tag{15.4}$$

Proof Suppose $L_\varepsilon = \{|\zeta - z| = \varepsilon\}$, $G_\varepsilon = G \setminus \{|\zeta - z| \le \varepsilon\}$, and the boundary of G_ε is $L \setminus L_\varepsilon$. We have

$$\frac{\partial}{\partial\overline{\zeta}}\Big(\frac{W(\zeta)}{\zeta - z}\Big) = \frac{1}{\zeta - z}\Big(\frac{\partial W(\zeta)}{\partial\overline{\zeta}}\Big) = \frac{\partial}{\partial\overline{\zeta}}\Big(\frac{\overline{\zeta - z}}{\zeta - z}\frac{\partial W}{\partial\overline{\zeta}}\Big), \quad \zeta \in G_\varepsilon.$$

Then

$$\frac{\partial}{\partial\overline{\zeta}}\Big(\frac{W(z)}{\zeta - z} - \frac{\overline{\zeta - z}}{\zeta - z}\frac{\partial W}{\partial\overline{\zeta}}\Big) = 0, \quad z \in G_\varepsilon, \tag{15.5}$$

or $\big(\frac{W(z)}{\zeta - z} - \frac{\overline{\zeta - z}}{\zeta - z}\frac{\partial W}{\partial\overline{\zeta}}\big) = \Phi(\zeta)$ is an analytic function in G_ε. By the Cauchy theorem we have

$$\frac{1}{2\pi i}\int_{L-L_\varepsilon}\Big(\frac{W(\zeta)}{\zeta - z} - \frac{\overline{\zeta - z}}{\zeta - z}\frac{\partial W}{\partial\overline{\zeta}}\Big)\,d\zeta = 0, \quad z \in G. \tag{15.6}$$

Consequently,

$$\begin{aligned}&\frac{1}{2\pi i}\int_{L}\Big(\frac{W(\zeta)}{\zeta - z} - \frac{\overline{\zeta - z}}{\zeta - z}\frac{\partial W}{\partial\overline{\zeta}}\Big)\,d\zeta \\ &= \frac{1}{2\pi i}\int_{L_\varepsilon}\frac{W(\zeta)}{\zeta - z}\,d\zeta - \int_{L_\varepsilon}\frac{\overline{\zeta - z}}{\zeta - z}\frac{\partial W}{\partial\overline{\zeta}}\,d\zeta, \quad z \in G.\end{aligned} \tag{15.7}$$

Without any difficulty we can prove that the first and the second integral of the right–hand side of (15.7) will just tend to $W(z)$ and zero respectively, when $\varepsilon \to 0$.

Theorem 1 is proved. Obviously, by formula (15.4) we can get a bianalytic function as soon as the boundary values $W(t)$ and $\partial W/\partial\overline{t}$ are known. (This just is the solution of the basic boundary value problem, see [1].)

15.3 INTEGRAL OF CAUCHY TYPE FOR BIANALYTIC FUNCTION

In formula (15.4) two boundary values $W(t)$ and $\partial W/\partial\overline{t}$ are contained. This is the difference from the Cauchy formula for analytic functions. Later for convenience we will simplify the integral of Cauchy type and consider it only for the bianalytic functions belonging to the class $\mathcal{D}_2^*(G)$.

Definition 2 *The integral of Cauchy type for bianalytic functions $w^*(z)$ belonging to the class $\mathcal{D}_2^*(G)$ will be*

$$\begin{aligned}w^*(z) &= \frac{\overline{z}}{2\pi i}\int_L\frac{\varphi(t)}{t - z}\,dt = \frac{1}{2\pi i}\int_L\frac{\overline{t}\varphi(t)}{t - z}\,dt - \frac{1}{2\pi i}\int_L\frac{\overline{t - z}}{t - z}\varphi(t)\,dt \\ &= \frac{1}{2\pi i}\int_L\frac{\omega(t)}{t - z}\,dt - \frac{1}{2\pi i}\int_L\frac{\overline{t - z}}{\overline{t}(t - z)}\omega(t)\,dt, \ z \notin L,\end{aligned} \tag{15.8}$$

where $\varphi(t)$ is an arbitrary continuous function and $\omega(t) = \overline{t}\varphi(t)$.

The first and the second term on the right–hand side of (15.8) are an analytic function and a bianalytic function, respectively. Consequently, we have

$\partial^2 w/\partial \bar{z}^2 = 0$, for $z \notin L$. Obviously, $w^*(z)$ is a bianalytic function belonging to $\mathcal{D}_2^*(G)$, if $z \notin L$.

Theorem 2 *If $W(z)$ is a bianalytic function in a bounded domain G, then $W(z) = w^*(z) + \Phi(z)$, $z \in G$, where $w^*(z)$ is a function as in* (15.8) *and $\Phi(z)$ is an arbitrary analytic function.*

Proof We know that $W(z) = \bar{z}\varphi_1(z) + \varphi_2(z)$, see [1]. As soon as we take

$$\varphi_1(z) = \frac{1}{2\pi i}\int_L \frac{\varphi(t)}{t-z}\,dt, \quad \Phi(z) = \varphi_2(z)\,,$$

we at once get Theorem 2.

Definition 3 *A piecewise bianalytic function with jump curve L will be a function $\Psi(z)$ which is bianalytic in any bounded domain not containing the curve L, having a finite k-th order at infinity and being continuously extendable from left and right to L. Here L is a closed smooth curve.*

Definition 4 *A piecewise bianalytic function $w^*(z)$ is called chracteristic bianalytic, if it belongs to $\mathcal{D}_2^*(G)$ in any bounded domain D not containing the curve L, has a finite k-th order at infinity and is continuously extendable from left and right to L. Here L is a closed smooth curve.*

Theorem 3 *Suppose L is a closed smooth curve, and $\omega(t)$ belongs to class H, (i.e. class of Hölder continuous functions). Then, the integral of Cauchy type $w^*(z)$* (15.8) *is a piecewise bianalytic function , which is bounded at infinity and L is the jump curve .*

Using the properties of the Cauchy type integral for analytic functions, see [6], we can easily prove Theorem 3.

Furthermore, the following theorem holds.

Theorem 4 *Suppose L is a closed smooth curve, and $\omega(t) = \bar{t}\varphi(t)$ belongs to the class H (i.e. to the class of Hölder continuous functions); then for the left boundary value $w^{*+}(t)$ and the right boundary value $w^{*-}(t)$, $t \in L$, of the integral of Cauchy type* (15.8) *(for characteristic bianalytic functions) $w^*(z)$, we have the following formula.*

Plemelj formulas:

$$\begin{aligned} w^{*+}(t) &= \frac{1}{2}\omega(t) + \frac{1}{2\pi i}\int_L \frac{\omega(\tau)}{\tau - t}\,d\tau - \frac{1}{2\pi i}\int_L \frac{\overline{\tau - t}}{\bar{\tau}(\tau - t)}\omega(\tau)\,d\tau, \\ w^{*-}(t) &= -\frac{1}{2}\omega(t) + \frac{1}{2\pi i}\int_L \frac{\omega(\tau)}{\tau - t}\,d\tau - \frac{1}{2\pi i}\int_L \frac{\overline{\tau - t}}{\bar{\tau}(\tau - t)}\omega(\tau)\,d\tau,\ t \in L. \end{aligned} \tag{15.9}$$

The Plemelj formulas (15.9) can be written in the form

$$\begin{aligned} w^{*+}(t) + w^{*-}(t) &= \frac{1}{\pi i}\int_L \frac{\omega(\tau)}{\tau - t}\,d\tau - \frac{1}{\pi i}\int_L \frac{\overline{\tau - t}}{\bar{\tau}(\tau - t)}\omega(\tau)\,d\tau, \\ w^{*+}(t) - w^{*-}(t) &= \omega(t), \quad t \in L. \end{aligned} \tag{15.10}$$

15.4 HILBERT PROBLEM FOR BIANALYTIC FUNCTIONS

Suppose S^+ is a domain in the plane, L is the boundary of S^+. In general, S^+ may be a multiply connected domain. In this case L is composed by $p+1$ mutually disjoint closed Liapunov curves $L_0, L_1, \cdots, L_p$, $L = L_0 \cup L_1 \cup \cdots \cup L_p$, (where L_0 surrounds all the others). Without loss of generality, we can assume that the origin belongs to S^+. We set the positive orientation of L such that, S^+ is always on its left. We denote the complementary set of $S^+ \cup L$ by S^- and its connected component with boundary L_j by S_j^-, $j = 0, 1, \cdots, p$, respectively. Obviously, $S^- = S_0^- \cup S_1^- \cup \cdots \cup S_p^-$, and in general S_0^- contains the point of infinity, if L_0 exists. Moreover a_j is a definite point in $S_j^-, j = 1, \ldots, p$.

Hilbert Problem (Problem H) Find a piecewise bianalytic function with jump curve L which satisfies the boundary conditions

$$\begin{aligned} w^{*+}(t) &= G(t)w^{*-}(t) \\ w^{+}(t) &= G(t)w^{-}(t), \end{aligned} \quad t \in L, \quad \text{(homogeneous problem)} \tag{15.11}$$

$$\begin{aligned} w^{*+}(t) &= G(t)w^{*-}(t) + g_1(t), \\ w^{+}(t) &= G(t)w^{-}(t) + g_2(t), \end{aligned} \quad t \in L, \text{ (nonhomogeneous problem)} \tag{15.12}$$

where $G(t)$, $g_i(t)$ $(i = 1, 2)$ are given functions of class H and $G(t) \neq 0, \quad t \in L$. We will call $G(t)$ and $g_i(t)$ the coefficient and free terms of the Hilbert problem, respectively. Moreover $w^*(z)$ is a characterestic bianalytic function.

For simplicity, at first we discuss a simple case, namely the Hilbert problem for characteristic bianalytic functions.

Problem H^* Find a piecewise characterestic bianalytic function with jump curve L which satisfies the boundary conditions

$$w^{*+}(t) = G(t)w^{*-}(t) \quad t \in L, \quad \text{(homogeneous problem)}, \tag{15.11'}$$

$$w^{*+}(t) = G(t)w^{*-}(t) + g_1(t),\ t \in L, \text{(nonhomogeneous problem)}, \tag{15.12'}$$

where $G(t)$, $g_1(t)$ are given functions of class H and $G(t) \neq 0, \quad t \in L$.

By the results of the Hilbert problem for analytic functions we know that $G(t) = X^+(t)/\ X^-(t)$, where $X(z)$ is the so-called standard function of the Hilbert problem for analytic functions. It has order $-\kappa$ at the point at infinity, where κ is the index of Hilbert problem, see [6]. Later we will call κ also the index of the Hilbert problem for bianalytic functions, $\kappa \sum_{j=0}^{p} \lambda_j, 2\pi\lambda_j = [\arg G(t)]|_{L_j}, j = 0, \ldots, p$,

$$X(z) = \begin{cases} \dfrac{1}{\Pi(z)} e^{\Gamma(z)}, & z \in S^+, \\ z^{-\kappa} e^{\Gamma(z)}, & z \in S^-, \end{cases}$$

$$\Gamma(z) = \frac{1}{2\pi i} \int_L \frac{\ln G_0(t)}{t - z}\, dt, \quad G_0(t) = t^{-\kappa} G(t) \Pi(t), \Pi(z) = \prod_{j=1}^{p} (z - a_j)^{\lambda_j}.$$

Substituting this representation into (15.12′) we get

$$w^{*+}(t)=\frac{X^+(t)}{X^-(t)}w^{*-}(t)+g_1(t) \tag{15.13}$$

or

$$\frac{w^{*+}(t)}{X^+(t)}=\frac{w^{*-}(t)}{X^-(t)}+\frac{g_1(t)}{X^+(t)}. \tag{15.14}$$

Remark that the function $w^*(z)/X(z)$ is also a characteristic bianalytic function.

By (15.14) we have (see 15.3)

$$\frac{w^*(z)}{X(z)}=\frac{1}{2\pi i}\int_L\frac{g_1(t)}{X^+(t)(t-z)}dt-\frac{1}{2\pi i}\int_L\frac{\overline{t-z}}{\bar{t}\,(t-z)}\frac{g_1(t)}{X^+(t)}dt.$$

This is a characteristic bianalytic function which is bounded at the point at infinity. Consequently, we have

$$w^*(z)=\frac{X(z)}{2\pi i}\int_L\frac{g_1(t)}{X^+(t)(t-z)}dt-\frac{X(z)}{2\pi i}\int_L\frac{\overline{t-z}}{\bar{t}(t-z)}\frac{g_1(t)}{X^+(t)}dt. \tag{15.15}$$

This is a characteristic bianalytic function which has order $-\kappa$ at infinity.

We consider the following different cases for the index κ.

1. $\kappa=0$. In this case (15.15) gives us a piecewise characteristic bianalytic function, which is bounded at the point at infinity.

2. $\kappa>0$. In this case $w^*(z)$ is also a characteristic bianalytic function, which is bounded at the point at infinity, and condition (15.13) is satisfied for $w^*(z)$.

3. $\kappa<0$. Now $w^*(z)$ has order $-\kappa>0$ at the point at infinity. In order that $w^*(z)$ is bounded at the point at infinity, we must also require:

$$\int_L\frac{g_1(t)}{X^+(t)}t^k\,dt=0,\quad k=0,\ 1,\ \cdots,\ -\kappa-1. \tag{15.16}$$

If conditions (15.16) are satisfied, then nonhomogeneous Hilbert problem (15.12′) has the unique solution (15.15).

Concerning the homogeneous Hilbert problem (15.11′) we have the following conclusions.

1. $\kappa>0$. Then by the condition $w^{*+}(t)/X^+(t)=w^{*-}(t)/X^-(t)$ we have $w^*(z)/X(z)=\bar{z}P_{\kappa-1}(z)$, which has order κ at the point at infinity.

2.$\kappa<0$. Then $w^*(z)/X(z)$ must have order $-\kappa>0$ at the point at infinity. Consequently, the homogeneous Hilbert problem has only the trivial solution.

3. $\kappa=0$. As $w^*(z)$ is not an analytic function, the homogeneous Hilbert problem also has only the trivial solution, i.e. in this case the homogeneous Hilbert problem may only have analytic solutions. We note that (15.8) is not an analytic function.

Finally we have the following theorems.

Theorem 5 *If $\kappa > 0$, the homogeneous Hilbert problem (Problem H^{*0}) has κ linearly independent solutions*

$$\bar{z}X(z),\ \bar{z}zX(z),\ \cdots,\ \bar{z}z^{\kappa-1}X(z). \tag{15.17}$$

If $\kappa \leq 0$, then Problem H^{*0} has only the trivial solution.

Theorem 6 *If $\kappa > 0$, the nonhomogeneous Hilbert problem (Problem H^*) has the general solution*

$$\begin{aligned} w^*(z) = \frac{X(z)}{2\pi i}\int_L \frac{g_1(t)}{X^+(t)(t-z)}\,dt \\ -\frac{X(z)}{2\pi i}\int_L \frac{\overline{t-z}}{\bar{t}(t-z)}\frac{g_1(t)}{X^+(t)}\,dt + \bar{z}X(z)P_{\kappa-1}(z), \end{aligned} \tag{15.18}$$

where $w^*(z)$ is a characteristic bianalytic function, which is bounded at the point at infinity, and condition (15.13) is satisfied for $w^*(z)$. In general the solution contains κ arbitrary constants.

If $\kappa \leq 0$, as soon as conditions (15.16) are satisfied, then Problem H^* has the unique solution

$$w(z) = \frac{X(z)}{2\pi i}\int_L \frac{g_1(t)}{X^+(t)(t-z)}\,dt - \frac{X(z)}{2\pi i}\int_L \frac{\overline{t-z}}{\bar{t}(t-z)}\frac{g_1(t)}{X^+(t)}\,dt. \tag{15.19}$$

By using the results of Problem H^{*0} and Problem H^*, we can solve Problem H^0 and Problem H. Indeed from (15.12) we obtain that

$$\Phi^+(t) = G(t)\Phi^-(t) + [g_2(t) - g_1(t)].$$

This is just Problem H for analytic function, it is well solved in [6]. Finally, we obtain the following theorems.

Theorem 7 *If $\kappa > 0$, the homogeneous Hilbert problem (Problem H^0) has the 2κ linearly independent solutions*

$$\begin{aligned} &\bar{z}X(z), \bar{z}zX(z), \cdots, \bar{z}z^{\kappa-1}X(z), \\ &w_0(z) + X(z), w_0(z) + zX(z), \cdots, w_0(z) + z^{\kappa-1}X(z), \end{aligned} \tag{15.20}$$

where $w_0(z)$ is the general solution of Problem H^{*0}.

If $\kappa < 0$, then Problem H^0 has only the trivial solution.

Theorem 8 *If $\kappa > 0$, the nonhomogeneous Hilbert problem (Problem H) has the general solution*

$$\begin{aligned} w(z) = \frac{X(z)}{2\pi i}\int_L \frac{g_1(t)}{X^+(t)(t-z)}\,dt - \frac{X(z)}{2\pi i}\int_L \frac{\overline{t-z}}{\bar{t}(t-z)}\frac{g_1(t)}{X^+(t)}\,dt \\ +\bar{z}X(z)P_{\kappa-1}(z) + \frac{X(z)}{2\pi i}\int_L \frac{g_1(t)}{X^+(t)(t-z)}\,dt + X(z)Q_{\kappa-1}(z), \end{aligned} \tag{15.21}$$

where $P_{\kappa-1}(z)$ and $Q_{\kappa-1}(z)$ are arbitrary polynomials of order $(\kappa - 1)$, and $w(z)$ is bounded at the point at infinity.

Obviously, condition (15.12) is satisfied for $w(z)$. In general the solution contains 2κ arbitrary constants.

Theorem 9 *When $\kappa < 0$, and the -2κ conditions*

$$\int_L \frac{g_2(t)}{X^+(t)} t^k dt = 0, \quad k = 0, 1, \cdots, -\kappa - 1, \tag{15.22}$$

and (15.16) *are satisfied, the nonhomogeneous Hilbert problem (Problem H) has the unique solution*

$$\begin{aligned} w(z) = \frac{X(z)}{2\pi i} \int_L \frac{g_1(t)}{X^+(t)(t-z)} dt \\ - \frac{X(z)}{2\pi i} \int_L \frac{\overline{t-z}}{\bar{t}(t-z)} \frac{g_1(t)}{X^+(t)} dt + \frac{X(z)}{2\pi i} \int_L \frac{g_2(t)}{X^+(t)(t-z)} dt. \end{aligned} \tag{15.23}$$

Obviously, condition (15.12) is satisfied for $w(z)$.

References

[1] Zhao Zhen: Bianalytic functions, complex harmonic functions and their basic boundary value problems. J. Beijing Normal Univ., Nat. Sci. 2 (1995) (Chinese).

[2] Zhao Zhen: Bianalytic functions and their applications. Proc. Second Asian Math. Conf. 1995, Thailand (to appear).

[3] Zhao Zhen: Riemann-Hilbert's problem for bianalytic functions. J. Beijing Normal Univ., Nat. Sci. 3 (1996), 316–320.

[4] Zhao Zhen: Schwarz's problems for complex partial differential equations of second order. Beijing Mathematics, part 1, no. 2 (1996), 131–137.

[5] Lu Jian-ke: Complex methods of plane elasticity. Wuhan Univ. Press, 1986 (Chinese).

[6] Muskhelishvili, N.I.: Singular integral equations. Noordhoof, Groningen, 1953.

[7] Vekua, I.N.: Generalized analytic functions. Pergomon, Oxford, 1962.

16 ON UNIQUENESS CONDITIONS FOR ENTIRE POLYHARMONIC FUNCTIONS

M.B. Balk
M.Ya.Mazalov

Department of Mathematics
Smolensk Pedagogical Institute
Przhevalski str. 4
Smolensk, 214000
Russia

Abstract: The main result of this paper is the answer — in the case of functions polyharmonic in the plane $\mathbb{R}^2$ — to a uniqueness problem for polyharmonic functions proposed recently by the well-known British mathematican W.K. Hayman [1]. Our reasoning is essentially based on the connection between polyharmonic functions of two real variables and polyanalytic functions of one complex variable [4].

16.1 INTRODUCTION

As it is known, a function $u(x, y)$ is called polyharmonic (in short: p. h.) of order n in the plane $\mathbb{R}^2$ (or *entire n-harmonic*) if it belongs to the class $C^{2n}(\mathbb{R}^2)$ and satisfies in $\mathbb{R}^2$ the condition

$$\Delta^n u = 0, \tag{16.1}$$

where Δ is the Laplace operator in $\mathbb{R}^2$. We suppose $u(x, y)$ to be real-valued and denote the set of all such functions as $PH_n(\mathbb{R}^2)$.

It is widely known that a p.h. function of order 1 (i.e. a harmonic function) in $\mathbb{R}^2$ vanishing on some oval (this means here and afterwards: in every point of a closed analytic curve without self-intersections) is identically equal to zero. Does something similar take place in the case of p.h. functions of order $n > 1$

H.G.W. Begehr et al. (eds.), Partial Differential and Integral Equations, 219–232.

vanishing on n ovals? W.K. Hayman and B. Korenblum obtained recently an unexpected — and even surprising — result (see [2], Theorem 5): There exists such **non-zero** (i.e. not identically equal to zero) entire p.h. function which vanishes on some **infinite** sequence of ovals. In other words, they detected a union of some *infinite* family of ovals which is *not a set of uniqeness* for the class $PH_n(\mathbb{R}^2)$. In the same paper [2] it was proved that the set L_n which is the union of arbitrary n pairwise distinct *circumferences* (or even of $n-1$ circumferences and one arbitrary oval having no common arc with any of these circumferences) is a set of uniqueness for the class $PH_n(\mathbb{R}^2)$ (in [2] a similar result was obtained for spheres and p.h. functions in regions G of $\mathbb{R}^p$, $p \geq 2$). The proof was essentially based (in the case which we discuss here) on such *specific* property of a *circumference*: if some entire p.h. function $U(x,y)$ vanishes on some *circumference* having the equation $p(x,y) = 0$ (where $p(x,y)$ is a polynomial of the form $p(x,y) = x^2 + y^2 + Ax + By + C$ and A, B, C are some constants) then $U \equiv pV$, where $V(x,y)$ is also some entire p.h. function.

This situation led W.K.Hayman to a problem (proposed by him in [1]) which admits in the case of p.h. functions in $\mathbb{R}^2$ the following formulation:

Does there - or does there not - exist for some $n > 1$ such set of n ovals $L_1, L_2, \ldots, L_n$ **all distinct from circumferences** *that their union is a set of uniqueness for the class $PH_n(\mathbb{R}^2)$?*

One can add the specific case (when $L_1 = L_2 = \ldots = L_n$):

Does there - or does there not - exist for some $n > 1$ such **(single)** *oval L_n in $\mathbb{R}^2$ that every entire p.h. function of order n vanishing on this oval is necessarily identically equal to zero?*

The answers to these questions are given below.

16.2 TERMINILOGY

A real-valued homogeneous polynomial $p_0(x,y)$ (with regard to the pair of variables x, y) is called a *homogeneous elliptic polynomial* [3] if it turns into zero only in the point $(0;0)$. Clearly the exact degree of p_0 with regard to the pair of variables x, y is even — say $2m$; the integer $m = (deg\, p_0)/2$ will be called afterwards the *ellipticity order* of this polynomial. Any real-valued (not necessarily homogeneous) polynomial $p(x,y)$ is called an *elliptic polynomial of ellipticity order m* (in short: an *m-elliptic polynomial*) if such is its leading part $p_0(x,y)$ (i.e. the sum of its terms of highest degree). Consider in the ring $\mathbb{R}[x,y]$ (of polynomials in x, y with real coefficients) some non-zero polynomial $p(x,y)$ (not necessarily elliptic). We call it *simple* if: either $p(x,y)$ is irreducible in $\mathbb{R}[x,y]$ and has a non-isolated zero in $\mathbb{R}^2$; or $p(x,y)$ is representable as a product of several mutually prime polynomials each possessing these two properties. A curve $L = \{(x,y) : p(x,y) = 0\}$ where $p(x,y)$ is a simple elliptic polynomial of ellipticity order m will be called afterwards an *m-elliptic curve.* E.g. such will be: the union of m pairwise distinct ellipses; the curve $x^{2m} + y^{2m} = 1$. Using the transformation $x = (z + \overline{z})/2$, $y = (z - \overline{z})/(2i)$ one can present any polynomial $p(x,y)$ in the form of a polynomial $P(z, \overline{z})$ with regard to the conjugate variables $z, \overline{z}$; there exists a unique polynomial in two independent complex variables z, w coinciding with $P(z,\overline{z})$ for $w = \overline{z}$; denote it as $P(z, w)$. Call $P(z,\overline{z})$ the *complex representation* of $p(x, y)$, call $P(z, w)$ the extension of $p(x, y)$ to the space $\mathbb{C}^2$, and call the function $P(z, 1/z)$ *coherent* with the function $p(x, y)$ (and with $P(z,\overline{z})$) on the unit circumference $\gamma = \{z : |z| = 1\}$ (or simply: the coherent function for $p(x, y)$).

16.3 FORMULATION OF THE MAIN RESULT

We intend to prove the following statement:

Theorem 1 *Every n-elliptic curve is a set of uniqueness for the class* $PH_n(\mathbb{R}^2)$.

In other words, *Every entire n-harmonic function* $u(x, y)$ *vanishing in every point of some n-elliptic curve vanishes identically in the entire plane* $\mathbb{R}^2$.

E.g. every entire n-harmonic function which turns into zero on n pairwise distinct ellipses is identically equal to zero; every entire n-harmonic function which vanishes in every point of the curve $x^{2n} + y^{2n} = 1$ is also identically equal to zero.

Theorem 1 can obviously be split into such two statements:

Theorem 1A. *There does not exist any non-zero p.h. polynomial* $u(x, y)$ *vanishing in every point of some n-elliptic curve.*

Theorem 1B. *There does not exist any entire transcendental (i.e. non-polynomial) n-harmonic function* $u(x, y)$ *vanishing (identically) on some n-elliptic curve.*

16.4 SEVERAL AUXILIARY FACTS

The principal idea which will be used here for the proof of Theorem 1 is the close connection between *polyharmonic* and *polyanalytic* functions. A function $f(z,\overline{z})$ defined in the complex plane $\mathbb{C}$ is called *entire polyanalytic* of order n (or entire n-analytic) if it can be presented in the form:

$$f(z,\overline{z}) = \sum_{k=0}^{n-1} f_k(z)\,\overline{z}^k, \tag{16.2}$$

where all functions $f_k(z)$ $(k = 0, 1, \ldots, n-1)$ are entire analytic functions.

We shall make use of some auxiliary facts.

Lemma 1 (see [4]) *The real part of any entire n-analytic function is an entire n-harmonic function; each entire n-harmonic function is the real part of some entire n-analytic function.*

Let $u(x, y)$ be some entire n-harmonic function; using the transformation $x = (z + \overline{z})/2$, $y = (z - \overline{z})/(2i)$ we can transfer to its complex presentation $U(z,\overline{z})$. As it is clear from Lemma 1, the function $U(z,\overline{z})$ can be extended to the entire complex space $\mathbb{C}^2$, where there exists such (unique) entire function (denote it as $U(z, w)$) which coincides with $U(z,\overline{z})$ for $w = \overline{z}$. The function

$U(z,\overline{z})$ admits such expansion:

$$U(z,\overline{z}) = \sum_{k=0}^{\infty} U_k(z,\overline{z}), \tag{16.3}$$

where all functions $U_k(z,\overline{z})$ ($k = 0,1,2,\ldots$) are homogeneous polynomials in $z, \overline{z}$:

$$U_k(z,\overline{z}) = \sum_{k_1+k_2=k} a_{k_1,k_2} z^{k_1} \overline{z}^{k_2}. \tag{16.4}$$

Lemma 1 implies the following statement.

Lemma 2 *If $U(z,\overline{z})$ is an entire n-harmonic function then in its representation* (16.3)–(16.4) *all homogeneous polynomials $U_k(z,\overline{z})$ are also n-harmonic functions and*

$$a_{k_1,k_2} = 0 \quad \text{if} \quad min(k_1,k_2) \geq n. \tag{16.5}$$

Notice that (for any k) the homogeneous polynomial $U_k(z,\overline{z})$ admits such a representation

$$U_k(z,\overline{z}) = \Psi_k(z,\overline{z}) + \overline{\Psi_k(z,\overline{z})},$$

where in the case $k = 2s+1$

$$\Psi_k(z,\overline{z}) = b_{2s+1,0} z^{2s+1} + b_{2s,1} z^{2s} \overline{z} + \cdots + b_{s+1,s} z^{s+1} \overline{z}^{s},$$

and in the case $k = 2s$

$$\Psi_k(z,\overline{z}) = b_{2s,0} z^{2s} + b_{2s-1,1} z^{2s-1} \overline{z} + \cdots + b_{s+1,s-1} z^{s+1} \overline{z}^{s-1} + 1/2 b_{s,s} z^s \overline{z}^s$$

(here all $b_{i,j}$ are complex constants).

Lemma 3 *If the homogeneous polynomial $U_k(z,\overline{z})$ is represented in the form*

$$U_k(z,\overline{z}) = \Psi_k(z,\overline{z}) + \overline{\Psi_k(z,\overline{z})}$$

then the function coherent with $U_k(z,\overline{z})$ on the unit circumference $\gamma = \{z : |z| = 1\}$ can be presented in the form:

$$U_k(z,1/z) = F_k^+(z) + F_k^-(z), \tag{16.6}$$

where

$$F_k^-(z) = \overline{F_k^+(1/\overline{z})},$$

and the following identities are true on γ:

$$F_k^-(z) = \overline{F_k^+(z)}, \quad U_k(z,1/z) = 2\Re F_k^+(z). \tag{16.7}$$

Corollary *If $z_0 \neq 0$ is some zero point for $U_k(z,1/z)$ then the same is true for $1/\overline{z}_0$.*

A function $F(z,\overline{z})$ is called an entire function of two conjugate variables $z, \overline{z}$ if its extension to $\mathbb{C}^2$ is an entire analytic function $F(z,w)$ of two independent complex variables.

Lemma 4 *If an entire function of two conjugate complex variables* $F(z,\overline{z})$ *vanishes in every point of some curve* $\Gamma = \{z : P(z,\overline{z}) = 0\}$, *generated by a simple polynomial* $P(z,\overline{z})$ $(\equiv p(x,y))$ *then there exists such an entire function of two conjugate variables* $\Phi(z,\overline{z})$ *that*

$$F(z,\overline{z}) \equiv P(z,\overline{z})\Phi(z,\overline{z}).$$

Proof We shall split the proof into several stages.
1) Obviously we may confine ourselves to the case when the polynomial $p(x,y)$ $(\equiv P(z,\overline{z}))$ is irreducible in the ring $\mathbb{R}[x,y]$.
2) Prove: If some real-valued polynomial $p(x,y)$ is irreducible in the ring $\mathbb{R}[x,y]$ then its extension (to $\mathbb{C}^2$) $P(z,w)$ is irreducible in the ring $\mathbb{C}[z,w]$ (of polynomials in z,w with complex coefficients).
Indeed, admit the opposite. Then

$$P(z,w) = P_1(z,w)P_2(z,w)$$

where none of the polynomials (with complex coefficients) P_1 or P_2 is a constant. Hence

$$P(z,\overline{z}) \equiv P_1(z,\overline{z})P_2(z,\overline{z}).$$

At least one of these two polynomials — let it be $P_1(z,\overline{z})$ — has a non-isolated zero in $\mathbb{C}$. As it is known (see e.g. [4], Lemma 4.4) P_1 admits such representation:

$$P_1(z,\overline{z}) = M(z,\overline{z})I(z,\overline{z})$$

where $M(z,\overline{z})$ is a real-valued polynomial and $I(z,\overline{z})$ is a polynomial which can possess only isolated zeros. We have:

$$P(z,\overline{z}) \equiv M(z,\overline{z})Q(z,\overline{z}) \quad \text{where } Q = IP_2.$$

Q is obviously not a constant; it is real-valued because so are P and M. Denoting $M(z,\overline{z})$ and $Q(z,\overline{z})$ as $r(x,y)$ and $q(x,y)$ we have: $p(x,y) = r(x,y)q(x,y)$, i.e. $p(x,y)$ is reducible in the ring $\mathbb{R}[x,y]$ — contrary to the condition.
3) Prove: Every polynomial $P(z,w)$ irreducible in the ring $\mathbb{C}[z,w]$ of polynomials with complex coefficients is also irreducible in the ring $H[z,w]$ of entire analytic functions in $\mathbb{C}^2$.

Indeed, admit the opposite: let $P(z,w) \equiv E_1(z,w)E_2(z,w)$, where E_1 and E_2 are irreversible entire functions (i.e. having zero points). Then the set S_k of all zeros of E_k $(k = 1,2)$ is a purely 1-dimensional analytic subset of $\mathbb{C}^2$ and is (simultaneously) a subset of some algebraic set (namely of the set of zeros of the polynomial $P(z,w)$).

Hence (see [6]) S_k $(k = 1,2)$ is an *algebraic* set. Therefore E_k is presentable in the form: $E_k = Q_k\Phi_k$, where Q_k is some polynomial and Φ_k is an entire function without zeros $(k = 1,2)$. Hence $P = Q\Phi$, where Q is a polynomial and Φ is an entire function without zeros. But $\Phi = P/Q$. Using Liouville's theorem

it is easy to conclude that $\Phi = \lambda = const, P = \lambda Q_1 Q_2$, and P is reducible in the ring $\mathbb{C}[z, w]$ — contrary to the condition.
4) Now let $F(z, w)$ be such entire function that $F(z, \bar{z})$ vanishes in every point of the simple curve $\Gamma = \{z = x + iy : p(x, y) = 0\}$ where the polynomial $p(x, y) (\equiv P(z, \bar{z}))$ is irreducible in the ring $\mathbb{R}[x, y]$. Prove that $F(z, w)$ is divisible by $P(z, w)$ in the ring $H[z, w]$ of entire analytic functions in $\mathbb{C}^2$.

In the proof we shall make use of the following theorem due to H. Cartan (see [6], p.67): If two entire analytic functions $F(z, w)$ and $G(z, w)$ are mutually prime (in the ring $H[z, w]$) then there exist such four **non-zero** entire functions $A(z, w)$, $B(z, w)$, $S(z)$, $T(w)$ that

$$A(z, w)\ F(z, w) + B(z, w)\ G(z, w) \equiv S(z)\ T(w).$$

Now admit that in our case $F(z, w)$ is not divisible by $P(z, w)$. Since $p(x, y)$ is irreducible (in the ring $\mathbb{R}[x, y]$), the polynomial $P(z, w)$ is irreducible in the ring $H[z, w]$. Let $F(z, w)$ and $P(z, w)$ be mutually prime in the ring $H[z, w]$. Then (according to Cartan's theorem) there must exist such entire analytic functions $A(z, w)$, $B(z, w)$, $S(z)$, $T(w)$ all not identically equal to zero, that

$$A(z, \bar{z}) F(z, \bar{z}) + B(z, \bar{z}) P(z, \bar{z}) \equiv S(z) T(\bar{z}).$$

Since the functions $F(z, \bar{z})$ and $P(z, \bar{z})$ vanish on a set of points having some finite cluster point, the same is true at least for one of the entire analytic functions $S(z)$ or $T(w)$ — which is therefore identically equal to zero. But this is impossible (according to Cartan's theorem) — if $F(z, w)$ and $P(z, w)$ are mutually prime in the ring $H[z, w]$. Hence $F(z, w)$ and $P(z, w)$ are not mutually prime in $H[z, w]$. Since $P(z, w)$ is irreducible in the ring $H[z, w]$, the function $F(z, w)$ must be divisible by $P(z, w)$ in $H[z, w]$, i.e. there exists such entire analytic function $\Phi(z, w)$ that

$$F(z, w) \equiv P(z, w) \Phi(z, w).$$

Lemma 5 *Suppose that:*
1)(M_k) is a sequence of non-negative numbers among which there is an infinity of non-zero numbers;
2)

$$\lim_{k \to \infty} \sqrt[k]{M_k} = 0; \tag{16.8}$$

3) m is a fixed positive integer.
Then for each (arbitrary small) number $\varepsilon > 0$ there exists an infinity of distinct positive integers k for which the following (strict) inequality is true:

$$M_{k+1} + M_{k+2} + \cdots + M_{k+m} < \varepsilon M_k. \tag{16.9}$$

Proof Put $\lambda = \varepsilon / m$; $\lambda_1 = \min(\lambda_1, 1)$ (thus $0 < \lambda_1 \leq 1$); $a_k = M_k (1/\lambda_1)^k$. Then (16.8) implies $a_k > 0$ for an infinity of indices k, and $\lim_{k \to \infty} a_k = 0$. It is easy to verify that *a sequence possessing such two properties* contains a maximal term. Moreover there can exist only a finite number of such maximal terms.

Now choose among the terms a_k all maximal terms, and among them choose the term with the greatest index; let it be a_{k_1}. After this choose among the terms (a_k) with indices $k > k_1$ the maximal term with the greatest index — let it be a_{k_2}. And so on. On the n-th step choose among the terms a_k with indices $k > k_{n-1}$ the maximal term with the greatest index — let it be a_{k_n}. We obtain a monotonically decreasing sequence of numbers

$$a_{k_1}, a_{k_2}, \dots, a_{k_n}, a_{k_{n+1}}, \dots$$

with the following property: for every positive integer i

$$a_{k_n+i} < a_{k_n}.$$

Denoting k_n as k we have:

$$M_{k+i}(1/\lambda_1)^{k+i} < M_k(1/\lambda_1)^k,$$

whence $M_{k+i} < \lambda_1 M_k$, or $M_{k+i} < (\varepsilon/m)M_k$.

Adding these inequalities for $i = 1, \dots, m$ we obtain (16.9).

Lemma 6 *Suppose that:*

1) $P_k(z,\overline{z})$ $(k = 0, 1, \dots, 2n)$ is a homogeneous polynomial of degree k with respect to the conjugate variables $z, \overline{z}$; $P_{2n}(z,\overline{z})$ is an elliptic polynomial of ellipticity order n;

2) $\Phi_k(z,\overline{z})$ $(k = 0, 1, 2 \dots)$ is a homogeneous polynomial of degree k with regard to $z, \overline{z}$; among these polynomials there exist infinitely many which are not identically equal to zero;

3) the numbers

$$M_k = \max_{|z|=1} |\Phi_k(z,\overline{z})| \quad (k = 0, 1, 2, \dots) \tag{16.10}$$

satisfy the condition

$$\lim_{k\to\infty} \sqrt[k]{M_k} = 0.$$

4) $U_k(z,\overline{z})$ $(k = 0, 1, 2, \dots)$ is a homogeneous polyharmonic polynomial of polyharmonicity order n;

5) for each number $k \geq 0$ the following identity is true:

$$U_{2n+k} \equiv P_{2n}\Phi_k + \mathbb{R}_{2n+k}, \tag{16.11}$$

where

$$\mathbb{R}_{2n+k} = P_0\Phi_{2n+k} + P_1\Phi_{2n+k-1} + \dots + P_{2n-1}\Phi_{k+1}. \tag{16.12}$$

Then there exist such number k and such circular ring

$$V' = \{z : 1/\rho \leq |z| \leq \rho\} \quad (\rho > 1) \tag{16.13}$$

that on its boundary the following inequality is true:

$$|\mathbb{R}_{2n+k}(z,1/z)| < |U_{2n+k}(z,1/z)|. \tag{16.14}$$

Proof Since $P_{2n}(z,\overline{z})$ is an elliptic polynomial it has no zero points on the unit circumference $\gamma = \{z : |z| = 1\}$; the same is true for the rational function $P_{2n}(z,1/z)$, coherent with $P_{2n}(z,\overline{z})$ on γ. Hence there exists such closed ring

$$V = \{z : 1/(1+\delta) \le |z| \le 1+\delta\} \quad (0 < \delta < 1),$$

that $P_{2n}(z,1/z)$ has no zero points in V.

Introduce the notations

$$c_0 = \min_{z\in V} |P_{2n}(z,1/z)| \quad (\text{obviously } c_0 > 0), \tag{16.15}$$

$$A_k = \max_{z\in\gamma} |P_k(z,1/z)|; \tag{16.16}$$

$$c_p = \max_{0\le k\le 2n-1} |A_k|/c_0. \tag{16.17}$$

Let $\varepsilon > 0$ be some number, $0 < \varepsilon < 1/2$ (it will be more precisely defined afterwards).
According to Lemma 5, there exist infinitely many of such indices k for which the inequality

$$M_{k+1} + M_{k+2} + \cdots + M_{k+2n} < (\varepsilon/c_p)M_k$$

is true. Denote the set of such indices as $\widetilde{N}$. We have on γ

$$|\mathbb{R}_{2n+k}(z,1/z)| \le (\varepsilon/c_p)M_k(c_0c_p) = \varepsilon c_0 M_k < (c_0/2)M_k\ . \tag{16.18}$$

On the other hand, we have on γ:

$$\max_{z\in\gamma}\{|P_{2n}(z,\overline{z})\Phi_k(z,\overline{z})|\} \ge c_0 M_k,$$

therefore (see (16.11))

$$\max_{z\in\gamma} |U_{2n+k}(z,1/z)| > (c_0/2)M_k. \tag{16.19}$$

Since $U_{2n+k}(z,\overline{z})$ is a p.h. polynomial of polyharmonicity order n, it can be represented in the form (see Lemma 2 and 3):

$$U_{2n+k}(z,\overline{z}) \equiv \Psi(z,\overline{z}) + \overline{\Psi(z,\overline{z})}$$

where

$$\Psi(z,\overline{z}) = b_{2n+k,0}z^{2n+k} + b_{2n+k-1,1}z^{2n+k-1}\overline{z} + \cdots + b_{n+k+1,n-1}z^{n+k+1}\overline{z}^{n-1}$$

(here all coefficients $b_{i,j}$ are some complex constants). Put

$$F_+(z) = \Psi(z,1/z); \quad F_-(z) = \overline{F_+(z)}.$$

Cleary $F_+(z)$ can be presented in the form

$$F_+(z) = z^{k+2}Q(z),$$

where $Q(z)$ is some non-zero polynomial of degree less than or equal to $2n-2$.

We have on the unit circumference γ (and hence everywhere):

$$U_{2n+k}(z, 1/z) = F_+(z) + F_-(z).$$

Take also into account that on γ

$$|F_+(z)| = |F_-(z)|,$$
$$|U_{2n+k}(z, 1/z)| \leq |2F_+(z)|.$$

Therefore (see (16.19))

$$\max_{z\in\gamma} |F_+(z)| \geq (c_0/4)M_k. \tag{16.20}$$

Consider the set K of all polynomials (with complex coefficients)

$$q(z) = q_0 + q_1 z + \cdots q_{2n-2} z^{2n-2}$$

of degree not exceeding $2n-2$ and satisfying the condition

$$\max_{z\in\gamma} |q(z)| = 1. \tag{16.21}$$

Introduce in K the norm

$$||q|| = |q_0| + |q_1| + \cdots + |q_{2n-2}|.$$

Cleary K forms (with respect to this norm) a compact normed space. Choose in the circular ring

$$V_2 = \{1 + \delta/2 < |z| < 1 + \delta\} \tag{16.22}$$

$2n-1$ distinct circumferences with the centre $z = 0$:

$$\gamma_1, \gamma_2, \ldots, \gamma_{2n-1}.$$

Let

$$c(q) = \max_{1\leq j\leq 2n-1} \{\min_{z\in\gamma_j} |q(z)|\}, \tag{16.23}$$

$$c_1 \equiv c_1(K) = \inf_{q\in K} c(q). \tag{16.24}$$

Cleary $c_1 \geq 0$. We intend to prove that

$$c_1 \neq 0. \tag{16.25}$$

Indeed, admit the opposite: let $c_1 = 0$. Since K is compact there must exist in K such polynomial $\widetilde{q(z)}$ that

$$c(\widetilde{q}) = \min_{q\in K} c(q) = c_1 = 0.$$

Hence, on each circumference γ_k, $(k = 1, \ldots, 2n-1)$ the polynomial $\widetilde{q(z)}$ has at least one root. Since the exact degree of $\widetilde{q(z)}$ does not exceed $2n-2$ we conclude: $\widetilde{q(z)} \equiv 0$. But this contradicts the condition (16.21). The obtained contradiction proves that

$$c_1 > 0.$$

Now go back to the polynomial $Q(z)$ and consider the auxiliary polynomial

$$q(z) = Q(z)/B,$$

where $B = \max_{z \in \gamma} |Q(z)|$. Since the polynomial $q(z)$ belongs to the set K, there exists at least one of the circumferences $\gamma_1, \ldots, \gamma_{2n-1}$ (denote it as Γ_k) on which $|q(z)| \geq c_1 > 0$, i.e. $|Q(z)|_{z \in \Gamma_k} > c_1 B$.

Since $F_+(z) = z^{k+2} Q(z)$, we have

$$|F_+(z)| \geq c_1 B |z|^{k+2} \quad \text{for every } z \in \Gamma_k\,. \tag{16.26}$$

The function $F_-(z)$ is holomorphic for $|z| \geq 1$ and has a finite limit in the point ∞. Hence (as it follows from the maximum modulus principle) in every point of the circumference Γ_k we have

$$|F_-(z)| \leq \max_{|z| \in \gamma} |F_-(z)| = \max_{z \in \gamma} |F_+(z)| = B.$$

Therefore we have for any $z \in \Gamma_k$

$$|U_{2n+k}(z, 1/z)| \geq |F_+(z)| - |F_-(z)| \geq B(c_1 |z|^{k+2} - 1) = Bc_1 |z|^k \left(|z|^2 - \frac{1}{c_1 |z|^k} \right).$$

But $|z| > 1 + \delta/2$, and k can be chosen arbitrary large. Hence for a sufficiently large number k we have: $|z|^2 - 1/(c_1|z|^2) > 1/2(1+\delta/2)^2$. Hence

$$\min_{z \in \Gamma_k} |U_{2n+k}(z, 1/z)| > M_k c_2 |z|^k, \tag{16.27}$$

where c_2 is some constant, $c_2 > 0$.

Consider on Γ_k the rational function $R_{2n+k}(z, 1/z)$ (coherent with the function $R_{2n+k}(z, \bar{z})$ on the unit circumference L).

The coherent function $R(z) = R_{2n+k}(z, 1/z)/z^{2n+k}$ is holomorphic for $|z| \geq 1$ and has a finite limit in the point ∞. Hence (by the maximum modulus principle) we obtain (with account of (16.18)):

$$\max_{z \in \Gamma_k} |R_{2n+k}(z, 1/z)|/z^{2n+k} \leq \max_{z \in \gamma} |R_{2n+k}(z, \bar{z})| \leq \varepsilon c_0 M_k,$$

$$|R_{2n+k}(z, 1/z)| \leq \varepsilon c_0 M_k |z|^{2n+k} \leq \varepsilon c_0 M_k (1+\delta)^{2n} |z|^k$$

for all $z \in \Gamma_k$. Hence

$$\max_{z \in \Gamma_k} |R_{2n+k}(z, 1/z)| \leq \varepsilon c_3 M_k |z|^k, \tag{16.28}$$

where c_3 is some positive constant (independent of k).

It follows from (16.28) and (16.27) that for $\varepsilon \leq c_2/c_3$ the inequality (16.14) is true for $\rho = R_k$.

Hence Lemma 6 is proved.

16.5 THE PROOF OF THEOREM 1

Proof of Theorem 1A Admit the opposite: let there exist such a non-zero n-harmonic polynomial $u(x,y)$ and such an n-elliptic curve $L = \{(x,y) : p(x,y) = 0\}$ that $u(x,y) \equiv 0$ on L. Then there must exist such a real-valued polynomial $q(x,y)$ of some (exact) degree k that

$$u(x,y) \equiv p(x,y)q(x,y). \tag{16.29}$$

Denoting by $p_{2n}(x,y)$, $q_k(x,y)$ and $u_{2n+k}(x,y)$ the terms of highest degree of the polynomials p, q and u, we arrive at the identity

$$u_{2n+k}(x,y) \equiv p_{2n}(x,y)q_k(x,y);$$

here u_{2n+k} is some non-zero n-harmonic polynomial and p_{2n} is some homogeneous n-elliptic polynomial. Transferring from the real variables x, y to the conjugate complex variables $z, \overline{z}$, we arrive at an identity of the form

$$U_{2n+k}(z,\overline{z}) = P_{2n}(z,\overline{z})Q_k(z,\overline{z}).$$

Since $\overline{z} = 1/z$ on the unit circumference $\gamma = \{z : |z| = 1\}$ the following equality is true (on γ — and hence everywhere in the punctured plane $\mathbb{C} \setminus \{0\}$):

$$U_{2n+k}(z,1/z) = P_{2n}(z,1/z)Q_k(z,1/z). \tag{16.30}$$

Estimate the number of roots (with counting multiplicities) on the unit circumference of the rational function presented on the right–hand side of (16.30). Since the elliptic polynomial $p_{2n}(x,y)$ has no zero points on γ the same is true for $P_{2n}(z,1/z)$; the rational function $Q_k(z,1/z)$ has not more than $2k$ roots in the plane $\mathbb{C}$; hence $P_{2n}(z,1/z)Q_k(z,1/z)$ *has not more than* $2k$ *roots on* γ. On the other hand, the homogeneous polynomial $u_{2n+k}(x,y)$ is some p.h. function of order n; therefore (see Lemmas 1,2) there must exist such polyanalytic polynomial $F_{2n+k}(z,\overline{z})$ of polyanalyticity order n that

$$u_{2n+k}(x,y) = \Re F_{2n+k}(z,\overline{z}),$$

$$F_{2n+k}(z,\overline{z}) = a_{2n+k,0}z^{2n+k} + a_{2n+k-1,1}z^{2n+k-1}\overline{z} + \cdots + a_{n+k+1,n-1}z^{n+k+1}\overline{z}^{n-1}$$

(here all $a_{i,j}$ are complex numbers).

We have therefore on the unit circumference γ

$$U_{2n+k}(z,1/z) = \Re F_{2n+k}(z,1/z), \tag{16.31}$$

$$F_{2n+k}(z,1/z) = a_{2n+k,0}z^{2n+k} + \cdots + a_{n+k+1,n-1}z^{k+2} = z^{k+2}H(z),$$

where $H(z)$ is some polynomial.

Consider first the case when $H(z)$ *has not any root on* γ.

When the point z describes the circumference γ once the point $w = z^{k+2}H(z)$ makes (according to the argument principle) not less than $k+2$ revolutions around the point $w = 0$; hence the function $w = F_{2n+k}(z, 1/z)$ maps γ onto a closed curve γ_1 which cuts the axis $\Re w = 0$ not less than $2(k+2)$ times. This means (see (16.31)) that the function $U_{2n+k}(z, 1/z)$ *has not less than* $2k+4$ *roots on* γ.

Now consider the case when the polynomial $H(z)$ *has some roots on* γ.

Then choose some circumference $\gamma_0 = \{|z| = r_0 < 1\}$ containing no zeros of $H(z)$. Let

$$d_0 = \min_{\gamma_0} |F_{2n+k}(z, 1/z)|.$$

Clearly, $d_0 > 0$. Denote $F_{2n+k}(z, 1/z)$ as $\Phi(z)$ and consider the auxiliary function $\Psi(z, d) = \Phi(z) - id$, $0 < d < d_0$. Notice that there must exist such number d $(0 < d < d_0)$ for which $\Psi(z, d)$ has no zero point on γ. Indeed, in the opposite case the polynomial $w = \Phi(z)$ maps an *infinite* set of points z, belonging to γ onto the interval $(0, id)$ of the axis $\Re w = 0$. But the function $\Re \Phi(z) = U_{2n+k}(z, 1/z)$ is a harmonic polynomial, and since it must be identically equal to zero.

Hence, for some number d, $0 < d < d_0$, $\Psi(z, d)$ has no zeros on γ. By the Rouché theorem the function $\Psi(z, d)$ has within γ_0 (and therefore also within γ) not less than $k+2$ zeros (with counting multiplicities). Reasoning as above we conclude that $\Re \Psi(z, d) = U_{2n+k}(z, 1/z)$ has on γ not less then $2k+4$ zero points.

The obtained two conclusions concerning the right–hand and left–hand sides of the identity (16.30) contradict one another; hence our assumption is wrong, and Theorem 1A is true.

Proof of Theorem 1B Assume the opposite: let there exist such an n-elliptic curve $\Gamma = \{z : P(z, \overline{z}) = 0\}$ in the plane and such an entire transcendental n-harmonic function $U(z, \overline{z})$ that $U(z, \overline{z})$ vanishes in every point of Γ. Denote as $U(z, w)$ and $P(z, w)$ the extensions of $U(z, \overline{z})$ and $P(z, \overline{z})$ to the space $\mathbb{C}^2$. According to Lemma 4 there exists such entire analytic function $\Phi(z, w)$ that

$$U(z, w) \equiv P(z, w)\Phi(z, w). \tag{16.32}$$

The functions U, P, Φ can be represented in the form

$$U(z, w) = \sum_{k=0}^{\infty} U_k(z, w),$$

$$P(z, w) = \sum_{k=0}^{2n} P_k(z, w),$$

$$\Phi(z, w) = \sum_{k=0}^{\infty} \Phi_k(z, w)$$

where all functions U_k, P_k, Φ_k are homogeneous polynomials in z, w of degree k. It follows from (16.32) that for every integer k

$$U_{2n+k}(z, \overline{z}) \equiv P_{2n}(z, \overline{z})\Phi_k(z, \overline{z}) + R_{2n+k}(z, \overline{z}), \tag{16.33}$$

$$R_{2n+k} = P_0\Phi_{2n+k} + P_1\Phi_{2n+k-1} + P_{2n-1}\Phi_{k+1}, \tag{16.34}$$

$$U_{2n+k}(z, 1/z) \equiv P_{2n}(z, 1/z)\Phi_k(z, 1/z) + R_{2n+k}(z, 1/z). \tag{16.35}$$

Since $U(z, \overline{z})$ is not a polynomial (in z, $\overline{z}$) among the polynomials Φ_k $(0 \le k < +\infty)$ infinitely many are not identically equal to zero.

Put

$$M_k = \max_{|z|=1} |\Phi_k(z, \overline{z})|.$$

Then obviously (since $\Phi(z, w)$ is an entire function)

$$\lim_{k\to\infty} |M_k|^{1/k} = 0.$$

Reasoning exactly as in the proof of Theorem 1A one can easily detect that for infinitely many indices k the function $P_{2n}(z, 1/z)\Phi_k(z, 1/z)$ has not more than $\underline{2k}$ zero points (with counting multiplicities) on the unit circumference γ; but them the same is true in some sufficiently narrow (closed) ring

$$D = \{z : (1+d)^{-1} \le |z| \le 1+d\}, \quad d > 0.$$

The rational function $U_{2n+k}(z, 1/z)$ has for every k in every ring of the form

$$V = \{z : (1+2b)^{-1} \le |z| \le 1+2b\} \quad (0 < b < 0.5)$$

not less than $2k + 4$ zeros (with counting multiplicities).

As it follows from Lemma 6 such a number k and such a (narrow) circular ring

$$V' = \{z : 1/\rho \le |z| \le \rho\}, \quad V' \subset V$$

can be chosen that in every point of its boundary the strict inequality

$$|R_{2n+k}(z, 1/z)| < |U_{2n+k}(z, 1/z)|$$

is true. Hence (according to Roucheé theorem) the function $U_{2n+k}(z, 1/z) - \mathbb{R}_{2n+k}(z, 1/z)$ has the same number of zero points as $U_{2n+k}(z, 1/z)$, — thus not less then $\underline{2k+4}$ zero points. This contradicts (16.35). Thus our assumption is wrong, and $U(z, \overline{z})$ cannot be a transcendental entire p.h. function.

Theorem 1B is proved.

16.6 A GENERALIZATION

Call a polynomial $p(x, y)$ from the ring $\mathbb{R}[x, y]$ (not necessarily an elliptic one!) a "polynomial of ellipticity order not less than n" if its highest order term $p_0(x, y)$ admits a representation

$$p_0(x, y) \equiv p_{01}(x, y)p_{02}(x, y)$$

where p_{01} and p_{02} belong to $\mathbb{R}[x, y]$ and at least one of them is an elliptic polynomial of exact degree not less than $2n$.

E.g. the polynomial $p = x^6 - y^6 - 1$ is not an elliptic polynomial; but its highest order term admits the representation $(x^4 + y^4 - x^2y^2)(x^2 - y^2)$ (where the first factor is an elliptic polynomial). The polynomial p is a "polynomial of ellipticity order 2".

Looking through the proof of Theorems 1A and 1B one can easily prove the following statement generalizing Theorem 1.

Theorem 2. *Suppose that*
1) $p(x, y)$ is a non-zero real-valued polynomial (not necessarily an elliptic one!) of ellipticity order not less than n;
2) $p(x, y)$ is a "simple" polynomial in the ring $\mathbb{R}[x, y]$ (see def. in section 16.2).
3) $u(x, y)$ is a n-harmonic function in $\mathbb{R}^2$, vanishing in every point of the curve $\Gamma = \{(x, y) : p(x, y) = 0\}$.
Then $u(x, y)$ vanishes in every point of the plane $\mathbb{R}^2$.

E.g. the polynomial $p(x, y) = x^6 - y^6 - 1$ is a simple polynomial in the ring $\mathbb{R}[x, y]$ (moreover, it is irreducible in this ring); its leading term is a polynomial of ellipticity order 2. According to Theorem 2 the only entire biharmonic function vanishing in every point of the (hyperbola-like) curve $x^6 - y^6 = 1$ is the function $u(x, y) \equiv 0$.

Remark Let D be the region bounded by some n-elliptic oval L. Among the functions n-harmonic in D and continious in $D \cup L$ there exist an infinite number of functions vanishing on L; but — only one of them can be extended as a n-harmonic function in the entire plane $\mathbb{R}^2$ — this is the function $u(x, y) \equiv 0$.

References

[1] Hayman, W.K.: A uniqueness problem for polyharmonic functions. In: Linear and Complex Analysis. Problem Book 3, Part II. Eds.:V.P.Havin, N.K.Nikolski. Lecture Notes in Mathematics 1574, Springer-Verlag, Berlin-Heidelberg, 1994, Problem 16.16, pp. 326-327.

[2] Hayman, W.K., Korenblum, B.: Representation and uniqueness theorems for polyharmonic functions. J. Anal. Math. 60 (1993), 113-133.

[3] Hörmander, L.: Analysis of linear partial differential operators, 1. Springer-Verlag, Berlin-Heidelberg, 1981.

[4] Balk, M.B.: Polyanalytic Functions. Mathematical Research 63. Akademie Verlag, Berlin, 1991.

[5] Behnke, H., Thullen, P.: Theorie der Funktionen mehrerer komplexen Veränderlichen. Springer–Verlag, Berlin, 1934.

[6] Ronkin, L.A.: Some topics on zero points distribution of entire functions of several variables. Mat. Sbornik 87 (1972), 351—368 (Russian).

17 BI-ANALYTIC FUNCTIONS AND THEIR APPLICATIONS IN ELASTICITY

Wei Lin †
Yuqiu Zhao

Department of Mathematics
Zhongshan University
Guangzhou 510275
People's Republic of China
stslw@zsu.edu.cn

Abstract: This paper is a review of some of our relevant work on bi-analytic functions and their applications in elasticity. The main basis lies in a mechanical interpretation of bi-analytical functions. First we provide a brief introduction of (λ, k) bi-analytical functions and the relation between the basic problems of plane elasticity, and (λ, k) bi-analytical functions. A mixed-contact problem solved by using (λ, k) bi-analytical functions and singular integral operators and some basic problems of elasticity reduced to BVP on some Riemann surfaces are demonstrated. Moreover, we restrict our attention to quasistatic linear thermoelasticity. We discuss the problems in general simply-connected domains. We first use the Bergman kernel function and the Green function of the domain to obtain integral representations for the BVPs and then decouple the thermoelastic systems on the bounded simply-connected domain. By utilizing the contractive mapping principle the decoupled temperature equation is studied. The representations of solutions of the field equations are obtained and their solvability results are proved. Finally their numerical solutions are given by means of the Galerkin finite element method and wavelet method respectively.

17.1 INTRODUCTION

Function theoretic methods for partial differential equations and integral equations has undergone long lasting development since L. Bers and I. Vekua. Nowadays, browsing through the papers on this subject, one can see that it still keeps

†This work is supported by the the National Natural Science Foundation of China and the Guangdong Natural Science Foundation

H.G.W. Begehr et al. (eds.), Partial Differential and Integral Equations, 233–247.

active. There are numerous application of methods in various fields such as mechanics and mathematics physics (see, e.g. some of the authors working in this field are R.P. Gilbert, H. Begehr, W. Wendland and A. Dzhuraev). Among those application is a systematic study of plane elasticity, initiated by Russian mathematicians in the middle of this century and systemalized and generalized thereafter, by applied mathematicians both from the West and the East.

This paper is a review of some of our relevant work on this topic. The main basis lies in a mechanical interpretation of bi–analytical functions developed by Hua, Lin and Wu [1] in the sixties. A bi-analytical function is an analogue of an analytical function in that both components of them satisfy some elliptic equation of canonical form and possess pararell properties. It comes to know that bi-analytical functions are essentially and closely related to orthotropic plane elasticity.

In the eighties, R.P. Gilbert and W. Lin [5] first brought to light the relations between (λ, k) bi-analytical functions and plane elasticity, both in isotropic and orthotropic cases. Several years later (1992), a deeper understanding of such relations lead to Begehr and Lin [4], where a mix-contact problem is solved by using (λ, k) bi-analytical functions and singular integral operators. Almost at the same time, the first author of the present paper, along with a Russian mathematician, developed an approach to deal with elasticity problems. In their treatment, basic problems of elasticity are reduced equivalently to the boundary value problem on some Riemann surface. This is achieved through generalizing the symmetry point and Schwarz principle about the circle to algebraic curves on Riemann surfaces of genus zero .

The main content of this paper devotes to the application of function theoretic methods to thermal elastic problems. Motivated by a paper of Shi and Xu [13], we have worked out a series of papers which aim at solving basic problems of plane thermoelasticity ([9, 10, 11, 12, 8]). Using (λ, k) bi-analytical functions and canonical operators in mathematical physics, the thermal elastic system is decoupled, and the stress problem and displacement problem are solved, both the isotropic and orthotropic cases are taken into consideration, and further, numerical solutions are constructed.

The present paper is arranged as follows. In section 2, we provide some brief introduction of (λ, k) bi-analytical functions. In section 3, we demonstrate the relation between the basic problems of plane elasticity with (λ, k) bi-analytical functions. In the remaining sections, we restrict our attention to quasi-static linear thermoelasticity, mainly in the unit disk, intending in this way to explain our approach to deal with general thermal problems in simply-connected domains from our above cited papers.

17.2 BI-ANALYTIC FUNCTIONS

In 1960's, Hua, Lin and Wu (the readers are referred to [1] for details) considered the system

$$\left(A\frac{\partial^2}{\partial x^2}+2B\frac{\partial^2}{\partial x\partial y}+C\frac{\partial^2}{\partial y^2}\right)\begin{pmatrix} u \\ v \end{pmatrix}=0 \tag{17.1}$$

where A, B and C are 2×2 matrices. System (17.1) may be reduced to some canonical form by applying three types of transformations: linear combinations of equations, linear transform of unknown functions and linear transform of variables. As an example, for an elliptic system, its characteristic quadratic form

$$Q(\xi,\eta)=|A\xi^2+2B\xi\eta+C\eta^2|=(\xi^2+\eta^2)(\xi^2+k^2\eta^2)$$

yields two pairs of conjugate complex roots (where $0<k\leq 1$). In this case, we obtain the canonical form

$$\Bigg[\begin{pmatrix} 1 & 0 \\ 0 & \lambda/k^2 \end{pmatrix}\frac{\partial^2}{\partial x^2} + \begin{pmatrix} 0 & (\lambda-k^2)/k \\ (\lambda-1)/k & 0 \end{pmatrix}\frac{\partial^2}{\partial x\partial y}+\begin{pmatrix} \lambda & 0 \\ 0 & 1 \end{pmatrix}\frac{\partial^2}{\partial y^2}\Bigg]\begin{pmatrix} u \\ v \end{pmatrix}=0\,, \tag{17.2}$$

which can be factorized into

$$\left[\begin{pmatrix} k & 0 \\ 0 & -\lambda \end{pmatrix}\frac{\partial}{\partial x}+\begin{pmatrix} 0 & \lambda \\ k & 0 \end{pmatrix}\frac{\partial}{\partial y}\right]\left[\begin{pmatrix} 1/k & 0 \\ 0 & 1/k \end{pmatrix}\frac{\partial}{\partial x}+\begin{pmatrix} 0 & 1 \\ 1 & 0 \end{pmatrix}\frac{\partial}{\partial y}\right]\begin{pmatrix} u \\ v \end{pmatrix}=0\,.$$

Setting

$$\begin{pmatrix} \theta \\ \omega \end{pmatrix}=\left[\begin{pmatrix} 1/k & 0 \\ 0 & 1/k \end{pmatrix}\frac{\partial}{\partial x}+\begin{pmatrix} 0 & 1 \\ 1 & 0 \end{pmatrix}\frac{\partial}{\partial y}\right]\begin{pmatrix} u \\ v \end{pmatrix},$$

we get

$$k\frac{\partial\theta}{\partial x}+\lambda\frac{\partial\omega}{\partial y}=0\;;\;\; k\frac{\partial\theta}{\partial y}-\lambda\frac{\partial\omega}{\partial x}=0\,.$$

Obviously, $\varphi(z)=k\theta-i\lambda\omega$ is analytic with respect to $z=x+iy$. We refer $f(z)=u+iv$ as the bi-analytic function of (λ,k) type, if u,v satisfy

$$\frac{1}{k}\frac{\partial u}{\partial x}-\frac{\partial v}{\partial y}=\theta\;;\;\;\frac{\partial u}{\partial y}+\frac{1}{k}\frac{\partial v}{\partial x}=\omega.$$

Derivative and integral are introduced, and a function theory of (λ,k) bi-analytic functions are developed in [1].

17.3 BASIC BOUNDARY VALUE PROBLEMS FOR PLANE ELASTICITY

In homogeneous and isotropic plane elasticity, the following stress-strain relation holds

$$\begin{cases} \sigma_x = (\lambda + 2\mu)\dfrac{\partial u}{\partial x} + \lambda\dfrac{\partial v}{\partial y} \\ \sigma_y = \lambda\dfrac{\partial u}{\partial x} + (\lambda + 2\mu)\dfrac{\partial v}{\partial y} \\ \tau_{xy} = \mu(\dfrac{\partial v}{\partial x} + \dfrac{\partial u}{\partial y}) \end{cases} \tag{17.3}$$

where u, v are displacements in x and y direction, respectively. σ_x, σ_y and τ_{xy} are respectively direct stresses and shear stress. λ and μ are *Lamé* constants

$$\lambda = \frac{E\nu}{(1+\nu)(1-2\nu)} \ , \ \mu = \frac{E}{2(1+\nu)} \ ,$$

where $E(> 0)$ is the elastic modulus and $\nu(0 < \nu < \frac{1}{2})$ is *Poission*'s ratio. Equilibrium equation (the field equation, without body force) is

$$\begin{cases} \dfrac{\partial \sigma_x}{\partial x} + \dfrac{\partial \tau_{xy}}{\partial y} = 0 \\ \dfrac{\partial \tau_{xy}}{\partial x} + \dfrac{\partial \sigma_y}{\partial y} = 0 \end{cases} \tag{17.4}$$

Assuming that the elastic body is with reference configuration Ω and with boundary Γ, then generally there are two types of basic boundary value problems

17.3.1 Stress boundary value problem

The load on Γ is known, determine the equilibrium inside Ω subject to

$$\begin{cases} X_n = \alpha\sigma_x + \beta\tau_{xy} = T_1 \\ Y_n = \alpha\tau_{xy} + \beta\sigma_y = T_2 \end{cases} \qquad (x,\ y) \in \Gamma\ . \tag{17.5a}$$

where $(\alpha,\ \beta)$ is the unit outer normal at $(x,\ y) \in \Gamma$, T_1 and T_2 are components of the known edge force. Of course they should obey equilibrium conditions on force and moments.

Sometimes this boundary value problem is termed *the first basic problem of plane elasticity.*

17.3.2 Displacement boundary value problem

The displacement at each boundary point is given, that is

$$\begin{cases} u(x,y) = u_0 \\ v(x,y) = v_0 \end{cases} \qquad (x,\ y) \in \Gamma\ , \tag{17.5b}$$

determine the equilibrium. This is sometimes called *the second basic problem of plane elasticity.*

Instead of (17.3), in orthotropic case, we have

$$\begin{cases} \sigma_x = \dfrac{1}{1-\nu_{12}\nu_{21}}(E_{11}\dfrac{\partial u}{\partial x} + \nu_{12}E_{22}\dfrac{\partial v}{\partial y}), \\ \sigma_y = \dfrac{1}{1-\nu_{12}\nu_{21}}(\nu_{21}E_{11}\dfrac{\partial u}{\partial x} + E_{22}\dfrac{\partial v}{\partial y}), \\ \tau_{xy} = G_{12}(\dfrac{\partial v}{\partial x} + \dfrac{\partial u}{\partial y}), \end{cases} \tag{17.6}$$

where E_{11} and E_{22} are respectively the *Young*'s moduli along x and y, and G_{12} is the shear modulus. ν_{12} is a *Poission*'s ratio, which describes the constriction in y direction caused by the stretch in x. ν_{21} has an analogous meaning with only x and y interchanged. These constants obey the following symmetric property: $\nu_{21}E_{11} = \nu_{12}E_{22}$, and $0 < \nu_{12} < 1/2$, $0 < \nu_{21} < 1/2$.

We use (17.4) and (17.6) as an example to show the relation between the elastic model and $(\lambda,\ k)$ bi-analytic functions.

Substituting (17.6) into (17.4), setting with

$$\nu = \sqrt{\nu_{12}\nu_{21}},\ E = \sqrt{E_{11}E_{22}},\ \delta^2 = \sqrt{E_{11}/E_{22}},\ K = \frac{E}{2G_{12}} - \nu,\ L = \frac{G_{12}}{E}(1-\nu^2),$$

we obtain

$$\Big[\begin{pmatrix} \delta^2 & 0 \\ 0 & L \end{pmatrix}\frac{\partial^2}{\partial x^2} + \begin{pmatrix} 0 & \nu+L \\ \nu+L & 0 \end{pmatrix}\frac{\partial^2}{\partial x\partial y} + \begin{pmatrix} L & 0 \\ 0 & \delta^{-2} \end{pmatrix}\frac{\partial^2}{\partial y^2}\Big]\begin{pmatrix} u \\ v \end{pmatrix} = 0 \tag{17.7}$$

Since $K > -\nu > -1$, we can divide our discussion into three cases: $K > 1$, $K = 1$ and $|K| < 1$, $K = 1$ corresponding to the isotropic case.

When $K > 1$, letting $k = K - \sqrt{K^2 - 1}$ and $\lambda = kL$, carrying out transformations of coordinates and unknowns and linear combination of equations as follows

$$\begin{pmatrix} x \\ y \end{pmatrix} = \begin{pmatrix} \delta & 0 \\ 0 & 1/\sqrt{k} \end{pmatrix}\begin{pmatrix} X \\ Y \end{pmatrix},$$

$$\begin{pmatrix} u \\ v \end{pmatrix} = \begin{pmatrix} \dfrac{1+k\nu}{(k+\nu)\sqrt{k}} & 0 \\ 0 & -\delta \end{pmatrix}\begin{pmatrix} U \\ V \end{pmatrix}.$$

Multiplying (17.7) on the left by

$$\begin{pmatrix} \dfrac{(k+\nu)\sqrt{k}}{1+k\nu} & 0 \\ 0 & -\delta \end{pmatrix},$$

we have thus turned (17.7) into

$$\left[\begin{pmatrix} 1 & 0 \\ 0 & \lambda/k \end{pmatrix} \frac{\partial^2}{\partial X^2} + \begin{pmatrix} 0 & (\lambda-k^2)/k \\ \lambda-1 & 0 \end{pmatrix} \frac{\partial^2}{\partial X \partial Y} + \begin{pmatrix} \lambda & 0 \\ 0 & k \end{pmatrix} \frac{\partial^2}{\partial Y^2}\right] \begin{pmatrix} U \\ V \end{pmatrix} = 0. \tag{17.8}$$

This is the same as the elliptic canonical form (17.2). Similar derivation is valid when $|K| < 1$:

$$\rho = \frac{\sqrt{2}-\sqrt{1-K}}{\sqrt{1+K}}, \quad k = \rho^2, \quad \lambda = \rho^2 \left(\frac{1+\nu+2L}{1-\nu}\right),$$

$$\begin{pmatrix} x \\ y \end{pmatrix} = \begin{pmatrix} \delta/2 & \delta/(2\rho) \\ 1/2 & -1/(2\rho) \end{pmatrix} \begin{pmatrix} X \\ Y \end{pmatrix}$$

$$\begin{pmatrix} u \\ v \end{pmatrix} = \begin{pmatrix} 1 & \dfrac{1-\nu}{2\rho(1-L)} \dfrac{\lambda-k^2}{k} \\ 0 & -\delta \end{pmatrix} \begin{pmatrix} U \\ V \end{pmatrix}$$

Multiplying (17.4) on the left by

$$\frac{1}{2}\begin{pmatrix} \dfrac{1}{1-\nu} & -\dfrac{\delta}{1-\nu} \\ \dfrac{\lambda-1}{2\rho(1-L)} & \dfrac{(\lambda-1)\delta}{2\rho(1-L)} \end{pmatrix}$$

then the equation turns to the same canonical form (17.8). That is to say, $U + iV$ is a (λ, k) bi-analytic function with respect to $Z = X + iY$.

Along this way, the basic boundary value problems of plane elasticity can be solved through solving the corresponding boundary value problems of (λ, k) bi-analytic fuctions. We have thus shown that (λ, k) bi-analytic fuctions provide a tool to deal with orthotropic plane elasticity while $(\lambda, 1)$ bi-analytic functions are closely related to isotropic elasticity.

17.4 QUASI–STATIC THERMOELASTICITY

For homogeneous and isotropic elasticity, by neglecting inertia term, the field equation in the quasi-static approximation of thermoelasticity is

$$\lambda \frac{\partial}{\partial x_i}\left(\sum_{j=1}^{2} \frac{\partial u_j}{\partial x_j}\right) + \mu \left(\sum_{j=1}^{2} \frac{\partial^2 u_i}{\partial x_j^2} + \frac{\partial}{\partial x_i}\left(\sum_{j=1}^{2} \frac{\partial u_j}{\partial x_j}\right)\right) = \beta \frac{\partial \theta}{\partial x_i} \quad i = 1,\ 2, \tag{17.9}$$

$$q\Delta\theta = \rho s \frac{\partial \theta}{\partial t} + \beta c \frac{\partial}{\partial t} \text{div} \mathbf{u}, \tag{17.10}$$

where $\mathbf{u} = (\mathbf{u_1}, \ \mathbf{u_2})$ is the displacement vector, θ is the temperature, and λ and μ are *Lamé* constants, β denotes the interaction constant, q, ρ, s and c denote, respectively, the heat conductivity, the density, the specific heat and the absolute reference temperature. All the above constants are positive.

Suppose the reference configuration of the elastic body be a bounded simply connected domain Ω, $\Gamma = \partial\Omega \in C^2$. Sometimes we use the notation $\Omega_T = \Omega \times [0, \ T]$ and $S_T = \partial\Omega \times [0, \ T]$.

The first initial-boundary value problem for (17.9) and (17.10) is

$$\mathbf{u} = \mathbf{u_0} \ , \quad \theta = \theta_0 \ \text{ on } \ \partial\Omega \times (0, \ T) \ ; \quad \theta(\cdot \ , \ 0) = \varphi \ \ ; \text{on } \ \Omega \ . \tag{17.11}$$

The second initial-boundary value problem for (17.9) and (17.10) is

$$T_{ij} n_j = T_i^0 \ , \quad \theta = \theta_0 \ \text{ on } \ \partial\Omega \times (0, \ T) \ ; \quad \theta(\cdot \ , \ 0) = \varphi \ \text{ on } \Omega \ . \tag{17.12}$$

where T_{ij} are stresses, T_1^0, T_2^0 are the components of external stress along the axes, n_1 and n_2 are direction cosines of the exterior unit normal to $\partial\Omega$.

The quasi-static theory of thermoelasticity in one space dimension is thoroughly studied in W.A. Day [14] via the decoupled temperature equation. In contact problems of one dimensional thermoelasticity, the decoupled temperature equation enjoys an interesting nonlocal nonlinearality (Shi and Shillor, see [15]). However, the decoupling in two space dimension is much difficult because of the structure of (17.9) and (17.10). P. Shi and Y. Xu [13] investigate the first initial boundary value problem on the unit disk. In our papers, the results obtained on this special case are generalized to all simply-connected domains with twice continuously differentiable boundary, and both the first and second initial boundary value problems are dealt with. By using function theoretic methods, we give a representation formula of solutions for general bounded simply-connected domains. We decouple the systems and prove the existence and uniqueness of solution to the first and second initial-boundary value problems.

Making a change of variables

$$\begin{cases} x = -x_1 \\ y = x_2 \end{cases} ; \qquad \begin{cases} u(x, \ y) = u_1(-x_1, \ x_2) \\ v(x, \ y) = u_2(-x_1, \ x_2) \end{cases} ,$$

and setting $\hat{\lambda} = \mu/(\lambda + 2\mu)$, $\hat{\beta} = \beta/(\lambda + 2\mu)$, we get

$$\left[\begin{pmatrix} 1 & 0 \\ 0 & -\hat{\lambda} \end{pmatrix} \frac{\partial^2}{\partial x^2} + \begin{pmatrix} 0 & \hat{\lambda} - 1 \\ 1 - \hat{\lambda} & 0 \end{pmatrix} \frac{\partial^2}{\partial x \partial y} + \begin{pmatrix} \hat{\lambda} & 0 \\ 0 & -1 \end{pmatrix} \frac{\partial^2}{\partial y^2} \right] \begin{pmatrix} u \\ v \end{pmatrix} = -\beta \begin{pmatrix} \theta_x \\ \theta_y \end{pmatrix} .$$

Let $z = x + iy$, $w(z) = u(x, y) + iv(x, y)$, $\partial/\partial z = 1/2\,(\partial/\partial x - i\partial/\partial y)$, $\partial/\partial\overline{z} = 1/2\,(\partial/\partial x + i\partial/\partial y)$.

$$\mathbf{D}w = (1-\hat{\lambda})\frac{\partial^2 w}{\partial\overline{z}^2} + (1+\hat{\lambda})\frac{\partial^2\overline{w}}{\partial z\partial\overline{z}},$$

then (17.9) and (17.10) turn to

$$\mathbf{D}w = -\hat{\beta}\frac{\partial\theta}{\partial\overline{z}}, \tag{17.13}$$

$$q\Delta\theta = \rho s\frac{\partial\theta}{\partial t} - 2\beta c\frac{\partial}{\partial t}Re\left\{\frac{\partial w}{\partial\overline{z}}\right\}. \tag{17.14}$$

The usual notations of function spaces, see, e.g. [3], $D(\Omega)$, $W^{1,2}(\Omega)$, $W_0^{1,2}(\Omega)$, $W^{1,2}(\Omega_T)$ and $W_0^{1,2}(\Omega_T)$,

$$W_2^{2,1}(\Omega_T) = \{u\ |u \in W^{1,2}(\Omega_T); u_{xx}, u_{xy}, u_{yy} \in L^2(\Omega_T)\}$$

with norm

$$\|u\|_{W_2^{2,1}(\Omega_T)}$$

$$= \left[\iint_{\Omega_T}\left(|u|^2 + |u_x|^2 + |u_y|^2 + |u_t|^2 + |u_{xx}|^2 + |u_{xy}|^2 + |u_{yy}|^2\right)dxdydt\right]^{1/2}$$

are used. $W_0^{1,2}(\Omega_T)$ and $W_{2,0}^{2,1}(\Omega_T)$ denote the completions of the family of $C^\infty(\Omega_T)$ functions vanishing in a neighbourhood of S_T in $W^{1,2}(\Omega_T)$ and $W_2^{2,1}(\Omega_T)$, respectively.

If $\partial\Omega \in C^2$, then $W_{2,0}^{2,1}(\Omega_T)$ possesses an equivalent norm

$$|u|_{W_{2,0}^{2,1}(\Omega_T)} = \left[\iint_{\Omega_T}\left(|u_t|^2 + |a\Delta u|^2\right)dxdydt\right]^{1/2}, \quad (a \neq 0).$$

Without loss of generality, one can pose the following problems.

The first initial boundary value problem:

$$w|_{S_T} = 0\ , \tag{17.15}$$

$$\theta|_{S_T} = 0\ ,\quad \theta(\cdot,\ 0) = \varphi(\cdot). \tag{17.16}$$

The second initial boundary value problem:

$$T_{ij}n_j|_{S_T} = 0\ , \tag{17.17}$$

$$\theta|_{S_T} = 0\ ,\quad \theta(\cdot,\ 0) = \varphi(\cdot). \tag{17.18}$$

On the unit disk

The representation of solutions of Dirichlet problem (17.15) of equation (17.13) is

$$w(z)=\hat{\beta}\left\{\frac{1}{2\pi}\int_\Omega\frac{\theta(\zeta)}{\zeta-z}d\sigma_\zeta\right.$$
$$\left.+\frac{1}{2\pi}\int_\Omega\frac{\bar{z}\theta(\zeta)}{1-\bar{z}\zeta}d\sigma_\zeta-\frac{\hat{\lambda}-1}{\hat{\lambda}+1}\frac{(1-z\bar{z})}{2\pi}\int_\Omega\frac{\bar{\zeta}(2-z\bar{\zeta})}{(1-z\bar{\zeta})^2}\theta(\zeta)d\sigma_\zeta\right\}. \tag{17.19}$$

Substituting (17.19) into (17.14), one gets the decoupled equation

$$(\rho s+\beta\hat{\beta}c)\frac{\partial\theta}{\partial t}-q\Delta\theta$$
$$=2\beta c\frac{\partial}{\partial t}\operatorname{Re}\left\{\frac{\hat{\beta}}{2\pi}\int_\Omega\left[\frac{1}{(1-\bar{z}\zeta)^2}+\frac{\hat{\lambda}-1}{\hat{\lambda}+1}\frac{z\bar{\zeta}(2-z\bar{\zeta})}{(1-z\bar{\zeta})^2}\right]\theta(\zeta,t)d\sigma_\zeta\right\}. \tag{17.20}$$

Setting

$$(\mathbf{M}\theta)(z)=\frac{1}{\pi}\int_\Omega\operatorname{Re}\left[\frac{1}{(1-\bar{z}\zeta)^2}+\frac{\hat{\lambda}-1}{\hat{\lambda}+1}\frac{z\bar{\zeta}(2-z\bar{\zeta})}{(1-z\bar{\zeta})^2}\right]\theta(\zeta)d\sigma_\zeta \tag{17.21}$$

we obtain

$$\frac{\partial\theta}{\partial t}-a\Delta\theta=b\frac{\partial}{\partial t}(\mathbf{M}\theta), \tag{17.22}$$

where $a=q/(\rho s+\beta\hat{\beta}c)$, $b=\beta\hat{\beta}c/(\rho s+\beta\hat{\beta}c)$. Shi and Xu [13] proved

Theorem 1 *Assume* $\varphi\in W_0^{1,2}(\Omega)$, *then there is a unique solution in* $W_{2,0}^{2,1}(\Omega_T)$ *to* (17.22) *and* (17.16), *provided that* $|b|<C_\lambda=[1+2(\hat{\lambda}-1)/(\hat{\lambda}+1)\sqrt{4/\pi+1}$. *The solution is analytic in the space variables for almost all fixed* $t\in[0,\ 1]$.

Shi and Xu suggest that the restriction to b in Theorem 4.1 is unnatural and removable. We remove it and improved above result. Infact, we can prove

Lemma 1 $||\mathbf{M}\theta||_{L^2(\Omega)}\leq||\theta||_{L^2(\Omega)}$.

Easily follows from this estimate and the principle of contractive mapping

Theorem 2 *Assume* $\varphi\in W_0^{1,2}(\Omega)$, *then there is a unique solution in* $W_{2,0}^{2,1}(\Omega_T)$ *to* (17.22) *and* (17.16). *The solution is analytic in the space variables for almost all fixed* $t\in[0,\ 1]$.

17.5 ON A BOUNDED SIMPLY–CONNECTED DOMAIN

Bergman function

For Ω, we may introduce the Bergman function

$$\overline{k_z(\zeta)}=K_\omega(z,\bar{\zeta})\ ,\ z,\ \zeta\in\Omega,$$

with the well-known property

$$\int_{\Omega} K_{\Omega}(z,\overline{\zeta})f(\zeta)d\sigma_{\zeta} = f(z)\ ,\quad f \in L^2(\Omega),\ f \text{ analytic},$$

With the Green function $g(\zeta, z) = -\ln\overline{|\zeta - z|} + h(\zeta, z)$, then

$$K_{\Omega}(z,\overline{\zeta}) = -\frac{2}{\pi}\frac{\partial^2 g(\zeta, z)}{\partial\overline{\zeta}\partial z} = -\frac{2}{\pi}\frac{\partial^2 h(\zeta, z)}{\partial\overline{\zeta}\partial z}.$$

For the unit disk,

$$K(z,\overline{\zeta}) = \frac{1}{\pi}\sum_{k=1}^{\infty} k(z\overline{\zeta})^{k-1} = \frac{1}{\pi(1 - z\overline{\zeta})^2}.$$

Denoting $\hat{\theta} = -\hat{\beta}\theta$, we get

$$\begin{cases} \mathbf{D}w = \partial\hat{\theta}/\partial\overline{z}, \\ w|_{\partial\Omega} = 0, \end{cases} \qquad \begin{cases} A\partial w/\partial\overline{z} + \overline{B(\partial w/\partial\overline{z})} = \hat{\theta} + \psi(z), \\ w|_{\partial\Omega} = 0, \end{cases} \tag{17.23}$$

where $A = 1 - \hat{\lambda}$, $B = 1 + \hat{\lambda}$. ψ is analytic. Denoting

$$\Psi(z) = \frac{1}{2}\hat{\theta} + \frac{A}{\delta}\psi(z) - \frac{B}{\delta}\overline{\psi(z)}\ ,$$

where $\delta = A^2 - B^2 = -4\hat{\lambda}$, one gets

$$w(z) = -\frac{2}{\pi}\int_{\Omega} g(\zeta, z)\frac{\partial\Psi(\zeta)}{\partial\zeta}d\sigma_{\zeta} = \frac{2}{\pi}\int_{\Omega}\Psi(\zeta)\frac{\partial g(\zeta, z)}{\partial\zeta}d\sigma_{\zeta}.$$

At last we obtain that

$$w(z) = (\mathbf{T}_{\Omega}\Psi)(z) + \frac{2}{\pi}\int_{\Omega}\Psi(\zeta)\frac{\partial h(\zeta, z)}{\partial\zeta}d\sigma_{\zeta}.$$

Here we have utilized Vekua's $\mathbf{T}$ operator

$$(\mathbf{T}_{\Omega}f)(z) = -\frac{1}{\pi}\int_{\Omega}\frac{f(\zeta)}{\zeta - z}d\sigma_{\zeta} \text{ such that } \frac{\partial(\mathbf{T}_{\Omega}f)(z)}{\partial\overline{z}} = f(z)\ .$$

Hence it suffices to determine Ψ or, in other word, ψ. In fact, ψ satisfies the integral equation

$$\psi(z) - \frac{A}{B}\overline{\mathbf{M}_1\psi} - \frac{\delta}{2B}\overline{\mathbf{M}_1(\hat{\theta})} = 0 \tag{17.24}$$

where $(\mathbf{M}_1\psi)(z) = -2/\pi\int_{\Omega}\psi(\zeta)\overline{K_{\Omega}(z,\overline{\zeta})}d\sigma_{\zeta}$, which satisfies $\mathbf{M}_1^2 = \mathbf{M}_1$, and

Lemma 2 *If* $\psi \in L^2(\Omega)$, *then*

$$||\mathbf{M}_1\psi||_{L^2(\Omega)} \leq ||\psi||_{L^2(\Omega)} \tag{17.25}$$

Theorem 3 *The integral equation* (17.24) *is uniquely solvable. If* $\psi \in L^2(\Omega)$ *is a solution, then*

$$\|\psi\|_{L(\Omega)} \le \|\hat{\theta}\|_{L^2(\Omega)}. \tag{17.26}$$

Decoupled equation
Substituting (17.23) to (17.14), we get

$$\frac{\partial \theta}{\partial t} - a_1 \Delta \theta = b_1 Re[\mathbf{M}_2 \theta_t]$$

where $a_1 = q/(\rho s + \beta\hat{\beta}c)$, $b_1 = -\beta\hat{\beta}c/(\rho s + \beta\hat{\beta}c)$,

$$\mathbf{M}_2 = \frac{\delta}{2B}(I - \frac{A}{B}\overline{\mathbf{M_1}})^{-1}\overline{\mathbf{M_1}}.$$

Noticing (17.26), we have

$$\|Re(\mathbf{M_2}\theta)\|_{L^2(\Omega)} \le \|\mathbf{M_2}\theta\|_{L^2(\Omega)} \le \|\theta\|_{L^2(\Omega)}.$$

Since $-1 < b_1 < 0$, we may prove the next result.

Theorem 4 *Assume* $\varphi \in W_0^{1,2}(\Omega)$, *then there exists a unique solution in* $W_{2,0}^{2,1}(\Omega_T)$ *to* (17.13), (17.14), (17.15) *and* (17.16), *and the solution is analytic in the space variables for almost all fixed* $t \in [0,\ T]$.

17.6 STRESS BOUNDARY VALUE PROBLEM

That is

$$\begin{aligned} \mathbf{D}w &= -\hat{\beta}\frac{\partial \theta}{\partial z}, \\ &\frac{\partial}{\partial t} Re\left\{\frac{\partial w}{\partial \bar{z}}\right\}, \\ \alpha_1 \sigma_{x_1} + \alpha_2 \tau_{x_1 x_2} &= T_1, \\ \alpha_1 \tau_{x_1 x_2} + \alpha_2 \sigma_{x_2} &= T_2. \end{aligned} \tag{17.27}$$

Noticing $z'(s) = i(\alpha_1 + i\alpha_2)$, we get the complex form of the boundary condition

$$\mu(z'(s))^2 \frac{\partial w}{\partial z} - (\lambda + \mu) Re \frac{\partial w}{\partial \bar{z}} = -z'(s)\frac{T_1 + iT_2}{2}.$$

For simplicity, assume $T_1 = T_2 \equiv 0$,

$$\mu(z'(s))^2 \frac{\partial w}{\partial z} - (\lambda + \mu) Re \frac{\partial w}{\partial \bar{z}} = -z'(s)\frac{T_1 + iT_2}{2}\ , \quad z \in \partial D. \tag{17.28}$$

Also

$$\theta|_{S_T} = 0\ , \quad \theta(z,0) = \theta_0(z)\ , \quad z \in D. \tag{17.29}$$

Defining $\hat{\theta} = -\hat{\beta}\theta/2$, we obtain

$$bf Dw = 2\frac{\partial \hat{\theta}}{\partial z}, \tag{17.30}$$

$$\frac{\partial w}{\partial \overline{z}} = \hat{\theta} + \frac{2A}{\delta}\varphi - \frac{2B}{\delta}\overline{\varphi}, \tag{17.31}$$

$$\frac{\partial^2 w}{\partial \overline{z}\partial z} = \frac{\partial \hat{\theta}}{\partial z} + \frac{2A}{\delta}\varphi'(z) = \frac{\partial \hat{\theta}}{\partial z} + \frac{2A}{\delta}(2\frac{\partial Re\varphi(z)}{\partial z}) = \frac{\partial}{\partial z}(\hat{\theta} - \frac{\lambda+\mu}{\mu}Re\varphi).$$

Hence

$$\frac{\partial w}{\partial z} = -F_0(z) + \psi(z) \tag{17.32}$$

where

$$F_0(z) = -\mathbf{T}_D\left(\frac{\partial}{\partial z}(\hat{\theta} - \frac{\lambda+\mu}{\mu})Re\varphi\right) \;;\; (\mathbf{T}_D f)(z) \equiv -\frac{1}{\pi}\int_D \frac{f(\zeta)d\sigma_\zeta}{\zeta - z}.$$

Obviously (by means of (17.31)), (17.30) yields a particular solution $w_0 = (\mathbf{T}_D\hat{\theta})(z)$. And the solvability of (17.27) and (17.28) is equivalent to the solvability of the problem

$$\begin{cases} \mathbf{D}w = 0, \\ \alpha_1\sigma_x + \alpha_2\tau_{xy} = \tilde{T}_1 = -2\mu Re\left(iz'(s)\mathbf{T}_D(\partial\hat{\theta}/\partial z)\right), \\ \alpha_1\tau_{xy} + \alpha_2\sigma_y = \tilde{T}_2 = -2\mu Im\left(iz'(s)\mathbf{T}_D(\partial\hat{\theta}/\partial z)\right). \end{cases}$$

For the stress boundary value problem, the boundary data have to fulfil exterior force and total moment being zero. In fact

$$\int_{\partial\Omega}(T_1 + iT_2)ds = -2i\mu\int_{\partial D} z'(s)\mathbf{T}_D(\frac{\partial\hat{\theta}}{\partial z})ds = -2i\mu\int_{\partial D}\mathbf{T}_D(\frac{\partial\hat{\theta}}{\partial z})dz = 0,$$

$$\begin{aligned}\int_{\partial\Omega}(x_1T_2 - x_2T_1)ds &= -\int_{\partial\Omega}(xT_2 + yT_1)ds = \mathrm{Im}\left[2i\mu\int_{\partial D} z'(s)\mathbf{T}_D(\frac{\partial\hat{\theta}}{\partial z})zds\right] \\ &= \mathrm{Im}\left[-2i\mu I_R\right],\end{aligned}$$

$$\begin{aligned} I_R &= \int_{|z|=R} z\mathbf{T}_D(\frac{\partial\hat{\theta}}{\partial z})dz \\ &= \int_{|z|=R} z\sum_{n=0}^{-\infty}\left[\frac{1}{\pi}\int_D \tau^n\frac{\partial\hat{\theta}}{\partial\tau}d\sigma_\tau\right]z^{-n-1}dz = \frac{1}{\pi}\int_D\frac{\partial\hat{\theta}}{\partial\tau}d\sigma = 0\end{aligned}$$

Decoupling

Substituting (17.31), (17.32) into (17.28), and noticing $z'(s)\overline{z'(s)} = 1$, we get

$$\psi(z) - F_0(z) = \frac{\lambda+\mu}{\mu}\overline{z'(s)}^2\,\mathrm{Re}\,\varphi\;\;, z\in\partial D. \tag{17.33}$$

Introducing

$$\chi(z) = \frac{1}{2\pi i}\int_D \frac{\overline{\zeta'(s)}\,\mathrm{Re}\,\varphi}{\zeta - z}d\zeta,$$

then

$$\psi(z) = \frac{\lambda+\mu}{\mu}\chi^+(z) \quad , \quad z \in D, \tag{17.34}$$

$$F_0(z) = \frac{\lambda+\mu}{\mu}\chi^-(z) \quad , \quad z \in \mathbb{C}\overline{D} \tag{17.35}$$

By means of (17.15), we obtain a singular (Cauchy kernel) integral equation for determining $Re(\varphi)$.

The estimate for $||Re\varphi||_{L(\Omega)}$

It is easy to verify the Green identity

$$\int_{\partial\Omega}(u_1T_1 + u_2T_2)ds = \int_{\Omega}\left\{\left[u_1\left(\frac{\partial\sigma_1}{\partial x_1} + \frac{\partial\tau_{x_1x_2}}{\partial x_2}\right) + u_2\left(\frac{\partial\tau_{x_1x_2}}{\partial x_1} + \frac{\partial\sigma_{x_2}}{\partial x_2}\right)\right]\right.$$

$$\left. + \left[\sigma_{x_1}\frac{\partial u_1}{\partial x_1} + \frac{1}{\mu}\tau^2_{x_1x_2} + \sigma_{x_2}\frac{\partial u_2}{\partial x_2}\right]\right\} d\sigma.$$

Noticing $T_1 = T_2 \equiv 0$, and $\beta\theta = -2(\lambda+2\mu)\hat{\theta}$, we have

$$\int_D\left(4\mu\left|\frac{\partial w}{\partial z}\right|^2 + 4(\lambda+\mu)\left|Re\frac{\partial w}{\partial \overline{z}}\right|^2\right)d\sigma = \int_D\left[u(\beta\frac{\partial\theta}{\partial x}) - v(\beta\frac{\partial\theta}{\partial y})\right]d\sigma$$

$$= \operatorname{Re}\int_D \beta w(\frac{\partial\theta}{\partial x} + i\partial\theta\partial y)d\sigma = 4(\lambda+2\mu)\int_D \hat{\theta}\operatorname{Re}(\frac{\partial w}{\partial\overline{z}})d\sigma. \tag{17.36}$$

From (17.34) and (17.32) we get

$$\frac{\partial w}{\partial z} = \mathbf{\Pi}_D(\hat{\theta} - \frac{\lambda+\mu}{\mu}\operatorname{Re}\varphi) \ , \quad z \in D, \tag{17.37}$$

where

$$(\mathbf{\Pi}_D f)(z) = -\frac{1}{\pi}\int_D \frac{f(\zeta)}{(\zeta-z)^2}d\sigma_\zeta$$

and

$$||\mathbf{\Pi}_{\mathbb{C}} f||_{L^2(\mathbb{C})} = ||f||_{L^2(\mathbb{C})}.$$

From (17.35) and applying the Green formula, we have

$$(\mathbf{\Pi}_D\hat{\theta})(z) = \frac{\lambda+\mu}{\mu}(\mathbf{\Pi}_D(Re\varphi))(z) \ , \quad z \in \mathbb{C}\overline{D}.$$

From this and (17.37), we have

$$||\frac{\partial w}{\partial z}||_{L^2(D)} = ||\hat{\theta} - \frac{\lambda+\mu}{\mu}\operatorname{Re}\varphi||_{L^2(D)}, \tag{17.38}$$

here we assume $\hat{\theta} = \operatorname{Re}\varphi \equiv 0$, for $z \in \mathbb{C}\overline{D}$. Substituting (17.38) and (17.31) into (17.36), we get

$$\frac{\lambda+\mu}{\mu}\int_D |\operatorname{Re}\varphi|^2 d\sigma = \int_D |\hat{\theta}\operatorname{Re}\varphi| d\sigma \le ||\operatorname{Re}\varphi||_{L^2(D)}||\theta||_{L^2(D)}.$$

Hence $\|\operatorname{Re}\varphi\|_{L^2(D)} \leq \mu/(\lambda+\mu)\|\hat{\theta}\|_{L^2(D)}$. Finally we have the result.

Theorem 6 *Assume D be a bounded simply-connected domain with ∂D belonging to C^2. If $\theta_0 \in W_0^{1,2}(D)$, then for problem (17.27), (17.28) and (17.29), there exists a unique solution $\theta \in W_{2,0}^{2,1}(D_T)$. Moreover, θ is analytic with respect to the space variables.*

Numerical solution

$$\begin{cases} \dfrac{\partial u}{\partial t} - \Delta u = \mathbf{M}(\dfrac{\partial u}{\partial t}) \ , \quad (z,\ t) \in \Omega_T, \\ u(z,\ 0) = \varphi(z), \\ u|_{S_T} = 0, \end{cases}$$

where M is an integral operator of $L^2(\Omega)$ such that $\|M\| = r < 1$.

Using Galerkin finite element method and backward Euler-Galerkin difference quotient, we establish the algorithms and estimate the convergence rate and correponding error bounds, see for details, [11]. We also use Daubechis wavelet (tensor product) to deal with the same problem and establish fast algorithms.

The symmetric method

We have generalized the symmetry point and Schwarz principle for the circle to algebraic curves on Riemann surfaces of genus zero and have reduced the basic problems of elasticity to boundary value problems on these Riemann surfaces.

References

[1] Hua,L., Lin, W., Wu, Z.: Second order systems of equations on the plane. Pitman, London, 1985.

[2] Bergman, S.: Integral operators in the theory of linear partial differential equations. Ergeb. Math. u. Grenzgeb. 23, Springer, Heidelberg, 1969.

[3] Ladyzhenskaya, O.A.: The boundary value problems of mathematic physics. Springer, New York, 1985.

[4] Begehr, H., Lin, W.: A mixed-contact boundary problem in orthotropic elasticity. Research Notes in Math. 263, ed. H. Begehr, A. Jeffrey, Pitman, London, 1992, 219-239.

[5] Gilbert, R.P., Lin, W.: Function theoretic solution to problems of orthotropic elasticity. J. Elasticity 15 (1985), 143-154.

[6] Vekua, I.N.: Generalized analytic functions. Pergamon Press, Oxford, 1962.

[7] Dzhurajev, A.D.: Method of Singular Integral Equations. Longman, Harlow, 1992.

[8] Lin, W., Zhao, Y.Q.: On plane problem of orthotropic quasi-static thermoelasticity. J. Elasticity 41 (1995), 161-175.

[9] Zhao, Y.Q.: On plane displacement boundary value problem of quasi-static linear thermoelasticity. Applicable Analysis 56 (1995), 117-129.

[10] Zhao, Y.Q.: On the plane orthotropic stress problem of quasi-static thermoelasticity. J. Elasticity 46 (1997), 199-216.

[11] Zhao, Y.Q., Lin, W.: Numerical solution to problems concerning quasistatic thermoelasticity. Applicable Analysis 57 (1995), 221-234.

[12] Zhao, Y.Q., Lin, W.: On the plane stress boundary value problem of quasistatic linear thermoelasticity. Acta Math. Appl. Sinica 13, No. 4 (1997), 385-394.

[13] Shi, P., Xu, Y.: Quasi-static linear thermoelasticity on the unit disk. J. Elasticity 31 (1993), 209-218.

[14] Day, W.A.: Heat conduction within linear thermoelasticity. Springer, New York, 1985.

[15] Shi, P., Shillor, M.: Uniqueness and stability of the solution to a thermoelastic contact problem. European J. Appl. Math. 1 (1990), 371–387.

18 EFFECT OF PERIODIC GASKET ON A PERIODIC CONTACT PROBLEM

Xing Li

Ningxia University
Yinchuan 750021
P.R. China

li_x@nxu.edu.cn

Abstract: In this paper, the effect of periodic gasket on a periodic contact problem is investigated by approaches using singular integral equations with Hilbert kernel. The periodic solutions for the complex stress functions are obtained in closed form.

18.1 FORMULATION

The stress and displacement components are given by the following formulae, see [5],

$$\sigma_x + \sigma_y = 2[\Phi(z) + \overline{\Phi(z)}],$$

$$\sigma_x - \sigma_y + 2i\tau_{xy} = 2[\bar{z}\Phi(z) + \Psi(z)],$$

$$2\mu_k(u + iv) = \kappa_k\phi(z) - z\overline{\phi(z)} - \overline{\Psi(z)}, k = 1, 2,$$

where $\Phi(z) = \phi'(z)$, $\Psi(z) = \psi'(z), \mu_1, \mu_2$ are the shear moduli of the two elastic materials, $\kappa_k = (3 - \nu_k)/(1 + \nu_k)$ (under plane stress conditions), $\kappa_k = 3 - 4\nu_k$ (under plane strain conditions), and ν_k are the Poisson ratios of the two materials.

The periodic fundamental strip is denoted by P_0, the bound of the elastic gasket in P_0 is denoted by L_0, see Fig.1, which is a simple, closed, smooth contour. The positive direction of L_0 is taken to be anticlockwise. $\gamma_0 = (-l, l)$ is the contact interval of rigid punch with lower half plane in P_0. $l_0 = (-a\pi/2, a\pi/2)$. P_k $(k = \pm1, \pm2, \cdots)$, is the congruent strip formed by extending P_0 periodically with period $a\pi$ $(a > 0)$, $L_k(k = \pm1, \pm2, \cdots)$, is the congruent contour of L_0, and γ_k $(k = \pm1, \pm2, \cdots)$, is the congruent interval of γ_0. We denote

H.G.W. Begehr et al. (eds.), Partial Differential and Integral Equations, 249–254.

$\gamma = \cup_{k=0}^{\infty}\gamma_k$. The inner region of L_k $(k = 0, \pm1, \pm2, \cdots)$, is denoted by S_k^+, the complement of $\cup_{k=-\infty}^{\infty} S_k^+$ in the complex plane is denoted by S^-.

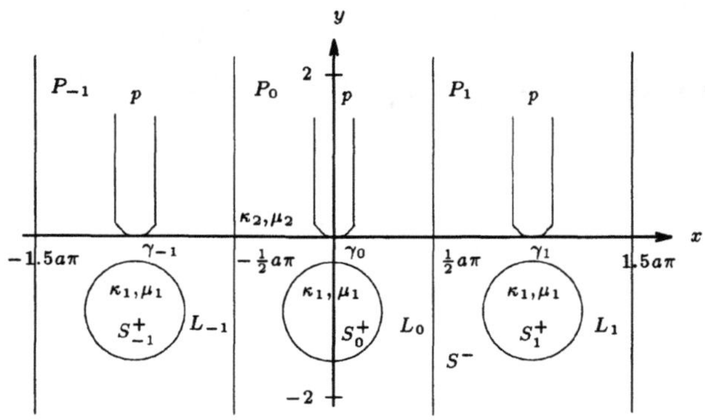

Figure 1: Contact model

Similar to [2] the complex stress functions can be written as follows after assuming that the stress vanishes as infinity.

$$\Phi(z) = \Phi_1(z) + \Phi_2(z), \tag{18.1}$$

$$\Psi(z) = \Psi_1(z) + \Psi_2(z) - \Phi_1(z) - \Phi_2(z) - z\Phi_1'(z) - z\Phi_2'(z), \tag{18.2}$$

in which $\Phi_1(z)$, $\Psi_1(z)$ are holomorphic in the lower half plane, continuous on the x–axis, $\Phi_2(z)$, $\Psi_2(z)$ are sectionally continuous in the complex plane, they are all periodic functions with period $a\pi$.

Now, we consider the periodic rigid punch with periodic profile $y = f(x)$ which is pressed vertically into frictionless contact with the boundary of the lower half-plane with periodic gasket. After contact, the vertical displacement of the region of contact will be given by

$$V_k^-(t) = f(t) + e, \ \ t \in \gamma_k.$$

Here e is a constant, we assume that $f(t)$ is periodic and satisfies a Hölder condition, the displacement at a point $t \in L_k$ on the positive and negative sides are given also respectively, which means that the functions $g_k^{\pm}(t) = u_k^{\pm}(t) + iv_k^{\pm}(t)(k = 1, 2)$ are given. We assume that these functions are periodic and satisfy a Hölder condition.

The condition of frictionless contact implies $\tau_{xy} = 0$ on γ_k, while on the remainder $\gamma' = \mathbb{R}\backslash\gamma$ of the x–axis, $\mathbb{R}$

$$\sigma_y - i\tau_{xy} = 0$$

So the external resultant force on γ will be

$$X + iY = -pi \text{ and } \sigma_y(-\infty i) = -\frac{p}{a\pi}, \ \ \tau_{xy}(-\infty i) = 0.$$

Here, p is the vertical well–distributed pressure on the rigid punch (see Fig. 1).

Then, similarly to [2], [4], [5], taking (18.1) and (18.2) into account we obtain the boundary value problems

$$\Phi_2^+(t) + (t-\bar{t})\overline{\Phi_2'^+(t)} + \overline{\Psi_2^+(t)} = \Phi_2^-(t) + (t-\bar{t})\overline{\Phi_2'^-(t)} + \overline{\Psi_2^-(t)},\ t \in L_k,$$

$$\kappa_1\Phi_2^+(t) - (t-\bar{t})\overline{\Phi_2'^+(t)} - \overline{\Psi_2^+(t)}$$
$$= \kappa_2\Phi_2^-(t) - (t-\bar{t})\overline{\Phi_2'^-(t)} - \overline{\Psi_2^-(t)} + \mu_2 g_k^+(t) - \mu_1 g_k^-(t),\ t \in L_k,$$

$$\Phi_1(t) + \Phi_2(t) - \Psi_1(t) - \Psi_2(t) - \overline{\Phi_1(t)} - \overline{\Phi_2(t)} + \overline{\Psi_1(t)} + \overline{\Psi_2(t)} = 0,\ t \in I\!R,$$

$$\kappa_1(\Phi_1(t) - \overline{\Phi_1(t)}) + \Psi_1(t) - \overline{\Psi_1(t)} + \kappa_1(\Phi_2(t) - \overline{\Phi_2(t)}) + \Psi_2(t) - \overline{\Psi_2(t)}$$
$$= 4i\mu_1 f'(t),\ t \in \gamma,$$

$$\Phi_1(t) + \overline{\Phi_1(t)}) + \Psi_1(t) + \overline{\Psi_1(t)} + \Phi_2(t) + \overline{\Phi_2(t)} + \Psi_2(t) + \overline{\Psi_2(t)} = 2\frac{P_0}{a\pi},\ t \in I\!R\backslash\gamma.$$

Because of the periodicity, we just need to get the solution of the problem in the periodic fundamental strip P_0. This means we must solve the boundary value problems, see [3],

$$\Phi_2^+(t) + (t-\bar{t})\overline{\Phi_2^{+\prime}(t)} + \overline{\Psi_2^+(t)} = \Phi_2^-(t) + (t-\bar{t})\overline{\Phi_2^{-\prime}(t)} + \overline{\Psi_2^-(t)}, t \in L_0, \quad (18.3)$$

$$\begin{aligned} &\kappa_1\Phi_2^+(t) - (t-\bar{t})\overline{\Phi_2^{+\prime}(t)} - \overline{\Psi_2^+(t)} \\ &= \kappa_2\Phi_2^-(t) - (t-\bar{t})\overline{\Phi_2^{-\prime}(t)} - \overline{\Psi_2^-(t)} + \mu_2 g_k^+(t) - \mu_1 g_k^-(t), t \in L_0, \end{aligned} \quad (18.4)$$

$$\Phi_1(t) + \Phi_2(t) - \Psi_1(t) - \Phi_2(t) - \overline{\Phi_1(t)} - \overline{\Phi_2(t)} + \overline{\Psi_1(t)} + \overline{\Psi_2(t)} = 0, t \in l_0, \quad (18.5)$$

$$\begin{aligned} &\kappa_1(\Phi_1(t) - \overline{\Phi_1(t)}) + \Psi_1(t) - \overline{\Psi_1(t)} - \kappa_1(\Phi_2(t) - \overline{\Phi_2(t)}) + \Psi_2(t) - \overline{\Psi_2(t)} \\ &= 4i\mu_1 f'(t), t \in \gamma_0, \end{aligned} \quad (18.6)$$

$$\Phi_1(t) + \overline{\Phi_1(t)} + \Psi_1(t) + \overline{\Psi_1(t)} + \Phi_2(t) + \overline{\Phi_2(t)} + \Psi_2(t) + \overline{\Psi_2(t)} = 2\frac{p}{a\pi}, t \in l_0\backslash\gamma_0, \quad (18.7)$$

18.2 SOLUTION

After adding (18.3) and (18.4), we obtain a classic Riemann-Hilbert boundary value problem in the form

$$\Phi_2^+(t) = \frac{\kappa_1+1}{\kappa_2+1}\Phi_2^-(t) + \frac{1}{\kappa_2+1}(\mu_2 g^+ - \mu_1 g^-) + \frac{\kappa_1-\kappa_2}{\kappa_2+1}\Phi_1(t), \quad (18.8)$$

so we get

$$\begin{aligned} \Phi_2(z) &= \tfrac{X(z)}{2a\pi i(\kappa_2+1)}\int_L R_1(t)\cot\left(\tfrac{t-z}{a}\right)dt + \frac{\kappa_1-\kappa_2}{\kappa_2+1}X(t)\Phi_1^*(t) \\ &= F_1(z) + GX(z)\Phi_1^*(z), \end{aligned} \quad (18.9)$$

where

$$\Phi_1^*(z) = \begin{cases} \Phi_1(z), & \text{when } z \in S^+, \\ 0, & \text{when } z \in S^-, \end{cases} \tag{18.10}$$

and

$$X(z) = \begin{cases} X^+(z) = G, & z \in S^+, \\ X^-(z) = 1, & z \in S^-, \end{cases}$$

$$R_1(t) = \frac{\mu_2 g^+ - \mu_1 g^-}{(\kappa_2 + 1)X^+(t)}, \quad G = \frac{\kappa_1 + 1}{\kappa + 1}.$$

From (18.3) we get the Riemann jump problem

$$\begin{aligned} \Psi_2^+(t) - \Psi_2^-(t) &= \overline{\Phi_2^-(t)} - \overline{\Phi_2^+(t)} \\ &+ (\bar{t} - t)(\Phi_2'^-(t) + \Phi_2'^-(t)) \equiv R_2(t). \end{aligned} \tag{18.11}$$

The solution of the boundary value problem (18.11) can be easily obtained as

$$\Psi_2(z) = \frac{1}{2\pi i}\int_L R_2(t)\cot(\frac{t-z}{a})dt \equiv F_2(z, \Phi_1(z)). \tag{18.12}$$

Here F is a known function, but $\Phi_1(z)$ is still an undetermined function. Let $z \to t \in \gamma_0$, substituting (18.9) into (18.5),(18.6) and (18.7) and considering (18.10), we get

$$\begin{aligned} &\Phi_1(t) - \overline{\Phi_1(t)} + \overline{\Psi_1(t)} - \Psi_1(t) \\ &= 2i(\mathrm{Im}F_2(t) - \mathrm{Im}F_1(t)) \equiv H_3(t), \quad t \in l_0, \end{aligned} \tag{18.5'}$$

$$\begin{aligned} &\kappa_1(\Phi_1(t) - \overline{\Phi_1(t)}) + \Psi_1(t) - \overline{\Psi_1(t)} \\ &= 2i[2\mu_1 f'(t) - \kappa_1 F_1(t) - \mathrm{Im}F_2(t))] \equiv H_1(t), \quad t \in \gamma_0, \end{aligned} \tag{18.6'}$$

$$\Phi_1(t) + \overline{\Phi_1(t)}) + \Psi_1(t) + \overline{\Psi_1(t)} + 2\left[\frac{p}{a\pi} - \mathrm{Re}F_1(t) - \mathrm{Re}F_2(t)\right] \equiv H_2(t), \; t \in l_0\backslash\gamma_0. \tag{18.7'}$$

Let $\Omega_1(z) = \Phi_1(z) - \Psi_1(z)$. Then (18.5′) can be rewritten as

$$\Omega_1^+(t) - \Omega_1^-(t) = H_1(t), \quad t \in l_0. \tag{18.13}$$

Thus

$$\Omega_1(z) = \Phi_1(z) - \Psi_1(z) = \frac{1}{2a\pi i}\int_{L_0} H_3(t)\cot(\frac{t-z}{a})dt, \quad \mathrm{Im}z < 0. \tag{18.14}$$

Substituting (18.14) into (18.6′), (18.7′), by using the Plemelj formula, we get

$$\begin{aligned} \Phi_1(t) - \overline{\Phi_1(t)} = \frac{1}{\kappa_1 + 1}\Big[H_1(t) + i\mathrm{Im}H_3(t) \\ + \frac{1}{2a\pi i}\int_{l_0}\Big(H_3(t_0)\cot(\frac{t-t_0}{a})dt_0 - \overline{H_3(t_0)}\overline{\cot(\frac{t-t_0}{a})d\bar{t}_0}\Big)\Big] \equiv H_4(t), \; t \in \gamma_0, \end{aligned} \tag{18.15}$$

$$\Phi_1(t) + \overline{\Phi_1(t)} = \frac{1}{2}\Big[H_2(t) + \mathrm{Re}H_3(t)$$

$$+\frac{1}{2a\pi i}\int_{l_0}\Big(H_3(t_0)\cot(\frac{t-t_0}{a})dt_0 - \overline{H_3(t_0)}\overline{\cot(\frac{t-t_0}{a})}d\bar{t}_0\Big)\Big] \equiv H_5(t),\ t \in l_0\backslash\gamma_0. \tag{18.16}$$

We introduce the new function

$$\Phi_{**}(z) = \begin{cases} -\Phi_1(z), & \mathrm{Im}z < 0, \\ \overline{\Phi_1(\bar{z})}, & \mathrm{Im}z > 0. \end{cases} \tag{18.17}$$

Substituting (18.17) into (18.16),(18.15), we obtain the Riemann-Hilbert boundary value problem with discontinuous coefficients

$$\Phi^+{**}(t) = G(t)\Phi^-_{**}(t) + g_{**}(t), \tag{18.18}$$

where

$$G(t) = \begin{cases} -1, & t \in \gamma_0, \\ 1, & t \in l_0\backslash\gamma_0, \end{cases} \qquad g_{**}(t) = \begin{cases} H_4(t), & t \in \gamma_0, \\ H_5(t), & t \in l_0\backslash\gamma_0. \end{cases}$$

The solution of the boundary value problem (18.18) can be obtained as, see [3],

$$\Phi_{**}(z) = \frac{Q(z)}{2a\pi i}\int_{L_0}\frac{g_{**}(t)}{Q^+(t)}(\cot\frac{t-z}{a} - \tan\frac{z}{a})dt, \tag{18.19}$$

where

$$Q(z) = \frac{1}{i\sqrt{R(z)}},\quad R(z) = \tan^2\frac{l}{a} - \tan^2\frac{z}{a}$$

taking the branch

$$\lim_{z\to a\pi/2}\tan\frac{z}{a}Q(z) = 1.$$

From (18.17),(18.19), we easily obtain

$$\Phi_1(z) = \frac{Q(z)}{2a\pi i}\int_{L_0}\frac{g_{**}(t)}{Q^+(t)}(\tan\frac{z}{a} - \cot\frac{t-z}{a})dt. \tag{18.20}$$

Substituting (18.20) into (18.14), gives

$$\Psi_1(z) = \frac{1}{2a\pi i}\Big[Q(z)\int_{L_0}\frac{g_{**}(t)}{Q^+(t)}(\tan\frac{z}{a} - \cot\frac{t-z}{a})dt$$
$$-\int_{L_0}H_3(t)\cot\frac{t-z}{a}dt\Big]. \tag{18.21}$$

Finally, substituting $\Phi_1(z), \Phi_2(z), \Psi_1(z), \Psi_2(z)$ into (18.1),(18.2), it is not difficult to obtain the complex stress functions $\Phi(z)$ and $\Psi(z)$.

References

[1] Li, Xing, Li, Zheng Xing: Effect of a periodic elastic gasket on periodic cracks. Engineering Fracture Mechanics 46(1993), 127-131.

[2] Liu, Shiqiang: Problem of the periodic contract of bounded plane materials. Acta Math. Appl. Sinica 15 (1993), No.3, 289-296 (Chinese).

[3] Lu, C.K., Cai, H.T.: Periodic problem of plane elasticity theory. Hunan Sci. Tech. Publ., 1986 (Chinese).

[4] Galin, L.A.: Contact problem in the theory of elasticity. Gostekhizdat, Moscow, 1953 (Russian).

[5] Muskhelishvili, N.I.: Some basic problems of the mathematical theory of elasticity. Noordhoff, Groningen, 1953.

19 SINGULAR INTEGRAL OPERATORS AND SINGULAR QUADRATURE OPERATORS FOR SINGULAR INTEGRAL EQUATIONS WITH HILBERT KERNEL

Jinyuan Du †

Wuhan University
Wuhan 430072
P.R. China
jydu@whu.edu.cn

Abstract: In this paper, we discuss some singular integral operators and singular quadrature operators for singular integral equations with Hilbert kernel, and obtain some useful properties of them. These results improve both the classical theory of singular integral equation with Hilbert kernel and the classical theory of singular quadrature with Hilbert kernel.

19.1 INTRODUCTION

In [1-3], the author discussed systematically the singular integral operators, singular quadrature operators and discretization matrices for singular integral equations with Cauchy kernel and using them gave a unified framework for various collocation methods of numerical solutions of singular integral equations with Cauchy kernel. In this paper, we shall consider the corresponding problems

†Supported by NNSF and Science Foundation of SEC of P.R. China

H.G.W. Begehr et al. (eds.), Partial Differential and Integral Equations, 255–266.

for singular integral equations (SIEs) with Hilbert kernel of the form

$$a(t)\varphi(t) + \frac{b(t)}{2\pi}\int_0^{2\pi}\varphi(\tau)\cot\frac{\tau-t}{2}d\tau + \frac{\lambda}{2\pi}\int_0^{2\pi}k(\tau,t)\varphi(\tau)d\tau = f(t),\ 0\leq t<2\pi. \quad (19.1)$$

In this equation, the input functions a, b, f, k are Hölder-continuous functions with period 2π of their arguments, denoted as $a, b, f, k \in H_{2\pi}$, λ is a given constant. It is required to find the solution φ in the class $H_{2\pi}$, see [4,5].

In 1978, S. Krenk discussed the numerical method for SIE (19.1) in case that $a(t)$ and $b(t)$ are constants [6]. In 1983, for the same case, N.I. Ioakimidis rediscussed it by using a somewhat different method [7]. He gave the coincidence theorem via concrete calculations. For general cases, in [8,9] the present author discussed the numerical methods for (19.1), and the existence and convergence are demonstrated on the basis of the coincidence theorem by applying specific computations again to establish the coincidence theorem. In this paper, we shall establish some singular integral operators and singular quadrature operators, which possess very similar and interesting properties. These results improve both the classical theory of singular integral equations and the theory of singular quadrature with Hilbert kernel. Using them we may give a unified framework for various collocation methods of (19.1). Under the framework, it is very simple and obvious to obtain the coincidence theoem of collocation methods (which is postponed to another paper in order to keep the present paper of reasonable length).

We need some special concepts and symbols used throughtout this paper (for more details, see [10]).

Let

$$\Theta(t) = \arg[a(t) - ib(t)] \quad (19.2)$$

be a real function continuous on $[0, 2\pi]$.

If we denote the index of (19.1) by 2κ, then

$$\kappa = \frac{1}{2\pi}[\Theta(2\pi) - \Theta(0)], \quad (19.3)$$

which is called the half index for convenience, is an integer because of the periodicity of Θ. The mean value of Θ

$$\theta = \frac{1}{2\pi}\int_0^{2\pi}\Theta(\tau)d\tau \quad (19.4)$$

is called the characteristic number of (19.1).

The canonical function of (19.1) is

$$X(z) = \begin{cases} \exp\left\{\kappa iz + \frac{1}{2\pi}\int_0^{2\pi}[\Theta(\tau) - \kappa\tau]\cot\frac{\tau-z}{2}d\tau\right\}, & \text{if } \operatorname{Im} z > 0, \\ \exp\left\{-\kappa iz + \frac{1}{2\pi}\int_0^{2\pi}[\Theta(\tau) - \kappa\tau]\cot\frac{\tau-z}{2}d\tau\right\}, & \text{if } \operatorname{Im} z < 0. \end{cases} \quad (19.5)$$

Obviously (cf. [4,5]), X is a sectionally holomorphic function in the half-planes with period 2π, and possesses the upper boundary value $X^+(t) = \lim_{z\to t} X(z)$ (for $\operatorname{Im} z > 0$) and the lower boundary value $X^-(t) = \lim_{z\to t} X(z)$ (for $\operatorname{Im} z < 0$) for every $t \in (-\infty, +\infty)$. Such a function is said to be in class $AH_{2\pi}$. If $\Phi \in AH_{2\pi}$, by using the mapping $w = \exp(iz)$, we easily see that its sectionally restricted functions Φ^+ on the upper half-plane and Φ^- on the lower half-plane, respectively, have the following series expansions

$$\Phi^+(z) = \sum_{-\infty}^{+\infty} a_j e^{ijz} \ (\operatorname{Im} z > 0), \quad \Phi^-(z) = \sum_{-\infty}^{+\infty} b_j e^{ijz} \ (\operatorname{Im} z < 0). \tag{19.6}$$

Set

$$Q.P(\Phi)(z) = \frac{a_0 + b_0}{2} + \sum_{j=1}^{+\infty} \left(a_{-j} e^{-ijz} + b_j e^{ijz}\right), \tag{19.7}$$

which is called the quasi-principal part of Φ. The number

$$\Delta(\Phi) = \frac{a_0 - b_0}{2} \tag{19.8}$$

is called the deficiency of Φ.

The fundamental function of (19.1) is

$$Z(t) = \exp\left\{\frac{1}{2\pi}\int_0^{2\pi} [\Theta(\tau) - \kappa\tau] \cot\frac{\tau - t}{2} d\tau\right\}, \quad \operatorname{Im} t = 0, \tag{19.9}$$

which is in class $H_{2\pi}$ with understanding the above principal value integral at 0 or 2π as

$$\int_0^{2\pi} f(\tau)\cot\frac{\tau}{2} d\tau = \lim_{\delta\to 0}\int_\delta^{2\pi-\delta} f(\tau)\cot\frac{\tau}{2} d\tau. \tag{19.10}$$

Let

$$r(t) = \sqrt{a^2(t) + b^2(t)} \neq 0 \qquad \text{(normal type)}. \tag{19.11}$$

The functions

$$w_1(t) = \frac{Z(t)}{r(t)}, \ w_2(t) = \frac{1}{r(t)Z(t)} \tag{19.12}$$

are called, respectively, the weight function of first kind and that of second kind associated with (19.1).

From [10], we know that the normalized equation of (19.1) is

$$a(t)w_1(t)y(t) + \frac{b(t)}{2\pi}\int_0^{2\pi} w_1(\tau)y(\tau)\cot\frac{\tau - t}{2} d\tau + \frac{\lambda}{2\pi}\int_0^{2\pi} w_1(\tau)k(\tau, t)y(\tau)d\tau = f(t). \tag{19.13}$$

With the relation

$$\varphi(t) = w_1(t)y(t) \tag{19.14}$$

the solutions of (19.1) in $H_{2\pi}$ are equivalent to those of (19.13) in the class $H_{2\pi}$, see [10]. Therefore we shall discuss the latter.

19.2 SINGULAR INTEGRAL OPERATORS

Introduce the singular integral operators (SIOs)

$$(\mathbf{A}y)(t) = a(t)w_1(t)y(t) + \frac{b(t)}{2\pi}\int_0^{2\pi} w_1(\tau)y(\tau)\cot\frac{\tau - t}{2}d\tau,$$

$$(\mathbf{B}y)(t) = a(t)w_2(t)y(t) - \frac{b(t)}{2\pi}\int_0^{2\pi} w_2(\tau)y(\tau)\cot\frac{\tau - t}{2}d\tau,$$

and their adjoint operators

$$(\mathbf{A}^*y)(t) = a(t)w_1(t)y(t) + \frac{1}{2\pi}\int_0^{2\pi} w_1(\tau)b(\tau)y(\tau)\cot\frac{\tau - t}{2}d\tau,$$

$$(\mathbf{B}^*y)(t) = a(t)w_2(t)y(t) - \frac{1}{2\pi}\int_0^{2\pi} w_2(\tau)b(\tau)y(\tau)\cot\frac{\tau - t}{2}d\tau.$$

Remark 1 For convenience we denote the adjoint operator of $\mathbf{A}$ as $\mathbf{B}^*$, of $\mathbf{B}$ as $\mathbf{A}^*$, not as $\mathbf{A}'$ and $\mathbf{B}'$ as in general texts.

Remark 2 $\mathbf{A}, \mathbf{B}, \mathbf{A}^*, \mathbf{B}^*$ all are SIOs from $H_{2\pi}$ to $H_{2\pi}$, i.e. $\mathbf{A} : H_{2\pi} \to H_{2\pi}$, $\mathbf{B} : H_{2\pi} \to H_{2\pi}$, $\mathbf{A}^* : H_{2\pi} \to H_{2\pi}$, $\mathbf{B}^* : H_{2\pi} \to H_{2\pi}$.

Again introduce the integral operators

$$(\mathbf{D}_1y)(t) = \frac{1}{2\pi}\int_0^{2\pi} w_1(\tau)y(\tau)\left\{[Q.P(X^{-1})(\tau) - Q.P(X^{-1})(t)]\cot\frac{\tau - t}{2} - i\Delta(X^{-1})\right\}d\tau,$$

$$(\mathbf{D}_2y)(t) = \frac{1}{2\pi}\int_0^{2\pi} w_2(\tau)y(\tau)\left\{[Q.P(X)(\tau) - Q.P(X)(t)]\cot\frac{\tau - t}{2} - i\Delta(X)\right\}d\tau.$$

Let H_n^T denote the class of all trigonometric polynomials of degree not greater than n, $H_n^T(\alpha)$ the family of all trigonometric polynomials of the form $a_n\sin(nt+\alpha)+T_{n-1}(t)$ with $T_{n-1}\in H_{n-1}^T, 0\le\alpha<\pi$. If $n<0$ we regard $H_n^T=\{0\}$ and $H_n^T(\alpha) = \{0\}$, It is obvious that any trigonometric polynomial of degree n $(n > 0)$ cannot belong to two different classes $H_n^T(\alpha_1)$ and $H_n^T(\alpha_2)$ $(\alpha_1\neq\alpha_2)$, while $H_0^T(\alpha)$ is $\{0\}$ when $\alpha = 0$ and the set of all constants when $\alpha \neq 0$.

Remark 3 Let $[\theta]_\pi$ denote the number congruent to θ (mod π) in $[0, \pi)$. Via slightly evaluating, from (1.5) we get that $Q.P(X) \in H_{-\kappa}^T([\pi/2 - \theta]_\pi)$ and $Q.P(X^{-1}) \in H_\kappa^T([\pi/2 + \theta]_\pi)$, in particular, $Q.P(X) = Q.P(X^{-1}) = \cos\theta$ when $\kappa = 0$.

Remark 4 From Remark 2.3 and Lemma 13 in [11], we easily know that $\mathbf{D}_1 : H_{2\pi} \to H_\kappa^T([\theta]_\pi)$, $\mathbf{D}_2 : H_{2\pi} \to H_{-\kappa}^T([-\theta]_\pi)$. In particular, If $\kappa > 0$ then $\mathbf{D}_2 = \mathbf{0}$, if $\kappa < 0$ then $\mathbf{D}_1 = \mathbf{0}$, if $\kappa = 0$ then

$$(\mathbf{D}_1y)(t) = -\frac{\sin\theta}{2\pi}\int_0^{2\pi} w_1(\tau)y(\tau)d\tau, \quad (\mathbf{D}_2y)(t) = \frac{\sin\theta}{2\pi}\int_0^{2\pi} w_2(\tau)y(\tau)d\tau.$$

In [10] we have proved the following theorems.

Theorem 1 $\mathbf{AB}=\mathbf{I}+b\mathbf{D}_2, \mathbf{BA}=\mathbf{I}-b\mathbf{D}_1$, *where* $\mathbf{I}$ *is the identical operator.*

Theorem 2 *If* $T_n \in H_n^T$, *then* $\mathbf{A}^*T_n = Q.P(XT_n)$, $\mathbf{B}^*T_n = Q.P(X^{-1}T_n)$.

Corollary 1 $\mathrm{Ima}(\mathbf{D}_1) = H_\kappa^T([\theta]_\pi)$, $\mathrm{Ima}(\mathbf{D}_2) = H_{-\kappa}^T([-\theta]_\pi)$. *If* $\kappa \neq 0$ *or* $\kappa = 0$ *with* $[\theta]_\pi = \pi/2$ *then* $\ker(\mathbf{A}) = bH_\kappa^T([\theta]_\pi)$, $\ker(\mathbf{B}) = bH_{-\kappa}^T([-\theta]_\pi)$ *and* $\mathbf{D}_1 bT_\kappa = T_\kappa$ *when* $T_\kappa \in H_\kappa^T([\theta]_\pi)$, $\mathbf{D}_2 bT_{-\kappa} = -T_{-\kappa}$ *when* $T_{-\kappa} \in H_{-\kappa}^T([-\theta]_\pi)$. *If* $\kappa = 0$ *with* $[\theta]_\pi \neq \pi/2$, *then* $\ker(\mathbf{A}) = \ker(\mathbf{B}) = \{0\}$.

Proof For cases $\kappa \neq 0$ and $\kappa = 0$ with $[\theta]_\pi = \pi/2$, obviously, $Q.P(XT_{\kappa-1}) = 0$ if $T_{\kappa-1} \in H_{\kappa-1}^T$, again, if $T_\kappa(t) = \sin(\kappa t + \theta)$ via slightly evaluating we get $Q.P(XT_\kappa) = 0$. Therefore from Theorem 2 one has $bH_\kappa^T([\theta]_\pi) \subseteq \ker(\mathbf{A})$, and from Remark 4 we know $\mathrm{Ima}(\mathbf{D}_1) \subseteq H_\kappa^T([\theta]_\pi)$. On the other hand, quoting Theorem 1, $y = b\mathbf{D}_1 y \in bH_\kappa^T([\theta]_\pi)$ if $y \in \ker(\mathbf{A})$, i.e., $\ker(\mathbf{A}) \subseteq bH_\kappa^T([\theta]_\pi)$. Again if $T_\kappa \in H_\kappa^T([\theta]_\pi)$, then $bT_\kappa = b\mathbf{D}_1(bT_\kappa)$, noting that $b \neq 0$ (namely $b(t)$ does not vanish) in the cases discussed now, and $\mathbf{D}_1(bT_\kappa)$ is a trigonometric polynomial, hence $T_\kappa = \mathbf{D}_1(bT_\kappa)$, i.e.,$H_\kappa^T([\theta]_\pi) \subseteq \mathrm{Ima}(\mathbf{D}_1)$.

For case $\kappa = 0$ with $[\theta]_\pi \neq \pi/2$, if $\mathbf{A}y = 0$ then $y = b\mathbf{D}_1 y$ by Theorem 1, further, noting that $\mathbf{D}_1 y$ is a constant by using Theorem 2 and Remark 3, we get $\mathbf{A}(b\mathbf{D}_1 y) = \mathbf{D}_1 y \mathbf{A} b = \cos\theta\, b\mathbf{D}_1 y = 0$, hence $y = b\mathbf{D}_1 y = 0$, i.e., $\ker(\mathbf{A}) = \{0\}$. Similarly, the rest may be proved.

Remark 5 From Theorem 1, Remark 4 and Corollary 1, we know that if $\kappa > 0$ then $\mathbf{A}$ has the left inverse $\mathbf{B}$ but no right inverse, if $\kappa < 0$ then $\mathbf{A}$ has the right inverse $\mathbf{B}$ but no left inverse. It is very interesting that for $\kappa = 0$ two extreme cases occur: either both the right inverse and the left inverse of $\mathbf{A}$ exist if $[\theta]_\pi \neq \pi/2$ or neither of them exists if $[\theta]_\pi = \pi/2$, which will be seen later.

Corollary 2 *If* $\kappa \neq 0$ *or* $\kappa = 0$ *with* $[\theta]_\pi = \pi/2$, *then* $b\mathbf{D}_1$ *and* $-b\mathbf{D}_2$ *are idempotent,* $\mathbf{D}_1$ *and* $\mathbf{D}_2$ *are right zero divisors of* $\mathbf{B}$ *and* $\mathbf{A}$ *respectively,* $b\mathbf{D}_1$ *and* $b\mathbf{D}_2$ *are left zero divisors of* $\mathbf{A}$ *and* $\mathbf{B}$ *respectively, i.e.,* $\mathbf{D}_1\mathbf{B} = 0$, $\mathbf{D}_2\mathbf{A} = 0$, $\mathbf{A}b\mathbf{D}_1 = 0$ *and* $\mathbf{B}b\mathbf{D}_2 = 0$.

This corollary follows from Theorem 1, Remark 5 and Corollary 1.

Corollary 3 *If* $\kappa > 0$ *and* $N_\kappa \in H_\kappa^T([\theta]_\pi)$ *is a given trigonometric polynomial, then under the condition* $\mathbf{D}_1 y = N_\kappa, \mathbf{A}y = f$ *possesses the unique solution* $y = \mathbf{B}f + bN_\kappa$; *if* $\kappa < 0$, *the condition of solvability for* $\mathbf{A}y = f$ *is* $\mathbf{D}_2 f = 0$ *and it possesses the unique solution* $y = \mathbf{B}f$ *when it is fulfilled; if* $\kappa = 0$, *with* $[\theta]_\pi = \pi/2$, *the condition of solvability for* $\mathbf{A}y = f$ *is* $\mathbf{D}_2 f = 0$ *and when it is fulfilled it possesses the unique solution* $y = \mathbf{B}f + bN$ *under the condition* $\mathbf{D}_1 y = N$ *where* N *is a given constant; if* $\kappa = 0$ *with* $[\theta]_\pi \neq \pi/2$ *then* $\mathbf{A}y = f$ *possesses a unique solution* $y = \mathbf{B}f - b\sec\theta\mathbf{D}_2 f$.

So we call $\mathbf{D}_1$ the unisolving operator, $\mathbf{D}_2$ the restricting operator.

Proof When $\kappa > 0$, obviously $y = \mathbf{B}f + bN_\kappa$ is the solution of $\mathbf{A}y = f$ by Remark 5 and Corollary 1, as well as $\mathbf{D}_1 y = N_\kappa$ by Corollary 1 and Corollary 2. Conversely if $\mathbf{A}y = f$ and $\mathbf{D}_1 y = N_\kappa$ then by Theorem 1 $y = \mathbf{B}f + b\mathbf{D}_1 y = \mathbf{B}f + bN_\kappa$. When $\kappa < 0$, if $\mathbf{A}y = f$, then $\mathbf{D}_2 f = 0$ by Corollary 2 and $y = \mathbf{B}f$ by Remark 5, conversely the latter two equations implicate $\mathbf{A}y = f$ by Theorem 1. When $\kappa = 0$ with $[\theta]_\pi = \pi/2$, if $\mathbf{A}y = f$ with $\mathbf{D}_1 y = N$, then $\mathbf{D}_2 f = 0$ by Corollary 2 and $y = \mathbf{B}f + bN$ by Theorem 1, again, when $\mathbf{D}_2 f = 0$, $y = \mathbf{B}f + bN$ is just the solution of $\mathbf{A}y = f$ by Theorem 1 and Corollary 1, as well as $\mathbf{D}_1 y = N$ by Corollary 1 and Corollary 2. When $\kappa = 0$ with $[\theta]_\pi \neq \pi/2$, we directly check that $y = \mathbf{B}f - b\sec\theta\mathbf{D}_2 f$ is the solution of $\mathbf{A}y = f$. $\mathbf{A}y = \mathbf{AB}f - \mathbf{A}(b\sec\theta\mathbf{D}_2 f) = f + b\mathbf{D}_2 f - b\mathbf{D}_2 f = f$. Again, from Corollary 1 we know that this solution is unique.

Remark 6 When $\kappa < 0$ or $\kappa = 0$ with $[\theta]_\pi = \pi/2$, $\mathbf{D}_2 f = 0$ is equivalent to $\int_0^{2\pi} w_2(\tau)h_{-\kappa}(\tau)f(\tau)d\tau = 0$ for any $h_{-\kappa} \in H_{-\kappa}^T([-\theta]_\pi)$, which is denoted as $f \perp_2 H_{-\kappa}^T([-\theta]_\pi)$.

Remark 7 Obviously, for the SIE $\mathbf{B}y = f$ we also have the results similar to those for the SIE $\mathbf{A}y = f$ discussed above. For example, its condition of solvability when $\kappa > 0$ or $\kappa = 0$ with $[\theta]_\pi = \pi/2$ is $\int_0^{2\pi} w_1(\tau)h_\kappa(\tau)f(\tau)d\tau = 0$ for any $h_\kappa \in H_\kappa^T([\theta]_\pi)$, denoted as $f \perp_1 H_\kappa^T([\theta]_\pi)$.

Remark 8 From the above last corollary, we now know that if $\kappa = 0$ with $[\theta]_\pi = \pi/2$ $\mathbf{A}$ has neither a left inverse nor a right inverse and if $\kappa = 0$ with $[\theta]_\pi \neq \pi/2$ then $\mathbf{A}$ has the inverse $\mathbf{A}^{-1} = \mathbf{B} - b\sec\theta\mathbf{D}_2$.

19.3 SINGULAR QUADRATURE OPERATORS

In this section, we construct some singular quadrature operators (SQOs) possessing properties similar to those of the SIOs $\mathbf{A}$ and $\mathbf{B}$ in the last section.

First we need a kind of singular integral operators with cosecant kernel. For the Hölder-continuous function y with $y(\tau + 2\pi) = -y(\tau)$, denoted as $y \in \overline{H}_{2\pi}$, we introduce

$$(\overline{\mathbf{A}}y)(t) = a(t)w_1(t)y(t) + \frac{b(t)}{2\pi}\int_0^{2\pi} w_1(\tau)y(\tau)\csc\frac{\tau - t}{2}d\tau,$$

$$(\overline{\mathbf{B}}y)(t) = a(t)w_2(t)y(t) - \frac{b(t)}{2\pi}\int_0^{2\pi} w_2(\tau)y(\tau)\csc\frac{\tau - t}{2}d\tau.$$

Obviously, they map $\overline{H}_{2\pi}$ into $\overline{H}_{2\pi}$ and possess the properties very similar to those of the SIOs $\mathbf{A}$ and $\mathbf{B}$ (to be omitted).

Definition 1 *Let* $\Delta_n(\tau) = \prod_{j=1}^n \sin\frac{1}{2}(\tau - \alpha_{n,j})$ *where* $\alpha_{n,j}$ $(j = 1, 2, \cdots, n)$ *are* n *different points lying in* $[0, 2\pi)$, $\nabla_m(\tau) = \prod_{j=1}^m \sin\frac{1}{2}(\tau - \beta_{m,j})$ *where* $\beta_{m,j}$ $(j = 1, 2, \cdots, m)$ *are* m *different points lying in* $[0, 2\pi)$. *We call* (Δ_n, ∇_m)

a pair of associated half-trigonometric polynomials of the SIOs $(\mathbf{A}, \mathbf{B})$, *or simply, a pair of associated HTPs if* $n = m + 2\kappa$ *and* $\mathbf{A}\Delta_n = \nabla_m, \mathbf{B}\nabla_m = \Delta_n$ *when* n *is even as well as* $\overline{\mathbf{A}}\Delta_n = \nabla_m, \overline{\mathbf{B}}\nabla_m = \Delta_n$ *when* n *is odd.*

There are many such pairs of associated HTPs. Some examples may be found in [3,4].

If $h_n(\tau) = \prod_{j=1}^{n} \cot \frac{1}{2}(\tau - t_j)$, a half-trigonometric polynomial, we denote the discretization operator at the set of its zeros t_j $(j = 1, 2, \cdots, n)$ by

$$\mathbf{r}_n^h f = (f(t_1), \cdots, f(t_n))^T. \tag{19.15}$$

If (Δ_n, ∇_m) is a pair of associated HTPs, from [12] we may obtain that the quasi-interpolation type singular quadrature operators $\mathbf{Q}_n^{\Delta A}$ for $\mathbf{A}$ and $\mathbf{Q}_m^{\nabla B}$ for $\mathbf{B}$ are as follows.

$$(\mathbf{Q}_n^{\Delta A} f)(t) = \frac{\nabla_m(t)}{\Delta_n(t)} f(t) + b(t) E_n^{\Delta D} d_n^{\Delta}\left(\cot \frac{1}{2}(\cdot - t)\right) \mathbf{r}_n^{\Delta} f \tag{19.16}$$

where the diagonal matrix

$$d_n^{\Delta}\left(\cot \frac{1}{2}(\cdot - t)\right) = \mathrm{diag}\left(\cot \frac{1}{2}(\alpha_{n,1} - t), \cdots, \cot \frac{1}{2}(\alpha_{n,n} - t)\right), \tag{19.17}$$

and

$$E_n^{\Delta D} = (u_{n,1}, \cdots, u_{n,n}) \tag{19.18}$$

with

$$u_{n,j} = \frac{\Delta_n^*(\alpha_{n,j})}{4\pi\, \Delta_n'\, (\alpha_{n,j})}, \tag{19.19}$$

$$\Delta_n^*(z) = \begin{cases} \displaystyle\int_0^{2\pi} w_1(\tau)\, \Delta_n\, (\tau) \cot \frac{\tau - z}{2} d\tau, & \text{if } n \text{ is even,} \\ \displaystyle\int_0^{2\pi} w_1(\tau)\, \Delta_n\, (\tau) \csc \frac{\tau - z}{2} d\tau, & \text{if } n \text{ is odd,} \end{cases} \tag{19.20}$$

which is called the associated function of $\Delta_n(\tau)$ with repect to the weight $w_1(\tau)$.

$$(\mathbf{Q}_m^{\nabla B} f)(t) = \frac{\Delta_n(t)}{\nabla_m(t)} f(t) - b(t) E_m^{\nabla R} d_m^{\nabla}\left(\cot \frac{1}{2}(\cdot - t)\right) \mathbf{r}_m^{\nabla} f \tag{19.21}$$

where the diagonal matrix

$$d_m^{\nabla}\left(\cot \frac{1}{2}(\cdot - t)\right) = \mathrm{diag}\left(\cot \frac{1}{2}(\beta_{n,1} - t), \cdots, \cot \frac{1}{2}(\beta_{m,m} - t)\right), \tag{19.22}$$

and

$$E_m^{\nabla R} = (v_{m,1}, \cdots, v_{m,m}) \tag{19.23}$$

with

$$v_{m,j} = \frac{\nabla_m^*(\beta_{m,j})}{4\pi\, \nabla_m'\, (\beta_{m,j})}, \tag{19.24}$$

$$\nabla_n^*(z) = \begin{cases} \displaystyle\int_0^{2\pi} w_2(\tau)\, \nabla_n\,(\tau) \cot \frac{\tau - z}{2} d\tau, & \text{if } n \text{ is even}, \\ \displaystyle\int_0^{2\pi} w_2(\tau)\, \nabla_n\,(\tau) \csc \frac{\tau - z}{2} d\tau, & \text{if } n \text{ is odd}, \end{cases} \tag{19.25}$$

which is called the associated function of $\nabla_m(\tau)$ with repect to the weight $w_2(\tau)$.

Definition 2 *If* (Δ_n, ∇_m) *is a pair of associated HTPs of the SIOs* $(\mathbf{A}, \mathbf{B})$, *then* $(\mathbf{Q}_n^{\Delta A}, \mathbf{Q}_m^{\nabla B})$ *given by* (19.16) *and* (19.21) *is said to be a pair of associated SQOs relative to* $(\mathbf{A}, \mathbf{B})$.

Remark 9 We easily obtain that $\text{diag}(u_{n,1}, \cdots, u_{n,n})\mathbf{r}_n^{\Delta} b = \mathbf{r}_n^{\Delta}(\nabla_m/2\Delta_n')$ and $\text{diag}(v_{m,1}, \cdots, v_{m,m})\mathbf{r}_m^{\nabla} b = \mathbf{r}_m^{\nabla}(\Delta_n/2\nabla_m')$. For example,

$$\begin{aligned} 2\,\Delta_n'\,(\alpha_{n,j}) b(\alpha_{n,j}) u_{n,j} &= \frac{b(\alpha_{n,j})\,\Delta_n^*\,(\alpha_{n,j})}{2\pi} \\ &= \begin{cases} (\mathbf{A}\Delta_n)(\alpha_{n,j}), & \text{if } n \text{ is even}, \\ (\overline{\mathbf{A}}\Delta_n)(\alpha_{n,j}), & \text{if } n \text{ is odd}, \end{cases} \\ &= \nabla_m(\alpha_{n,j}). \end{aligned}$$

In particular, if $\alpha_{n,j}$ coincides with some $\beta_{m,i}$ then $b(\alpha_{n,j})u_{n,j} = 0$.

We also use Δ_n and ∇_m to denote the sets of their zeros respectively.

Remark 10 $(u_{n,j}b)'(\alpha_{n,j})$ $(j = 1, \cdots, n)$ and $(v_{m,j}b)'(\beta_{m,j})$ $(j = 1, \cdots, m)$ exist. In fact, for the case $u_{n,j} = 0$ this is trivial, if $u_{n,j} \neq 0$, i.e.,$\Delta_n^*(\alpha_{n,j}) \neq 0$ from definition .1, we know

$$\nabla_m(\tau) = a(\tau) w_1(\tau)\,\Delta_n\,(\tau) + \frac{b(\tau)\,\Delta_n^*\,(\tau)}{2\pi}, \tag{19.26}$$

in which both the first and the second term are differentiable at $\alpha_{n,j}$, and so is $b\Delta_n^*$, hence $b'(\alpha_{n,j})$ exists by Lemma 4.2 in [12]. In particular, $b'(\alpha_{n,j})$ exists when $\alpha_{n,j} \in \Delta_n \backslash \nabla_m$ since $u_{n,j} \neq 0$ in this case by (19.26).

Let $C_{2\pi}'(h_n)$ denote the family of continuous functions with period 2π and possessing derivatives at the zeros of h_n. By this remark we may obtain the following conclusion.

Remark 11 $\mathbf{Q}_n^{\Delta A} : C_{2\pi}'(\Delta_n) \to C_{2\pi}'(\nabla_m)$ and $\mathbf{Q}_m^{\nabla B} : C_{2\pi}'(\nabla_m) \to C_{2\pi}'(\Delta_n)$, or $\mathbf{Q}_n^{\Delta A} : C_{2\pi}'(\Delta_n \backslash \nabla_m) \to C_{2\pi}'(\nabla_m \backslash \Delta_n)$, $\mathbf{Q}_m^{\nabla B} : C_{2\pi}'(\nabla_m \backslash \Delta_n) \to C_{2\pi}'(\Delta_n \backslash \nabla_m)$.

Remark 12 $p(\mathbf{Q}_n^{\Delta A}) \geq [\frac{1}{2}(n-1)]$, $p(\mathbf{Q}_m^{\nabla B}) \geq [\frac{1}{2}(m-1)]$, where $p(\mathbf{Q}_n^{\Delta A})$ and $p(\mathbf{Q}_m^{\nabla B})$ denote the trigonometric precisions of $\mathbf{Q}_n^{\Delta A}$ and $\mathbf{Q}_m^{\nabla B}$ respectively, $[x]$ the integral part of x, see [12].

We point out an important fact to be used thereafter: $E_n^{\Delta D}$ is just the matrix of the trigonometric interpolation type quadrature formula for the integral operator

$$\mathbf{D}f = \frac{1}{2\pi}\int_0^{2\pi} w_1(\tau)f(\tau)d\tau \tag{19.27}$$

and $E_m^{\nabla R}$ is just the the matrix of the trigonometric interpolation type quadrature formula for the integral operator

$$\mathbf{R}f = \frac{1}{2\pi}\int_0^{2\pi} w_2(\tau)f(\tau)d\tau. \tag{19.28}$$

Remark 13 Set $\mathbf{Q}_n^{pD}f = E_n^{\Delta D}\mathbf{r}_n^{\Delta}f$ and $\mathbf{Q}_m^{\nabla R}f = E_m^{\nabla R}\mathbf{r}_m^{\nabla}f$, then we have $p(\mathbf{Q}_n^{\Delta D}) \geq [\frac{1}{2}(n-1)]$ and $p(\mathbf{Q}_m^{\nabla R}) \geq [\frac{1}{2}(m-1)]$.

Using $E_n^{\Delta D}$ we set the quadrature formula of $\mathbf{D}_1$

$$(\mathbf{Q}_n^{\Delta D_1}f)(t) = E_n^{\Delta D}d_n^{\Delta}(X_1(\cdot,t))\mathbf{r}_n^{\Delta}f \tag{19.29}$$

where

$$d_n^p(X_1(\cdot,t)) = \mathrm{diag}(X_1(\alpha_{n,1},t),\cdots,X_1(\alpha_{n,n},t)) \tag{19.30}$$

which originates from the kernel X_1 in $\mathbf{D}_1$, and

$$X_1(\tau,t) = \left[Q.P\left(X^{-1}\right)(\tau) - Q.P\left(X^{-1}\right)(t)\right]\cot\frac{\tau-t}{2} - i\Delta\left(X^{-1}\right). \tag{19.31}$$

From Definition 2 and Remark 7 we know $\Delta_n \perp_2 H_{-\kappa}^T([-\theta]_\pi)$ unless $\kappa = 0$ with $[\theta]_\pi \neq \pi/2$. Now we further assume $\Delta_n \perp_2 H_{-\kappa}^T$. In this case, the quadrature formula (19.29) is also a trigonometric interpolation type quadrature formula. This fact is not obvious.

We use $\mathbf{L}_n^h f$ to denote the trigonometric interpolation polynomial of normal form for f at the zeros $t_j (j = 1,2,\cdots,n)$ of a trigonometric polynomial h_n of degree n, then, see [12,13],

$$\left(\mathbf{L}_n^h f\right)(\tau) = \begin{cases} \displaystyle\sum_{j=1}^n \frac{h_n(\tau)}{2h_n'(t_j)}\csc\frac{\tau-t_j}{2}f(t_j), & \text{if } n \text{ is odd,} \\ \displaystyle\sum_{j=1}^n \frac{h_n(\tau)}{2h_n'(t_j)}\cot\frac{\tau-t_j}{2}f(t_j), & \text{if } n \text{ is even.} \end{cases} \tag{19.32}$$

Lemma 1 *$p(\mathbf{Q}_n^{\Delta D_1}) \geq [\frac{1}{2}(n-1)]$, i.e., $\mathbf{Q}_n^{\Delta D_1}f = \mathbf{D}_1\mathbf{L}_n^{\Delta}f$.*

Proof We only prove the case when n is even.

$$(\mathbf{D}_1\mathbf{L}_n^p f)(t) = \sum_{j=1}^n \frac{f(\alpha_{n,j})}{4\pi\,\Delta_n'(\alpha_{n,j})}\int_0^{2\pi} w_1(\tau)\Delta_n(\tau)X_1(\tau,t)\cot\frac{\tau-\alpha_{n,j}}{2}d\tau. \tag{19.33}$$

Using the equality $\cot x \cot y = \cot(y-x)(\cot x - \cot y) - 1$ and observing $\Delta_n \perp_2 H^T_{-\kappa}$, we have

$$\begin{aligned}
&\int_0^{2\pi} w_1(\tau)\, \Delta_n\, (\tau) X_1(\tau,t) \cot \frac{\tau-\alpha_{n,j}}{2} d\tau \\
&= \cot \frac{\alpha_{n,j}-t}{2} \left(\int_0^{2\pi} w_1(\tau)\, \Delta_n\, (\tau)\, [X_1(\tau,\alpha_{n,j}) - X_1(\tau,t)]\, d\tau \right) \\
&\quad + X_1(\alpha_{n,j},t) \int_0^{2\pi} w_1(\tau)\, \Delta_n\, (\tau) \cot \frac{\tau-\alpha_{n,j}}{2} d\tau \\
&= X_1(\alpha_{n,j},t)\, \Delta_n^*\, (\alpha_{n,j})\, d\tau.
\end{aligned}$$

Now noting (19.24), (19.19) and (19.32), we get $\mathbf{Q}_n^{\Delta D_1} f = \mathbf{D}_1 \mathbf{L}_n^{\Delta} f$.

Similarly, using $E_m^{\nabla R}$ we set the quadrature formula of $\mathbf{D}_2$

$$(\mathbf{Q}_m^{\nabla D_2} f)(t) = E_m^{\nabla R} d_m^{\nabla}(X_2(\cdot,t)) \mathbf{r}_m^{\nabla} f \tag{19.34}$$

with

$$d_m^{\nabla}(X_2(\cdot,t)) = \operatorname{diag}(X_2(\beta_{n,1},t),\cdots,X_2(\beta_{m,m},t)) \tag{19.35}$$

which originates from the kernel X_2 in $\mathbf{D}_2$, and

$$X_2(\tau,t) = [Q.P(X)(\tau) - Q.P(X)(t)] \cot \frac{\tau-t}{2} - i\Delta(X). \tag{19.36}$$

Lemma 2 $p(\mathbf{Q}_m^{\nabla D_2}) \geq [\frac{1}{2}(m-1)]$, *i.e.*, $\mathbf{Q}_n^{\nabla D_2} f = \mathbf{D}_2 \mathbf{L}_n^{\nabla} f$.

Now we assume that $b(t)$ in (19.2) is a trigonometric polynomial of degree μ and n is sufficiently large.

The following theorem is analogous to Theorem 1.

Theorem 3 $\mathbf{Q}_n^{\Delta A} \mathbf{Q}_m^{\nabla B} = \mathbf{I} + b\mathbf{Q}_m^{\nabla D_2}$, $\mathbf{Q}_m^{\nabla B} \mathbf{Q}_n^{\Delta A} = \mathbf{I} - b\mathbf{Q}_n^{\Delta D_1}$.

Proof Noting Remark 3, we know that the compositions of $\mathbf{Q}_n^{\Delta A}$ and $\mathbf{Q}_m^{\nabla B}$ are reasonable. First suppose $f \in \ker(\mathbf{r}_n^{\Delta})$, we get $\mathbf{Q}_m^{\nabla B} \mathbf{Q}_n^{\Delta A} f = f$, for a general function f. Therefore we have

$$\begin{aligned}
\mathbf{Q}_m^{qB} \mathbf{Q}_n^{pA} f &= f - L_n^p f + \mathbf{BA} L_n^p f \\
&= f - L_n^p f + (\mathbf{I} - b\mathbf{D}_1) L_n^p f \\
&= f - b\mathbf{Q}_n^{pD_1} f.
\end{aligned}$$

The proof of the first equality is similar.

Introduce the adjoints of the SQOs $\mathbf{Q}_n^{\Delta A}$ and $\mathbf{Q}_m^{\nabla B}$, which are the SQOs for the SIOs $\mathbf{A}^*$ and $\mathbf{B}^*$ respectively, then

$$(\mathbf{Q}_n^{\Delta A^*} f)(t) = \frac{\nabla m(t)}{\Delta_n(t)} f(t) + E_n^{\Delta D} d_n^{\Delta} \left(\frac{1}{\cdot - t} \right) \mathbf{r}_n^{\Delta} b f, \tag{19.37}$$

$$(\mathbf{Q}_n^{\nabla B^*} f)(t) = \frac{\Delta_n(t)}{\nabla m(t)} f(t) - E_n^{\nabla R} d_m^{\nabla} \left(\frac{1}{\cdot - t} \right) \mathbf{r}_m^{\nabla} b f. \tag{19.38}$$

Theorem 4

$$(\mathbf{Q}_n^{\triangle A^*} f)(t) = \frac{\nabla_m(t)f(t) - (\mathbf{L}_n^{\triangle} \nabla_m f)(t)}{\Delta_n(t)},$$

$$(\mathbf{Q}_m^{\nabla B^*} f)(t) = \frac{\Delta_n(t)f(t) - (\mathbf{L}_m^{\nabla} \Delta_n f)(t)}{\nabla_m(t)}.$$

Proof We prove the first equality only.

$$\begin{aligned}(\mathbf{Q}_n^{\triangle A^*} f)(t) &= \frac{\nabla_m(t)}{\Delta_n(t)} f(t) + d_n^{\triangle}\left(\frac{1}{\cdot - t}\right) \mathbf{r}_n^{\triangle}\left(\frac{\nabla_m f}{\Delta_n'}\right) \\ &= \frac{1}{\Delta_n(t)}\left[\nabla_m(t)f(t) - (\mathbf{L}_n^{\triangle} \nabla_m f)(t)\right].\end{aligned}$$

Using Theorem 1 and Theorem 2 we may get the following corollaries, which are analogous to Corollaries 1-3. Their proofs follow trivially in a completely analogous way to that used for Corollaries 1-3.

Corollary 4 Ima $\left(\mathbf{Q}_n^{\triangle D_1}\right) = H_\kappa^T([\theta]_\pi)$ *and* Ima $\left(\mathbf{Q}_m^{\nabla D_2}\right) = H_{-\kappa}^T([-\theta]_\pi)$. *If $\kappa \neq 0$ or $\kappa = 0$ with $[\theta]_\pi = \pi/2$ then* ker $\left(\mathbf{Q}_n^{\triangle A}\right) = bH_\kappa^T([\theta]_\pi)$, *more precisely, $\mathbf{Q}_n^{\triangle D_1} bT_\kappa = T_\kappa$ when $T_\kappa \in H_\kappa^T([\theta]_\pi)$;* ker $\left(\mathbf{Q}_m^{\nabla B}\right) = bH_{-\kappa}^T([-\theta]_\pi)$, *or $\mathbf{Q}_m^{\nabla D_2} bT_{-\kappa} = -T_{-\kappa}$ when $T_{-\kappa} \in H_{-\kappa}^T([-\theta]_\pi)$. If $\kappa = 0$ with $[\theta]_\pi \neq \pi/2$, then* ker $\left(\mathbf{Q}_n^{\triangle A}\right) =$ ker $\left(\mathbf{Q}_m^{\nabla B}\right) = \{0\}$.

Corollary 5 *If $\kappa \neq 0$ or $\kappa = 0$ with $[\theta]_\pi = \pi/2$, then $b\mathbf{Q}_n^{\triangle D_1}$ and $-b\mathbf{Q}_m^{\nabla D_2}$ are idempotent, $\mathbf{Q}_n^{\triangle D_1}$ and $\mathbf{Q}_m^{\nabla D_2}$ are right zero divisors of $\mathbf{Q}_m^{\nabla B}$ and $\mathbf{Q}_n^{\triangle A}$ respectively, $b\mathbf{Q}_n^{\triangle D_1}$ and $b\mathbf{Q}_m^{\nabla D_2}$ are left zero divisors of $\mathbf{Q}_n^{\triangle A}$ and $\mathbf{Q}_m^{\nabla B}$ respectivly, i.e., $\mathbf{Q}_n^{\triangle D_1}\mathbf{Q}_m^{\nabla B} = 0$, $\mathbf{Q}_m^{\nabla D_2}\mathbf{Q}_n^{\triangle A} = 0$, $\mathbf{Q}_n^{\triangle A} b\mathbf{Q}_n^{\triangle D_1} = 0$ and $\mathbf{Q}_m^{\nabla B} b\mathbf{Q}_m^{\nabla D_2} = 0$.*

Corollary 6 *If $\kappa > 0$ and $N_\kappa \in H_\kappa^T([\theta]_\pi)$ is a given trigonometric polynomial, then under the condition $\mathbf{Q}_n^{\triangle D_1} y = N_\kappa$, $\mathbf{Q}_n^{\triangle A} y = f$ possesses the unique solution $y = \mathbf{Q}_m^{\nabla B} f + bN_\kappa$. If $\kappa < 0$, the condition of solvability for $\mathbf{Q}_n^{\triangle A} y = f$ is $\mathbf{Q}_m^{\nabla D_2} f = 0$ and it possesses the unique solution $y = \mathbf{Q}_m^{\nabla B} f$ when this condition is satisfied. If $\kappa = 0$, with $[\theta]_\pi = \pi/2$, the condition of solvability for $\mathbf{Q}_n^{\triangle A} y = f$ is $\mathbf{Q}_m^{\nabla D_2} f = 0$ and when it is satisfied it possesses the unique solution $y = \mathbf{Q}_m^{\nabla B} f + bN$ under the condition $\mathbf{Q}_n^{\triangle D_1} y = N$ where N is a given constant. If $\kappa = 0$ with $[\theta]_\pi \neq \pi/2$ then $\mathbf{Q}_n^{\triangle A} y = f$ possesses the unique solution $y = \mathbf{Q}_m^{\nabla B} f - b \sec\theta \mathbf{Q}_m^{\nabla D_2} f$.*

References

[1] Du Jinyuan: Singular integral operators and singular quadrature operators associated with singular integral equations of the first kind and their applications. Acta Math. Sci. 15 (1995), 2, 219–234.

[2] Du Jinyuan: Singular integral operators and singular quadrature operators associated with singular integral equations. Submitted to J. Math. Anal. Appl.

[3] Du Jinyuan: The classical collocation methods for singular integral equations with Cauchy kernels. Submitted to J. Math. Anal. Appl.

[4] Jian-ke Lu: Boundary value problems for analytic functions. World Scientific, Singapore, 1993.

[5] Muskhelishvili: Singular integral equations. 2nd ed. Noordhoff, Groningen 1968.

[6] Krenk, S.: Numerical quadrature of periodic singular integral equations. J. Inst. Math. Appl. 21 (1978), 181–187.

[7] Ioakimidis, N.I.: A natural interpolation formula for the numerical solution of singular integral equations with Hilbert kernel. BIT 23 (1983), 92–104.

[8] Du Jinyuan: On the numerical solution for singular integral equations with Hilbert kernel. Chinese J. Num. Math. Appl. 11 (1989), 2, 9–27.

[9] Du Jinyuan: On the trigonometric polynomials interpolating approximate solutions of singular integral equations with Hilbert kernel. In: Integral equations and boundary value problems, eds. G.C. Wen and Z. Zhao. World Scientific, Singapore, 1991.

[10] Du Jinyuan: Some systems of orthogonal trigonometric polynomials associated with singular integral equations with Hilbert kernel. J. Wuhan Univ., Sci. ed. 2 (1988), 2, 17–35.

[11] Du Jinyuan: Quadrature formulas for singular integrals with Hilbert kernel. J. Comp. Math. 6 (1988), 3, 205–225.

[12] Du Jinyuan: Quadrature formulas of quasi-interpolation type for singular integrals with Hilbert kernel. J. Approximation Theory, to appear.

[13] Schoenberg, I.J.: Notes on spline functions. Indagationes Math. 34 (1972), 412–422.

20 SOME CO–CIRCULAR CRACK PROBLEMS WITH CYCLIC SYMMETRY

Jianke Lu †

Wuhan University
Wuhan 430072
P.R. China

jydu@whu.edu.cn

Abstract: Some co-circular crack problems with cyclic symmetry in the infinite plane are considered. They are reduced to singular integral equations along a single crack, whose solutions may be expressed in terms of elementary functions with certain coefficients determined by some integrals.

The cyclically symmetric problems in plane elasticity were discussed in [1], [2] and general problems for cyclically symmetric cracks, in [3]. Some first fundamental problems for co-circular cracks with cyclic symmetry in the infinite plane will be studied in detail in the present paper, whose solutions may be obtained in finite analytic form so that they may be conveniently used in practice.

20.1 FORMULATION OF THE PROBLEMS

An isotropic infinite elastic plane is weakened by a set of cyclically symmetric co-circular cracks $L_k = a_k b_k$, $k = 0, 1, \cdots, n-1$, on the unit circle $|t| = 1$ with end-points $a_k = \exp[(-\alpha + 2k\pi)i/n]$ and $b_k = \exp[(\alpha + 2k\pi)i/n]$ $(0 < \alpha < \pi)$, oriented from a_k to b_k. The weakened plane is denoted by S and n is called the *order* of cyclic symmetry. A cyclically symmetric problem means the boundary condition is also cyclically symmetric with the same order n. $\omega = \omega_n = \exp(2\pi i/n)$ is called the *cyclic constant.*

†Supported by NNSF and Science Foundation of SEC of P.R. China

H.G.W. Begehr et al. (eds.), Partial Differential and Integral Equations, 267–276.

We consider the first fundamental problems. For simplicity, assume that there is no external stress on each crack. In general, the Kolosov functions $\Phi(z)$ and $\Psi(z)$ holomorphic in S are finite at infinity: $\Phi(\infty) = \Gamma$, $\Psi(\infty) = \Gamma'$, see [4]. Also assume that there is no rotation at infinity so that Γ is real, denoting tension ($\Gamma > 0$) or compression ($\Gamma < 0$) in all directions. Γ' denotes tension or compression in a single direction and hence Γ' must be zero by cyclic symmetry unless $n = 2$. We want to determine the elastic equilibrium, i.e., search for the stresses σ_ρ, σ_θ, $\tau_{\rho\theta}$ at any point $z = \rho e^{i\theta}$ in polar coordinates, which may be expressed by, see [4],

$$\left.\begin{aligned} \sigma_\rho + \sigma_\theta &= 4\mathrm{Re}\Phi(z), \\ \sigma_\theta - \sigma_\rho + 2i\tau_{\rho\theta} &= 2[\bar{z}\Phi'(z) + \Psi(z)]e^{2i\theta}. \end{aligned}\right\} \tag{20.1}$$

By cyclic symmetry of the stresses, $\Phi(z)$ and $\Psi(z)$ are respectively cyclically symmetric and quasi-symmetric, see [2],

$$\Phi(\omega z) = \Phi(z), \quad \Psi(\omega z) = \bar{\omega}^2\Psi(z). \tag{20.2}$$

In case $n = 2$, $\Psi(z)$ is also cyclically symmetric since $\omega = -1$.

20.2 REDUCTION TO SINGULAR INTEGRAL EQUATIONS ON A SINGLE CRACK

As shown in [2], Section 30, in the general case, by introducing a new unknown functon $\Omega(t)$ on $L = \sum_{k=0}^{n-1} L_k$, $\Phi(z)$ and $\Psi(z)$ may be respectively expressed as

$$\Phi(z) = \frac{1}{2\pi i}\int_L \frac{\Omega(t)}{t - z}dt + \Gamma, \quad z \in S, \tag{20.3}$$

$$\Psi(z) = -\frac{1}{2\pi i}\int_L \frac{\overline{\Omega(t)}}{t - z}d\bar{t} - \frac{1}{2\pi i}\int_L \frac{\bar{t}\Omega(t)}{(t - z)^2}dt + \Gamma', \quad z \in S, \tag{20.4}$$

where $\Omega(t)$ is the solution of the singular integral equation

$$\begin{aligned} &\frac{1}{\pi i}\int_L \frac{\Omega(t)}{t - t_0}dt + \frac{1}{2\pi i}\int_L \frac{\partial}{\partial t_0}\log\frac{t - t_0}{\bar{t} - \bar{t}_0}\Omega(t)dt \\ &+\frac{1}{2\pi i}\int_L \frac{\partial}{\partial t_0}\frac{t - t_0}{\bar{t} - \bar{t}_0}\overline{\Omega(t)}d\bar{t} = -2\Gamma - \bar{\Gamma}'\frac{d\bar{t}_0}{dt_0}, \quad t_0 \in L, \end{aligned} \tag{20.5}$$

in the class h_0, see [5], [6], i.e., possibly with integrable singularities at all the tips of L, fulfilling the supplementary condition

$$\int_L \Omega(t)dt = 0. \tag{20.6}$$

The unique existence of such a solution is guaranteed by the general theory.

In case of co-circular cracks, on account of $t\bar{t} = 1$, $t_0\bar{t}_0 = 1$ and so

$$\frac{t - t_0}{\bar{t} - \bar{t}_0} = -tt_0,\ d\bar{t} = -\frac{dt}{t^2},\ \frac{d\bar{t}_0}{dt_0} = -\frac{1}{t_0^2},$$

(20.4) becomes

$$\Psi(z) = \frac{1}{2\pi i}\int_L \frac{\overline{\Omega(t)}}{t^2(t-z)}dt - \frac{1}{2\pi i}\int_L \frac{\Omega(t)}{t(t-z)^2}dt + \Gamma',\ z \in S, \tag{20.7}$$

and the equation (20.5) is reduced to

$$\frac{1}{\pi i}\int_L \frac{\Omega(t)}{t - t_0}dt + \frac{1}{2\pi i}\int_L \frac{\overline{\Omega(t)}}{t}dt = 2\Gamma - \frac{\bar{\Gamma}'}{t_0^2},\ t_0 \in L. \tag{20.8}$$

In case of cyclic symmetry, by the cyclically symmetric property of $\Phi(z)$, it is easily seen by (20.3) that $\Omega(t)$ possesses the same property: $\Omega(\omega t) = \Omega(t)$.

In order to simplify (20.8) to an equation on a single circular arc, we make the mapping $\zeta = z^n$ and denote $\tau = t^n$ for $t \in L$ with the restriction $|\arg \tau| \leq \alpha$ and $|\arg \tau^{1/n}| \leq \alpha/n$. The arc $L_0 = a_0b_0$ is mapped to the arc $L_0^* = a_0^*b_0^*$ on the unit circle $|\tau| = 1$, where $a_0^* = e^{-i\alpha}$, $b_0^* = e^{i\alpha}$. Then S is mapped onto a region S^* in the ζ-plane cut by L_0^*. Denote $\Omega(t) = \Omega(\tau^{1/n}) = \Omega^*(\tau)$. As

$$\begin{aligned} \frac{1}{\pi i}\int_L \frac{\Omega(t)}{t - t_0}dt &= \frac{1}{\pi i}\sum_{k=0}^{n-1}\int_{L_0} \frac{\Omega(t)}{t - \bar{\omega}^k t_0}dt \\ &= \frac{1}{\pi i}\int_{L_0} \frac{nt^{n-1}\Omega(t)}{t^n - t_0^n}dt \\ &= \frac{1}{\pi i}\int_{L_0^*} \frac{\Omega^*(\tau)}{\tau - \tau_0}d\tau,\ \tau_0 \in L_0^*, \end{aligned} \tag{20.9}$$

$$\frac{1}{2\pi i}\int_L \frac{\overline{\Omega(t)}}{t}dt = \frac{1}{2\pi i}\int_{L_0} \frac{n\overline{\Omega(t)}}{t}dt = \frac{1}{2\pi i}\int_{L_0^*} \frac{\overline{\Omega^*(\tau)}}{\tau}d\tau, \tag{20.10}$$

the equation (20.6) is reduced to

$$\frac{1}{\pi i}\int_{L_0^*} \frac{\Omega^*(\tau)}{\tau - \tau_0}d\tau + \frac{1}{2\pi i}\int_{L_0^*} \frac{\overline{\Omega^*(\tau)}}{\tau}d\tau = 2\Gamma - \frac{\bar{\Gamma}'}{\tau_0},\ \tau_0 \in L_0^* \tag{20.11}$$

(remember that Γ' occurs only when $n = 2$). The supplementary condition (20.6) becomes now

$$\int_{L_0^*} \frac{\Omega^*(\tau)}{\tau^{1-1/n}}d\tau = 0. \tag{20.12}$$

(20.11) should be solved again in the class h_0. Its solution with the restriction (20.12) is known to be unique as before.

After $\Omega^*(\tau)$ is obtained, $\Phi(z)$ and $\Psi(z)$ may be also expressed as integrals along L_0^*. Denote $\Phi(z) = \Phi(\zeta^{1/n}) = \Phi^*(\zeta)$. Similar to (20.9), it is obvious

$$\Phi^*(\zeta) = \frac{1}{2\pi i}\int_{L_0^*}\frac{\Omega^*(\tau)}{\tau-\zeta}d\tau + \Gamma. \tag{20.13}$$

As for $\Psi(z)$, we have, by (20.7),

$$\Psi(z) = \sum_{k=0}^{n-1}\frac{1}{2\pi i}\int_{L_0}\frac{\overline{\Omega(t)}}{\omega^k t^2(\omega^k t - z)}dt - \sum_{k=0}^{n-1}\frac{1}{2\pi i}\int_{L_0}\frac{\Omega(t)}{t(\omega^k t - z)^2}dt + \Gamma'.$$

It is easily verified that

$$\sum_{k=0}^{n-1}\frac{1}{\omega^k(\omega^k t - z)} = \frac{ntz^{n-2}}{t^n - z^n}, \tag{20.14}$$

$$\sum_{k=0}^{n-1}\frac{1}{(\omega^k t - z)^2} = nz^{n-2}\frac{z^n + (n-1)t^n}{(t^n - z^n)^2}. \tag{20.15}$$

Thus, by substitution, we have

$$\frac{\Psi(z)}{z^{n-2}} = \frac{1}{2\pi i}\int_{L_0}\frac{n\overline{\Omega(t)}dt}{t(t^n - z^n)} - \frac{1}{2\pi i}\int_{L_0}\frac{n[z^n - (n-1)t^n]}{t(t^n - z^n)^2}\Omega(t)dt + \Gamma', \ z \in S.$$

By (20.2), we see that $\Psi_1(z) = \Psi(z)/z^{n-2}$ is cyclically symmetric. $\Psi_1(z)$ can not possess a singularity at $z = 0$ by the cyclic quasi-symmetry of $\Psi(z)$ (when $n > 2$) and so it is holomorphic in S and cyclically symmetric.

Denote $\Psi_1(z) = \Psi_1(\zeta^{1/n}) = \Psi_1^*(\zeta)$ which is holomorphic in S^* (when $n = 2$, $\Psi_1(z) = \Psi(z) = \Psi^*(\zeta)$). Then, we obtain

$$\Psi_1^*(\zeta) = \frac{1}{2\pi i}\int_{L_0^*}\frac{\overline{\Omega^*(\tau)}d\tau}{\tau(\tau-\zeta)} - \frac{1}{2\pi i}\int_{L_0^*}\frac{[\zeta + (n-1)\tau]}{\tau(\tau-\zeta)^2}\Omega^*(\tau)d\tau + \Gamma', \zeta \in S^*. \tag{20.16}$$

Thus, our problem is thoroughly solved when $\Omega^*(\tau)$ is obtained.

Remark Equation (20.11) is independent of n and so is its solution $\Omega^*(\tau)$, which is advantageous to solutions for the case of different n's. $\Omega^*(\tau)$ depends upon n just because of (20.12).

20.3 PROPERTIES OF CERTAIN INTEGRALS

Before solving (20.11), we give some properties of certain class of integrals. Let

$$R(\zeta) = (\zeta - a_0^*)(\zeta - b_0^*) = (\zeta - e^{-i\alpha})(\zeta - e^{i\alpha}) = \zeta^2 - 2\zeta\cos\alpha + 1, \ \zeta \in S^*, \tag{20.17}$$

and $\sqrt{R(\zeta)}$ be taken as any continuous branch in S^*, e.g., for definiteness, that branch with

$$\sqrt{R(0)} = +1 \text{ or } \lim_{\zeta\to\infty}\frac{\sqrt{R(\zeta)}}{\zeta} = -1 \tag{20.18}$$

(cf. [4], Section124a). Denote $\sqrt{R(\tau)}$ ($\tau \in L_0^*$) as the left (positive) side boundary value of $\sqrt{R(\zeta)}$. Then,

$$\sqrt{R(\tau)} = 2e^{i\theta/2}\sqrt{\sin\frac{\alpha+\theta}{2}\sin\frac{\alpha-\theta}{2}} = \frac{1}{\sqrt{2}}e^{i\theta/2}\sqrt{\cos\theta-\cos\alpha},$$

$$\tau = e^{i\theta}, -\alpha \le \theta \le \alpha, \tag{20.19}$$

where the radicals on the right–hand sides take non-negative values since $\arg\sqrt{R(\zeta)}$ remains unchanged when ζ varies from $\zeta = 0$ to $\zeta = 1$ along the real axis and $\arg\sqrt{R(0)} = 0$ by (20.18).

Consider the integral

$$I_\lambda = \frac{1}{\pi i}\int_{L_0}\frac{d\tau}{\tau^\lambda\sqrt{R(\tau)}}, \tag{20.20}$$

where λ is a real parameter. When λ is not an integer, for $\tau = e^{i\theta} \in L_0^*$, τ^λ means $e^{i\lambda\theta}(-\alpha \le \theta \le \alpha)$. We note some of its properties.

1° I_λ is real for any (real) λ,

$$I_\lambda = \frac{1}{\pi}\int_0^\alpha \frac{\cos(1/2-\lambda)\theta}{\sqrt{\sin(\alpha+\theta)/2\sin(\alpha-\theta)/2}}d\theta = \frac{\sqrt{2}}{\pi}\int_0^\alpha \frac{\cos(1/2-\lambda)}{\sqrt{\cos\theta-\cos\alpha}}d\theta, \tag{20.21}$$

which follows at once from (20.19).

2° $I_{1-\lambda} = I_\lambda$ or $I_{\frac{1}{2}+\mu} = I_{\frac{1}{2}-\mu}$.

3° $I_\lambda \le I_{\frac{1}{2}}$ and the equality occurs only when $\lambda = 1/2$.

4° I_λ is positive and monotonic increasing (decreasing) when $0 \le \lambda \le 1/2$ $(1/2 \le \lambda \le 1)$.

5° $I_\lambda + I_{-\lambda} > 0$ when $|\lambda| \le 1/2$, $I_\lambda - I_{-\lambda} > 0$ when $0 < \lambda < 1$.

6° $I_0 = I_1 = 1$, $I_2 = I_{-1} = \cos\alpha$.

Properties 2°–5° are immediate consequences of (20.21). By definition,

$$I_0 = \frac{1}{\pi i}\int_{L_0^*}\frac{d\tau}{\sqrt{R(\tau)}},\ I_{-1} = \frac{1}{\pi i}\int_{L_0^*}\frac{\tau d\tau}{\sqrt{R(\tau)}}.$$

Let Λ be a closed contour surrounding L_0^*, oriented clockwisely. By the residue theorem and (20.18), it is readily seen that

$$\frac{1}{2\pi i}\int_\Lambda \frac{d\zeta}{\sqrt{R(\zeta)}} = 1.$$

Shrinking Λ to L_0^*, we see that $I_0 = 1$.

In the neighborhood of $\zeta = \infty$, we may easily verify

$$\frac{\zeta}{\sqrt{R(\zeta)}} = -1 - \frac{\cos\alpha}{\zeta} + \cdots$$

and so

$$\frac{1}{2\pi i}\int_{\Lambda}\frac{\zeta}{\sqrt{R(\zeta)}}d\zeta = \cos\alpha,$$

from which $I_{-1} = \cos\alpha$ follows.

We shall also evaluate certain Cauchy principal value integrals for later needs. First, we establish

$$\frac{1}{\pi i}\int_{L_0^*}\frac{\sqrt{R(\tau)}}{\tau-\tau_0}d\tau = \tau_0 - \cos\alpha,\ \ \tau_0 \in L_0^*. \tag{20.22}$$

In fact, let

$$F(w) = \frac{1}{2\pi i}\int_{\Lambda}\frac{\sqrt{R(\zeta)}}{\zeta-w}d\zeta,\ \ w \notin \Lambda,$$

where Λ is the contour as before but leaves w in its exterior. In the neighborhood of $\zeta = \infty$, we have the expansion

$$\begin{aligned}\frac{\sqrt{R(\zeta)}}{\zeta-w} &= -(1+\frac{w}{\zeta}+\cdots)(1-\frac{\cos\alpha}{\zeta}+\cdots) \\ &= -1-\frac{w-\cos\alpha}{\zeta}+\cdots.\end{aligned}$$

Then, by the residue theorem,

$$F(w) = \sqrt{R(w)} + w - \cos\alpha.$$

Let Λ shrink to L_0^*, we get

$$F(w) = \frac{1}{\pi i}\int_{L_0^*}\frac{\sqrt{R(\tau)}}{\tau-w}d\tau = \sqrt{R(w)} + w - \cos\alpha. \tag{20.23}$$

Then, (20.22) follows from the Plemelj formla,

$$\frac{1}{\pi i}\int_{L_0^*}\frac{\sqrt{R(\tau)}}{\tau-\tau_0}d\tau = \frac{F^+(\tau_0)+F^-(\tau_0)}{2}.$$

In a similar manner, we may verify that

$$\frac{1}{\pi i}\int_{L_0^*}\frac{d\tau}{\sqrt{R(\tau)}(\tau-\zeta)} = \frac{1}{\sqrt{R(\zeta)}},\ \ \zeta\notin L_0^*, \tag{20.24}$$

and hence

$$\frac{1}{\pi i}\int_{L_0^*}\frac{d\tau}{\sqrt{R(\tau)}(\tau-\tau_0)} = 0,\ \ \tau \in L_0^*. \tag{20.25}$$

20.4 SOLUTION FOR THE CASE $N > 2$ OR $N = 2$ WITH $\Gamma' = 0$

For solving (20.11), we set

$$A = \frac{1}{2\pi i}\int_{L_0^*} \frac{\overline{\Omega^*(\tau)}}{\tau} d\tau \quad \text{or} \quad \bar{A} = \frac{1}{2\pi i}\int_{L_0^*} \frac{\Omega^*(\tau)}{\tau} d\tau. \tag{20.26}$$

Then, (20.11) becomes

$$\frac{1}{\pi i}\int_{L_0^*} \frac{\Omega^*(\tau)}{\tau - \tau_0} d\tau = 2\Gamma - A, \; \tau_0 \in L_0^*. \tag{20.27}$$

By the inverse formula for Cauchy principal value integral in the class h_0, see [5] or [6], we have

$$\Omega^*(\tau_0) = \frac{1}{\sqrt{R(\tau_0)}}\left[\frac{2\Gamma - A}{\pi i}\int_{L_0^*} \frac{\sqrt{R(\tau)}}{\tau - \tau_0} d\tau + D\right],$$

where D is an undetermined constant. Therefore, by (20.22), we obtain

$$\Omega^*(\tau) = \frac{1}{\sqrt{R(\tau)}}\left[(2\Gamma - A)(\tau - \cos\alpha) + D\right], \; \tau \in L_0^*. \tag{20.28}$$

The constants A and D should be determined by (20.26) and (20.12). By substituting (20.28) into them, we get respectively, on account of properties 6° and 2° in the previous section,

$$\left.\begin{aligned} 2\bar{A} &= D + (2\Gamma - A)(1 - \cos\alpha), \\ [D - (2\Gamma - A)\cos\alpha] I_{1/n} &+ (2\Gamma - A) I_{-1/n} = 0. \end{aligned}\right\} \tag{20.29}$$

Eliminating D, we have

$$\left[2\bar{A} - (2\Gamma - A)\right] I_{1/n} + (2\Gamma - A) I_{-1/n} = 0. \tag{20.30}$$

Taking its imaginary part

$$\text{Im}A(I_{1/n} + I_{-1/n}) = 0,$$

we see that, by 5°, Section 20.3, A is real, and so is D. Thus, by (20.30),

$$A = \frac{2\Gamma(I_{1/n} - I_{-1/n})}{3I_{1/n} - I_{-1/n}}, \tag{20.31}$$

where $3I_{1/n} - I_{-1/n} > 0$ by 3° and 6°, Section 20.3. (20.31) may be rewritten as

$$A = \frac{2\Gamma(1 - J_{1/n})}{3 - J_{1/n}}, \; J_{1/n} = I_{-1/n}/I_{1/n} < 1, \tag{20.31'}$$

and then, by (20.29).

$$D = \frac{4\Gamma(\cos\alpha - J_{1/n})}{3 - J_{1/n}}. \tag{20.32}$$

We obtain finally, by (20.28),

$$\Omega^*(\tau) = \frac{4\Gamma}{3 - J_{1/n}} \frac{\tau - J_{1/n}}{\sqrt{R(\tau)}}, \quad \tau \in L_0^*. \tag{20.33}$$

Substituting (20.33) into (20.13) and (20.16), we obtain $\Phi^*(\zeta)$, $\Psi_1^*(\zeta)$ and then $\Phi(z)$, $\Psi(z)$ by noting that

$$\overline{\sqrt{R(\tau)}} = \sqrt{R(\tau)}/\tau, \tag{20.34}$$

and finally the stresses at any point by (20.1).

20.5 SOLUTION FOR THE CASE $N = 2$ WITH $\Gamma = 0$.

Two subcase will be discussed:

(i) $\Gamma' = B'$ is real, i.e., there is a tension or compression along the x-axis. Define A again as (20.26). Then (20.11) and (20.12) become respectively

$$\frac{1}{\pi i}\int_{L_0^*} \frac{\Omega^*(\tau)}{\tau - \tau_0} d\tau = -\frac{B'}{\tau_0} - A, \tag{20.35}$$

$$\int_{L_0^*} \frac{\Omega^*(\tau)}{\sqrt{\tau}} d\tau = 0. \tag{20.36}$$

The general solution of (20.35) in class h_0 is

$$\begin{aligned}
\Omega^*(\tau_0) &= \frac{1}{\sqrt{R(\tau_0)}}\left[-\frac{1}{\pi i}\int_{L_0^*} \frac{\sqrt{R(\tau)}}{\tau - \tau_0}\left(A + \frac{B'}{\tau}\right) d\tau + D\right] \\
&= \frac{1}{\sqrt{R(\tau_0)}}\left[\frac{B'}{\tau_0}\frac{1}{\pi i}\int_{L_0^*} \frac{\sqrt{R(\tau)}}{\tau} d\tau - \left(\frac{B'}{\tau_0} + A\right)\frac{1}{\pi i}\int_{L_0^*} \frac{\sqrt{R(\tau)}}{\tau - \tau_0} d\tau + D\right] \\
&= \frac{1}{\sqrt{R(\tau_0)}}\left[\frac{B'}{\tau_0}(1 - \cos\alpha) - \left(\frac{B'}{\tau_0} + A\right)(\tau_0 - \cos\alpha) + D\right]
\end{aligned}$$

by (20.23) an (20.22). Thus,

$$\Omega^*(\tau) = \frac{1}{\sqrt{R(\tau)}}\left[B'\left(\frac{1}{\tau} - 1\right) - A(\tau - \cos\alpha) + D\right]. \tag{20.37}$$

Substituting this into (20.26) and (20.36), we get respectively

$$\begin{aligned}
2\bar{A} &= -AI_o + B'I_2 + (D - B' + A\cos\alpha)I_1 \\
&= D - (A + B')(1 - \cos\alpha), \\
0 &= -AI_{-1/2} + B'I_{3/2} + (D - B' + A\cos\alpha)I_{1/2}
\end{aligned}$$

or

$$(B' - A)J_{1/2} + D - B' + A\cos\alpha = 0.$$

Eliminating D, we obtain

$$2\bar{A} = A(J_{1/2} - 1) - B'(J_{1/2} - \cos\alpha),$$

from which it is readily seen that A is real since $J_{1/2} > 0$, and so

$$A = \frac{B'(J_{1/2} - \cos\alpha)}{J_{1/2} - 3}, \quad D = \frac{B'(4J_{1/2} - 2J_{1/2}\cos\alpha + \cos^2\alpha - 3)}{J_{1/2} - 3}. \tag{20.38}$$

Substituting this into (20.37), we obtain

$$\Omega^*(\tau) = \frac{B'}{\sqrt{R(\tau)}}\left[\frac{1}{\tau} - \tau + \frac{(3 - \cos\alpha)(\tau - J_{1/2})}{3 - J_{1/2}}\right]. \tag{20.39}$$

(ii) $\Gamma' = iC'$ is pure imaginary, i.e., there is a tension or compression along the y-axis. Then (20.11) and (20.12) respectively become

$$\frac{1}{\pi i}\int_{L_0^*} \frac{\Omega^*(\tau)}{\tau - \tau_0} d\tau = -\frac{iC'}{\tau_0} - A \tag{20.40}$$

and (20.36), where A is defined by (20.26) again. As before, its general solution in the class h_0 is

$$\Omega^*(\tau) = \frac{1}{\sqrt{R(\tau)}}\left[iC'\left(\frac{1}{\tau} - 1\right) - A(\tau - \cos\alpha) + D\right]. \tag{20.41}$$

Then, by (20.26) and (20.36), it is easily seen that

$$\begin{aligned} 2\bar{A} &= -A + iC'\cos\alpha + D - iC' + A\cos\alpha, \\ 0 &= (iC' - A)J_{1/2} + D - iC' + A\cos\alpha \end{aligned}$$

so that

$$2\bar{A} = A(J_{1/2} - 1) - iC'(J_{1/2} - \cos\alpha),$$

by which it follows that $A = iA'$ and also $D = iD'$ are pure imaginary, and

$$A' = C'\frac{J_{1/2} - \cos\alpha}{J_{1/2} + 1}, \quad D' = C'\left[\frac{(1 + \cos\alpha)^2}{J_{1/2} + 1} - 2\cos\alpha\right]. \tag{20.42}$$

Substituting this in (20.41), we obtain

$$\Omega^*(\tau) = \frac{iC'}{\sqrt{R(\tau)}}\left[\frac{1}{\tau} - \frac{J_{1/2} - \cos\alpha}{J_{1/2} + 1} - \frac{J_{1/2}(1 + \cos\alpha)}{J_{1/2} + 1}\right]. \tag{20.43}$$

Remark For the general case where $\Gamma \neq 0$ and $\Gamma' = B' + iC'$ is complex, the solution may be obtained by superposition.

References

[1] Lu Jianke: On mathematical problems for the elastic plane with cyclic symmetry. J. Wuhan Univ. **2** (1964), 1-13 (Chinese, English abstract).

[2] Lu Jianke: Complex variable methods in plane elasticity. World Scientific, Singapore, 1995.

[3] Lu Jianke: On cyclically symmetric crack problems. Wuhan Univ. J. Nat. Sci. **2** (1997), No.2, 132-138.

[4] Muskhelishvili N.I.: Some basic problems of the mathematical theory of elasticity. Noordhoff, Groningen, 1953.

[5] Muskhelishvili N.I.: Singular integral equations. Noordhoff, Leyden, 1977.

[6] Lu Jianke: Boundary value problems for analytic functions. World Scientific, Singapore, 1993.

21 THE OBLIQUE DERIVATIVE PROBLEM FOR NONLINEAR PARABOLIC EQUATIONS OF SECOND ORDER WITH MEASURABLE COEFFICIENTS IN A HIGHER DIMENSIONAL DOMAIN

Guo–Chun Wen

Peking University
Beijing 100871
P.R. China

wengc@pku.edu.cn

Abstract: In [1], Yu. A. Alkhutov and I. T. Mamedov discussed the solvability of the Dirichlet problem for linear uniformly parabolic equations

$$\sum_{i,k=1}^{n} a_{ik}(x,t)u_{x_i x_k} - u_t = f(x,t),$$

with measurable coefficients satisfying the conditions

$$\sup \sum_{i,k=1}^{n} a_{ik}^2(x,t)/\inf[\sum_{i=1}^{n} a_{ii}(x,t)]^2 < \frac{1}{n-1/2}.$$

In the present paper, we discuss the solvability of oblique derivative problem for nonlinear uniformly parabolic equations of second order with measurable coefficients.
AMS No: 35K20, 35K55, 35K60.

21.1 FORMULATION OF THE OBLIQUE DERIVATIVE PROBLEM FOR NONLINEAR PARABOLIC EQUATIONS

Let Ω be a bounded domain in $I\!R^n$ and the boundary $\partial\Omega \in C^2$. Denote $Q = \Omega \times I$, $I = \{0 < t \leq T\}$, $0 < T < \infty$, $\partial Q = S_1 \cup S_2$ is the parabolic boundary,

H.G.W. Begehr et al. (eds.), Partial Differential and Integral Equations, 277–286.

where $S_1 = \Omega \times \{t=0\}$ is the bottom and $S_2 = \partial\Omega \times \bar{I}$ is the lateral boundary. We consider the nonlinear parabolic equation of second order

$$F(x,t,u,D_xu,D_x^2u) - Hu_t = 0 \text{ in } Q, \tag{21.1}$$

namely

$$Lu = \sum_{i,j=1}^{n} a_{ij}u_{x_ix_j} + \sum_{i=1}^{n} b_iu_{x_i} + cu - Hu_t = d \text{ in } Q, \tag{21.2}$$

where $D_xu = (u_{x_i}), D_x^2u = (u_{x_ix_j}), 0 < H < \infty$, and

$$a_{ij} = \int_0^1 F_{\tau r_{ij}}(x,t,u,p,\tau r)d\tau,\ b_i = \int_0^1 F_{\tau p_i}(x,t,u,\tau p,0)d\tau,$$

$$c = \int_0^1 F_{\tau u}(x,t,\tau u,0,0)d\tau,\ d = -F(t,x,0,0,0),$$

$$r = D_x^2u,\ p = D_xu,\ r_{ij} = \frac{\partial^2 u}{\partial x_i \partial x_j},\ p_i = \frac{\partial u}{\partial x_i}.$$

Suppose that (21.1) (or (21.2)) satisfies Condition C, i.e. for arbitrary functions $u_1(x,t),\ u_2(x,t) \in C^{1,0}_{\beta,\beta/2}(\overline{Q}) \cap \tilde{W}_2^{2,1}(Q)$, $F(x,t,u,D_xu,D_x^2u)$ satisfy the condition

$$\begin{aligned} &F(x,t,u_1,D_xu_1,D_x^2u_1) - F(x,t,u_2,D_xu_2,D_x^2u_2) \\ &\quad = \sum_{i,j=1}^{n} \tilde{a}_{ij}u_{x_ix_j} + \sum_{i=1}^{n} \tilde{b}_iu_{x_i} + \tilde{c}u, \end{aligned} \tag{21.3}$$

where $\tilde{W}_2^{2,1}(Q) = W_2^{2,0}(Q) \cap W_2^{0,1}(Q),\ 0 < \beta < 1, u = u_1 - u_2$ and

$$\tilde{a}_{ij} = \int_0^1 F_{u_{x_ix_j}}(x,t,\tilde{u},\tilde{p},\tilde{r})d\tau,\ \tilde{b}_i = \int_0^1 F_{u_{x_i}}(x,t,\tilde{u},\tilde{p},\tilde{r})d\tau,\ \tilde{u} = u_2 + \tau(u_1 - u_2),$$

$$\tilde{c} = \int_0^1 F_u(x,t,\tilde{u},\tilde{p},\tilde{r})d\tau,\ \tilde{p} = D_x[u_2 + \tau(u_1 - u_2)],\ \tilde{r} = D_x^2[u_2 + \tau(u_1 - u_2)],$$

and $\tilde{a}_{ij}, \tilde{b}_i, \tilde{c}, d$ satisfy the conditions

$$\alpha_0 \sum_{j=1}^{n} |\xi_j|^2 \le \sum_{i,j=1}^{n} \tilde{a}_{ij}\xi_i\xi_j \le \alpha_0^{-1} \sum_{j=1}^{n} |\xi_j|^2, 0 < \alpha_0 < 1, \tag{21.4}$$

$$\sup \sum_{i,j=1}^{n} \tilde{a}_{ij}^2 / \inf[\sum_{i=1}^{n} \tilde{a}_{ii}]^2 \le q_0 < \frac{1}{n - 1/2}. \tag{21.5}$$

$$|\tilde{a}_{ij}| \le k_0,\ |\tilde{b}_i| \le k_0,\ |\tilde{c}| \le k_0,\ L_p(d,Q) \le k_1, i,j = 1,...,n, \tag{21.6}$$

where $k_0, k_1, p\,(> n+2)$ are nonnegative constants. Moreover, for almost every point $(x,t) \in Q$ and $D_x^2 u$, $\tilde{a}_{ij}(x,t,u,D_x u, D_x^2 u)$, $\tilde{b}_i(x,t,u,D_x u), \tilde{c}(x,t,u)$ are continuous in $u \in I\!R$, $D_x u \in I\!R^n$.

The so-called oblique derivative problem(Problem O) is to find a continuously differentiable solution $u = u(x,t) \in C^{1,0}_{\beta,\beta/2}(\bar{Q}) \cap \tilde{W}_2^{2,1}(Q)$ satisfying the initial-boundary conditions

$$u(x,0) = f(x), \quad x \in \Omega, \tag{21.7}$$

$$\begin{aligned} lu &= a\frac{\partial u}{\partial \nu} + bu = g(x,t), \quad (x,t) \in S_2, \text{ i.e.} \\ lu &= \sum_{i,j=1}^{n} a_{ij}\cos(\nu, x_i)\frac{\partial u}{\partial x_j} + bu = g(x,t), \quad (x,t) \in S_2, \end{aligned} \tag{21.8}$$

in which $d_j(x,t) = \sum_{i=1}^{n} a_{ij}\cos(\nu, x_i)$, $f(x)$, $g(x,t), b(x,t)$ satisfy the conditions

$$\begin{aligned} &C_\alpha^2[f(x), \bar{\Omega}] \le k_2, \ C^{1,1}_{\alpha,\alpha/2}[g(x,t), S_2] \le k_2, \\ &C^{1,1}_{\alpha,\alpha/2}[d_j(x,t), S_2] \le k_0\,(j = 1,\cdots,n), \ C^{1,1}_{\alpha,\alpha/2}[b(x,t), S_2] \le k_0, \\ &\cos(\nu, \hat{n}) \ge d_0 > 0, \ \textstyle\sum_{j=1}^{n} d_j\cos(\nu, x_j) \ge d_0 > 0, \end{aligned} \tag{21.9}$$

where ν is some unit vector and $\hat{n}$ is the unit outward normal on S_2, $\alpha\,(0 < \beta \le \alpha < 1), k_0, k_2, d_0\,(0 < d_0 < 1)$ are nonnegative constants. In particular, Problem O with the conditions $a = 1, \nu = \hat{n}$, $b = 0$ on S_2 is the initial-Neumann boundary value problem, which will be called Problem N.

In the following, we give a priori estimates of solutions for Problem O. Afterwards, by using the method of parameter extension and the Schauder fixed-point theorem, the existence and uniqueness of solutions for Problem O are proved.

21.2 A PRIORI ESTIMATES OF SOLUTIONS FOR PROBLEM O FOR (21.1)

We first prove the following theorem.

Theorem 1 *If the equation* (21.1) *satisfies Condition C, then the solution of Problem O is unique.*

Proof Let $u_1(x,t), u_2(x,t)$ be two solutions of Problem O. It is easily seen that $u = u_1 - u_2$ is a solution of the initial-boundary value problem

$$\sum_{i,j=1}^{n} \tilde{a}_{ij} u_{x_i x_j} + \sum_{i=1}^{n} \tilde{b}_i u_{x_i} + \tilde{c}u - Hu_t = 0 \text{ in } Q, \tag{21.10}$$

$$u(x,0) = 0, \quad x \in S_1, \tag{21.11}$$

$$lu(x,t)=0 \text{ i.e. } a\frac{\partial u}{\partial \nu}+bu=0,\ (x,t)\in S_2, \tag{21.12}$$

where $\tilde{a}_{ij}, \tilde{b}_i, \tilde{c}$ are as stated in (21.3). Introducing the transformation of function $u = v\exp(Bt)$, B is an appropriately large number, such that $HB > \tilde{c}$, the initial-boundary value problem (21.10)-(21.12) is reduced to

$$\sum_{i,j=1}^{n} \tilde{a}_{ij}v_{x_ix_j}+\sum_{i=1}^{n}\tilde{b}_iv_{x_i}-[HB-\tilde{c}]v-Hv_t=0 \text{ in } Q, \tag{21.13}$$

$$v(x,0)=0 \text{ on } S_1, \tag{21.14}$$

$$lv(x,t)=0, \text{ i.e. } a\frac{\partial v}{\partial \nu}+bv=0 \text{ on } S_2. \tag{21.15}$$

Noting that $HB-\tilde{c}>0$, $(x,t)\in Q$, by the maximum principle of solutions for (21.13), it is derived that $v(x,t)=u(x,t)=0$, i.e. $u^1(x,t)=u^2(x,t)$, $(x,t)\in Q$.

In the following, we shall give estimates in $C^{1,0}(\bar{Q})$ and $C^{1,0}_{\beta,\beta/2}(\overline{Q})$ of solutions $u(x,t)$ of Problem O.

Theorem 2 *Under the same condition as in Theorem* 1, *any solution* $u(x,t)$ *of Problem O for* (21.1) *satisfies the estimate*

$$\| u \|_{C^{1,0}(\bar{Q})}=\| u \|_{C^{0,0}(\bar{Q})}+\sum_{i=1}^{n}\| u_{x_i} \|_{C^{0,0}(\overline{Q})}\le M_1, \tag{21.16}$$

where M_1 *is a nonnegative constant only dependent on* q_0,α,k,p,Q, *i.e.* $M_1 = M_1(q_0,\alpha,\ k,p,Q)$, $k=(k_0,k_1,k_2)$.

Proof Suppose that (21.16) is not true. There exist sequences of functions $\{a^m_{ij}\},\{b^m_i\},\{c^m\},\{d^m\}$ and $\{f^m(x)\},\{g^m(t,x)\},\{a^m(x,t)\},\{b^m(x,t)\}$, which satisfy Condition C and (21.7)-(21.8), and $\{a^m_{ij}\}$, $\{b^m_i\}$, $\{c^m\},\{d^m\}$ weakly converge to a^0_{ij},b^0_i,c^0,d^0, and $\{f^m\},\{g^m\}$, $\{a^m\},\{b^m\}$ uniformly converge to f^0,g^0,a^0,b^0 on S_1 or S_2 respectively, and the initial-boundary value problem

$$\sum_{i,j=1}^{n} a^m_{ij}u_{x_ix_j}+\sum_{i=1}^{n}b^m_iu_{x_i}+c^mu-Hu_t=d^m \text{ in } Q, \tag{21.17}$$

$$u^m(x,0)=f^m(x),\ x\in S_1, \tag{21.18}$$

$$lu^m(x,t)=g^m(x,t), \text{ i.e. } a^m\frac{\partial u^m}{\partial \nu}+b^mu^m=g^m(x,t),\ (x,t)\in S_2 \tag{21.19}$$

has a solution $u^m(x,t)$, such that $\| u^m \|_{C^{1,0}(\bar{Q})}=A_m(m=1,2,...)$ is unbounded. There is no harm in assuming that $A_m\ge 1$, and $\lim_{m\to\infty}A_m=+\infty$. It is easy to see that $U^m=u^m/A_m$ is a solution of the initial-boundary value problem

$$\tilde{L}U^m=\sum_{i,j=1}^{n} a^m_{ij}U^m_{x_ix_j}-HU^m_t=B^m,\ B^m=-\sum_{i=1}^{n}b^m_iU^m_{x_i}-c^mU^m+d^m/A_m, \tag{21.20}$$

$$U^m(x,0) = \frac{f^m(x)}{A_m}, \quad x \in S_1,$$

$$lU^m(x,t) = \frac{g^m}{A_m}, \text{ i.e. } a^m \frac{\partial U^m}{\partial \hat{n}} + b^m U^m = \frac{g^m}{A_m}, \quad (x,t) \in S_2. \tag{21.21}$$

Noting that $\sum_{i=1}^n b_i^m U_{x_i}^m + c^m U^m$ in (21.20) is bounded, by using the result in Theorem 3 below, we can obtain the estimate

$$\| U^m \|_{C^{1,0}_{\beta,\beta/2}(\bar{Q})} = \| U^m \|_{C^{0,0}_{\beta,\beta/2}(\bar{Q})} + \sum_{i=1}^n \| U^m_{x_i} \|_{C^{0,0}_{\beta,\beta/2}(\bar{Q})} \leq M_2, \tag{21.22}$$

$$\| U^m \|_{\tilde{W}^{2,1}_{p_0}(Q)} \leq M_3 = M_3(q_0, \alpha, k, Q, p), \; m = 1, 2, \ldots, \tag{21.23}$$

where $\beta \, (0 < \beta \leq \alpha)$, $p_0 \, (2 \leq p_0 < p)$, $M_j = M_j(q_0, \alpha, k, p, Q)$ $(j = 2, 3)$ are nonnegative constants. Hence from $\{U^m\}$, $\{U^m_{x_i}\}$, we can choose a subsequence $\{U^{m_k}\}$ such that $\{U^{m_k}\}$, $\{U^{m_k}_{x_i}\}$ uniformly converge to $U^0, U^0_{x_i}$ on $\bar{Q}$ respectively, and $\{U^{m_k}_{x_i x_j}\}$, $\{U^{m_k}_t\}$ weakly converge to $U^0_{x_i x_j}, U^0_t$ on Q respectively, and U^0 is a solution of the initial-boundary value problem

$$\sum_{i,j=1}^n a^0_{ij} \tilde{U}^0_{x_i x_j} + \sum_{i=1}^n b^0_i \tilde{U}^0_{x_i} + c^0 U^0 - H U^0_t = 0, \tag{21.24}$$

$$U^0(x,0) = 0, \quad x \in S_1, \tag{21.25}$$

$$U^0(x,t) = 0, \text{ i.e. } a \frac{\partial U^0}{\partial \nu} + b U^0 = 0, \quad (x,t) \in S_2. \tag{21.26}$$

According to Theorem 1, we know that $U^0(x,t) = 0$, $(x,t) \in \overline{Q}$. However, from $\| U^m \|_{C^{1,0}(\bar{Q})} = 1$, it can be derived that there exists a point $(x^*, t^*) \in \bar{Q}$, such that $|U^0(x^*,t^*)| + \sum_{i=1}^n | U^0_{x_i}(x^*,t^*)| > 0$. This contradiction proves that (21.16) is true.

Theorem 3 *Under the same condition as in Theorem* 1, *any solution* $u(x,t)$ *of Problem O satisfies the estimates*

$$\| u \|_{C^{1,0}_{\beta,\beta/2}(\bar{Q})} \leq M_4 = M_4(q_0, \alpha, k, p, Q), \tag{21.27}$$

$$\| u \|_{\tilde{W}^{2,1}_{p_0}(Q)} \leq M_5 = M_5(q_0, \alpha, k, p, Q), \tag{21.28}$$

where $\beta \, (0 < \beta \leq \alpha), p_0 \, (2 \leq p_0 < p), M_4, M_5$ *are nonnegative constants.*

Proof First of all, we find a solution $\hat{u}(x,t)$ of the equation

$$\Delta \hat{u} - H \hat{u}_t = 0 \tag{21.29}$$

with the initial-boundary conditions (21.7) and (21.8), which satisfies the estimate

$$\| \hat{u} \|_{C^{2,1}(\bar{Q})} \leq M_6 = M_6(q_0, \alpha, k, Q, p). \tag{21.30}$$

(see [2]). Thus the function

$$\tilde{u}(x,t) = u(x,t) - \hat{u}(x,t) \tag{21.31}$$

is a solution of the equation

$$L\tilde{u} = \sum_{i,j=1}^{n} a_{ij}\tilde{u}_{x_i x_j} + \sum_{i=1}^{n} b_i \tilde{u}_{x_i} + c\tilde{u} - H\tilde{u}_t = \tilde{d}, \tag{21.32}$$

$$\tilde{u}(x,0) = 0, \ \ x \in S_1, \tag{21.33}$$

$$l\tilde{u}(x,t) = 0, \ \ (x,t) \in S_2, \tag{21.34}$$

where $\tilde{d} = d - L\hat{u}$. Introduce a local coordinate system in the neighborhood G of the surface $S_0 \in \partial\Omega$:

$$x_i = h_i(\xi_1, \cdots, \xi_{n-1})\xi_n + g_i(\xi_1, \cdots, \xi_{n-1}), \ i = 1, \cdots, n, \tag{21.35}$$

where $\xi_n = 0$ is just the surface $S_0 : x_i = g_i(\xi_1, \cdots, \xi_{n-1})$ $(i = 1, \cdots, n)$, and

$$h_i(\xi) = \left.\frac{d_i(x)}{d(x)}\right|_{x_i = g_i(\xi)}, \ i = 1, \cdots, n, d^2(x) = \sum_{i=1}^{n} d_i^2(x),$$

the boundary condition (21.34) can be reduced to the form

$$\frac{\partial \tilde{u}}{\partial \xi_n} + \tilde{b} = 0 \ \text{ on } \ \xi_n = 0, \tag{21.36}$$

where $\tilde{u} = \tilde{u}[x(\xi), t]$, $\tilde{d} = d[x(\xi), t]$. Secondly, we find a solution $v(x,t)$ of Problem N for the equation (21.29) with the boundary condition

$$\frac{\partial v}{\partial \xi_n} = -\tilde{b} \ \text{ on } \ \xi_n = 0, \tag{21.37}$$

which satisfies the estimate

$$\| v \|_{C^{2,1}(\bar{Q})} \le M_7 = M_7(q_0, \alpha, k, Q, p) < \infty, \tag{21.38}$$

and the function

$$V(x,t) = \tilde{u}e^{v(x,t)} \tag{21.39}$$

is a solution of the following initial-boundary value problem

$$\sum_{i,j=1}^{n} \tilde{a}_{ij} V_{\xi_i \xi_j} + \sum_{i=1}^{n} \tilde{b}_i V_{x_i} + \tilde{c}V - HV_t = \tilde{d}, \tag{21.40}$$

$$\frac{\partial V}{\partial \xi_n} = 0, \ \ \xi_n = 0. \tag{21.41}$$

On the basis of Theorem 4 below, we can derive the estimates

$$\| V \|_{C^{1,0}_{\beta,\beta/2}(\bar{Q})} \leq M_8 = M_8(q_0, \alpha, k, p, Q), \tag{21.42}$$

$$\| V \|_{\tilde{W}^{2,1}_{p_0}(Q)} \leq M_9 = M_9(q_0, \alpha, k, p, Q), \tag{21.43}$$

of $V(\xi, t)$, where $\beta\,(0 < \beta \leq \alpha), p_0\,(2 \leq p_0 < p), M_8, M_9$ are nonnegative constants. Combining (21.30),(21.38),(21.42) and (21.43), the estimates (21.27) and (21.28) are obtained.

Now, we shall give some estimates of solutions of Problem N for (21.1).

Theorem 4 *Suppose that the equation* (21.1) *satisfies Condition C. Then any solution* $u(x,t)$ *of Problem N satisfies the estimates*

$$C^{1,0}_{\beta,\beta/2}[u, \bar{Q}] \leq M_{10} = M_{10}(q_0, \alpha, k, p, Q), \tag{21.44}$$

$$\| u \|_{\tilde{W}^{2,1}_{p_0}(Q)} \leq M_{11} = M_{11}(q_0, \alpha, k, p, Q), \tag{21.45}$$

where $p_0\,(2 \leq p_0 < p), M_{10}, M_{11}$ *are nonnegative constants,* β *is a constant as stated in* (21.27).

Proof First of all, we extend the function $f(x)$ into the whole space $\mathbb{R}^n$, such that $C^2_\alpha(f, \mathbb{R}^n) \leq 2k_0$, and find a solution $\hat{u}$ of the Cauchy problem

$$\Delta u - H u_t = 0, \quad (t, x) \in \mathbb{R}^n \times \bar{I}, \tag{21.46}$$

$$u(x, 0) = f(x), \quad x \in \mathbb{R}^n, \tag{21.47}$$

which satisfies the estimate

$$C^{2,1}[\hat{u}, \bar{Q}] \leq M_{12} = M_{12}(q_0, \alpha, k, Q, p). \tag{21.48}$$

Thus the function $\tilde{u}(x,t) = u(x,t) - \hat{u}(x,t)$ is a solution of the following initial-boundary value problem

$$L\tilde{u} = \sum_{i,j=1}^{n} a_{ij}\tilde{u}_{x_i x_j} + \sum_{i=1}^{n} b_i \tilde{u}_{x_i} + c\tilde{u} - H\tilde{u}_t = \tilde{d}, \tag{21.49}$$

$$\tilde{u}(x, 0) = 0, \quad x \in S_1, \tag{21.50}$$

$$\frac{\partial \tilde{u}}{\partial n} = g(x,t) - \frac{\partial \hat{u}}{\partial n}, \quad (x,t) \in S_2, \tag{21.51}$$

where $\tilde{d} = d - L\hat{u}$. Afterwards, choosing an arbitrary point x^* in Ω and a small positive number ε, we construct a function $J(x) \in C^2_\alpha(\bar{\Omega})$ such that

$$J(x) = \begin{cases} 1, & x \in \Omega_\varepsilon, \\ 0, & x \in \Omega \setminus \Omega_{2\varepsilon}, \end{cases} \quad 0 \leq J(x) \leq 1, \ x \in \Omega_{2\varepsilon} \setminus \Omega_\varepsilon,$$

in which $\Omega_\varepsilon = \{|x - x^*| \leq \varepsilon\}$ and $J(x)$ satisfies the estimate

$$C^2[J(x), \Omega] \leq M_{13} = M_{13}(\varepsilon, \Omega). \tag{21.52}$$

Denote $U(x,t) = J(x)\tilde{u}(x,t)$, obviously, $U = U(x,t)$ is a solution of the following initial-boundary value problem

$$\sum_{i,j=1}^{n} a_{ij}U_{x_ix_j} + \sum_{i=1}^{n} b_iU_{x_i} + cU - HU_t = d^*, \ \ (x,t) \in \bar{Q}, \tag{21.53}$$

$$U(x,0) = 0, \ \ x \in S_1, \tag{21.54}$$

$$U(x,t) = 0, \ \ (x,t) \in S_2, \tag{21.55}$$

where $d^* = JLu + \sum_{i,j=1}^{n} a_{ij}[J_{x_i}u_{x_j} + J_{x_ix_j}u] + \sum_{i=1}^{n} b_iJ_{x_i}u$. By the method of inner estimate from [1], we obtain

$$C^{1,0}_{\beta,\beta/2}[U,\bar{Q}] \le M_{14}, \ \ C^{1,0}_{\beta,\beta/2}[\tilde{u},Q_\varepsilon] \le M_{15}, \tag{21.56}$$

where $Q_\varepsilon = \Omega_\varepsilon \times \bar{I}$, $M_j = M_j(q_0,\alpha,k,p,Q_\varepsilon)$, $j = 14, 15$. Combining (21.48), (21.52) and (21.56), we obtain the estimate

$$C^{1,0}_{\beta,\beta/2}[u,Q_\varepsilon] \le M_{16} = M_{16}(q_0,\alpha,k,Q,p). \tag{21.57}$$

Next, we choose any point $x^* \in S_2$, a small positive number d, and denote $S_3 = S_2 \cap \{|x - x^*| \le d, \ 0 \le t \le T\}$. Then we can find out a solution $\hat{u}(x,t)$ of (21.46) on $\bar{Q}$, such that $\hat{u}(x,t)$ satisfies the boundary condition

$$\frac{\partial \hat{u}}{\partial n} = g(x,t), \ \ (x,t) \in S_3.$$

Thus $\tilde{u} = u - \hat{u}$ is a solution of equation (21.49), which satisfies the boundary condition

$$\frac{\partial \tilde{u}}{\partial n} = 0, \ \ (x,t) \in S_3. \tag{21.58}$$

Making a non-singular transformation by a twice continuously differentiable function $\zeta = \zeta(x)$, such that S_3 is mapped onto S_4 in the plane $\zeta_n = 0$, and the domain Q onto the domain Q_1 in the half space $\zeta_n < 0$, so that the equation (21.2) and the boundary condition (21.58) are reduced to the equation and boundary condition

$$\sum_{i,j=1}^{n} A_{ij}\tilde{u}_{\zeta_i\zeta_j} + \sum_{i=1}^{n} B_i\tilde{u}_{\zeta_i} + C\tilde{u} - H\tilde{u}_t = D \text{ in } Q_1, \tag{21.59}$$

$$\frac{\partial \tilde{u}}{\partial n} = 0 \text{ on } S_4, \tag{21.60}$$

respectively. Now, we extend the function $\tilde{u}$ to a symmetric domain Q_2 of Q_1 about S_4, i.e. we construct the function

$$U = \begin{cases} \tilde{u}(\zeta,t), & (\zeta,t) \in Q_1, \\ \tilde{u}(\zeta^*,t), & (\zeta,t) \in Q_2, \end{cases} \tag{21.61}$$

where $\zeta^* = (\zeta_1, ..., \zeta_{n-1}, -\zeta_n)$, and $U(\zeta, t)$ is a solution of the equation

$$\sum_{i,j=1}^{n} \tilde{A}_{ij} U_{\zeta_i \zeta_j} + \sum_{i=1}^{n} \tilde{B}_i U_{\zeta_i} + \tilde{C} U - H U_t = \tilde{D} \text{ in } Q_1 \cup Q_2, \tag{21.62}$$

where

$$\tilde{A}_{ij} = \begin{cases} A_{ij}(\zeta, t), \\ (-1)^k A_{ij}(\zeta^*, t), \end{cases} k = \begin{cases} 1, i \neq j, \ i \text{ or } j = n, \\ 0, \text{ other cases}, \end{cases} \tilde{B}_i = \begin{cases} B_i(\zeta, t), \\ (-1)^k B_i(\zeta^*, t), \end{cases}$$

$$k = \begin{cases} 1, \ i \neq j, \\ 0, \ i = n, \end{cases} \tilde{C} = \begin{cases} C(\zeta, t), \\ C(\zeta^*, t), \end{cases} \tilde{D} = \begin{cases} D(\zeta, t), & (\zeta, t) \in Q_1, \\ D(\zeta^*, t), & (\zeta, t) \in Q_2. \end{cases}$$

By using a similar method as stated in the proof of (21.57), we can derive that $U(\zeta, t)$ and $u(x, t)$ satisfy the estimate

$$C^{1,0}_{\beta,\beta/2}[U, Q_1 \cup Q_2] \leq M_{17}, \ C^{1,0}_{\beta,\beta/2}[u, Q_1] \leq M_{18}, \tag{21.63}$$

where $M_j = M_j(q_0, \alpha, k, p, Q), j = 17, 18$. Combining (21.57) and (21.63), the estimates (21.44) and (21.45) are obtained.

21.3 THE SOLVABILITY OF PROBLEM O FOR THE PARABOLIC EQUATION (21.1)

In this section, we shall prove the solvability of Problem O for the equation (21.1) by using the estimates of solutions as stated in Section 21.2 and the method of parameter extension.

Theorem 5 *Let the equation* (21.64) *satisfy Condition C. Then Problem O for* (21.64) *is solvable.*

Proof To use the method of parameter extension, we introduce the equation with the parameter $\tau \in [0, 1]$:

$$\Delta u - H u_t + \tau(Lu - \Delta u) = D(x, t), \tag{21.64}$$

$$\frac{\partial u}{\partial \tilde{n}} + \tau[a\frac{\partial u}{\partial \nu} - \frac{\partial u}{\partial \tilde{n}} + bu] = B(x, t), \ (x, t) \in S_2, \tag{21.64'}$$

where $D(x, t) \in L_p(\overline{Q})$, $p > n+2$. When $\tau = 0$, by the result in [1] and [2], it is easy to see that Problem O is solvable. Suppose that when $\tau = \tau_0$, Problem O for (21.64) is solvable, we shall verify that there exists a positive number $\delta < 1$, such that for any $\tau \in E = \{|\tau - \tau_0| \leq \delta, \ \tau \in [0, 1]\}$, Problem O for (21.64) is solvable. For this, we rewrite (21.64) in the form

$$\Delta u - H u_t + \tau_0(Lu - \Delta u) = (\tau_0 - \tau)(Lu - \Delta u) + D(x, t). \tag{21.65}$$

We are free to choose a function $u_0(x, t) \in B = C^{1,0}_{\beta,\beta/2}(\bar{Q}) \cap \tilde{W}^{2,1}_{p_0}(Q)$, and substitute it into the position of u in the right-hand side of (21.65), obviously

$(\tau_0 - \tau)(Lu_0 - \Delta u_0) + D(x,t) \in L_p(\bar{Q})$, hence Problem O for such an equation (21.65) has a solution $u^1(x,t) \in B$. Thus by using successive iteration, we can obtain a sequence of functions $\{u^m(x,t)\}$, which satisfy

$$\begin{aligned} &\Delta u^m - Hu_t^m + \tau_0(Lu^m - \Delta u^m) \\ &= (\tau_0 - \tau)(Lu^{m-1} - \Delta u^{m-1}) + D(x,t), \end{aligned} \tag{21.66}$$

and the boundary conditions (21.7)-(21.8). According to the proof in Section 21.2, it can be proved that $U^m = u^m(x,t) - u^{m-1}(x,t)$ satisfies the estimate

$$\| U^m \|_{C^{1,0}(\bar{Q})} \leq |\tau_0 - \tau| M_{19} \| U^{m-1} \|_{C^{1,0}(\bar{Q})}, \tag{21.67}$$

where the constant $M_{19} = M_{19}(q_0, \alpha, k, p, Q) \geq 0$. By the completeness of the space $C^{1,0}(\bar{Q})$, there exists a function $u^*(x,t) \in C^{1,0}(\bar{Q})$, such that

$$\| u^m - u^* \|_{C^{1,0}(\bar{Q})} \to 0 \text{ as } n \to \infty.$$

It is not difficult to see that $u^*(x,t)$ is a solution of Problem O for (21.64). Thus, from the solvability of Problem O for (21.64) with $\tau = 0$, we can derive the solvability of Problem O for (21.64) with $\tau = \delta, 2\delta, ..., [1/\delta]\delta, 1$ successively, see e.g. [3]. In particular, Problem O for (21.64) with $\tau = 1, D(x,t) = 0$ is solvable. This completes the proof.

References

[1] Alkhutov, Yu.A., Mamedov, I.T.: The first boundary value problem for non-divergence second order parabolic equations with discontinuous coefficients. Math. USSR Sbornik 59(1988), 471-495.

[2] Ladyzhenskaya, O.A., Solonnikov, V.A., Ural'ceva, N.N.: Linear and Quasi-linear Equations of Parabolic Type. Trans. Math. Monographs 23, Amer. Math. Soc., Providence, RI, 1968.

[3] Wen , G.C., Begehr, H.: Boundary Value Problems for Elliptic Equations and Systems. Longman Scientific and Technical, Harlow, 1990.

22 EXISTENCE AND UNIQUENESS FOR BOUNDARY VALUE CONTACT PROBLEMS

Ivan V. Andronov

Department of Mathematical Physics
St.Petersburg University
St. Petersburg, Russia

Boris P. Belinskiy †

Department of Mathematics
University of Tennessee at Chattanooga
Chattanooga, Tennessee 37403-2598, USA
bbelinsk@cecasun.utc.edu

Abstract: Boundary value contact problems (BVCP) are the problems of mathematical physics, whose statement contains some conditions at particular points or lines of the boundary. Such problems appear when describing acoustic scattering on elastic plates or shells with any non-homogeneities. For a very wide class of the BVCP, we prove uniqueness theorems. We also prove the existence for one particular model.

†The second author was supported in part by The University of Tennessee at Chattanooga Center of Excellence in Computer Applications Scholarship and The Faculty Development Grant. This is an expanded version of a lecture given at the ISAAC '97 Conference, University of Delaware, June 3–7, 1997

H.G.W. Begehr et al. (eds.), Partial Differential and Integral Equations, 287–301.

22.1 INTRODUCTION

The problem of description of an oscillating continuum is one of the classical problems of mathematical physics. If this continuum is an acoustic medium, the boundary value problem for the Helmholtz equation appears. The Dirichlet and Neumann problems describe the scattering from an acoustically soft and rigid body correspondingly. The radiation conditions should be also stated at infinity. The solution for the scattering problems above exists and is unique at least in case of a compact scatterer body [1]. The physical models we consider are characterized by three peculiarities. First, the boundary operator contains derivatives of the order greater or equal than the order of the operator in the domain. Second, the boundary is non–homogeneous, and as a result, the statement of the problem contains some conditions at particular points or lines of the boundary. Third, the scatterer is non–compact (a finite perturbation of a half–plane). Though there are many papers in applied mathematical literature devoted to the solution of particular, nowadays very complex, problems, e.g., [2-7], the rigorous consideration of the scattering from a non–compact arbitrary shaped elastic body can hardly be found in the literature. The proof of uniqueness is certainly of independent interest. Also, uniqueness is closely connected with existence. Indeed, for many models, the scattering problem can be reduced to a Fredholm type equation in an appropriate functional space, e.g. [1,7–9]. The complete justification of any numerical algorithm for a Fredholm equation requires the proof of existence. The Fredholm alternative implies existence if uniqueness is proved.

The paper is organized as follows. Section 2 contains the statement of the boundary value contact problem. We also formulate the "optical" theorem [10-13] which is used in the next sections. The uniqueness for some "simple" problems is proved in Section 3. Section 4 contains the proof of Sommerfeld formula [14-16]. The uniqueness for a rather general BVCP is proved in Section 5. The problem of acoustic scattering from a finite crack in an infinite elastic plate was reduced in [7] to the integral equation perturbed by a finite–dimensional operator. The uniqueness of the solution for this model follows from the results of Section 5, whereas the existence is proved in Section 6. In Appendix the representations for Green's functions are discussed.

The paper is partially based on the recent authors' results [7, 16-18].

22.2 THE STATEMENT OF BVCP FOR THE ACOUSTIC SCATTERING

Let the acoustic medium be located in the half–plane $\{-\infty < x < \infty,\ y > 0\}$ which is compactly perturbed to form the domain $\Omega^* = \mathbf{R}^2_+ \setminus \Omega$. The scatterer Ω can be a compact body, a finite set of cracks in the plane, a finite set of supporting ribs, etc. For example, let the scatterer be the crack, $\Omega = \{|x| < a,\ y = 0\}$. We are looking for the solution of the following BVCP. Find the acoustic pressure $u : \Omega^* \to \mathbf{C}$ satisfying the Helmholtz equation

$$(\nabla^2 + k^2)u(\mathbf{r}) = 0,\ \mathbf{r} \in \Omega^*, \tag{22.1}$$

subject to the boundary conditions

$$\left(D\frac{\partial^4}{\partial x^4} - m\omega^2\right)\frac{\partial u}{\partial y}(x,0) + \rho\omega^2 u(x,0) = 0, \quad |x| > a, \; y = 0, \tag{22.2}$$

$$u(\mathbf{r}) = 0, \quad |x| < a, \; y = 0, \tag{22.3}$$

radiation conditions

$$u(\mathbf{r}) = \sqrt{\frac{2}{\pi k r}} e^{i(kr - \frac{\pi}{4})}\Psi_u(\varphi) + o(r^{-\frac{1}{2}}), \quad r \to \infty, \tag{22.4a}$$

uniformly with respect to $\varphi \in [0, \pi]$, and

$$u(\mathbf{r}) = A_u^{\pm} e^{\pm i\kappa x - \sqrt{\kappa^2 - k^2} y} + 0(1), \quad x \to \pm\infty, \tag{22.4b}$$

for any finite $y > 0$, and the contact conditions

$$\xi_u''(x) = \xi_u'''(x) = 0, \quad x = a^+, -a^-. \tag{22.5}$$

Here

$$\xi_u(x) = \frac{1}{\rho\omega^2}\frac{\partial u}{\partial y}(x, 0). \tag{22.6}$$

The parameters k, D, m, ρ, and ω characterize the model and are supposed to be positive. The function $\Psi_u(\varphi)$ is called the scattering amplitude, its analysis represents the central problem of the scattering theory [1,11]. The quantities $A_u^{\pm}$ are called the amplitudes of the surface waves. The parameter κ is the only positive solution of the dispersion equation

$$\left(D\kappa^4 - m\omega^2\right)\sqrt{\kappa^2 - k^2} - \rho\omega^2 = 0. \tag{22.7}$$

The direct analysis of the equation (22.7) implies the following result.

Lemma 1 $\kappa > \max\left(k, \left(m\omega^2/D\right)^{1/4}\right)$.

The field $u(\mathbf{r}) \in H^1(W_1)$ for any compact domain $W_1 \in \mathbf{R}_2^+$, whereas $\xi(x) \in H^2(W_2)$ for any compact domain $W_2 \in \mathbf{R} \setminus [-a, a]$ [19, 20].

Above, the statement of the homogeneous scattering problem is given because our main goal, the proof of the uniqueness theorem, requires the consideration of such a problem.

One of possible non–homogeneous problems is the scattering of the plane wave

$$u_i(\mathbf{r}) = B\exp(ik(x\cos\varphi_0 - y\sin\varphi_0)). \tag{22.8}$$

Here B is the given amplitude of the plane wave. If the plate were homogeneous, the total field would be just the sum of two plane waves

$$u_1(\mathbf{r}) = u_i(\mathbf{r}) + u_r(\mathbf{r}), \tag{22.9}$$

$$u_r(\mathbf{r}) = BR\exp(ik(x\cos\varphi_0 + y\sin\varphi_0)).$$

Here R is the reflection coefficient whose exact value can be found easily but is not important for us. In case of an arbitrary non–homogeneity, the field $u_1(\mathbf{r})$ is not the solution of the BVCP. For example, for the model above, $u_1(\mathbf{r})$ does not satisfy the conditions (22.3)-(22.5). In this case we postulate the following representation for the solution:

$$u(\mathbf{r}) = u_i(\mathbf{r}) + u_r(\mathbf{r}) + u_s(\mathbf{r}). \tag{22.10}$$

Here the scattering field $u_s(\mathbf{r})$ satisfies the radiation conditions (22.4a), (22.4b).

Note, the boundary conditions (22.2) and (22.3) are not the only possible once. The most general conditions would include differential operators of an arbitrary order. We choose the model (22.1)-(22.5) because it describes an important applied problem, i.e., the acoustic scattering from a thin elastic plate with a finite crack in it. For this model, the contact conditions (22.5) describe "free edges" of the plate. Other contact conditions are possible as well. If the BVCP in a finite domain is considered, they should guarantee selfadjointness of the corresponding operator [19].

Two basic problems appear in connection with any BVCP. Those are the **uniqueness** and **existence** of the solution.

Note immediately that the elastic part of the boundary (where the boundary condition (22.2) is stated) in our problem is non–compact. If it were compact, the uniqueness could be proved in the same way as for the Dirichlet and Neumann problems. Indeed, the proof for those problems is based on the separation of variables for the Helmholtz equation in the exterior of any circle, containing the scatterer, and completeness of the system of separated solutions $\{H_n^{(1)}(kr)\exp(in\varphi)\}$ there. For a non–compact elastic boundary, this proof is not applicable.

In the following, we need the "optical" theorem [10-13].

Lemma 2 *("optical" theorem). Let the scattering of a plane wave* (22.8) *be considered and the representation* (22.10) *for the field be postulated. Then the following identity holds for any solution of the BVCP* (22.1)-(22.10)

$$\frac{2\pi}{k}\int_0^\pi |\Psi_u(\varphi)|^2 d\varphi + \frac{\kappa}{2k\rho\omega^2}\Big(5\kappa^4 D - 4\kappa^2 k^2 D - m\omega^2\Big)\Big(|A_u^+|^2 + |A_u^-|^2\Big)$$

$$= -\frac{4\pi}{k}\Re\Big(\bar{R}B\Psi(\varphi_0)\Big). \tag{22.11}$$

The identity (22.11) is the energy conservation law for the model if the scattering of a plane wave (22.8) is considered. In case of the scattering of other fields, similar identities appear. They have the same left hand–side as (22.11), and the right–hand side is always linear with respect to the amplitude of the incident field.

The scattering amplitude $\Psi_u(\varphi)$ is defined for (real) angles $\varphi \in [0, \pi]$. It appears that the scattering amplitude allows the analytic continuation to the complex φ–plane and the amplitudes of the surface waves $A_u^{\pm}$ are closely connected with the amplitude [18].

Lemma 3 *The scattering amplitude* $\Psi_u(\varphi)$ *is a meromorphic function of* φ *with no poles on the real axis. Let* $\varphi_{\pm} = \cos^{-1}(\kappa/k)$. *The following identities hold*

$$A_u^{\pm} = \mp 2\pi i Res_{\varphi=\varphi_{\pm}} \Psi_u(\varphi). \tag{22.12}$$

Lemma 4 *Let* $u(\mathbf{r})$ *be a solution of a homogeneous BVCP* (22.1)-(22.6). *Then*

$$\Psi_u(\varphi) = 0, \ \varphi \in [0, \pi] \ \text{and} \ A_u^{\pm} = 0.$$

Proof There are no exterior sources for the homogeneous problem, and hence $B = 0$ in the identity (22.11). Lemma 1 shows that the expression $5\kappa^4 D - 4\kappa^2 k^2 D - m\omega^2$ is positive. Hence $\Psi_u(\varphi) = 0$ a.e. on $[0, \pi]$ and $A_u^{\pm} = 0$. Lemma 3 shows that in fact $\Psi_u(\varphi) = 0$ for any φ. The identity (22.12) completes the proof. □

Lemma 4 and the radiation conditions (22.4) imply the following asymptotic representations for the scattered field in case of the homogeneous BVCP:

$$u(\mathbf{r}) = o(r^{-\frac{1}{2}}), \ r \to \infty, \tag{22.13a}$$

uniformly with respect to $\varphi \in [0, \pi]$, and

$$u(\mathbf{r}) = o(1), \ x \to \pm\infty, \tag{22.13b}$$

for any finite y.

The scattering amplitude and the amplitudes of the surface waves are responsible for the energy transfer. We shall prove that their absence (see Lemma 4) implies the uniqueness, i.e., the identity $u(\mathbf{r}) = 0$.

22.3 UNIQUENESS OF THE SOLUTION FOR SOME PARTICULAR BVCP

Consider the particular case of the BVCP above, i.e., we assume that $a = 0$. In this case, the boundary condition (22.2) is satisfied for all $x \in \mathbf{R} \setminus \{0\}$ and the boundary condition (22.3) disappears. We will call this problem BVCP1. It models the scattering on a narrow crack in an elastic plate. In [2, 21] such boundary conditions are treated by adding a linear combination (with unknown factors) of the Dirac delta function and its derivatives to the right–hand side of (22.2):

$$\left(D\frac{\partial^4}{\partial x^4} - m\omega^2\right)\frac{\partial u}{\partial y}(x,0) + \rho\omega^2 u(x,0)$$

$$= \sum_{n\leq 3} c_n \delta^{(n)}(x), \ -\infty < x < \infty, \ y = 0. \tag{22.14}$$

Applying Fourier transform yields the explicit form of the solution for the problem (22.1), (22.4), (22.14) [2]

$$u(\mathbf{r}) = -\frac{1}{2\pi}\int_{\Lambda} \exp\Big(i\lambda x - \gamma(\lambda)\Big) \sum_{n\leq 3} \frac{c_n(i\lambda)^n}{l(\lambda)} d\lambda, \tag{22.15}$$

$$l(\lambda) = \Big(D\lambda^4 - m\omega^2\Big)\gamma(\lambda) - \rho\omega^2, \quad \gamma(\lambda) = \sqrt{\lambda^2 - k^2}.$$

Here c_n, $n \leq 3$ are arbitrary constants to be determined from the contact conditions (22.5) (note, $a = 0$ for the BVCP1). The function $\gamma(\lambda)$ is defined on the complex λ–plane cut along the vertical lines $\lambda = k + it$ and $\lambda = k - it$, $t \in [0, \infty)$, and $\gamma(\lambda) > 0$ for $\lambda > k$. The integrand in (22.15) is analytic on this plane except the finite set of poles located outside Λ and the segment $[-k, k]$. The contour of integration Λ coincides with the real axis except small semicircles around the real poles $\lambda = \pm\kappa$.

The representation (22.15) contains arbitrary constants c_n, $n \leq 3$, linearly. Applying the contact conditions (22.5) yields the 4×4 homogeneous linear algebraic system for the constants c_n, $n \leq 3$. The problem of solvability of this system is closely connected to the problem of uniqueness. The determinant of the system is expressed in terms of some special integrals, and its direct analysis is not simple. We shall show in this section, that the determinant is distinct from zero and the problem has a unique solution. Lemma 4, i.e., the absence of the scattering amplitude and the amplitudes of the surface waves in the solution of any homogeneous BVCP, is very important for our proof.

Lemma 5 *The determinant of the linear algebraic system for constants c_n, $n \leq 3$ is not equal to zero.*

Proof Let the determinant be zero. Then there exists a non–trivial solution of the homogeneous algebraic system for constants c_n. Analytic properties of the function $l(\lambda)$ allow one to apply the method of steepest descents [22] to the integral (22.15). As a result, asymptotic representation (22.4a) appears with the following scattering amplitude:

$$\Psi_u(\varphi) = k \sin\varphi \sum_{n\leq 3} \frac{c_n(k\cos\varphi)^n}{l(k\cos\varphi)}. \tag{22.16}$$

Since not all the constants c_n are equal to zero, the scattering amplitude is not identically zero. This result contradicts Lemma 4. Thus, the determinant is not equal to zero. □

Theorem 1 *The BVCP1 has a unique solution.*

Proof Lemma 5 implies solvability of the linear algebraic system for any right–hand side. In particular, there exist (at least one!) Green's function $G(\mathbf{r}, \mathbf{r}_0)$ satisfying all the boundary and contact conditions, the radiation conditions, and the non–homogeneous Helmholtz equation

$$(\nabla^2 + k^2)G(\mathbf{r}, \mathbf{r}_0) = \delta(\mathbf{r} - \mathbf{r}_0), \quad \mathbf{r} \in \Omega^*, \quad \mathbf{r}_0 \in \Omega^*. \tag{22.17}$$

Here essential use is made of the fact that we are able to construct Green's function for the BVCP1 explicitly and thus prove its existence (but not uniqueness!). Applying Green's second formula to the functions $u(\mathbf{r})$ and $G(\mathbf{r})$ in the domain $\Omega \cap \{r < R\}$ yields

$$u(\mathbf{r}) = R \int_0^\pi \left(\frac{\partial G}{\partial r_0} u - \frac{\partial u}{\partial r_0} G \right) \Big|_{r_0 = R} d\varphi_0$$

$$+ D(g\xi_u''' - \xi_u g''' - g'\xi_u'' + \xi_u' g'')\Big|_{x_0=-R}^{x_0=R}. \tag{22.18}$$

Here $g(\mathbf{r}, x_0) = -\frac{\partial G}{\partial y_0}(\mathbf{r}, x_0, 0)$. The radiation condition (22.4a) for the field G implies $g(\mathbf{r}, \pm R) = O(1)$ as $R \to \infty$. The representation (22.18) is derived, for example, in [23]. Using radiation conditions for the fields G and $u(\mathbf{r})$ (see (22.4) and (22.13) correspondingly) and letting $R \to \infty$ yields from (22.18) $u(\mathbf{r}) = 0$. Hence, the BVCP1 has a unique solution. □

22.4 SOMMERFELD'S FORMULA

The proof of uniqueness in the previous section was based on the closed form solution for the scattering problem BVCP1. The closed form solution can be constructed for some specific BVCP only. For some other models, the scattering problem can be reduced to a Fredholm type equation in an appropriate functional space, e.g., [7-9]. For those problems, the proof of uniqueness requires a new method to be used. Lemma 4 shows that any solution of the homogeneous BVCP does not contain the scattering amplitude and the surface waves. We shall construct the representation for the scattered field which contains the scattering amplitude linearly. Such a representation obviously implies uniqueness.

The representation of this type is known for the problem of scattering from a compact obstacle as Sommerfeld's formula [14, 15]. We derive here the similar formula for the BVCP (22.1)-(22.6). Let the scattering of the plane wave (22.8) be considered and the representation (22.10) be postulated. Then the scattered field $u_s(\mathbf{r})$ satisfies the equation (22.1), the homogeneous boundary conditions (22.2) and (22.6), the radiation conditions (22.4), the boundary condition

$$u(\mathbf{r}) = f(\mathbf{r}), \quad |x| < a, \; y = 0, \tag{22.3a}$$

and the contact conditions

$$\xi_u''(x) = p^\pm, \; \xi_u'''(x) = q^\pm, \quad x = a^+, -a^-. \tag{22.5a}$$

Here the function $f(\mathbf{r})$ and the numbers $p^\pm$, $q^\pm$ are known, but their exact form is not important for us.

In what follows, we use two special functions which describe oscillations of an acoustic medium in presence of an infinite elastic plate without any non-homogeneities. Spatial Green's function $G_0(\mathbf{r}, \mathbf{r}_0)$ is defined as the solution of the Helmholtz equation with the delta function in the right-hand side subject to the homogeneous boundary condition, whereas surface Green's function

$g_0(\mathbf{r}, x_0)$ is the solution of the homogeneous Helmholtz equation subject to the boundary condition with the delta function in the right–hand side (see Appendix).

Applying Green's second formula to the functions $u(\mathbf{r})$ and $G_0(\mathbf{r}, \mathbf{r}_0)$ in the domain $\mathbf{R}^2_+ \cap \{r < R\}$ with sufficiently large R, using all boundary, contact, and radiation conditions, and letting $R \to \infty$ yields the representation

$$u(\mathbf{r}) = \int_{-a}^{a} E(\mathbf{r}_0) G_0(\mathbf{r}, \mathbf{r}_0)\Big|_{y_0=0} dx_0 + F(x_0) g_0(\mathbf{r}, x_0)\Big|_{-a}^{a}. \tag{22.19}$$

Here E and F are the following differential operators

$$E(\mathbf{r}_0) = -f(\mathbf{r})\frac{\partial}{\partial y_0} + \frac{\partial u}{\partial y_0},$$

$$F(x_0) = D(\xi_u'''(x_0) - \xi_u''(x_0)\frac{\partial}{\partial x_0} + \xi_u'(x_0)\frac{\partial^2}{\partial x_0^2} - \xi_u(x_0)\frac{\partial^3}{\partial x_0^3}). \tag{22.20}$$

The representation (22.19) and the radiation conditions for Green's functions $G_0(\mathbf{r}, \mathbf{r}_0)$ and $g_0(\mathbf{r}, x_0)$ allow one to justify the radiation conditions (22.4) and to find the following expression for the scattering amplitude

$$\Psi_u(\varphi) = \int_{-a}^{a} E(\mathbf{r}_0)\Psi_{G_0}(\varphi, \mathbf{r}_0)\Big|_{y_0=0} dx_0 + F(x_0)\Psi_{g_0}(\varphi, x_0)\Big|_{-a}^{a}. \tag{22.21}$$

The scattering amplitudes in both left–hand side and right–hand side of the identity (22.21) are defined on $\varphi \in (0, \pi)$. Lemma 3 and analytic properties of scattering diagrams $\Psi_{G_0}(\varphi, \mathbf{r}_0)$ and $\Psi_{g_0}(\varphi, x_0)$ show that the identity (22.21) is valid on the complex φ–plane except some complex poles.

In Appendix, we construct the representation (22.38a) for Green's functions $G_0(\mathbf{r}, \mathbf{r}_0)$ and $g_0(\mathbf{r}, x_0)$ in terms of their scattering diagrams. Substituting (22.38a) into (22.19) and changing the order of integration yields

$$u(\mathbf{r}) = -\int_C e^{ikr\cos(\varphi-\psi)}\Big(\int_{-a}^{a} E(\mathbf{r}_0)\Psi_{G_0}(\psi, \mathbf{r}_0)\Big|_{y_0=0} dx_0 + F(x_0)\Psi_{g_0}(\psi, x_0)\Big|_{-a}^{a}\Big) d\psi. \tag{22.22}$$

The identities (22.21) and (22.22) imply

$$u(\mathbf{r}) = -\int_C e^{ikr\cos(\varphi-\psi)}\Psi_u(\psi)d\psi. \tag{22.23}$$

The identity (22.23) represents the desired connection between the scattered field and the analytic continuation of the scattering amplitude for this field. Note, this connection for Green's functions is given by (22.38a).

22.5 UNIQUENESS OF THE SOLUTION FOR THE BVCP

Sommerfeld's formula allows the uniqueness of the solution for a general boundary value contact problem to be proved.

Theorem 2 *(Uniqueness theorem) The BVCP* (22.1), (22.2), (22.3*a*), (22.4), (22.5*a*), (22.6) *has a unique solution.*

Proof If two different solutions for the problem exist, then their difference $u(\mathbf{r})$ solves the homogeneous problem (22.1)-(22.6). Lemma 4 implies that the scattering amplitude $\Psi_u(\varphi)$ is equal to zero on $[0, \pi]$. According to Lemma 3, the scattering amplitude is a meromorphic function of φ. Hence, $\Psi_u(\varphi) = 0$ identically. Sommerfeld's formula (22.23) implies that $u(\mathbf{r}) = 0$ identically. Thus the solution of the BVCP is unique. □

Note, the uniqueness theorem for the BVCP can be also proved in case of a more complex scatterer than a crack considered here. For example, it can be a compact body attached to the plate with a minimum requirements on its boundary (Green's formula should be applicable). Also, other contact conditions than (22.5) can be stated. It can be also shown that the uniqueness theorem is valid if the Kirchhoff model of the plate (boundary condition (22.2)) is substituted for Timoshenko—Mindlin model [17].

22.6 EXISTENCE OF THE SOLUTION FOR THE BVCP

The scattering of the plane wave from the crack in an infinite elastic plate was considered in [7]. The uniqueness of the solution for this model follows from the results of Section 22.5. The existence is proved here by means of the Fredholm alternative and Galerkin method.

The notation $\mathbf{C}[-a, a]$ is used below for the set of complex–valued functions $f : [-a, a] \to \mathbf{C}$.

In [7], the BVCP (22.1), (22.2), (22.3a), (22.4), (22.5a), (22.6) is reduced to the integro–algebraic (IA) system

$$G * \phi + T_1\mathbf{c} = f,$$

$$T_2\phi + T_3\mathbf{c} = \mathbf{v}. \tag{22.24}$$

Here the function $\phi \in \mathbf{C}[-a, a]$ and the vector $\mathbf{c} = [c_1, c_2, c_3, c_4]^T \in \mathbf{C}^4$ are unknown. The function $f \in \mathbf{C}[-a, a]$ and the vector $v \in \mathbf{C}^4$ are given. The operators in (22.24) are acting as follows:

$$(G * \phi)(x) = \int_{-a}^{a} G(x - t)\phi(t)dt,$$

$$T_1\mathbf{c} = \sum_{m=1}^{m=4} \eta_{1,m}(x)c_m, \quad (T_2\phi)_m = \int_{-a}^{a} \eta_{2,m}(t)\phi(t)dt \quad m = 1, 2, 3, 4,$$

and T_3 is 4×4 complex–valued matrix. The functions G, $\eta_{p,m} \in \mathbf{C}[-a,a]$ are given. The IA system is a finite–dimensional perturbation of the convolution type integral equation with the kernel G.

It can be shown that the IA system (22.24), in general, is not equivalent to the BVCP (22.1), (22.2), (22.3a), (22.4), (22.5a), (22.6). Hence, though the BVCP has a unique solution, the solution of the IA system might be non–unique.

Lemma 6 *There exists a set of parameters $k, D, \rho,$ and ω such that the solution of the IA system* (22.24) *is not unique.* We do not consider this specific case of non–uniqueness in our brief text.

Assumption 1 *The solution of the IA system* (22.24) *is unique.*

Some standard notations should be introduced. $F[v]$ denotes the Fourier transform of a function v; x and λ are dual variables. $H^s(\mathbf{R})$ is the set of all distributions $u : \mathbf{R} \to \mathbf{C}$ such that $F[u] \in L^2_{loc}(\mathbf{R})$ and $(1+|\lambda|)^s F[u] \in L^2(\mathbf{R})$. Further, $C^{k+\alpha}(\mathbf{R})$ with an integer positive k and an $\alpha \in (0,1)$ is the set of all $u \in C^k(\mathbf{R})$ such that $u^{(k)}$ is Hőlder continuous with the index α. The following functional space is very important for our considerations.

Definition 1 $A_{\delta,\alpha} = \{\phi : [-a,a] \to \mathbf{C} \mid \phi(t) = \eta(t)/(a^2-t^2)^\delta,\ \eta \in C^\alpha[-a,a]\}$ *with some $\delta, \alpha \in (0,1)$. It is assumed everywhere that η is continued by zero on $|t| > a$.*

Lemma 7 *(The structure of the kernel) The following representations and inclusions hold*

a. $G(x) = C_1 \ln|x| + G_0(x)$ *where* $G_0 \in C^{1+\alpha}(\mathbf{R})$.

b. $G_0''(x) \in L^p(D)$ *for any* $p > 0$ *and any compact set* $D \in \mathbf{R}$.

c. $G = C_2 \cos \kappa x + v.p. \int_R F[G](\lambda) e^{i\lambda x} d\lambda$.

Proof is based on the analysis of $F[G](\lambda)$and requires the special consideration of the real poles $\pm\kappa$. □

The next four lemmas describe some properties of the class $A_{\delta,\alpha}$.

Lemma 8 *Let* $\phi \in A_{\frac{1}{2},\alpha}$, *then*

$$F[\phi] = \sqrt{\frac{\pi}{2\lambda a}}(e^{i\lambda a}\eta(a) + e^{-i\lambda a}\eta(-a)) + O(\lambda^{-1}) \text{ as } \lambda \to \pm\infty. \qquad (22.25)$$

Proof can be found in [22]. □

Lemma 9 *Let* $\phi \in A_{\delta,\alpha}$, *then* $F[\phi]$ *is an entire analytic function of* $\lambda \in \mathbf{C}$.

Proof This is Wiener–Paley–Schwartz theorem (see [25]). □

Lemma 10 *Let* $\phi \in A_{\delta,\alpha}$, *then* $l(x) = \int_{-a}^{a} \ln|x-t|\phi(t)dt \in L^1_{loc}(\mathbf{R})$.

Lemma 11 *Let* $\phi \in A_{\delta,\alpha}$, *then* $\int_{R\setminus[-\mu,\mu]} |F[\phi]|^2 \frac{d\lambda}{|\lambda|} < \infty$ *for any* $\mu > 0$.

The proof is based on Haussdorff–Young theorem [25] and lemmas 8 and 9. □

Lemma 12 *(Refinement of the properties of the solution) If $f \in C^{1+\varepsilon}(\mathbf{R})$ and the IA system* (22.24) *has a solution $\phi \in L^p[-a,a]$ for some $p \in (1,2)$, then this solution is "better", i.e., $\phi \in A_{\delta,\alpha}$, $\delta = \frac{1}{2}$, $\alpha = \min(\frac{1}{2}, \varepsilon)$.*

Proof Considering first the properties of the convolution $\Phi = G_0 * \phi$ and using the previous lemmas yield $\Phi \in C^{1+\varepsilon}(R)$. We now have the equation with logarithmic kernel and can use the well-known results on this equation and analytic properties of Cauchy principal value integrals (see [26]). □

The solution of the IA system can be constructed by means of Galerkin method [7]. Lemma 12 implies the following structure of the solution:

$$\phi(t) = \frac{\eta(t)}{\sqrt{a^2 - t^2}} \tag{22.26}$$

with an unknown function $\eta(t) \in C^{1+\alpha}[-a,a]$. We look for this function in the form of the series with respect to Chebyshev polynomials

$$\eta(t) = \sum_{n \geq 0} p_n T_n(t/a). \tag{22.27}$$

For the unknown coefficients p_n and $\mathbf{c}$ the linear algebraic system (LS) appears

$$\sum_{n \geq 0} B^1_{mn}\, p_n + \sum_{n=1}^{4} B^2_{mn}\, c_n = f_m,\ m \geq 0, \tag{22.28}$$

$$\sum_{n \geq 0} B^3_{mn}\, p_n + \sum_{n=1}^{4} B^4_{mn}\, c_n = v_m,\ 1 \leq m \leq 4.$$

For what follows, the matrix $\{B^1_{mn}\}$ and the vector $\{f_m\}$ are especially important for us. They are given by the following integrals:

$$B^1_{mn} = \int_{-a}^{a}\int_{-a}^{a} G(x-t)\frac{T_m(x/a)}{\sqrt{a^2-x^2}}\frac{T_n(t/a)}{\sqrt{a^2-t^2}}dxdt, \tag{22.29a}$$

$$f_m = \int_{-a}^{a} f(x)\frac{T_m(x/a)}{\sqrt{a^2-x^2}}dx. \tag{22.29b}$$

We should imbed the LS (22.28) in an appropriate Hilbert space. The LS (22.28) is the finite-dimensional perturnation of the linear system with the matrix $\{B^1_{mn}\}$. Below, we ignore details caused by that perturbation.

Lemma 13 *(Asymptotic behavior of the vector $\mathbf{f} = \{f_m\}$)*

a. *If $f \in C^\infty[-a,a]$, then $f_m = O(m^{-N})$ for any $N > 0$ as $m \to \infty$.*

b. *If $f \in L^2_{1/\sqrt{a^2-x^2}}$, then $\mathbf{f} \in l^2$.*

c. *If $f \in C^N[-a,a]$ with some integer $N > 0$, then $f_m = O(m^{-N})$ as $m \to \infty$.*

d. *If $f' \in L^p[-a,a]$ with some $p > 4$, then $\{m f_m\} \in l^2$.*

Proof is based on integration by parts in the representation (22.29b), Parseval's identity, and Hölder's inequality. □

Lemma 7 implies the decomposition of the matrix $\mathbf{B} = \{B^1_{mn}\}$, $\mathbf{B} = \mathbf{U} + \mathbf{V}$ with $\mathbf{U}$ and $\mathbf{V}$ originating from the first and second terms of the representation of G given by that Lemma. The identity (see [27])

$$-\frac{1}{\pi}\int_{-1}^{1} \ln|x-t| \frac{T_n(t)}{\sqrt{1-t^2}} dt = \gamma_n T_n(x),$$

$$\gamma_0 = \ln 2, \; \gamma_n = \frac{1}{n}, \; n > 0$$

shows that Chebyshev polynomials are eigenfunctions of the integral operator with logarithmic kernel. Hence, the matrix $\mathbf{U}$ is diagonal and can be inverted. This allows to rewrite the LS (22.28) as a finite–dimensional perturbation of the LS

$$p_m + \sum_{n>0} W_{mn}\, p_n = F_m, \; m > 0. \tag{22.30}$$

Lemma 14 *The LS* (22.29) *can be considered as an equation of the Hilbert–Schmidt type in* l^2*,i.e., the matrix* $\mathbf{W} = \{W_{mn}\}$ *generates the Hilbert– Schmidt operator in* l^2 *and the vector* $\mathbf{F} = \{F_m\} \in l^2$.

Proof is based on integration by parts in the representation (22.29a), Parseval's identity, and the smoothness properties for the function G_0 (see Lemma 7). □

Lemma 15 *The solution of the LS* (22.28) *in* l^2 *exists.*

Proof is based on the Fredholm alternative and Assumption 1. □

Theorem 3 *(Existence theorem) Let* $f \in C^{1+\varepsilon}[-a,a]$. *Then the solution of the IA system* (22.24) *from the class* $A_{\delta,\alpha}$ *with* $\alpha = \min(\frac{1}{2}, \varepsilon)$ *exists and is unique.*

In [7], the solution $u(x,y)$ of the BVCP (22.1), (22.2), (22.3a), (22.4), (22.5a), (22.6) is constructed in terms of the solution of the IA system (22.24). We finally find the following result.

Theorem 4 *The solution* $u(x,y)$ *for the problem of scattering of a plain incident wave, constructed by means of the solution* $\{\phi(x),\ \mathbf{c}\}$ *of the IA system* (22.24), *satisfies* **pointwise** *Helmholtz equation* (22.1), *boundary conditions* (22.2), (22.3a), *radiation conditions* (22.4), *contact conditions* (22.5a), (22.6). *Also,* $u(\mathbf{r}) \in H^1(W_1)$ *for any compact domain* $W_1 \in \mathbf{R}^+_2$, *whereas* $\xi(x) \in H^2(W_2)$ *for any compact domain* $W_2 \in \mathbf{R} \setminus [-a,a]$.

22.7 APPENDIX: GREEN'S FUNCTIONS

Consider an acoustic medium in presence of an infinite elastic plate without any non–homogeneities. Spatial Green's function $G_0(\mathbf{r}, \mathbf{r}_0)$ for this model is defined as the solution of the non–homogeneous Helmholtz equation

$$(\nabla^2 + k^2) G_0(\mathbf{r}, \mathbf{r}_0) = \delta(\mathbf{r} - \mathbf{r}_0), \; \mathbf{r}, \mathbf{r}_0 \in \mathbf{R}^2_+ \tag{22.31}$$

subject to the boundary condition

$$\left(D\frac{\partial^4}{\partial x^4}-m\omega^2\right)\frac{\partial G_0}{\partial y}(x,0)+\rho\omega^2 G_0(x,0)=0,\quad -\infty<x<\infty,\ y=0\,. \tag{22.32}$$

Surface Green's function $g_0(\mathbf{r},x_0)$ is the solution of the homogeneous Helmholtz equation subject to the boundary condition (22.31) with $\delta(x-x_0)$ instead of zero in the right hand–side and radiation conditions. Their explicit representation is well known (e.g, [2]):

$$G_0(\mathbf{r},\mathbf{r}_0)=-\frac{1}{4\pi}\int_\Lambda \frac{1}{\gamma(\lambda)}e^{i\lambda(x-x_0)}\left(e^{-\gamma(\lambda)|y-y_0|}+\frac{l_0(\lambda)}{l(\lambda)}e^{-\gamma(\lambda)(y+y_0)}\right)d\lambda, \tag{22.33a}$$

$$g_0(\mathbf{r},x_0)=-\frac{\rho\omega^2}{2\pi}\int\frac{1}{l(\lambda)}e^{i\lambda(x-x_0)-\gamma(\lambda)y}d\lambda. \tag{22.33b}$$

The definition of the contour Λ and functions $\gamma(\lambda)$ and $l(\lambda)$ is given in Section 22.3. Also, $l_0(\lambda)=l(\lambda)+2\rho\omega^2$. Note the elementary identity

$$g_0(\mathbf{r},x_0)=-\frac{\partial G_0}{\partial y_0}(\mathbf{r},x_0). \tag{22.34}$$

Both Green's functions satisfy radiation conditions similar to (22.4). Applying the method of steepest descents to the representations (22.33a,b) yields the following expressions for scattering amplitudes:

$$\Psi_{G_0}(\varphi,\mathbf{r}_0)=\frac{1}{4\pi i}e^{-ikx_0\cos\varphi}(e^{-iky_0\sin\varphi}+\frac{l_0(k\cos\varphi)}{l(k\cos\varphi})e^{iky_0\sin\varphi}), \tag{22.35a}$$

$$\Psi_{g_0}(\varphi,x_0)=-\frac{\rho\omega^2}{2\pi}e^{-ikx_0\cos\varphi}\frac{k\sin\varphi}{l(k\cos\varphi)}. \tag{22.35b}$$

Another representation for Green's functions appears by introducing the new variable $\lambda=k\cos\psi$:

$$G_0(\mathbf{r},\mathbf{r}_0)=-\frac{1}{4\pi i}\int_C e^{ikr\cos(\varphi-\psi)}e^{-ikx_0\cos\psi}\Big(e^{-iky_0\sin\varphi}$$

$$+\frac{l_0(k\cos\psi)}{l(k\cos\psi}\big)e^{iky_0\sin\psi}\Big)d\psi, \tag{22.36a}$$

$$g_0(\mathbf{r},x_0)=\frac{\rho\omega^2}{2\pi}\int_C e^{ikr\cos(\varphi-\psi)}e^{-ikx_0\cos\psi}\frac{k\sin\psi)}{l(k\cos\psi)}d\psi. \tag{22.36b}$$

Note, the representation (22.36a) is valid for $y>y_0$. The contour C coincides with the contour $C^*=(\pi-i\infty,\pi)\cup(\pi,0)\cup(0,i\infty)$ except small neighborhoods of the points (see (22.12))

$$\varphi_+=\cos^{-1}\left(\frac{\kappa}{k}\right)=i\ln\left(\frac{\kappa}{k}+\sqrt{\left(\frac{\kappa}{k}\right)^2-1}\right),\ \varphi_-=\pi-\varphi_+.$$

It avoids $\varphi_\pm$ by following small semicircles centered at those points. The semicircles are located to the right of φ_+ and to the left of φ_-. The contour C^* was introduced by Sommerfeld (e.g., [14]).

The identity (22.34) implies

$$\Psi_{g_0}(\varphi, x_0) = -\frac{\partial \Psi_{G_0}}{\partial y_0}(\varphi, x_0, 0). \tag{22.37}$$

The scattering amplitudes $\Psi_{G_0}(\varphi, \mathbf{r}_0)$ and $\Psi_{g_0}(\varphi, x_0)$ are defined for all (real) angles $\varphi \in [0, \pi]$. The direct analysis of (22.35) implies that those amplitudes allow the analytic continuation for the complex φ–plane as meromorphic functions. The points $\psi = \varphi_\pm$ are the only poles located on the contour C^*. The direct comparison of the representations (22.35) and (22.36) implies the formulas

$$G_0(\mathbf{r}, \mathbf{r}_0) = -\int_C e^{ikr\cos(\varphi-\psi)}\Psi_{G_0}(\psi, \mathbf{r}_0)d\psi, \tag{22.38a}$$

$$g_0(\mathbf{r}, x_0) = -\int_C e^{ikr\cos(\varphi-\psi)}\Psi_{g_0}(\psi, x_0)d\psi. \tag{22.38b}$$

The formulas (22.38) represent Green's functions in terms of analytic continuation of their scattering amplitudes.

References

[1] Colton, D., Kress, R.: Integral equation methods in scattering theory. A Wiley–Interscience Publication, John Wiley & Sons, New York, 1983.

[2] Kousov, D.P.: On a solution of the H equation for a half–plane with boundary conditions containing high order derivatives. J. Appl. Math. Mech. 31 (1967), 164–170.

[3] Rebinsky, D.A., Norris, A.N.: Acoustic and flexural wave scattering from a three–member junction. J. Acoust. Soc. America 98 (1995), 3309–3319.

[4] Photiadis, D.M.: The effect of irregularity on the scattering of acoustic waves from a ribbed plate. J. Acoust. Soc. America 91 (1992), 1897–1903.

[5] Engineer, J.G., Abrahams, I.D.: The radiation of sound waves from a lightly loaded finite elastic shell, I: duct and shell resonances for linear motions. J. Sound Vibration 153 (1992), 291–310.

[6] Kriegsmann, G.A.: Acoustic pulse scattering by baffled membranes. J. Acoust. Soc. America 79 (1986), 1–8.

[7] Andronov, I.V., Belinskiy, B.P.: Acoustic scattering by a narrow crack in an elastic plate. Wave Motion 24 (1996), 101–115.

[8] Belinskiy, B.P.: Sound radiation from a plate with protruding stiffener. Soviet Physics–Acoustics 24 (1978), 183–187.

[9] Andronov, I.V.: The propagation of sound in a fluid under the elastic plate with a crack. J. Appl. Math. Mech. 57 (1993), 363–369.

[10] Reed, M., Simon, B.: Methods of modern math. physics 3: Scattering theory. Academic Press, New York, 1979.

[11] Lax, P.D., Phillips, R.S.: Scattering Theory. Academic Press, New York, 1967.

[12] Belinskiy, B.P., Kousov, D.P.: The optical theorem for the plate–liquid system. Soviet Physics–Acoustics 26 (1980), 8–11.

[13] Kriegsmann, G.A., Norris, A.N., Reiss, E.L.: An "optical" theorem for acoustic scattering by baffled flexible surfaces. J. Sound Vibration 99 (1985), 301–307.

[14] Sommerfeld, A.: Vorlesungen über theoretische Physik VI. Dieterich, Wiesbaden, 1950.

[15] Apelcin, V.F., Kyurkchan, A.G.: Analytical properties of wave fields. Moscow State Univ., Moscow, 1990 (Russian).

[16] Andronov, I.V., Belinskiy, B.P.: Sommerfeld's formula and uniqueness for the boundary value contact acoustic problems. Preprint, 1997.

[17] Andronov, I.V., Belinskiy, B.P.: Acoustic scattering on an elastic plate described by the Timoshenko model: contact conditions and the uniqueness of the solution. J. Acoust. Soc. America, to appear.

[18] Andronov, I.V., Belinskiy, B.P., Dauer, J.P.: The connection between the scattering diagram and the amplitudes of the surface waves for acoustic scattering by a baffled flexible plate. J. Sound Vibration 195 (1996), 667–673.

[19] Belinskiy, B.P.: Fredholm property of boundary–contact problems of acoustics. Soviet Phys. Doklady 29 (1984), 797–799.

[20] Belinskiy, B.P.: Boundary–contact problem of acoustics in an infinite domain. Soviet Phys. Doklady 30 (1985), 560–562.

[21] Andronov, I.V.: Application of zero–range models to the problem of diffraction by a small hole in elastic plate. J. Math. Phys. 34 (1993), 2226–2241.

[22] Erdélyi, A.: Asymptotic expansions. Dover Publ., Inc., New York, 1956.

[23] Belinskiy, B.P., Kousov, D.P.: Green's type formulas for flexurally vibrating plates. Soviet Physics–Acoustics 27 (1981), 394–398.

[24] Taylor, M.E.: Pseudodifferential operators. Princeton Univ. Press, Princeton, 1981.

[25] Hörmander, L.: The analysis of linear partial differential operators, I. Distribution theory and Fourier analysis. Springer–Verlag, New York, 1983.

[26] Gakhov, F.D.: Boundary value problems. Nauka, Moscow, 1977 (Russian); Pergamon, Oxford, 1966.

[27] Abramowitz, M., Stegun, I.A.: Handbook of mathematical functions with formulas, graphs, and mathematical tables. National Bureau of Standards. Applied Mathematics Series 55, U.S. Government Pr. Office, Washington, D.C., 1964.

23 WEAKLY HYPERBOLIC EQUATIONS — A MODERN FIELD IN THE THEORY OF HYPERBOLIC EQUATIONS

M. Dreher
M. Reissig

Faculty for Mathematics and Computer Sciences
Freiberg University of Mining and Technology
Bernhard–von–Cotta–Str. 2
09596 Freiberg, Germany

(dreher,reissig)@mathe.tu-freiberg.de

Abstract: In the last two decades the theory of weakly hyperbolic equations was developed in an astonishing way. The goal of the paper is to give new directions which were discussed in recent years. In section 2 the Cauchy problem is discussed for a model quasilinear weakly hyperbolic equation with time and spatial degeneracy. Sharp Levi conditions, i.e. relations between the lower order terms and the main part, are supposed. The tools are classical ones (energy method, Gronwall type inequalities, fixed point arguments), but one has to find essential modifications of them to prove the existence of Sobolev solutions. Section 3 is devoted to the discussion of oscillations in the coefficients. The above mentioned tools are not applicabel. Only very careful microlocal considerations allow to characterize suitable function spaces which give a so-called strictly hyperbolic type property for corresponding linear equations. Open problems motivate further considerations.

23.1 INTRODUCTION

In this paper we give an overview of problems and new trends in the theory of *weakly hyperbolic equations.*

We start with the Cauchy problem

$$u_{tt} - a(t)u_{xx} = 0, \quad u(x,0) = u_0(x), \quad u_t(x,0) = u_1(x).$$

H.G.W. Begehr et al. (eds.), Partial Differential and Integral Equations, 303–318.

In the strictly hyperbolic theory ($a(t) \geq c > 0$) we have the correctness in the classes of Sobolev spaces and the estimate

$$\sup_{[0,T]} \left(\|u(t)\|_{H^{m+1-\beta t}} + \|u'(t)\|_{H^{m-\beta t}} \right) \\ \leq C_0(1 + e^{C_m T}) \left(\|u_0\|_{H^{m+1}} + \|u_1\|_{H^m} \right)$$

for a log–Lipschitz coefficient $a(t)$, see [2].

In the weakly hyperbolic theory ($a(t) \geq 0$) the situation is completely different. C^∞–regularity of $a(t)$ is not sufficient for well–posedness of the Cauchy problem even in the classes of distributions. In [3] it was shown that for any $T > 0$ a coefficient $a(t) \in C^\infty([0,\infty))$ and C^∞ data u_0, u_1 exist, such that the solution belongs to $C^\infty([0,T), C^\infty(\mathbb{R}))$, but not to $C([0,T], \mathcal{D}'(\mathbb{R}))$. This function $a(t)$ is positive for $t < T$, oscillating for $t \to T - 0$ and identically zero for $t \geq T$.

We list three ways to restrict the influence of *oscillations*:

- One assumes a nonlocal condition of the form

  $$\int_0^T \frac{|a'(\tau)|}{a(\tau) + \varepsilon} d\tau \leq C |\ln \varepsilon|,$$

  see [1].

- The coefficient $a(t)$ is assumed to be analytic, see [4].

- One supposes $a = a(x,t) = \sigma^2(x)\lambda^2(t)$ with $\lambda(0) = 0$, $\lambda'(t) > 0$, see [6].

In opposite to the strictly hyperbolic case, lower order terms have an influence on the C^∞–well–posedness, too. From [6] we know that the Cauchy problem for

$$u_{tt} - t^{2l} x^{2n} u_{xx} + t^k x^m u_x = 0$$

is C^∞– well–posed if $k \geq l - 1$, $m \geq n$. The necessity of this conditions was proved in [5]. These conditions are called *Levi conditions of C^∞-type* which are used to ensure the well–posedness in C^∞. This example shows that one has to distinguish between Levi conditions for time degeneracy and Levi conditions for spatial degeneracy.

One possibility to formulate Levi conditions of C^∞–type for the equation

$$u_{tt} - \sigma^2(x)\lambda^2(t)u_{xx} + b(x,t)u_x + c(x,t)u_t + d(x,t)u = g(x,t) \tag{23.1}$$

is to assume

$$|b(x,t)| \leq C|\sigma(x)|\lambda'(t),$$

compare with $\sigma(x) = x^n$, $\lambda(t) = t^l$, $b(x,t) = t^k x^m$.

Now let us turn to quasilinear problems, e.g.

$$u_{tt} - \sigma^2(x)\lambda^2(t)u_{xx} = f(x,t,u,u_t,u_x). \tag{23.2}$$

In this case we have the Levi condition of C^∞–type

$$\left|\frac{\partial f}{\partial p}(x,t,u,v,p)\right| \le C|\sigma(x)|\lambda'(t).$$

In the case of weakly hyperbolic equations another phenomenon occurs – the *loss of Sobolev regularity*, even if the coefficients are smooth and have no oscillations. Let us consider the Cauchy problem

$$u_{tt} - t^2 u_{xx} = a u_x, \quad u(x,0) = \varphi(x), \quad u_t(x,0) = 0.$$

As it was shown in [7], the solution can be represented in the form

$$u(x,t) = \sum_{k=0}^{n} c_k t^{2k} \partial_x^k \varphi\left(x + \frac{t^2}{2}\right)$$

if $a = 4n+1$, $n \in \mathbb{N}_0$. In other words, if $\varphi \in H^s$, then $u(.,t) \in H^{s-(a-1)/4}$.

Typical problems for the linear equation (23.1) are:

- well–posedness of the Cauchy problem in Sobolev spaces, C^∞ or Gevrey spaces; well–posedness means existence and cone of dependence as usual;
- propagation of singularities;
- construction of parametrix;
- influence of oscillations in the coefficients.

Typical problems for the quasilinear equation (23.2) are:

- local existence of solutions in Sobolev spaces, C^∞ or Gevrey spaces;
- uniqueness of solutions of the Cauchy problem and existence of a cone of dependence;
- global regularity of solutions of the Cauchy problem;
- stability of solutions.

A natural way to attack the problem to prove the existence of Sobolev solutions for the Cauchy problem for (23.2) is to linearize the equation and to apply fixed points arguments. A so–called *strictly hyperbolic type estimate* for the linearized equation is essential for this purpose. That means:

It exists a Banach space B such that the following holds:

If $f(x,t)/\lambda'(t) \in B$, then the solution of

$$u_{tt} - \sigma^2(x)\lambda^2(t) u_{xx} = f(x,t), \quad u(x,0) = u_t(x,0) = 0$$

satisfies u/λ, u_t/λ, $\sigma u_x \in B$ with

$$\left\|\frac{u}{\lambda}\right\|_B + \left\|\frac{u_t}{\lambda}\right\|_B + \|\sigma u_x\|_B \le C\|f\|_B .$$

In the strictly hyperbolic case this property holds with $B = C([0,T], H^s)$, $\sigma(x) = 1$ and $\lambda(t) = 1 + t$.

If such a *strictly hyperbolic type estimate* holds for the linear problem, then one can mostly construct a sequence of approximate solutions of (23.2) in the usual way which will converge for small t to a solution.

In the case of weakly hyperbolic equations it is not obvious how to derive such an estimate and how to fix the space B, since we have to expect a loss of Sobolev regularity.

23.2 WEAKLY HYPERBOLIC EQUATIONS WITH TIME AND SPATIAL DEGENERACY, BUT WITHOUT OSCILLATIONS

In this section we consider the model problem with *time and spatial degeneracy*, but without *oscillations* of the coefficients of the main part

$$u_{tt} - \sum_{i=1}^{n} \partial_{x_i} \left(\lambda^2(t)\sigma^2(x)u_{x_i}(x,t)\right) = f(b_1(x,t)u_{x_1}(x,t), \ldots, b_n(x,t)u_{x_n}(x,t)),$$
$$u(x,0) = \varphi_0(x), \quad u_t(x,0) = \varphi_1(x) \tag{23.3}$$

under the following assumptions:

$$\sigma \in C_b^\infty(\mathbb{R}^n), \quad \varphi_0 \in H^{N+1}(\mathbb{R}^n), \quad \varphi_1 \in H^N(\mathbb{R}^n), \tag{23.4}$$

$$\exists M > 0 : \quad \text{supp}\,(|\varphi_0| + |\varphi_1|) \subset [-M, M], \tag{23.5}$$

$$b_i \in C([0,T_0], C_b^\infty(\mathbb{R}^n)), \quad f \in C^\infty(\mathbb{R}^n), \tag{23.6}$$

$$\lambda \in C^1([0,T_0]), \quad \lambda(0) = \lambda'(0) = 0, \quad \lambda'(t) > 0 \quad (t > 0), \tag{23.7}$$

$$|b_i(x,t)| \leq C_b|\sigma(x)|\lambda'(t), \quad (x,t) \in \mathbb{R}^n \times [0,T_0], \quad i = 1, \ldots, n, \tag{23.8}$$

$$\left|\frac{\partial f}{\partial p_i}(p_1, \ldots, p_n)\right| \leq C_f, \quad |p_1| + \ldots + |p_n| \leq \delta, \quad i = 1, \ldots, n, \tag{23.9}$$

$$f(0) = 0 \tag{23.10}$$

The functions σ and λ are used to formulate the *degeneration*, which occurs at the points (x_0, t_0) with $\sigma(x_0) = 0$ or $\lambda(t_0) = 0$. The assumption (23.8) describes the Levi condition and the assumption (23.7) excludes oscillations with respect to t in the coefficients.

We want to prove the local existence of a solution in Sobolev spaces and the finite speed of propagation of this solution. As one can see from the example of Qi Min-You, we have to expect a loss of regularity, i.e. the solution will belong to a space $C^1([0,T], H^{N-\gamma})$. The number γ of lost derivatives will be proportional to the product of the constants C_b, C_f from (23.8), (23.9). The assumption $f(0) = 0$ will only be used in the proof of the finite speed of propagation.

Our approach will be divided into the following steps:

- reduction of the starting problem (23.3) to a family of Cauchy problems for ordinary differential equations and for a quasilinear weakly hyperbolic equation whose right-hand side has a *suitable asymptotic behaviour* (in this step we feel the loss of derivatives);
- derivation of a strictly hyperbolic type property for the solutions of the linearized weakly hyperbolic Cauchy problems by using the energy method, a *Gronwall type lemma for differential inequalities with a singular coefficient* and an approximation argument;
- using the results from the previous step, a fixed point argument gives a solution of the auxiliary problem, starting problem, respectively.

a) We write

$$u(x,t) = u^0(x,t) + \ldots + u^d(x,t) + v(x,t),$$

where the functions $u^1, \ldots, u^n$ are solutions of ordinary differential equations in t with the parameter x:

$$\begin{aligned}
&u^0_{tt} = f(0), \quad u^0(x,0) = \varphi_0(x), \quad u^0_t(x,0) = \varphi_1(x),\\
&\ldots,\\
&u^j_{tt} = f\left(b_1 \sum_{l=0}^{j-1} u^l_{x_1}, \ldots, b_n \sum_{l=0}^{j-1} u^l_{x_n}\right) - f\left(b_1 \sum_{l=0}^{j-2} u^l_{x_1}, \ldots, b_n \sum_{l=0}^{j-2} u^l_{x_n}\right)\\
&\qquad + \sum_{i=1}^{n} \partial_{x_i}\left(\lambda^2\sigma^2 u^{j-1}_{x_i}\right),\\
&u^j(x,0) = u^j_t(x,0) = 0, \quad j = 1, \ldots, d, \quad \sum_{l=0}^{-1} = 0.
\end{aligned}$$

This system can be solved step by step, we get solutions

$$u^j \in C^2\left([0,T_j], H^{N-2j}\right).$$

By Hadamard's formula and (2.6) we see that

$$\left\|u^j(t)\right\|_{H^{N-2j}} + \left\|u^j_t(t)\right\|_{H^{N-2j}} \le C\lambda^j(t).$$

The function u is a solution of (23.3), if and only if v solves the Cauchy problem

$$\begin{aligned}
v_{tt} - \sum_{i=1}^{n} \partial_{x_i}\left(\lambda^2\sigma^2 v_{x_i}\right) &= f\left(b_1\left(\sum_{l=0}^{d} u^l_{x_1} + v_{x_1}\right), \ldots, b_n\left(\sum_{l=0}^{d} u^l_{x_n} + v_{x_n}\right)\right)\\
&- f\left(b_1 \sum_{l=0}^{d-1} u^l_{x_1}, \ldots, b_n \sum_{l=0}^{d-1} u^l_{x_n}\right) + \sum_{i=1}^{n} \partial_{x_i}\left(\lambda^2\sigma^2 u^d_{x_i}\right)\\
&=: f_d(x,t,b_1 v_{x_1}, \ldots, b_n v_{x_n}),
\end{aligned} \tag{23.11}$$

$v(x,0) = v_t(x,0) = 0$.

The right-hand side satisfies

$$\|f_d(.,t,0)\|_{H^{N-2d-2}} \le C_d\lambda^d(t)\lambda'(t),$$
$$\left\|\frac{\partial f_d}{\partial p_i}(.,t,p_1(.,t),\ldots,p_n(.,t))\right\|_\infty \le C_f, \quad |p_1|+\ldots+|p_n| \le \delta'$$

for small t. We write $N_0 := N - 2d - 2$ and from Hadamard's formula we deduce that

$$\begin{aligned}
&\|f_d(x,t,b_1v_{x_1},\ldots,b_nv_{x_n})\|_{H^{N_0}} \\
&\quad \le \|f_d(x,t,b_1v_{x_1},\ldots,b_nv_{x_n}) - f_d(.,t,0)\|_{H^{N_0}} + \|f_d(.,t,0)\|_{H^{N_0}} \\
&\quad \le C_f \textstyle\sum_{i=1}^n \sum_{|\alpha|=N_0} \|\partial_x^\alpha(b_iv_{x_i})\|_2 + C\lambda'(t)\,\|v\|_{H^{N_0}} + C_d\lambda^d(t)\lambda'(t) \\
&\quad \le C_f \textstyle\sum_{i=1}^n \sum_{|\alpha|=N_0} \|b_i\partial_x^\alpha v_{x_i}\|_2 + C_d'\lambda'(t)\,\|v\|_{H^{N_0}} + C_d\lambda^d(t)\lambda'(t)
\end{aligned}$$

for $\sum_{i=1}^n |b_i|(\sum_{l=0}^d |u_{x_i}^l| + |v_{x_i}|) \le \delta$. With $a(x,t) := \sigma^2(x)\lambda^2(t)$ we define the energies

$$\begin{aligned}
e_j(v)(t)^2 &= \sum_{|\alpha|=j} \int a(x,t) \sum_{i=1}^n (\partial_x^\alpha \partial_{x_i} v)^2 + (\partial_x^\alpha v_t)^2 + (\partial_x^\alpha v)^2\,dx, \\
E_N(v)(t) &= \sum_{j=0}^N e_j(v)(t)
\end{aligned}$$

and obtain

$$\begin{aligned}
\|f_d(.,t,b_1v_{x_1},\ldots,b_nv_{x_n})\|_{H^{N_0}} &\le \left(\sqrt{n}C_fC_b\tfrac{\lambda'(t)}{\lambda(t)} + C_d'\lambda'(t)\right) E_{N_0}(v)(t) \\
&\quad + C_d\lambda^d(t)\lambda'(t).
\end{aligned}$$

b) For further considerations we need an estimate for solutions of

$$w_{tt} - \sum_{i=1}^n \partial_{x_i}\left(\lambda^2\sigma^2 w_{x_i}\right) = g(x,t), \quad w(x,0) = w_t(x,0) = 0, \tag{23.12}$$

$$\|g(t)\|_{H^{N_0}} \le C_g\lambda^{d-1}(t)\lambda'(t) + C_g'\lambda^d(t)\lambda'(t). \tag{23.13}$$

We suppose that the function $g \in C\left([0,T_0], H^{N_0}\right)$ has compact support with respect to x and that there exists a solution w with compact support with respect to x which satisfies

$$w, w_t, \sigma w_x \in C\left([0,T_0], H^{N_0}\right). \tag{23.14}$$

Differentiating $e_j(w)(t)^2$ and using $a_t = 2a\lambda'(t)/\lambda(t)$ and the differential equation we get

$$e_j(w)(t)e_j(w)'(t) = \sum_{|\alpha|=j} \int \frac{\lambda'(t)}{\lambda(t)} a \sum_{i=1}^{n} \left(\partial_x^\alpha \partial_{x_i} w\right)^2 + a \sum_{i=1}^{n} \left(\partial_x^\alpha \partial_{x_i} w\right)\left(\partial_x^\alpha \partial_{x_i} w_t\right)$$

$$+ \left(\partial_x^\alpha w_t\right)\left(\partial_x^\alpha g\right) + \left(\partial_x^\alpha w_t\right) \sum_{i=1}^{n} \partial_x^\alpha \partial_{x_i}(a w_{x_i}) + \left(\partial_x^\alpha w\right)\left(\partial_x^\alpha w_t\right) dx.$$

By the aid of partial integration and the Leibniz formula we get

$$\int \left(\partial_x^\alpha w_t\right) \sum_{i=1}^{n} \partial_x^\alpha \partial_{x_i}(a w_{x_i}) dx$$

$$= \int \sum_{\substack{\beta \le \alpha, \\ \beta \ne 0}} \binom{\alpha}{\beta} \left(\partial_x^\alpha w_t\right) \sum_{i=1}^{n} \partial_{x_i}\left(\left(\partial_x^\beta a\right)\left(\partial_x^{\alpha-\beta} w_{x_i}\right)\right) - \left(\partial_x^\alpha \partial_{x_i} w_t\right) a \sum_{i=1}^{n} \partial_x^\alpha w_{x_i} dx,$$

which leads to

$$e_j(w)'(t) \le \frac{\lambda'(t)}{\lambda(t)} e_j(w)(t) + \sum_{|\alpha|=j} \|\partial_x^\alpha g\|_2 + C_j \sum_{k=0}^{j} e_k(w)(t).$$

Finally we get

$$E_{N_0}(w)'(t) \le \left(\frac{\lambda'(t)}{\lambda(t)} + C_{N_0}\right) E_{N_0}(w)(t) + \|g(t)\|_{H^{N_0}}. \tag{23.15}$$

We can not use Gronwall's Lemma here, since the factor $\lambda'(t)/\lambda(t) + C_{N_0}$ is unbounded. We apply the Lemma of Nersesyan (Gronwall type lemma for differential inequalities with a singular coefficient) instead:

Lemma 1 (Nersesyan) *Let $y(t) \in C([0,T]) \cap C^1(0,T)$ be a solution of the differential inequality*

$$y'(t) \le K(t)y(t) + f(t), \quad 0 < t < T,$$

where the functions $K(t)$ and $f(t)$ belong to $C(0,T)$. We assume that for every $t \in (0,T)$ and every $\delta \in (0,t)$

$$\int_0^\delta K(\tau)\,d\tau = \infty, \quad \int_\delta^T K(\tau)\,d\tau < \infty,$$
$$\lim_{\delta \to +0} \int_\delta^t \exp\left(\int_s^t K(\tau)\,d\tau\right) f(s)\,ds \quad \textit{exists}, \tag{23.16}$$
$$\lim_{\delta \to +0} y(\delta) \exp\left(\int_s^t K(\tau)\,d\tau\right) = 0.$$

Then it holds

$$y(t) \le \int_0^t \exp\left(\int_s^t K(\tau)\,d\tau\right) f(s)\,ds.$$

Using this lemma we obtain with (23.13)

$$\begin{aligned} E_{N_0}(w)(t) &\le \int_0^t \exp\left(\int_s^t \frac{\lambda'(\tau)}{\lambda(\tau)} + C_{N_0}\,d\tau\right) \|g(\tau)\|_{H^{N_0}}\,ds \\ &\le C_g e^{C_{N_0}t}\frac{1}{d-1}\lambda^d(t) + C_g' e^{C_{N_0}t}\frac{1}{d}\lambda^{d+1}(t). \end{aligned} \tag{23.17}$$

During our computations we needed more regularity for w than (23.14). The existence of a solution w with compact support with respect to x and the final estimate (23.17) can be seen as follows: We replace $a(x,t) = \sigma^2(x)\lambda^2(t)$ by $a_\varepsilon(x,t) = (\sigma^2(x)+\varepsilon)(\lambda(t)+\varepsilon)^2$ and replace g by a smooth compactly supported function g_ε, such that $\{g_\varepsilon\}$ converges to g in the norm of $C\left([0,T_0], H^{N_0}\right)$ if ε tends to 0. This modified problem is strictly hyperbolic and has a smooth solution w_ε which is compactly supported and fulfils (23.17). The sequence $\{w_{\varepsilon_k}\}$ ($\varepsilon_k \to 0$) converges to a compactly supported solution w of (23.12) satisfying (23.17).

c) We apply this estimate for w to a linearized version of (23.11): Let $\{v^k\}$ be a sequence with

$$\begin{aligned} &v^0(x,t) \equiv 0, \\ &v^k_{tt} - \sum_{i=1}^n \partial_{x_i}\left(\lambda^2\sigma^2 v^k_{x_i}\right) = f_d(x,t,b_1 v^{k-1}_{x_1},\dots,b_n v^{k-1}_{x_n}), \\ &v^k(x,0) = v^k_t(x,0) = 0. \end{aligned}$$

We have $E_{N_0}(v^0)(t) \equiv 0 \le \lambda^d(t)$, hence

$$\left\|f_d(x,t,b_1 v^0_{x_1},\dots,b_n v^0_{x_n})\right\|_{H^{N_0}} \le \sqrt{n}C_f C_b \lambda^{d-1}(t)\lambda'(t) + (C_d' + C_d)\lambda^d(t)\lambda'(t).$$

This implies

$$E_{N_0}(v^1)(t) \le \sqrt{n}C_f C_b e^{C_{N_0}t}\frac{1}{d-1}\lambda^d(t) + (C_d + C_d')e^{C_{N_0}t}\frac{1}{d}\lambda^{d+1}(t).$$

We suppose $2C_f C_b\sqrt{n}/(d-1) \le 1/3$ and restrict the interval for t such that

$$e^{C_{N_0}t} \le 2, \quad 2(C_d + C_d')\frac{1}{d}\lambda(t) \le \frac{1}{3}$$

and conclude

$$E_{N_0}(v^1)(t) \le \frac{2}{3}\lambda^d(t) \le \lambda^d(t).$$

By induction we get the uniform bound $E_{N_0}(v^k)(t) \le \lambda^d(t)$, $k \in \mathbb{N}$. In order to show the convergence of the sequence $\{v^k\}$ we consider the sequence of differences $w^k = v^{k+1} - v^k$. The function w^k satisfies

$$\begin{aligned} &w^k_{tt} - \sum_{i=1}^n \partial_{x_i}\left(\lambda^2\sigma^2 w^k_{x_i}\right) = \sum_{i=1}^n g_{dki}(x,t) b_i(x,t) w^{k-1}_{x_i} \\ &:= f_d(x,t,b_1 v^k_{x_1},\dots,b_n v^k_{x_n}) - f_d(x,t,b_1 v^{k-1}_{x_1},\dots,b_n v^{k-1}_{x_n}), \\ &w^k(x,0) = w^k_t(x,0) = 0. \end{aligned}$$

The L^2–norm of the right–hand side can be bounded by

$$\sqrt{n}C_f C_b \frac{\lambda'(t)}{\lambda(t)} E_0(w^{k-1})(t).$$

From $E_0(w^0)(t) \leq \lambda^d(t)$ and the special choice of d we see by induction that

$$E_0(w^k)(t) \leq \left(\frac{2}{3}\right)^k \lambda^d(t).$$

This implies that $\{v^k\}$ is a Cauchy sequence in $C^1([0,T], L^2)$. By interpolation and the uniform estimate of $E_{N_0}(v^k)(t)$ we deduce that v^k converges in $C^1\left([0,T], H^{N_0-1}\right)$. The limit is a solution of (23.11) and satisfies

$$\text{supp }(v(.,t)) \subset [-M - Ct, M + Ct].$$

This proves the finite propagation speed of the constructed solution. Thus we have shown:

Theorem 1 *Let the assumptions* (23.4) *to* (23.10) *be satisfied. Then the Cauchy problem* (23.3) *has a solution u with $u, u_t, \sigma u_x \in C\left([0,T], H^{N-\gamma}\right)$, where*

$$\gamma = 2d + 2, \qquad \frac{2C_f C_b \sqrt{n}}{d-1} \leq \frac{1}{3}.$$

This solution has a finite speed of propagation.

Open problems:

1. Uniqueness of the solution in $C^1\left([0,T], H^{N-\gamma}\right)$;
2. A local existence result in Gevrey spaces under Gevrey–type Levi conditions.

23.3 WEAKLY HYPERBOLIC EQUATIONS WITH TIME DEGENERACY AND FAST OSCILLATIONS, BUT WITHOUT SPATIAL DEGENERACY

In this section we consider the model problem with *time degeneracy and oscillations* in the coefficients of the main part, but without *spatial degeneracy*

$$u_{tt} - \lambda^2(t)a^2(t) \,\Delta\, u = f(b_1(x,t)u_{x_1}(x,t), \ldots, b_n(x,t)u_{x_n}(x,t)), \qquad (23.18)$$

$$u(x,0) = \varphi_0(x), \quad u_t(x,0) = \varphi_1(x).$$

The function $\lambda = \lambda(t)$ describes the degeneracy with respect to t as in the previous section. To understand the influence of oscillations which are produced by the function $a = a(t)$, let us remember a result from [9]. There it was shown that the Cauchy problem for

$$u_{tt} - \exp\left(-\frac{2}{t^\alpha}\right) a^2\left(\frac{1}{t}\right) u_{xx} = 0 \qquad (23.19)$$

is C^∞–well–posed if and only if $\alpha \geq 1/2$. Here $a = a(t)$ is a non–constant, 1–periodic positive function from C^∞. This example shows that the product between amplitudes and frequencies of the oscillations is important. An interpretation of this example and a definition of so–called fast oscillations by the condition

$$|D_t a(t)| \leq C\frac{\lambda'(t)}{\lambda(t)}|\ln\lambda(t)|, \quad t \in (0, T_0] \tag{23.20}$$

were given in [10]. There it was proved that (23.20) is sufficient for C^∞–well–posedness (necessity for (23.19) is clear). We suppose:

- $\lambda = \lambda(t) \in C^2([0, T_0])$ satisfies (23.7);
- there exist positive constants $d_0 \geq \frac{1}{2}$ and d_1 such that

$$d_0\frac{\lambda(t)}{\Lambda(t)} \leq \frac{\lambda'(t)}{\lambda(t)} \leq d_1\frac{\lambda(t)}{\Lambda(t)}, \quad |\lambda''(t)| \leq d_1\lambda(t)\left(\frac{\lambda(t)}{\Lambda(t)}\right)^2 \tag{23.21}$$

  for all $t \in (0, T_0]$, where $\Lambda(t) := \int_0^t \lambda(\tau)\,d\tau$;
- the function $a = a(t) \in C^1((0, T_0])$ satisfies (23.20) and

$$0 < c \leq a(t) \leq C, \quad t \in (0, T_0]. \tag{23.22}$$

Our approach will be divided into the following steps:

- derivation of a strictly hyperbolic type property for the solutions of linear weakly hyperbolic Cauchy problems by using microlocal considerations for explicit solutions of linear ordinary differential equations depending on parameters;
- using the a–priori estimate and a fixed point argument to construct a Sobolev solution of the starting problem (23.18).

a) An essential point to handle quasilinear equations is the strictly hyperbolic type property for corresponding linear equations (compare with step b) of the previous section). If we try to follow the reasoning from the previous section, the conditions (23.20) and (23.22) imply for the energies E_{N_0} an inequality of the type (compare with (23.15))

$$E_{N_0}(u)'(t) \leq C\frac{\lambda'(t)}{\lambda(t)}|\ln\lambda(t)|E_{N_0}(u)(t) + \dots \quad .$$

Consequently, Lemma 1 is not applicable and it is not clear how to apply the energy method.

To obtain a strictly hyperbolic type property for the Cauchy problem

$$u_{tt} - \lambda^2(t)a^2(t)\,\Delta\,u = -\lambda'(t)f(x,t), \quad u(x,0) = u_t(x,0) = 0 \tag{23.23}$$

we need another approach, which was developped in [8]. One can prove:

Lemma 2 *There exists a Banach space $B_{M,Q,T}$ such that for every f belonging to $B_{M,Q,T}$ one has a uniquely determined solution of (23.23) with the property that u/λ, u_t/λ and u_x belong to $B_{M,Q,T}$, too. Moreover, the a-priori estimate*

$$\left\|\frac{u}{\lambda}\right\|_{M,Q,T}+\left\|\frac{u_t}{\lambda}\right\|_{M,Q,T}+\|\nabla_x u\|_{M,Q,T}\leq C_{apr}\,\|f\|_{M,Q,T} \tag{23.24}$$

holds with a constant C_{apr} independent of $T\in(0,T_0]$ and f.

The idea of the construction of $B_{M,Q,T}$ is to describe the elements $u=u(x,t)$ by the behaviour of its partial Fourier transform $\hat{u}=\hat{u}(t,\xi)$ in the cotangent space. Therefore we study the Cauchy problem

$$D_t^2\hat{u}-\lambda^2(t)a^2(t)|\xi|^2\hat{u}=\lambda'\hat{f},\quad \hat{u}(0,\xi)=\hat{u}_t(0,\xi)=0 \tag{23.25}$$

under the assumptions $\hat{f}\in C([0,T],L^\infty(\mathbb{R}^n))$ and $\lambda(t)\in C^2([0,T_0])$, (23.21), (23.22).

In the light of (23.24), it is sufficient to show which properties of $\hat{f}$ are transferred to $\hat{u}/\lambda$, $\hat{u}_t/\lambda$ and $\xi_j\hat{u}$. This can be done by following the next steps:

- we divide $\mathbb{R}^n\times[0,T]$ into the so-called pseudo-differential zone Z_{pd} and the hyperbolic zone Z_{hyp} ([10]);
- we transform the equation from (23.25) to a first order system in each zone;
- we find a fundamental solution $X=X(t,s,\xi)$ for the first order system in each zone and an estimate of the form ([10])

$$\|X(t,s,\xi)\|\leq\exp\left(C\int_s^t g(\tau,\xi)d\tau\right);$$

- we represent the solution of a suitable Cauchy problem by the aid of the fundamental solution;
- using these explicit representations leads to assumptions for $\hat{f}=\hat{f}(t,\xi)$ of the form

$$|\hat{f}(t,\xi)|\leq C(\xi)\exp\left(Q\int_{a(\xi)}^t g(\tau,\xi)d\tau\right),$$

  where $C(\xi)$ appears because ξ is a parameter and $a(\xi)$ depends on the zone;
- the right-hand sides of these estimates determine the weights in the definition of $B_{M,Q,T}$.

b) Let us go into detail. By t_ξ we denote the solution of the equation $\Lambda(t_\xi)\langle\xi\rangle=N\ln\langle\xi\rangle$, where $\langle\xi\rangle=(c+|\xi|^2)^{\frac{1}{2}}$ and N, c are sufficiently large positive numbers.

The *pseudo–differential zone* $Z_{pd}(c,N)$ is defined by (see [10])

$$Z_{pd}(c,N) := \{(x,t) \in \mathbb{R}^n \times [0,T] : \Lambda(t)\langle\xi\rangle \le N \ln\langle\xi\rangle\}.$$

Substituting $w_2 := D_t\hat{u}$, $w_1 := \varrho(t,\xi)\hat{u}$, where $\varrho = \varrho(t,\xi)$ is the positive root of $\varrho^2 - 1 - (\lambda^2(t)/\Lambda(t))\langle\xi\rangle \ln\xi = 0$, (23.25) reads as

$$D_tW - A(t,\xi)W = F, \quad W(0,\xi) = 0, \tag{23.26}$$

$$W = \begin{pmatrix} w_1 \\ w_2 \end{pmatrix}, \quad F = \begin{pmatrix} 0 \\ \lambda'(t)\hat{f} \end{pmatrix}, \quad A(t,\xi) = \begin{pmatrix} \frac{D_t\varrho}{\varrho} & \varrho \\ \lambda^2(t)\alpha^2(t)|\xi|^2 & 0 \end{pmatrix}.$$

If $X = X(t,s,\xi)$ is the fundamental matrix, that is, it solves

$$D_tX - A(t,\xi)X = 0, \quad X(s,s,\xi) = I, \quad 0 \le s \le t \le t_\xi,$$

then $W(t,\xi) = \int_0^t X(t,s,\xi)F(s,\xi)\,d\xi$ solves (23.26).

The assumptions concerning λ, a and the definition of $Z_{pd}(c,N)$ imply $\|A(t,\xi)\| \le C_{pd}g(t,\xi)$, where $g(t,\xi) = \varrho(t,\xi) + \varrho_t(t,\xi)/\varrho(t,\xi)$. Consequently, using the matrizant we obtain

$$\|X(t,s,\xi)\| \le \exp\left(\int_s^t \|A(\tau,\xi)\|\,d\tau\right) \le \exp\left(C_{pd}\int_s^t g(\tau,\xi)d\tau\right).$$

Thus, on the one hand, we have the representations

$$\begin{aligned} \varrho(t,\xi)\hat{u}(t,\xi) &= \int_0^t X_{12}(t,s,\xi)\lambda'(s)\hat{f}(s,\xi)\,ds, \\ D_t\hat{u}(t,\xi) &= \int_0^t X_{22}(t,s,\xi)\lambda'(s)\hat{f}(s,\xi)\,ds, \end{aligned}$$

on the other hand it is clear how to estimate $|X_{12}(t,s,\xi)|$, $|X_{22}(t,s,\xi)|$. Both results together yield a *strictly hyperbolic type result* for the solution of (23.25) in $Z_{pd}(c,N)$.

Lemma 3 *If $\hat{f}(t,\xi) \in C([0,T], L^\infty(\mathbb{R}^n))$ satisfies the inequality*

$$|\hat{f}(t,\xi)| \le A(\xi)\exp\left(Q\int_{t_\xi}^t g(\tau,\xi)\,d\tau\right), \quad Q > C_{pd},$$

in $Z_{pd}(c,N)$, then the solution $\hat{u}(t,\xi)$ of (23.25) satisfies in $Z_{pd}(c,N)$ the estimates

$$\begin{aligned} |\xi_j\hat{u}(t,\xi)| &\le C\frac{A(\xi)}{Q - C_{pd}}\exp\left(Q\int_{t_\xi}^t g(\tau,\xi)\,d\tau\right), \\ |\hat{u}_t(t,\xi)| &\le CA(\xi)\lambda(t)\exp\left(Q\int_{t_\xi}^t g(\tau,\xi)\,d\tau\right). \end{aligned}$$

Remark 1 *The Cauchy problem* (23.25) *feels neither oscillations nor hyperbolicity in* $Z_{pd}(c,N)$.

c) Now let us devote to the *hyperbolic zone* ([10]):

$$Z_{hyp}(c,N) := \{(\xi,t) \in \mathbb{R}^n \times [0,T] : \Lambda(t)\langle\xi\rangle \geq N \ln\langle\xi\rangle\}.$$

Let us denote the characteristic roots by

$$\tau_i(t,\xi) := (-1)^i \lambda(t) a(t)|\xi|.$$

There exists a well-known approach for the construction of fundamental solutions in the strictly hyperbolic case. One knows how to apply the perfect diagonalizer. It is not clear how to use this tool for equations with fast oscillations. Nevertheless in [10] only the first step of perfect diagonalization was done to derive C^∞-well-posedness for linear weakly hyperbolic equations with fast oscillations. One can use this idea to derive a corresponding result to Lemma 1.19 from [10] in $Z_{hyp}(c,N)$. Using suitable transformations, diagonalization and one step of perfect diagonalization transforms (23.25) to

$$D_t W - \begin{pmatrix} \tau_1 & 0 \\ 0 & \tau_2 \end{pmatrix} W + FW + BW = N^{-1}\frac{\lambda'(t)\hat{f}}{2a(t)}\begin{pmatrix} -1 \\ 1 \end{pmatrix} \tag{23.27}$$

in $Z_{hyp}(c,N)$, where $F := (1/2)(D_t a(t)/a(t))I$, N is the diagonalizer and B satisfies

$$\|B(t,\xi)\| \leq C\left(\frac{\lambda(t)}{\Lambda(t)} + \frac{\lambda(t)|\ln\lambda(t)|^2}{\langle\xi\rangle\Lambda^2(t)}\right) =: K(t,\xi),$$

see [10]. If $X(t,s,\xi)$ is a fundamental solution of (23.27), then

$$W(t,\xi) = \int_{t_\xi}^t X(t,s,\xi)F(s,\xi)\,ds + X(t,t_\xi,\xi)W(t_\xi,\xi)$$

solves (23.27) with a Cauchy condition on $t = t_\xi$. Using the ansatz $X(t,s,\xi) = E(t,s,\xi)Q(t,s,\xi)$, where E is the fundamental matrix of the left-hand side of (23.27) without BW, the norms of $Q(t,s,\xi)$, $X(t,s,\xi)$, respectively, can be estimated by $C\exp(C_{hyp}\int_s^t K(\tau,\xi)d\tau)$, $t_\xi \leq s \leq t$. This is exactly what we need to prove a *strictly hyperbolic type estimate* in $Z_{hyp}(c,N)$.

Lemma 4 *If* $\hat{f} \in C([0,T], L^\infty(\mathbb{R}^n))$ *satisfies the inequality*

$$|\hat{f}(t,\xi)| \leq C\langle\xi\rangle^{-M} \exp\left(Q\int_{T_0}^t K(\tau,\xi)\,d\tau\right), \quad Q > C_{hyp},$$

in $Z_{hyp}(c,N)$, *then*

$$|\xi_j \hat{u}(t,\xi)| \leq C\langle\xi\rangle^{-M} \exp\left(Q\int_{T_0}^t K(\tau,\xi)\,d\tau\right),$$
$$|\hat{u}_t(t,\xi)| \leq C\langle\xi\rangle^{-M}\lambda(t) \exp\left(Q\int_{T_0}^t K(\tau,\xi)\,d\tau\right).$$

Remark 2 *Contrary to $Z_{pd}(c,N)$ the Cauchy problem* (23.25) *feels oscillations as well as hyperbolicity in $Z_{hyp}(c,N)$.*

d) Now we are able to construct $B_{M,Q,T}$. Therefore we introduce the *weight function* ($(\xi,t) \in \mathbb{R}^n \times [0,T]$)

$$\mathcal{N}_{Q,M}(t,\xi) := \begin{cases} \langle\xi\rangle^M \exp\left(Q\int_t^{T_0} K(\tau,\xi)\,d\tau\right) & : t \in [t_\xi, T_0], \\ \langle\xi\rangle^M \exp\left(Q\int_{t_\xi}^{T_0} K(\tau,\xi)\,d\tau + Q\int_t^{t_\xi} g(\tau,\xi)\,d\tau\right) & : t \in [0, t_\xi]. \end{cases}$$

For a given positive number M we denote by B^M the normed linear space

$$B^M = \left\{ u \in \mathcal{S}'(\mathbb{R}^n) : \hat{u} \in L^\infty_{loc} \text{ is a function with } \sup_{\xi\in\mathbb{R}^n} \langle\xi\rangle^M |\hat{u}(\xi)| < \infty \right\}.$$

For given positive numbers M, Q and T, $T \le T_0$, we denote by $B_{M,Q,T}$ the linear space

$$B_{M,Q,T} = \Big\{ u \in C([0,T], B^M) : \hat{u}(\xi,t) \ \textit{satisfies} \ \mathcal{N}_{Q,M}(t,\xi)|\hat{u}(\xi,t)| \le C \ \forall(\xi,t) \in \mathbb{R}^n \times [0,T] \Big\}.$$

One can show that $B_{M,Q,T}$ is a Banach space with the norm

$$||u||_{M,Q,T} := \max_{[0,T]} \sup_{\xi\in\mathbb{R}^n} \mathcal{N}_{Q,M}(t,\xi)|\hat{u}(\xi,t)|.$$

All results of this section together make it possible to prove Lemma 2.

e) In the last point let us study (23.18) by the aid of the spaces $B_{M,Q,T}$.

- To have a reasonable space for the data we can use the imbedding $C([0,T], W_1^r(\mathbb{R}^n)) \subset B_{M,Q,T}$ if $r \ge r_0$, where r_0 depends on the given M and Q.
- To have a reasonable space for the solution we can use the imbedding $B_{M,Q,T} \subset C([0,T], H^l(\mathbb{R}^n))$ if $M - l > n/2$, $l > 0$ and $Q > 0$.
- Up to now we are not able to study superposition operators in $B_{M,Q,T}$.
- The only way for us to include nonlinearities is to prove that $B_{M,Q,T}$ is even a Banach algebra (see [8]).
- Thus one can suppose that f depends on $b_i(x,t)u_{x_i}(x,t)$ analytically. To simplify the calculations in [8] we restricted ourselves to the assumption that f be entire.

Now we can use the approach of the previous section to construct a uniquely determined solution of (23.18). We do not need a reduction process, we only use successive approximation and the a-priori estimate (23.24). To obtain a Cauchy sequence in a suitable space $B_{M,Q,T}$ we have to suppose a Levi condition which is close to the sharp Levi condition $O(\lambda'(t))$. Let us formulate a final result, where we renounce to represent the most general version.

Theorem 2 *Let us consider the Cauchy problem*

$$u_{tt} - \lambda^2(t)a^2(t) \bigtriangleup u = f(b_1(x,t)u_{x_1}(x,t), \ldots, b_n(x,t)u_{x_n}(x,t)),$$
$$u(x,0) = \varphi_0(x), \quad u_t(x,0) = \varphi_1(x),$$

under the assumptions $\lambda \in C^2([0,T])$, (23.7), (23.21) *and* (23.22) *and the assumptions for* f

- f *is polynomial in its arguments,*
- $b_k = b_k(x,t) \in C([0,T_0], W_1^s(\mathbb{R}^n))$ *for every given* $s \in \mathbb{N}$, *where*

$$||b_k(.,t)||_{W_1^s(\mathbb{R}^n)} \leq o(\lambda'(t)), \quad t \in (0,T_0], \quad \textit{(Levi condition)}\ .$$

Then there exists a (in general sufficiently large) constant r *such that for arbitrary data* φ_0, φ_1 *belonging to* $W_1^r(\mathbb{R}^n)$, $W_1^{r-1}(\mathbb{R}^n)$, *respectively, we have a locally defined solution*

$$u \in C([0,T], H^{[n/2]+3}(\mathbb{R}^n)) \cap C^1([0,T], H^{[n/2]+2}(\mathbb{R}^n))$$
$$\cap C^2([0,T], H^{[n/2]+1}(\mathbb{R}^n)).$$

The solution is uniquely determined in $B_{M,Q,T}$.

Open problems:

1. Is it possible to apply the energy method to (23.18) under the assumptions of this section?

2. Find a strictly hyperbolic type result for the Cauchy problems for

$$u_{tt} - \lambda^2(t)a^2(t) \bigtriangleup u + \sum_{i=1}^{n} b_i(x,t)u_{x_i} = f$$

under the Levi condition

$$||b_i(x,t)||_{W_1^s(\mathbb{R}^n)} \leq O(\lambda'(t)), \quad t \in (0,T_0]$$

and

$$u_{tt} - \lambda^2(t)a^2(t)\sigma^2(x) \bigtriangleup u = f.$$

Remark 3 *We omit a long list of references. But any interested reader can ask the authors for a list of references concerning the subject of this paper.*

References

[1] Colombini, F., Jannelli, E., Spagnolo, S.: Well-posedness in the Gevrey classes of the Cauchy problem for a non-strictly hyperbolic equation with coefficients depending on time. Ann. Sc. Norm. Sup. Pisa 10 (1983), 291–312.

[2] Colombini, F., Lerner, N.: Hyperbolic operators with non-Lipschitz coefficients. Duke Math. J. 77 (1995), 657–698.

[3] Colombini, F., Spagnolo, S.: An example of a weakly hyperbolic Cauchy problem not well posed in C^∞. Acta Math. 148 (1982), 243–253.

[4] D'Ancona, P.: A note on a theorem of Jörgens. Math. Z. 218 (1995), 239–252.

[5] Ivrii, V.Ya., Petkov, V.M.: Necessary conditions for the well-posedness of the Cauchy problem for non-strictly hyperbolic equations. Usp. Mat. Nauk 29(5) (1974), 3–70 (Russian); Russ. Math. Surv. 29(5) (1974), 1-70.

[6] Oleinik, O.A.: On the Cauchy problem for weakly hyperbolic equations. Comm. pure appl. math. 23 (1970), 569–586.

[7] Qi Min-you: On the Cauchy problem for a class of hyperbolic equations with initial data on the parabolic degenerating line. Acta Math. Sinica 8 (1958), 521–529.

[8] Reissig, M., Yagdjian, K.: Weakly hyperbolic equations with fast oscillating coefficients. Preprint 7, Technische Universität Bergakademie Freiberg, 1997.

[9] Tarama, S.: On the second order hyperbolic equations degenerating in the infinite order. Math. Japonica 42 (1995), 523–534.

[10] Yagdjian, K.: The Cauchy problem for hyperbolic operators. Multiple characteristics, micro-local approach. Akademie Verlag, Berlin, 1997.

24 EIGENVALUE PROBLEMS FOR ELLIPTIC TYPE PARTIAL DIFFERENTIAL OPERATORS WITH SPECTRAL PARAMETER CONTAINED LINEARLY IN BOUNDARY CONDITIONS

Boris Belinskiy †

Department of Mathematics
University of Tennessee at Chattanooga
Chattanooga, Tennessee 37403-2598, USA
bbelinsk@cecasun.utc.edu

Abstract: It is shown that the eigenfunctions of an elliptic type partial differential operator in a compact domain with an eigenvalue parameter contained linearly in the boundary condition form a basis of a corresponding Hilbert space.

24.1 INTRODUCTION

Let $\Omega \subset \mathbb{R}^k$ be a compact domain. We consider the following eigenvalue problem.

Problem 1 *Find the eigenvalues λ such that there exists a non–trivial solution $u : \Omega \to \mathbb{C}$ for the equation*

$$-div(p(x) \nabla u) + q(x)u = \lambda r(x)u, \ x \in \Omega \tag{24.1}$$

†The author was supported in part by the University of Tennessee at Chattanooga Faculty Development Grant and Faculty Research Grant. This is an expanded version of a lecture given at the ISAAC 97 Conference, University of Delaware, June 3–7, 1997

H.G.W. Begehr et al. (eds.), Partial Differential and Integral Equations, 319–333.

with the boundary condition

$$-\beta_1(x)u + \beta_2(x)p(x)\frac{\partial u}{\partial n} = \lambda\beta_1'(x)u, \ x \in \partial\Omega. \tag{24.2}$$

Here n means the exterior normal vector to $\partial\Omega$. The coefficients of the equation (24.1) and the boundary condition (24.2) satisfy the following assumptions.

Assumption 1 *a. The boundary $\partial\Omega$ is such that the embedding theorems for Sobolev spaces hold, e.g., it is smooth.*

b. The functions $p, q, r : \Omega \to \mathbb{R}$ are measurable and bounded on Ω, p and r are positive definite there.

c. The functions $\beta_1, \beta_2, \beta_1' : \partial\Omega \to \mathbb{R}$ are measurable and bounded on $\partial\Omega$ and the function $\rho(x) = \beta_1'(x)\beta_2(x)$ is positive definite there.

Problem 1 in the one–dimensional case is the Sturm–Liouville problem. Its statement is the following.

Problem 2 *Find the eigenvalues λ such that there exists a non–trivial solution $u : [a,b] \to \mathbb{C}$ for the equation*

$$-(p(x)u')' + q(x)u = \lambda r(x)u, \ x \in (a,b) \tag{24.3}$$

with the boundary conditions

$$\cos\alpha u(a) + \sin\alpha(pu')(a) = 0, \tag{24.4a}$$

$$-\beta_1 u(b) + \beta_2 p(b)u'(b) = \lambda(\beta_1{}' u(b) - \beta_2{}' p(b)u'(b)). \tag{24.4b}$$

Here $\alpha, \beta_1, \beta_2, \beta_1{}', \beta_2{}' \in \mathbb{R}$, $\alpha \in [0,\pi)$.

Assumption 2 *a. The functions $p, q, r : [a,b] \to \mathbb{R}$ are measurable on $[a,b]$, p and r are positive definite there.*

b. $\rho = \beta_1'\beta_2 - \beta_1\beta_2{}' > 0$.

Problem 2 has been studied by many authors (see, e.g., [1, 2, 3, 4]). The main **goal** of those papers is the analysis of the **spectrum** and justification of the **eigenfunction expansion.** The considerations of [1, 4] are based on the operator–theoretic formulation of Problem 2. In [2, 3], the residue calculus in a manner similar to Titchmarsh [5] is employed.

There is rather extensive literature associated with Problem 2. The most detailed review of the results, published before 1980, was, to our best knowledge, written by Fulton [2]. In [6] the problem of eigenoscillations of various mechanical systems is formulated as the Sturm–Liouville problem with the eigenvalue parameter (frequency of oscillations) appearing in the boundary conditions. It is shown that the eigenfunctions form a basis of a corresponding Hilbert space.

In [7, 8, 9] the Laplace equation $\Delta u = 0$ in a two–dimensional domain Ω with the boundary condition $\partial u/\partial n = -(\lambda p_1(x) - q_1(x))u$, $x \in \partial\Omega$, is considered. In [7] Ω is a unit circular disk. The separation of variables allows to reduce the problem to a Sturm–Liouville problem and to use the Titchmarsh technique [5]. In [8, 9] more general domains are considered. The problem is treated with the help of conformal mappings. The distribution function for eigenvalues is analyzed asymptotically, the convergence of the eigenfunction expansion is discussed.

The eigenfunction expansion for an arbitrary elliptic type equation (24.1) with no spectral parameter in the boundary condition (24.2), i.e., $\beta_1'(x) = 0$, is well known. Its justification is based on the concept of a weak solution of the spectral problem and the Riesz representation theorem (see, e.g., [10, 11]). As a result, the spectral problem is reduced to an operator equation

$$u = \lambda A u, \; u \in \Xi, \tag{24.5}$$

with a compact self-adjoint operator A in an appropriate Hilbert space Ξ. Fredholm theory implies that the eigenfunctions $\{u_k\}$ form an orthogonal basis in Ξ. Note, the dimension of the domain Ω is not important for this construction.

Both Problems 1 and 2 can be reduced to operator equations containing spectral parameter in a more complicated way. Keldysh [12] considered a wide class of operator equations of the form

$$L(\lambda)\psi = 0, \; L(\lambda) = \sum_{i=0}^{n} \lambda^i A_i \tag{24.6}$$

in a Hilbert space Ξ. Such an operator is called an operator polynomial or operator pencil. With some appropriate assumptions, Keldysh proved that the spectrum of the problem (24.6) is discrete and the system $\{\psi_n\}$ of eigenfunctions and associated functions is complete in Ξ. An abstract theory of the operator polynomials has been developed extensively in [13, 14, 15, 16, 17, 18]. In our paper the operator polynomial is of the form

$$L(\lambda) = A_0 + \lambda A_1. \tag{24.7}$$

Note, in [19, 20] operator polynomials of the form $L(\lambda) = A_0 + \lambda A_1 + \lambda^2 A_2$ are considered. An important assumption in those papers is the existence of a constant $c \in \mathbb{R}$ such that the operator $L(c)$ is positive definite. In the following this property will be shown to hold for the operator polynomials that are considered.

In [16] an abstract operator polynomial (24.7) is studied. The forms associated with the operators A_0, A_1 are supposed to be hermitian and at least one of them has a finite number of negative squares. Problem 2 is considered as an application and the expansion theorem for it is proven. Assumptions 2 above (used systematically in [1, 2, 3, 4]) are partially removed. It is assumed instead that $\rho \neq 0$ and $r(x) \neq 0$ a.e. on $[a, b]$. It is assumed in addition that $\lambda = 0$ is not an eigenvalue of the Problem 2 (this condition does not appear in our paper).

We prove in this paper that the eigenfunctions of the Problem 1 form the basis of the Hilbert space $H^1(\Omega)$. Section 24.2 contains some auxiliary facts.

The main result is discussed in Sections 24.3 and 24.5. An estimation for the eigenvalues from below is given in Section 24.4. Section 24.6 contains the physical motivation of Problem 1.

24.2 SOME AUXILIARY FACTS

We use the standard notations for the Hilbert spaces $L^2(\Omega)$, $L^2(\partial\Omega)$, and $H^1(\Omega)$. The inner products in those spaces are given by

$$< u_1, u_2 >_0 = \int_\Omega u_1 \bar{u}_2 dx, \tag{24.8a}$$

$$< u_1, u_2 >_1 = \int_{\partial\Omega} u_1 \bar{u}_2 dx, \tag{24.8b}$$

and

$$< u_1, u_2 >_2 = \int_\Omega (\nabla u_1 \nabla \bar{u}_2 + u_1 \bar{u}_2) dx. \tag{24.8c}$$

The corresponding norms are

$$||u||_0 = < u, u >_0^{1/2}, \tag{24.9a}$$

$$||u||_1 = < u, u >_1^{1/2}, \tag{24.9b}$$

and

$$||u||_2 = < u, u >_2^{1/2}. \tag{24.9c}$$

The following result is a particular case of the embedding theorems for Sobolev spaces (see [10, 11]).

Lemma 1 *The trace of any function $u \in H^1(\Omega)$ is defined as an element of $L^2(\partial\Omega)$ and the inequality*

$$||u||_1^2 \leq C_1 \left(\frac{2}{\delta} ||u||_0^2 + \delta || \nabla u||_0^2 \right) \tag{24.10}$$

holds for sufficiently small positive numbers δ.

Here, and below, the symbol C_n is used to denote different positive constants which do not depend on the functions under consideration and whose exact values are not important for the proof.

Note, the normal derivative of a function $u \in H^1(\Omega)$ might not be defined as an element of $L^2(\partial\Omega)$.

Lemma 2 *If $q(x)/r(x)$ is bounded below, there exists a shift $d \in \mathbb{R}$ of the spectral parameter $\lambda \to \lambda - d$ such that the function $q(x)$ can be assumed to be positive on Ω.*

Proof The shift $\lambda \to \lambda - d$ transforms the differential equation (24.1) into an equation of the same structure with $q(x) \to q(x) + dr(x)$. By taking

$$d > \sup_{x \in \Omega} \frac{-q(x)}{r(x)}$$

the function $q(x)$ can be assumed to be positive on Ω. Note further that the function $\rho(x)$ is invariant with respect to the shift of the spectral parameter. Indeed, the boundary condition (24.2) preserves its form after the shift but $\beta_1(x) \to \beta_1(x) - d\beta_1'$. The coefficients $\beta_2(x)$, $\beta_1'(x)$ remains the same and therefore $\rho(x)$ remains the same. □

The next result deals with the Riesz basis (see [13].

Lemma 3 *Every bounded invertible operator transforms any orthonormal basis of a Hilbert space Ξ into another basis of Ξ, a so-called Riesz basis.*

24.3 MAIN RESULT

We prove below that, with Assumption 1.1, eigenfunctions of the Problem 1.1 form a basis of $H^1(\Omega)$. The analysis is based on the reduction of the Problem 1.1 to an operator polynomial equation

$$L(\lambda)u = 0, \quad L(\lambda) = A_0 + \lambda A_1 \tag{24.11}$$

with the property

$$L(-\lambda_0) > 0 \text{ for some real } \lambda_0. \tag{24.12}$$

The definition of a weak solutions of Problem 1 follows by the known procedure (see [10, 11]). Multiplying equation (24.1) by a conjugate to an arbitrary function $\eta \in C^1(\Omega)$, integrating the result over Ω, and integrating by parts yields the identity

$$-\int_{\partial\Omega} p(x)\frac{\partial u}{\partial n}\bar{\eta}dx + \int_{\Omega}\Big(p(x)\nabla u \nabla \bar{\eta} + q(x)u\bar{\eta}\Big)dx = \lambda\int_{\Omega} r(x)u\bar{\eta}dx. \tag{24.13}$$

The normal derivative should be excluded from this identity because it might be undefined on $\partial\Omega$ (see Section 24.2). The boundary condition (24.2) shows that

$$-p(x)\frac{\partial u}{\partial n} = -\frac{\beta_1(x)}{\beta_2(x)}u - \lambda\frac{\beta_1'(x)}{\beta_2(x)}u.$$

Note, the function $\beta_2(x)$ is positive definite on $\partial\Omega$ according to Assumption 1.1c.

Definition 1 *The weak solution of the eigenvalue problem* (24.1)-(24.2) *is an element $u \in H^1(\Omega)$ satisfying the identity*

$$\int_{\Omega}\Big(p(x)\nabla u \nabla \bar{\eta} + q(x)u\bar{\eta}\Big)dx - \int_{\partial\Omega}\frac{\beta_1(x)}{\beta_2(x)}u\bar{\eta}dx$$
$$= \lambda\int_{\Omega} r(x)u\bar{\eta}dx + \lambda\int_{\partial\Omega}\frac{\beta_1'(x)}{\beta_2(x)}u\bar{\eta}dx \tag{24.14}$$

for any $\eta \in H^1(\Omega)$.

Lemma 4 *If $u \in C^2(\Omega)$, $p \in C^1(\Omega)$, $q, r \in C(\Omega)$, β_α, $\alpha = 1, 2$, and $\beta_1' \in C(\partial\Omega)$, and the boundary $\partial\Omega$ is smooth, then a weak solution satisfies the eigenvalue problem* (24.1)–(24.2) *in the classical sense.*

The proof is based on integration by parts and the arbitrariness of the function η. □

If the coefficients p, q, r, β_α, $\alpha = 1, 2$, β_1' and the boundary $\partial\Omega$ satisfy only Assumption 1, but not the conditions of Lemma 1, then Definition 1 is an extension of the classical solution conception (see, e.g., [10] for discussion of this philosophy).

Now introduce a new inner product in $H^1(\Omega)$

$$< u_1, u_2 >_3 = \int_\Omega (p(x) \nabla u_1 \nabla \bar{u}_2 + q(x) u_1 \bar{u}_2) dx \tag{24.15}$$

and the corresponding norm

$$||u||_3 = < u, u >_3^{1/2} . \tag{24.16}$$

The norm (24.16) is equivalent to the original norm (24.8c), i.e., there exist some positive constants C_2, C_3, independent on u, such that

$$C_2||u||_2 \leq ||u||_3 \leq C_3||u||_2. \tag{24.17}$$

Lemma 1 implies

$$||u||_1 \leq C_4||u||_3. \tag{24.18}$$

The reduction of the identity (24.14) to an operator equation is based on the following result.

Lemma 5 *The following representations hold*

$$\int_\Omega r(x) u\bar{\eta} dx = < Ru, \eta >_3, \tag{24.19a}$$

$$\int_{\partial\Omega} \frac{\beta_1(x)}{\beta_2(x)} u\bar{\eta} dx = < Ku, \eta >_3, \tag{24.19b}$$

$$\int_{\partial\Omega} \frac{\beta_1'(x)}{\beta_2(x)} u\bar{\eta} dx = < Tu, \eta >_3 \tag{24.19c}$$

where the operators R, K, T : $H^1(\Omega) \to H^1(\Omega)$ are compact, self-adjoint, and R and T are positive.

Proof The proof is based on the Riesz representation theorem and the embedding theorems for Sobolev spaces (see [10]). Denote

$$l_1(u, \eta) = \int_\Omega r(x) u\bar{\eta} dx, \tag{24.20a}$$

$$l_2(u, \eta) = \int_{\partial\Omega} \gamma(x) u\bar{\eta} dx \tag{24.20b}$$

with a measurable bounded function $\gamma : \partial\Omega \to \mathbb{R}$. Note, the functions $\beta_1(x)/\beta_2(x)$ and $\beta_1'(x)/\beta_2(x)$ have the required properties of $\gamma(x)$.

We prove first that $l_\alpha(u,\eta)$, $\alpha = 1,2$, are linear functionals on $\eta \in H^1(\Omega)$. The proof is based on the inequalities

$$|l_1(u,\eta)| \le C_5\|u\|_3\|\eta\|_3, \tag{24.21a}$$

$$|l_2(u,\eta)| \le C_6\|u\|_3\|\eta\|_3. \tag{24.21b}$$

We first check (24.21a). Schwarz inequality, Assumption 1b, and (24.17) imply

$$|l_1(u,\eta)| \le C_7\|u\|_0\|\eta\|_0 \le C_7\|u\|_0\|\eta\|_3 \le C_7 C_4^2\|u\|_3\|\eta\|_3. \qquad (24.22a).$$

The inequality (24.21b) is checked in the same way. In particular, Assumption 1c and inequality (24.18) are used. We find

$$|l_2(u,\eta)| \le C_8\|u\|_1\|\eta\|_1 \le C_9\|u\|_1\|\eta\|_3 \le C_{10}\|u\|_3\|\eta\|_3. \qquad (24.22b).$$

Therefore $l_\alpha(u,\eta)$, $\alpha = 1,2$, are linear functionals on $\eta \in H^1(\Omega)$ for any given $u \in H^1(\Omega)$. The Riesz representation theorem shows that the representations (24.19) are valid for some bounded operators R, K and T. The selfadjointness of these operators is obvious. The positiveness of R and T is a consequence of Assumptions 1b and 1c correspondingly.

The compactness of R is a well–known fact (see [10]). Now the compactness of K will be proven. Let the sequence $\{u_n\}$ converge weakly in $H^1(\Omega)$ to an element u. The boundedness of K implies the weak convergence of $\{Ku_n\}$ to Ku in $H^1(\Omega)$. The embedding theorems for Sobolev spaces imply that the sequences $\{u_n\}$ and $\{Ku_n\}$ converge strongly in $L^2(\partial\Omega)$. The representations (24.19b) and (24.20b) and the second of the inequalities (24.22b) imply

$$\|K(u_n - u_m)\|_3^2 = l_2(u_n - u_m, K(u_n - u_m))$$

$$\le C_9\|u_n - u_m\|_1\ \|K(u_n - u_m)\|_3 \to 0 \text{ as } n,\ m \to \infty. \tag{24.23}$$

Hence, $\|K(u_n - u_m)\|_3 \to 0$ as n, $m \to \infty$, and the compactness of K is proven. The proof for T is similar. □

Explicit representations for the operators R, K and T can be given in terms of Green's functions for some boundary value problems. For example, the identity (24.19b) and the definition (24.15) show that the function $v = Ku$ satisfies the boundary value problem

$$-div(p(x)\nabla v) + q(x)v = 0,\ x \in \Omega, \tag{24.24a}$$

with the boundary condition

$$p(x)\frac{\partial u}{\partial n} = \frac{\beta_1(x)}{\beta_2(x)}u,\ x \in \partial\Omega, \tag{24.24b}$$

(the strong formulation is given). Introducing the "surface" Green function $g(x,x_0)$, $x \in \Omega$, $x_0 \in \partial\Omega$, as the solution of the boundary value problem

$$-div(p(x)\nabla g) + q(x)g = 0,\ x \in \Omega \tag{24.25a}$$

with the boundary condition

$$p(x)\frac{\partial g}{\partial n} = \delta(x - x_0), \; x \in \partial\Omega, \tag{24.25b}$$

and using Green's second formula yields

$$v(x_0) = (Ku)(x_0) = \int_{\partial\Omega} \frac{\beta_1(x)}{\beta_2(x)} g(x, x_0)u(x)dx.$$

The proof of the statement of Lemma 5 about compactness would require the analysis of the kernel of the integral operator K.

Using Lemma 5 the identity (24.14) can be rewritten as

$$< u, \eta >_3 - < Ku, \eta >_3 = \lambda(< Ru, \eta >_3 + < Tu, \eta >_3). \tag{24.26}$$

The arbitrariness of $\eta \in H^1(\Omega)$ in identity (24.26) implies

$$u - Ku = \lambda(Ru + Tu). \tag{24.27}$$

Hence, the following result is proven.

Lemma 6 *The generalized eigenfunctions of the eigenvalue problem* (24.1)-(24.2) *satisfy the following operator polynomial equation in* $H^1(\Omega)$

$$L(\lambda)u = 0, \;\; L(\lambda) = U - \lambda V, \tag{24.28a}$$

$$U = I - K, \;\; V = R + T. \tag{24.28b}$$

Here I is the identity.

The following result is very important for this paper.

Lemma 7 *The operator pencil $L(-\lambda_0)$, from* (24.28*a*), *is positive definite for sufficiently large positive values of λ_0.*

Proof The representations (24.19) imply

$$< L(-\lambda_0)u, u >_3 = ||u||_3^2 - \int_{\partial\Omega} \frac{\beta_1(x)}{\beta_2(x)}|u|^2dx$$

$$+\lambda_0\Big(\int_\Omega r(x)|u|^2dx + \int_{\partial\Omega}\frac{\beta_1'(x)}{\beta_2(x)}|u|^2dx\Big). \tag{24.29}$$

According to Assumption 1c, the function $\beta_1'(x)/\beta_2(x)$ is positive definite on $\partial\Omega$ and the function $-\beta_1(x)/\beta_2(x)$ is bounded below. Hence, there exists a positive number λ_0, such that

$$\lambda_0 \inf_{x\in\partial\Omega}\frac{\beta_1'(x)}{\beta_2(x)} + \inf_{x\in\partial\Omega}\Big(-\frac{\beta_1(x)}{\beta_2(x)}\Big) > 0. \tag{24.30}$$

Hence,

$$< L(-\lambda_0)u, u >_3 \geq \|u\|_3^2 \tag{24.31}$$

and the operator pencil $L(-\lambda_0)$ is positive definite. □

Substitute the new spectral parameter

$$\mu = \lambda_0 + \lambda \tag{24.32}$$

into the operator pencil equation (24.27), where λ_0 is the parameter from Lemma 7. The spectral problem then has the form

$$\Big(L(-\lambda_0) - \mu(R+T)\Big)u = 0. \tag{24.33}$$

Introduce a new unknown

$$\psi = L(-\lambda_0)^{1/2}u. \tag{24.34}$$

Here $L(-\lambda_0)^{1/2}$ is the positive square root of the positive definite self-adjoint operator $L(-\lambda_0)$.

Lemma 8 *The function $\psi = L(-\lambda_0)^{1/2}u$ satisfies the operator equation*

$$\psi - \mu Z\psi = 0, \tag{24.35}$$

$$Z = L(-\lambda_0)^{1/2}(R+T)L(-\lambda_0)^{1/2}$$

with a positive compact self-adjoint operator Z. The spectrum $\{\mu_k\}$ of this equation is discrete, positive and has the only point of accumulation at $+\infty$, and the system of eigenfunctions ψ_k forms an orthogonal basis of $H^1(\Omega)$.

Proof The results follow from the standard Fredholm theorems for compact selfadjoint operators. □

The positiveness of the operator $R+T$ follows from Assumptions 1a, c. Our technique does not allow to remove those assumptions (for the one–dimensional case see [16]). We assume below that the orthogonal basis $\{\psi_k\}$ is orthonormal.

Theorem 1 *With Assumptions 1, the system of the generalized eigenfunctions of the eigenvalue problems (24.1)–(24.2) forms a Riesz basis of $H^1(\Omega)$. The eigenvalues form a real discrete set with the only point of accumulation at $+\infty$. The number of negative eigenvalues is at most finite, namely all eigenvalues satisfy the inequality $\lambda_k \geq -\lambda_0$, where λ_0 is the number from Lemma 7.*

Proof We use Lemmas 3 and 8 and the connection (24.32) between the spectra of the problems (24.28) and (24.35). □

24.4 RAYLEIGH–RITZ FORMULA AND LOWER BOUND ESTIMATION FOR EIGENVALUES

The spectral problem (24.33) can be rewritten as the following variational problem. Find a non–trivial element $u \in H^1(\Omega)$ such that the ratio

$$\mu = \frac{< L(-\lambda_0)u, u >_3}{< (R+T)u, u >_3} \tag{24.36}$$

yields the minimum value. Here λ_0 is the positive number from Lemma 7 such that the operator $L(-\lambda_0)$ is positive definite.

Both quadratic forms in (24.36) are represented explicitly by (24.29). By using a variational approach, a Rayleigh–Ritz procedure can be developed for numerical analysis of the spectral problem (24.1)–(24.2) (e.g., see [21]). The equation (24.36) also allows us to find a lower bound estimation for eigenvalues. Indeed, representations (24.29) and (24.36) together with formula (24.32) imply

$$\lambda = \frac{||u||_3^2 - \int_{\partial\Omega} \frac{\beta_1(x)}{\beta_2(x)} |u|^2 dx}{\int_\Omega r(x)|u|^2 dx + \int_{\partial\Omega} \frac{\beta_1'(x)}{\beta_2(x)} |u|^2 dx}. \qquad (24.37)$$

Obviously the same representation appears by substituting $\eta = u$ into (24.26).

Let S be the subset of the points on $\partial\Omega$ where $\beta_1(x)/\beta_2(x) \geq 0$. Assumption 1 allows to state that the following quantities are finite

$$C_{11} = \sup_S \frac{\beta_1(x)}{\beta_2(x)}, \qquad (24.38a)$$

$$C_{12} = \sup_\Omega \frac{r(x)}{q(x)}, \qquad (24.38b),$$

$$C_{13} = \sup_{\partial\Omega} \frac{\beta_1'(x)}{\beta_2(x)}. \qquad (24.38c)$$

Now some elementary inequalities

$$\int_\Omega r(x)|u|^2 dx \leq C_{12} \int_\Omega q(x)|u|^2 dx \leq C_{12}||u||_3^2 \qquad (24.39)$$

are needed. Using them and the inequality (24.18) in (24.37) yields the desired lower bound estimation

$$\lambda_k \geq \frac{1 - C_{11}C_4^2}{C_{12} + C_{13}C_4^2}. \qquad (24.40)$$

If the subset S is empty, the numerator in (24.40) should be replaced by 1.

24.5 DISCUSSION: EIGENFUNCTIONS AS AN ORTHOGONAL BASIS

Eigenfunctions of the eigenvalue problem (24.1)–(24.2) form a Riesz basis in the space $H^1(\Omega)$ equipped with the inner product (24.15) which is natural for the problem. We construct here an inner product such that the eigenfunctions form an orthonormal basis in $H^1(\Omega)$.

Let

$$< u_1, u_2 >_4 = < L(-\lambda_0)^{1/2} u_1, L(-\lambda_0)^{1/2} u_2 >_3 \qquad (24.41)$$

be a new inner product in $H^1(\Omega)$. Here λ_0 is any (real) number from Lemma 7.

Theorem 2 *Let λ_0 be any number from Lemma 7. Then, with Assumptions 1, the system of the generalized eigenfunctions of the eigenvalue problems* (24.1)–(24.2) *forms an orthonormal basis of $H^1(\Omega)$ with the inner product* (24.41)

Proof According to Theorem 3, only orthogonality relation should be checked. Using the definition (24.41), the representation (24.34), and Lemma 8 yields

$$< u_k, u_n >_4 = < L(-\lambda_0)^{\frac{1}{2}} u_k, L(-\lambda_0)^{\frac{1}{2}} u_n >_3 = < \psi_k, \psi_n >_3 = \delta_{k,n}. \square \quad (24.42)$$

Lemma 9 *The generalized eigenfunctions u_k of the eigenvalue problem* (24.1)–(24.2) *satisfy, for any number $\lambda^* \in \mathbb{C}$, the orthogonality relations ($k \neq n$)*

$$< L(\lambda^*) u_k, u_n >_3 = 0, \quad (24.43a)$$

$$< V u_k, u_n >_3 = 0, \quad (24.43b)$$

$$< U u_k, u_n >_3 = 0. \quad (24.43c)$$

Proof The relation (24.43a) follows directly from (24.43b,c) and (24.28). The relation (24.43c) follows from (24.43b) by using $U u_k - \lambda_k V u_k = L(\lambda_k) u_k = 0$. The proof of (24.43b) is based on equation (24.28) and the selfadjointness of the operators U and V

$$(\lambda_k - \lambda_n) < V u_k, u_n >_3 = < \lambda_k V u_k, u_n >_3 - < V u_k, \lambda_n u_n >_3$$

$$= < U u_k, u_n >_3 - < u_k, \lambda_n V u_n >_3 = < U u_k, u_n >_3 - < u_k, U u_n >_3 = 0 \quad (24.44)$$

and (24.43b) follows. □

In the particular case $\lambda^* = -\lambda_0$ with λ_0 from Lemma 4, the orthogonality relation (24.43*a*) follows directly from (24.42), i.e.

$$< L(-\lambda_0) u_k, u_n >_3 = \delta_{k,n} \quad (24.45)$$

for sufficiently large λ_0.

The orthogonality relation (24.45) can be rewritten in an explicit form by means of (24.28) and (24.19)

$$< L(-\lambda_0) u_k, u_n >_3 = \int_\Omega (p(x) \nabla u_k \nabla \bar{u}_n + q(x) u_k \bar{u}_n) dx - \int_{\partial\Omega} \frac{\beta_1(x)}{\beta_2(x)} u_k \bar{u}_n dx$$

$$+ \lambda_0 \Big(\int_\Omega r(x) u_k \bar{u}_n dx + \int_{\partial\Omega} \frac{\beta_1'(x)}{\beta_2(x)} u_k \bar{u}_n dx \Big) = \delta_{k,n}. \quad (24.46)$$

Now a sufficient condition is formulated which allows to choose $\lambda_0 = 0$ in all the considerations of Section 3, and also in (24.41), (24.42), (24.45), (24.46).

Lemma 10 *Lemma 7 holds for $\lambda_0 = 0$ if the function $\beta_1(x)/\beta_2(x)$ is non-positive on $\partial\Omega$.*

Proof The result follows from (24.30) and Assumption 1c. □

The following is assumed in the rest of this section.

Assumption 3

$$\lambda_0 = 0. \quad (24.47)$$

Hence $L(-\lambda_0) = U$. In particular, the orthogonality relation (24.45) has the form

$$< Uu_k, u_n >_3 = \delta_{k,n}. \tag{24.48}$$

Lemma 8 and (24.32) imply now that all $\lambda_k = \mu_k > 0$. The representation (24.29) implies

$$< L(0)u, u >_3 = \|u\|_3^2 - \int_{\partial\Omega} \frac{\beta_1(x)}{\beta_2(x)} |u|^2 dx. \tag{24.49}$$

Define the Hilbert space $L^2(r;\Omega)$ with the inner product

$$< u_1, u_2 >_5 = < Vu_1, u_2 >_3$$

$$= \int_\Omega r(x) u_1 \bar{u}_2 dx + \int_{\partial\Omega} \frac{\beta_1'(x)}{\beta_2(x)} u_1 \bar{u}_2 dx. \tag{24.50}$$

It is convenient to normalize the eigenfunctions u_k according to the inner product (24.50). Let

$$\Gamma_k = \sqrt{\lambda_k} u_k.$$

Lemma 11 *The following orthogonality relation holds*

$$< \Gamma_k, \Gamma_n >_5 = \delta_{k,n}. \tag{24.51}$$

Proof The relation follows from the orthogonality relations (24.43b) and (24.48) and the equation (24.28a). Indeed

$$< \Gamma_k, \Gamma_n >_5 = < V\Gamma_k, \Gamma_n >_3 = \sqrt{\lambda_k \lambda_n} < Vu_k, u_n >_3 = 0 \text{ if } k \neq n, \tag{24.52a}$$

$$< \Gamma_k, \Gamma_k >_5 = < V\Gamma_k, \Gamma_k >_3 = \lambda_k < V\dot{u}_k, u_k >_3 = < Uu_k, u_k > = 1. \square \tag{24.52b}$$

Theorem 3 *With Assumptions 1 and 3, the system of generalized eigenfunctions $\{\Gamma_k\}$ of the eigenvalue problem (24.1)–(24.2) forms a basis of $L^2(r;\Omega)$ and $H^1(\Omega)$ which is orthonormal in $L^2(r;\Omega)$ and orthogonal in $H^1(\Omega)$.*

Proof The space $H^1(\Omega)$ is a dense subset of $L^2(r;\Omega)$. The system of eigenfunctions $\{\Gamma_k\}$ is a basis in $H^1(\Omega)$, which is orthonormal in $L^2(r;\Omega)$ (see Lemma 11). The eigenfunction expansion for an arbitrary element $F \in H^1(\Omega)$ with respect to the orthonormal basis $\{\Gamma_k\}$ has the form

$$F = \sum_k < F, \Gamma_k >_5 \Gamma_k \tag{24.53}$$

and converges in $H^1(\Omega)$. The density of $H^1(\Omega)$ in $L^2(r;\Omega)$ implies the convergence of the series (24.53) for any element $F \in L^2(r;\Omega)$, where the convergence takes place in $L^2(r;\Omega)$. $\square$

24.6 PHYSICAL MOTIVATION OF THE SPECTRAL PROBLEM (24.1)–(24.2)

The following initial boundary value problem is considered. Find the solution $u(x,t): \Omega \times (0,\infty) \rightarrow \mathbb{R}$ of the hyperbolic equation

$$div(p(x) \nabla u) - q(x)u = r(x)u_{tt}, \; x \in \Omega, \; t \in (0,\infty) \tag{24.54}$$

with the boundary condition

$$\beta_1(x)u - \beta_2(x)p(x)\frac{\partial u}{\partial n} = \beta_1'(x)u_{tt}, \; x \in \partial\Omega \tag{24.55}$$

and initial conditions

$$u(x,0) = u^0(x), \; u_t(x,0) = u^1(x). \tag{24.56}$$

The initial boundary value problem (24.54)–(24.55) describes the oscillations of an n–dimensional non–homogeneous continuum Ω whose boundary $\partial\Omega$ is attached to an $(n-1)$–dimensional continuum. Problem (24.1)–(24.2) appears naturally as a result of the Laplace transform of (24.54)–(24.55). We construct the **(formal)** solution for (24.54)–(24.56) by means of the eigenfunction expansion developed in the previous sections.

First, the energy conservation law is derived. Multiplying (24.54) by u_t, integrating over Ω and integrating by parts yields

$$\frac{d}{dt}\Big(\frac{1}{2}\int_\Omega (p(\nabla u)^2 + qu^2 + ru_t^2)dx\Big) = \int_{\partial\Omega} p\frac{\partial u}{\partial n}u_t dx. \tag{24.57}$$

Using the boundary condition (24.55) to simplify the right–hand side of (24.57) yields the energy conservation law

$$\frac{dE[u]}{dt} = 0, \tag{24.58a}$$

$$E[u] = \frac{1}{2}(< Uu, u >_3 + < Vu_t, u_t >_3). \tag{24.58b}$$

The functional $E[u]$ can be called the **energy** of the physical system. The first term in (24.58b) describes the potential energy and the second one describes the kinetic energy. Assumption 3 guarantees the positiveness of E for any (real valued) field u. Repeating the derivation of Section 24.3 yields the initial value problem (see (24.28a))

$$Uu + Vu_{tt} = 0, \tag{24.59a}$$

$$u(0) = u^0, u_t(0) = u^1. \tag{24.59b}$$

The solution of problem (24.59) can be constructed in terms of the eigenfunctions u_k of the operator polynomial (24.28a)

$$u(x,t) = \sum_k C_k(t)u_k(x). \tag{24.60}$$

Substituting (24.60) into (24.59a) yields

$$\sum_k \left(\frac{d^2}{dt^2}C_k + \lambda_k C_k\right) V u_k = 0. \tag{24.61}$$

The orthogonality relation (24.43b) implies the ordinary differential equation for C_k

$$\frac{d^2}{dt^2}C_k + \lambda_k C_k = 0. \tag{24.62}$$

The initial conditions for (24.62) can be found by means of (24.59b) and (24.60)

$$C_k(0) = \frac{< Vu^0, u_k >_3}{< Vu_k, u_k >_3}, \ \frac{d}{dt}C_k(0) = \frac{< Vu^1, u_k >_3}{< Vu_k, u_k >_3}. \tag{24.63}$$

Note, all the forms $< Vu_k, u_k >_3 > 0$. The field u can be represented finally as the following (**formal**) series

$$u(x,t) \tag{24.64}$$

$$= \sum_k \frac{1}{< Vu_k, u_k >_3} \left(< Vu^1, u_k >_3 \frac{\sin \sqrt{\lambda_k} t}{\sqrt{\lambda_k}} + < Vu^0, u_k >_3 \cos \sqrt{\lambda_k} t \right) u_k(x).$$

It is assumed here that all $\lambda_k \neq 0$.

References

[1] Walter, J.: Eigenvalue problems with eigenvalue parameter in the boundary conditions. Math. Z. 133 (1973), 301–312.

[2] Fulton, C.T.: Singular eigenvalue problems with eigenvalue parameter contained in the boundary conditions. Proc. Royal Soc. Edinburgh 87A (1980), 1–34.

[3] Fulton, C.T.: Two–point boundary value problems with eigenvalue parameter contained in the boundary conditions. Proc. Royal Soc. Edinburgh 77A (1977), 293–308.

[4] Hinton, D.: An expansion theorem for an eigenvalue problem with eigenvalue parameter in the boundary condition. Quarterly J. Math. 2 (1979), 33–42.

[5] Titchmarsh, E.C.: Eigenfunction expansions associated with second order differential equations, I. Univ. Press, Oxford, London, 1962.

[6] Belinskiy, B.P., Dauer, J.P.: Eigenoscillations of mechanical systems with boundary conditions containing the frequency. Quarterly Appl. Math., to appear.

[7] Eastham, M.S.P.: An eigenvalue problem with the parameter in the boundary condition. Quarterly J. Math. Oxford 13 (1962), 304–320.

[8] Eastham, M.S.P.: Eigenvalue problems with the parameter in the boundary condition. Quarterly J. Math. Oxford 14 (1963), 259–272.

[9] Eastham, M.S.P.: The distribution of the eigenvalues in certain eigenvalue problems. Quarterly J. Math. Oxford 15 (1964), 251–257.

[10] Ladyzhenskaia, O.A.: The boundary value problems of mathematical physics. Springer–Verlag, New York, 1985.

[11] Gilbarg, D., Trudinger, N.S.: Elliptic partial differential equations of second order, 2nd ed. Springer–Verlag, Berlin, 1983.

[12] Keldysh, M.V.: On the characteristic values and characteristic functions of certain classes of non–selfadjoint equations. Doklady Akad. Nauk SSSR 77 (1951), 11–14 (Russian).

[13] Gohberg, I.C., Krein, M.G.: Introduction to the theory of linear non–selfadjoint operators, Translation of Mathematical Monographs 18. Amer. Math. Soc., Providence, R.I., 1969.

[14] Rodman, L.: An introduction to operator polynomials. Birkhäuser Verlag, Boston, 1989.

[15] Markus, A.S.: Introduction to the spectral theory of polynomial pencils, Translation of Mathematical Monographs. Amer. Math. Soc., Providence, R.I., 1988.

[16] Langer, H., Schneider, A.: On spectral properties of regular quasidefinite pencils $F - \lambda G$. Results Math. 19 (1991), 89–109.

[17] Schäfke, F.W., Schneider, A.: S–hermitesche Rand–Eigenwert Probleme. I. Math. Ann. 162 (1965), 9–26.

[18] Schäfke, F.W., Schneider, A.: S–hermitesche Rand–Eigenwert Probleme. II. Math. Ann. 165 (1966), 236–260.

[19] Kostyuchenko, A.G., Shkalikov, A.A.: Self–adjoint quadratic operator pencils and elliptic problems. Func. Anal. Appl. 17 (1983), 109–128.

[20] Kostyuchenko, A.G., Shkalikov, A.A.: On the theory of self–adjoint quadratic operator polynomials. Vestnik Moscow State Univ. Ser. 1. Mathematics, Mechanics, 6 (1983), 40–51 (Russian).

[21] Stakgold, I.: Boundary value problem of mathematical physics, II. Macmillan Co., New York, 1971.

25 ON INDEFINITE NONLINEAR NEUMANN PROBLEMS

Klaus Pflüger

Freie Universität Berlin
Institut für Mathematik I
Arnimallee 3
14195 Berlin, Germany
pflueger@math.fu-berlin.de

Abstract: We study the subcritical nonlinear Neumann problem $-\Delta u = a(x)u^p$ in a bounded domain $\Omega \subset \mathbb{R}^n$ with boundary condition $\partial_n u = b(x)u^q$ on $\partial\Omega$, where a and b are continuous functions which may change sign. We derive existence and nonexistence results for positive solutions of this problem. In addition, we extend these results to a nonlinear interface problem in a domain $\Omega = \Omega_1 \cup \Omega_2$ with interface conditions $\partial_n u_1 = \partial_n u_2$, $b(x)|u_1|^{q-1}u_1 = \lambda(x)u_2$ on $\overline{\Omega}_1 \cap \overline{\Omega}_2$.

25.1 INTRODUCTION

In this paper we study existence of positive solutions to the following Neumann problem in a bounded domain $\Omega \subset \mathbb{R}^n$ ($n \geq 3$)

$$-\Delta u = a(x)u^p, \quad u > 0 \quad \text{in} \quad \Omega, \tag{25.1}$$

$$\partial_n u = b(x)u^q \quad \text{on} \quad \Gamma = \partial\Omega. \tag{25.2}$$

Here $\partial_n u$ denotes the outward normal derivative of u and we assume that a and b are bounded and continuous functions which may change sign, and the exponents p and q are subcritical, i. e. $1 < p < (n+2)/(n-2)$, $1 < q < n/(n-2)$.

This problem with constant coefficients a and b (and its parabolic analog) has recently been studied by Chipot, Fila and Quittner [5] as a model for heat conduction in a body where cooling and heating (according to the sign of a and b) appear inside and at the boundary at a rate proportional to some power

H.G.W. Begehr et al. (eds.), Partial Differential and Integral Equations, 335–346.

of u. If the exponents p and q are critical, this problem appears in conformal deformations of manifolds (cf. [4], [9]) and in the best constant for the Sobolev trace inequality (see [8]).

In the paper [5], Chipot et. al. considered the case $a = \text{const.} < 0$ and $b = 1$. We summarize their main results for general bounded domains in $\mathbb{R}^n$:

(i) If $p \leq q$ and $a < a_0$, then then there exists a nontrivial solution of (25.1), (25.2), where $a_0 = 0$ if $p < q$, $a_0 = a_\Omega := -|\Gamma|/|\Omega|$ if $p = q$.

(ii) If $p > q$, then there exists a constant $a_1 < 0$, such that (25.1), (25.2) has a nontrivial solution for $a < a_1$ and there is no solution for $a > a_1$.

(iii) If $p = q$ and $a > a_\Omega$, then there exists no nontrivial solution.

It should be remarked that the nonexistence result (iii) is used to prove that every solution of the corresponding evolution equation blows up in a finite time, while a global existence result is only obtained under more restrictive conditions on the geometry of Ω (in particular, if Ω is a ball) or on the growth of the nonlinearities (for instance if $q < (n+1)/(n-1)$ and $p > 2q - 1$).

For unbounded domains Ω, problems of the type (25.1), (25.2) have been studied earlier by the author in the definite case (see [13], [15]). Indefinite semilinear elliptic problems with homogeneous Dirichlet or Neumann (i. e. $b(x) = 0$) conditions have been studied recently in [2], [3] and [17].

Our aim here is to summarize some results for semilinear elliptic equations with nonlinear boundary- or transmission conditions in the presence of indefinite nonlinearities, thus extending the results in [3], [6] and [14] to this case. Our main results for problem (25.1), (25.2) are the following.

Theorem 1 *If $1 < p = q < n/(n-2)$, then problem (25.1), (25.2) has a nontrivial positive solution if and only if*

$$\int_\Omega a(x)\,dx + \int_\Gamma b(x)\,d\Gamma < 0 \quad \textit{and} \tag{25.3}$$

$$a(x) > 0 \quad \textit{or} \quad b(x) > 0 \quad \textit{somewhere in } \Omega \textit{ or } \Gamma.$$

For constant coefficients, this result recovers exactly (i) and (iii) in the case $p = q$. In the general case, we have the following.

Theorem 2 *If $p \neq q$, there is no nontrivial positive solution if*

$$b(x) \geq 0 \quad \textit{and} \quad \int_\Omega a(x) \geq 0 \qquad \textit{or} \qquad a(x) \geq 0 \quad \textit{and} \quad \int_\Gamma b(x) d\Gamma \geq 0\,.$$

If $p > q$, $\int_\Omega a(x)dx < 0$, $b(x) \leq 0$ and $a(x) > 0$ somewhere, then there exists a positive solution.

If $p < q$, $\int_\Gamma b(x)d\Gamma < 0$, $a(x) \leq 0$ and $b(x) > 0$ somewhere, then there exists a positive solution.

These theorems reflect some typical behaviour of equations with non-coercive linear part (the first eigenvalue of the linearized problem (25.1), (25.2) is zero) and indefinite nonlinearities. We remark that the theorems above can be generalized to problems of the type $-\Delta u + m(x)u = a(x)u^p$ in Ω, where the first eigenvalue of $-\Delta + m(x)$ with homogeneous Neumann boundary conditions is zero. The coercive case, where the first eigenvalue is strictly positive, is easier to handle and will not be considered explicitly in this paper. For example, one has the following existence results:

Theorem 3 *If the first eigenvalue of $-\Delta + m(x)$ with homogeneous Neumann condition is positive, then there exists a nontrivial positive solution of $-\Delta u + m(x)u = a(x)u^p$ in Ω with boundary condition (1.2) in the following cases:*
if $p = q$ and $a(x) > 0$ or $b(x) > 0$ somewhere;
if $p > q$, $b(x) \leq 0$ and $a(x) > 0$ somewhere;
if $p < q$, $a(x) \leq 0$ and $b(x) > 0$ somewhere.

This paper is organized as follows: In the next section we prove some necessary conditions, which will imply the nonexistence statements in Theorems 1 and 2, in Section 25.3 we consider the case $p = q$, in Section 25.4 the case $p \neq q$. The techniques we use here, are based on the paper [3]. Finally, in the last Section 25.5 we present some extensions to nonlinear transmission problems with indefinite character.

25.2 NECESSARY CONDITIONS

To obtain necessary conditions for the existence of a nontrivial solution, we assume that $u \geq 0$ is a solution of (25.1), (25.2). According to the regularity result of Cherrier [4] we can apply the maximum principle and claim that $u > 0$ in Ω. If $u(x_0) = 0$ for some x_0 in Γ, then the Hopf boundary lemma implies $\partial_n u(x_0) < 0$ which is clearly impossible in view of (25.2).

Therefore, we have $u > 0$ in $\overline{\Omega}$ and we may multiply (25.1) with u^{-p} to obtain by partial integration

$$\begin{aligned}
\int_\Omega a(x)\,dx &= \int_\Omega -\Delta u\, u^{-p} dx \\
&= \int_\Omega \nabla u \nabla (u^{-p})\,dx - \int_\Gamma \partial_n u\, u^{-p} d\Gamma \\
&= -p \int_\Omega |\nabla u|^2 u^{-p-1}\,dx - \int_\Gamma b(x) u^q u^{-p}\,d\Gamma\,.
\end{aligned}$$

This shows

$$\int_\Omega a(x)\,dx + \int_\Gamma b(x) u^{q-p}\,d\Gamma < 0\,. \tag{25.4}$$

In the same way, multiplying with u^{-q},we obtain

$$\int_\Omega a(x)u^{p-q}\,dx + \int_\Gamma b(x)\,d\Gamma < 0\,. \tag{25.5}$$

Therefore $\int_\Omega a(x)\,dx + \int_\Gamma b(x)\,d\Gamma < 0$ is a necessary condition in the case $p = q$.

On the other hand, direct integration of (25.1) leads to

$$\int_\Omega a(x)u^p\,dx = -\int_\Omega \Delta u\,dx = -\int_\Gamma \partial_n u\,d\Gamma,$$

which shows

$$\int_\Omega a(x)u^p\,dx + \int_\Gamma b(x)u^q\,d\Gamma = 0\,. \tag{25.6}$$

Therefore, a or b must be positive somewhere, which proves the "only if" statement in Theorem 1.

From the inequalities (25.4) and (25.5) we derive that

$$\text{if} \qquad b(x) \geq 0 \qquad \text{then} \qquad \int_\Omega a(x)\,dx < 0 \tag{25.7}$$

is necessary, while

$$\text{if} \qquad a(x) \geq 0 \qquad \text{then} \qquad \int_\Gamma b(x)\,d\Gamma < 0 \tag{25.8}$$

is a necessary condition for existence. This proves the first part of Theorem 2.

25.3 SUFFICIENT CONDITIONS IN THE CASE $p = q$

To obtain a nontrivial solution of (25.1), (25.2), we consider the following constrained maximization problem. Let

$$I(u) = \int_\Omega a(x)|u|^{q+1}\,dx + \int_\Gamma b(x)|u|^{q+1}\,d\Gamma\,.$$

Then we look for a solution of the variational problem

$$M = \max\{I(u) \mid u \in S\}\,, \quad S = \{u \in H^1(\Omega) \mid \int_\Omega |\nabla u|^2 = 1\}\,. \tag{25.9}$$

We assume that (25.3) from Theorem 1 holds. To show that $M = \sup\{I(u) \mid u \in S\}$ is finite, we use the following decomposition of $H^1(\Omega)$: for $u \in H^1(\Omega)$ we set $t = |\Omega|^{-1}\int_\Omega u(x)\,dx$, the avarage of u in Ω, and $v = u - t$. Then $\int_\Omega v(x)\,dx = 0$ and we set

$$H^1(\Omega) = \mathbb{R} \oplus V\,, \quad \text{where} \quad V = \{v \in H^1(\Omega) \mid \int_\Omega v\,dx = 0\}\,. \tag{25.10}$$

From the Poincaré inequality $||\varphi||^2_{H^1} \leq C\left(\int_\Omega |\nabla\varphi|^2 dx + |\int_\Omega \varphi\,dx|^2\right)$ for $\varphi \in H^1(\Omega)$, we see that

$$||u||^2 = |t|^2 + ||v||^2_V\,, \quad \text{with} \quad ||v||^2_V = \int_\Omega |\nabla v|^2 dx$$

defines an equivalent norm on $H^1(\Omega)$.

Under the assumption that a is positive somewhere, we get $M > 0$, since we can choose a function with compact support in $\{x \in \Omega \mid a(x) > 0\}$, $\int_\Omega |\nabla u|^2 dx = 1$. On the other hand, if $b(x) > 0$ somewhere, we can find a smooth function φ defined in $\overline{\Omega}$ with $\text{supp}(\varphi_{|\Gamma}) \subset \{x \in \Gamma \mid b(x) > 0\}$, such that $I(\varphi) > 0$. Then, for a suitable $t > 0$ we find that $\int_\Omega |\nabla(t\varphi)|^2 dx = 1$ and $I(\varphi) > 0$, which again shows that $M > 0$.

Now let $u_k \in S$ be a maximizing sequence, i. e. $I(u_k) \to M$, and let $u_k = t_k + v_k$ be the decomposition according to (25.10). Since $\int_\Omega |\nabla u_k|^2 dx = \int_\Omega |\nabla v_k|^2 dx = 1$, v_k converges weakly in V, and since the embedding $V \hookrightarrow L^{q+1}(\Omega)$ and trace $V \to L^{q+1}(\Gamma)$ are compact, v_k has a strongly convergent subsequence in $L^{q+1}(\Omega)$ and $L^{q+1}(\Gamma)$.

Next we show that $|t_k|$ is bounded. Assume that this is not true. Then, along a subsequence, $|t_k|^{-1} v_k \to 0$ in $L^{q+1}(\Omega)$ and in $L^{q+1}(\Gamma)$. Consequently, from (25.3) we get

$$\begin{aligned} \int_\Omega a(x)|1 + |t_k|^{-1} v_k|^{q+1} dx + \int_\Gamma b(x)|1 + |t_k|^{-1} v_k|^{q+1} d\Gamma & \longrightarrow \\ \int_\Omega a(x)\, dx + \int_\Gamma b(x) d\Gamma & < 0. \end{aligned}$$

From this we conclude

$$\begin{aligned} I(u_k) &= |t_k|^{q+1} \left(\int_\Omega a(x)|1 + |t_k|^{-1} v_k|^{q+1} dx + \int_\Gamma b(x)|1 + |t_k|^{-1} v_k|^{q+1} d\Gamma \right) \\ &\to -\infty, \end{aligned}$$

which is not possible. Therefore, t_k is bounded and contains a convergent subsequence. We may assume that u_k converges weakly in $H^1(\Omega)$ and strongly in $L^{q+1}(\Omega)$ and $L^{q+1}(\Gamma)$. The limit function u satisfies $I(u) = M$ and u is not a constant in view of (25.3). Hence $0 < \int_\Omega |\nabla u|^2 dx \le 1$ and there exists $\mu \ge 1$ such that $\mu u \in S$. Since $I(\mu u) = \mu^{q+1} M \le M$, we have $\mu = 1$. Further we claim $u \ge 0$, since we can choose a maximizing sequence of positive functions. Consequently there exists a Lagrange multiplier $\lambda \in \mathbb{R}$ such that

$$2\lambda \int_\Omega \nabla u \nabla \varphi = (q+1) \left(\int_\Omega a(x) u^q \varphi\, dx + \int_\Gamma b(x) u^q \varphi\, d\Gamma \right) \tag{25.11}$$

for every $\varphi \in H^1(\Omega)$. Setting $\varphi = u$, we get

$$\frac{2\lambda}{q+1} \int_\Omega |\nabla u|^2 dx = I(u) = M > 0\,,$$

which shows $\lambda > 0$. Let $\tilde{\lambda} = 2\lambda/(q+1)$. From (25.11) we obtain

$$-\tilde{\lambda} \Delta u = a(x) u^q \quad \text{in} \quad \Omega$$

and by partial integration

$$\tilde{\lambda} \int_\Gamma \partial_n u \, \varphi \, d\Gamma = \int_\Gamma b(x) u^q \varphi \, d\Gamma$$

for every $\varphi \in H^1(\Omega)$. Therefore $w = \tilde{\lambda}^{-1/(q-1)} u$ is a solution of (25.1), (25.2) and the maximum principle implies $u > 0$ in Ω. This proves the existence statement of Theorem 1.

25.4 SUFFICIENT CONDITIONS IN THE CASE $p \neq q$

If $p \neq q$ we cannot use constrained maximization as before, since rescaling is no longer possible in this case. Instead we shall use the mountain-pass lemma to obtain a nontrivial solution. Define

$$J(u) = \int_\Omega \left[\frac{1}{2} |\nabla u|^2 - \frac{1}{p+1} a(x) |u|^{p+1} \right] dx - \int_\Gamma \frac{1}{q+1} b(x) |u|^{q+1} d\Gamma .$$

We shall show that J satisfies the assumptions of the mountain-pass lemma (cf. [16]), that is:

(i) J satisfies the Palais-Smale condition.

(ii) There are numbers $\rho, \delta > 0$, such that $J(u) \geq \delta$ for every $u \in H^1(\Omega)$ with $\|u\| = \rho$.

(iii) $J(0) = 0$ and there exists $e \in H^1(\Omega)$, $\|e\| > \rho$, such that $J(e) < \delta$.

In the sequel we need the following estimates which easily follow from the continuity of the mappings $w \mapsto \int_\Omega a(x)|w|^{p+1} dx$ and $w \mapsto \int_\Gamma b(x)|w|^{q+1} d\Gamma$ from $L^{p+1}(\Omega) \to \mathbb{R}$ (resp. $L^{q+1}(\Gamma) \to \mathbb{R}$):

There exists a positive real number $\eta > 0$ such that for all $t \in \mathbb{R}$, $v \in V$ satisfying $\|v\|_V \leq \eta |t|$ we have

$$\int_\Omega a(x) |t + v(x)|^{p+1} dx \leq \frac{1}{2} |t|^{p+1} \int_\Omega a(x) dx \ \ (\text{if } \int_\Omega a(x)\, dx < 0), \tag{25.12}$$

$$\int_\Gamma b(x) |t + v(x)|^{q+1} d\Gamma \leq \frac{1}{2} |t|^{q+1} \int_\Gamma b(x) d\Gamma \ \ (\text{if } \int_\Gamma b(x)\, d\Gamma < 0), \tag{25.13}$$

for every $1 < p < (n+2)/(n-2)$, $1 < q < n/(n-2)$.

Now we prove (i)–(iii) in the case $p > q$.

To prove (i), let u_k be a Palais-Smale sequence in $H^1(\Omega)$ for J. Then for every $\varepsilon > 0$, $k \geq k_\varepsilon$ we have

$$\begin{aligned} |J(u_k)| &= \left| \int_\Omega \left[\tfrac{1}{2} |\nabla u_k|^2 - \tfrac{1}{p+1} a(x) |u_k|^{p+1} \right] dx - \int_\Gamma \tfrac{1}{q+1} b(x) |u_k|^{q+1} d\Gamma \right| \\ &\leq M \end{aligned}$$

and

$$|\langle J'(u_k), u_k\rangle| = \left|\int_\Omega \left[|\nabla u_k|^2 - a(x)|u_k|^{p+1}\right] dx - \int_\Gamma b(x)|u_k|^{q+1} d\Gamma\right|$$
$$\leq \varepsilon\|u_k\|. \tag{25.14}$$

Setting $\varepsilon = 1$, multiplying (25.14) with $1/(p+1)$ and adding, we obtain

$$\left(\frac{1}{2} - \frac{1}{p+1}\right)\int_\Omega |\nabla u_k|^2 dx + \left(\frac{1}{p+1} - \frac{1}{q+1}\right)\int_\Gamma b(x)|u_k|^{q+1} d\Gamma - \|u_k\| \leq M.$$

Now, by our assumptions, the middle term on the left-hand side is nonnegative. Setting $u_k = t_k + v_k \in \mathbb{R} \oplus V$ and $\vartheta = 1/2 - 1/(p+1) > 0$, we get

$$\vartheta\|v_k\|_V^2 - \|t_k + v_k\| \leq M. \tag{25.15}$$

We now want to show that $|t_k|$ and v_k are bounded. Without restrictions we may assume that $|t_k| \geq c > 0$ for all k. Then (25.15) implies

$$\vartheta\frac{\|v_k\|_V^2}{|t_k|} - \left(1 + \frac{\|v_k\|_V^2}{|t_k|^2}\right)^{1/2} \leq M/|t_k|,$$

or

$$\vartheta\frac{\|v_k\|_V^2}{|t_k|} - \frac{\|v_k\|_V}{|t_k|} \leq M_1. \tag{25.16}$$

Now suppose that t_k is not bounded. If $\|v_k\|_V \to \infty$, we obtain from (25.16)

$$\|v_k\|_V\left(\vartheta\frac{\|v_k\|_V}{|t_k|} - \frac{1}{|t_k|}\right) \leq M_1$$

and since $1/|t_k| \to 0$, we see that $\|v_k\|_V/|t_k| \to 0$. That is, to every $\eta > 0$ there is a k_η such that $\|v_k\|_V \leq \eta|t_k|$ holds for $k \geq k_\eta$.

On the other hand, if $\|v_k\|_V$ was bounded, the inequality $\|v_k\|_V \leq \eta|t_k|$ is trivially satisfied for k large enough.

In both cases we can apply (25.12) to obtain

$$J(u_k) \geq \frac{1}{2}\|v_k\|_V^2 - \frac{|t_k|^{p+1}}{2}\frac{\alpha}{p+1}, \quad \text{where} \quad \alpha = \int_\Omega a(x)\, dx.$$

Since $\alpha < 0$, we get $J(u_k) \to \infty$ which is a contradiction. Therefore, t_k must be bounded. Now from (25.15) it is easy to see that $\|v_k\|_V$ must also be bounded if t_k is. This shows that every Palais-Smale sequence for J is bounded. Now by standard arguments, using the compact embedding, resp. trace $V \hookrightarrow L^{p+1}(\Omega)$, $V \to L^{q+1}(\Gamma)$, we conclude that u_k has a convergent subsequence, thus (i) is proved.

To show that (ii) holds, fix $\rho > 0$ and let η be such that (25.12) holds. Choose $u = t + v \in H^1(\Omega)$ with $\|u\| = \rho$, i. e. $t^2 + \|v\|_V^2 = \rho^2$.

Assume first that $||v||_V \leq \eta\,|t|$. Then $t^2 \geq \rho^2(1+\eta^2)^{-1}$ and (25.12) implies

$$J(u) \geq \int_\Omega \left[\frac{1}{2}|\nabla v|^2 - \frac{1}{p+1}a(x)|t+v|^{p+1}\right] dx \geq \frac{-\alpha}{2}\frac{|t|^{p+1}}{p+1}.$$

Since $-\alpha > 0$, we obtain

$$J(u) \geq \frac{-\alpha}{2(p+1)}\left(\frac{\rho^2}{1+\eta^2}\right)^{(p+1)/2} > 0.$$

Now consider the case $||v||_V \geq \eta\,|t|$. We have $||u||^2 \leq t^2 + ||v||_V^2 \leq (\frac{1}{\eta^2}+1)||v||_V^2$. From the Sobolev embedding theorem we get

$$\left|\int_\Omega a(x)|u|^{p+1}dx\right| \leq C_1||u||^{p+1} \leq C_1(\frac{1}{\eta^2}+1)^{(p+1)/2}||v||_V^{p+1},$$

with a constant $C_1 > 0$. For v small enough, we infer

$$J(u) \geq \frac{1}{2}||v||_V^2 - \frac{C_2}{p+1}||v||_V^{p+1} \geq \frac{1}{4}||v||_V^2.$$

Since $||u|| = \rho$ and $||v||_V \geq \eta||u||(1+\eta^2)^{-1/2}$, we obtain $J(u) \geq \frac{1}{4}\frac{\eta^2\rho^2}{1+\eta^2} > 0$, hence the choice $\delta = \min\left\{\frac{1}{4}\frac{\eta^2\rho^2}{1+\eta^2}, \frac{-\alpha}{2(p+1)}\left(\frac{\rho^2}{1+\eta^2}\right)^{(p+1)/2}\right\}$ proves (ii).

To prove (iii), we note that there is an open subset $U \subset \Omega$, where $a(x) > 0$. Now choose a positive function $w \in H^1(\Omega)$ with compact support in U such that $J(w) \neq 0$ and observe that $J(\lambda w) \to -\infty$ as $\lambda \to +\infty$. For λ large enough, we infer that $e = \lambda w$ satisfies (iii).

Now the following variant of the mountain-pass lemma (see [3]) shows the existence of a positive solution.

Theorem 4 *Let X be a Banach space and $J \in C^1(X)$ a functional satisfying the conditions (i)-(iii) of the mountain-pass lemma. Assume that $\pi : X \to X$ is a continuous function satisfying $J(\pi(u)) \leq J(u)$ for every $u \in X$, $\pi(0) = 0$, $\pi(e) = e$ (with e defined in (iii)). Then, there exists a critical point $u \in \overline{\pi(X)}$ and $J(u) \geq \delta$.*

Setting $\pi(u) = |u|$, we obtain a positive solution, which is strictly greater than 0 by the maximum principle.

The case $p < q$ is proved in a similar way. Instead of (25.12) we use (25.13) in the proof of (i) and (ii), and use the trace-inequality

$$\left|\int_\Gamma b(x)|u|^{q+1}d\Gamma\right| \leq C_1||u||^{q+1} \leq C_1(\frac{1}{\eta^2}+1)^{(q+1)/2}||v||_V^{q+1}$$

(for $||v||_V \geq \eta\,|t|$) to obtain

$$J(u) \geq \frac{1}{2}||v||_V^2 - \frac{C_2}{q+1}||v||_V^{q+1} \geq \frac{1}{4}||v||_V^2.$$

Condition (iii) is proved by constructing a positive function w with support concentrated in a neighbourhood of $\{x \in \Gamma \mid b(x) > 0\}$.

Thus, Theorem 2 is proved.

25.5 A NONLINEAR TRANSMISSION PROBLEM

In this section we extend some of the above results to a nonlinear transmission problem, which was studied in a more general setting, but in the definite case in [14]. For a motivation and some physical background of such problems see [11] and [14], for the linear theory see [12]. Let $\Omega_1 \subset \Omega \subset \mathbb{R}^n$ $(n \geq 3)$ be an open domain such that $\Gamma = \partial\Omega_1 \subset \operatorname{int} \Omega$ and denote $\Omega_2 = \Omega \setminus \overline{\Omega}_1$, $\Gamma_2 = \partial\Omega$. The problem we want to study now is

$$-\Delta u_1 - a(x)|u_1|^{p-1}u_1 = 0 \quad \text{in} \quad \Omega_1 , \tag{25.17}$$

$$-\Delta u_2 + \lambda(x)u_2 = 0 \quad \text{in} \quad \Omega_2, \quad u_2 = 0 \quad \text{on} \quad \Gamma_2 , \tag{25.18}$$

$$\partial_{\mathsf{n}} u_1 = \partial_{\mathsf{n}} u_2 \qquad \text{on} \quad \Gamma , \tag{25.19}$$

$$b(x)|u_1|^{q-1}u_1 = \lambda(x)u_2 \quad \text{on} \quad \Gamma . \tag{25.20}$$

Here n is the normal vector on Γ, pointing outward of Ω_1 and λ is a continuous function on $\overline{\Omega}_2$, such that the first eigenvalue of $-\Delta u_2 + \lambda(x)u_2 = 0$ in Ω_2, $u_{2|\Gamma_2} = 0$ is strictly positive. Again we assume that a and b are bounded and continuous functions and $1 < p < (n+2)/(n-2)$, $1 < q < n/(n-2)$.

Note that, in contrast to linear transmission conditions, here it is not possible to define a linear subspace of $H^1(\Omega_1) \times H^1(\Omega_2)$ where condition (25.20) is satisfied. In order to handle the transmission conditions (25.19), (25.20) we define a suitable subspace of $H^1(\Omega_1) \times H^1(\Omega_2)$ in the following way. Set

$$\mathcal{D} = \{(\eta_1, \eta_2) \mid \eta_1 = \varphi_{|\Omega_1},\ \eta_2 = \psi_{|\Omega_2},\ \varphi,\ \psi \in C_0^\infty(\Omega)\}$$

and let $\mathcal{H}$ be the completion of

$$\mathcal{D}_0 = \{\eta = (\eta_1, \eta_2) \in \mathcal{D} \mid \partial_{\mathsf{n}}\eta_1 = \partial_{\mathsf{n}}\eta_2 \quad \text{on} \quad \Gamma\}$$

in the norm $\|\eta\| = \left(\|\eta_1\|^2_{H^1(\Omega_1)} + \|\eta_2\|^2_{H^1(\Omega_2)}\right)^{1/2}$. As a subspace of $H^1(\Omega_1) \times H^1(\Omega_2)$ the embedding, resp. trace operators

$$\mathcal{H} \hookrightarrow L^p(\Omega_1) \times L^r(\Omega_2) , \qquad \mathcal{H} \longrightarrow L^q(\Gamma) \times L^s(\Gamma)$$

are compact for $1 \leq p, r < 2n/(n-2)$, resp. $1 \leq q, s < 2(n-1)/(n-2)$. To prove existence of a (weak) solution to (25.17)-(25.20), we consider the functional

$$J_T(u) = \int_{\Omega_1} \left[\frac{1}{2}|\nabla u_1|^2 - \frac{a(x)}{p+1}|u_1|^{p+1}\right] dx + \int_{\Omega_2} \left[\frac{1}{2}(|\nabla u_2|^2 + \lambda(x)u_2^2\right] dx - \int_\Gamma \left[\frac{b(x)}{q+1}|u_1|^{q+1} - \frac{\lambda(x)}{2}u_2^2\right] d\Gamma .$$

However, it is a priori not clear that critical points of J_T satisfy the transmission conditions (25.19), (25.20) in a suitable sense. For this reason we define the closed subspace

$$\mathcal{H}_T = \{(u_1, u_2) \in \mathcal{H} \mid u_1 = u_2 \quad \text{on} \quad \Gamma\}$$

and, denoting $\gamma : H^1(\Omega_i) \to H^{1/2}(\Gamma)$ the usual trace operator,

$$H_T^{1/2}(\Gamma) := \gamma(\mathcal{H}_T) \qquad \text{and} \quad H_T^{-1/2} \; = \; \text{the dual of } H_{0,T}^{1/2}(\Gamma)\,.$$

Now we can prove the following theorem.

Theorem 5 *If $p \geq q$, $\int_\Omega a(x)dx < 0$, $b(x) \leq 0$ and $a(x) > 0$ somewhere, or if $p < q$, $\int_\Gamma b(x)d\Gamma < 0$, $a(x) \leq 0$ and $b(x) > 0$ somewhere, then there exists a nontrivial weak solution in $\mathcal{H}$ of Problem* (25.17)-(25.20), *where* (25.19) *is satisfied in $H_T^{-1/2}$ and* (25.20) *holds in $L^s(\Gamma)$, $s = 2(n-1)/n$.*

Proof For the existence of a critical point of J_T we follow the same arguments as in the proof of Theorem 2. As before, for $(u_1, u_2) \in \mathcal{H}$, we can split $u_1 = t+v$, where v has zero mean value and use $||u_1||^2 := |t|^2 + ||v||_V^2$ as an equivalent norm on $H^1(\Omega_1)$. Then, for a Palais-Smale sequence $u_k = (u_{1k}, u_{2k})$ in $\mathcal{H}$, we obtain the estimate

$$\vartheta(||v_k||_V^2 + ||u_{2k}||_{H^1}^2) - (||t_k + v_k||_{H^1}^2 + ||u_{2k}||_{H^1}^2)^{1/2} \leq M\,,$$

where $\vartheta = 1/2 - 1/(p+1) > 0$. The assumption $|t_k| \to \infty$ then implies that

$$||v_k||_V \left(\vartheta \frac{||v_k||_V}{|t_k|} - \frac{1}{|t_k|}\right) + ||u_{2k}||_{H^1} \left(\vartheta \frac{||u_{2k}||_{H^1}}{|t_k|} - \frac{1}{|t_k|}\right) \leq M_1\,.$$

If $||v_k||_V \to \infty$ (or $||u_{2k}||_{H^1} \to \infty$), we claim that $||v_k||_V/|t_k| \to 0$ (or $||u_{2k}||_{H^1}/|t_k| \to 0$), and again we can use (25.12) to claim

$$J_T(u_k) \geq \frac{1}{2}||v_k||_V^2 - \frac{|t_k|^{p+1}}{2} \frac{\alpha}{p+1}\,, \qquad \text{where} \quad \alpha = \int_\Omega a(x)\,dx\,,$$

which leads to a contradiction. It follows that $|t_k|$, $||v_k||_V$ and $||u_{2k}||_{H^1}$ are bounded and J_T satisfies the Palais-Smale condition.

The estimate $J_T(u) \geq \delta$ for $||u||_{\mathcal{H}} = \rho$ is proved as before. Finally we note that for a function $v = (v_1, 0) \in \mathcal{H}$, where $v_1 \not\equiv 0$ has compact support in $\{x \mid a(x) > 0\}$, we have $J_T(\ell v) \to -\infty$ as $\ell \to \infty$. Again the mountain pass lemma implies the existence of a nontrivial critical point.

The prove of the remaining assertions of Theorem 5 is exactly the same as in the definite case [14]. One can choose a sequence $\eta_k = (\eta_{1k}, \eta_{2k}) \in \mathcal{D}_0$ which converges to the critical point $u = (u_1, u_2)$. From $\partial_n \eta_{1k} = \partial_n \eta_{2k}$ on Γ and $J_T'(\eta_k) \to 0$, one can prove that $\partial_n u_{1k} = \partial_n u_{2k}$ holds in $H_T^{-1/2}(\Gamma)$ (see Lemma 4 of [14]). This can be used to show that the second interface condition $b(x)|u_1|^{q-1}u_1 = \lambda(x)u_2$ holds in $L^{2(n-1)/n}(\Gamma)$.

Finally, the case $p < q$ can be proved in a similar way.

Remark. The solution constructed above is in general not positive. Note that the refined version of the mountain pass lemma, Theorem 4, is not applicable here, since the subspace $\mathcal{H}$ is not invariant under the mapping $u \mapsto |u|$.

References

[1] Amann, H.: Parabolic evolution equations and nonlinear boundary conditions. J. Differ. Equations 72 (1988), 201–269.

[2] Alama, S., Tarantello, G.: On semilinear elliptic equations with indefinite nonlinearities. Calc. Var. Partial Differ. Eq. 1 (1993), 439–475.

[3] Berestycki, H., Capuzzo-Dolcetta, I., Nirenberg, L.: Variational methods for indefinite superlinear homogeneous elliptic problems. NoDEA 2 (1995), 553 – 572.

[4] Cherrier, P.: Problème de Neumann non linéaire sur les variétés Riemanniennes. J. Funct. Anal. 57 (1984), 154–206.

[5] Chipot, M., Fila, M., Quittner, P.: Stationary solutions, blow up and convergence to stationary solutions for semilinear parabolic equations with nonlinear boundary conditions. Acta Math. Univ. Comenianae 60 (1991), 35–103.

[6] Chipot, M., Voirol, F.: On a class of nonlinear elliptic problems with Neumann boundary conditions growing like a power. Z. Anal. Anwend. 14 (1995), 853-868.

[7] Chipot, M., Chafrir, I., Fila, M.: On the solutions to some elliptic equations with nonlinear Neumann boundary conditions. Adv. Differ. Equ. 1 (1996), 91–110.

[8] Escobar, J.F.: Sharp constant in a Sobolev trace inequality. Indiana Univ. Math. J. 37 (1988), 687–698.

[9] Escobar, J.F.: Conformal deformation of a Riemannian metric to a scalar flat metric with constant mean curvature. Ann. Math. 136 (1992), 1–50.

[10] Fila, M.: Boundedness of global solutions for the heat equation with nonlinear boundary condition. Comment. Math. Univ. Carolinae 30 (1989), 479–484.

[11] Fife, P.C.: Dynamics of internal layers and diffusive interfaces. CBMS-NFS Reg. Conf. Ser. Appl. Math. 53 (1988).

[12] Nicaise, S.: Polygonal interface problems. Peter Lang, Frankfurt/Main, 1993.

[13] Pflüger, K: Semilinear elliptic problems with nonlinear boundary conditions in unbounded domains. Z. Anal. Anwend. 14 (1995), 829–851.

[14] Pflüger, K.: Nonlinear transmission problems in bounded domains of $\mathbb{R}^n$. Appl. Anal. 62 (1996), 391–403.

[15] Pflüger, K.: Remarks on nonlinear Neumann problems in periodic domains. Result. Math. 31 (1997), 365–373.

[16] Rabinowitz, P.H.: Minimax methods in critical point theory with applications to differential equations. Reg. Conf. Ser. Math. 65 (1986).

[17] Tehrani, H.T.: On indefinite superlinear elliptic equations. Calc. Var. 4 (1996), 139–153.

[18] Terracini, S: Symmetry properties of positive solutions to some elliptic equations with nonlinear boundary conditions. Differ. Int. Eq. 8 (1995), 1911-1922.

26 HIGHER ORDER ORDINARY DIFFERENTIAL EQUATIONS WITH SUPER–SINGULAR POINTS

Nusrat Rajabov

Tajik State National University
Dushanbe 734025
Tajikistan
nrajabov@td.silk.org

Abstract: Integral representations are given for solutions to ordinary differential equations with different kind of singularities in endpoints of or inside the interval.

26.1 INTRODUCTION

Let $\Gamma_0 = \{a < x < b\}$ be a segment on the real axis, containing the points $x = b_1$, $x = b_2$, ..., $x = b_n$. Moreover, we assume that the points $b_1, b_2, \ldots, b_n$ satisfy $b_1 < b_2 < \ldots < b_n$, and allow b_1, b_n to coinside accordingly with the points a and b in particular. Further denote $\Gamma_b = \{b_1, b_2, \ldots, b_n\}$. In the set $\Gamma = \Gamma_0 \backslash \Gamma_b$ we will consider the equation

$$L_n^\alpha(y) = \sum_{j=0}^{m}(\prod_{k=1}^{n} |x - b_k|^\alpha)^{m-j} a_j(x) y^{(m-j)}(x) = f(x), \qquad (26.1)$$

where $a_0(x) = 1$, $a_j(x)$ $(1 \le j \le n)$, $f(x)$ are given functions in Γ, m is a natural number, α a constant > 0.

Many problems of applied character lead to different cases of this equation. Also problems in the theory of partial differential equations with singular or super–singular coefficients are reducible to the consideration of ordinary differential equations with singular or super–singular points, see [1], [2]. Up to now there is no universal method constructed as, for example, in the theory of ordinary differential equations with regular coefficients that enables us to

H.G.W. Begehr et al. (eds.), Partial Differential and Integral Equations, 347–358.

consider all the possible cases at once. In our works published in the last years, especially in the monographs [4], [5] and in a series of papers in proceedings of international conferences for linear systems a universal method was suggested that allow us to consider all the possible cases.

When $\alpha < 1$ equation (26.1) is called an ordinary linear differential equation of m–th order with n weak singular points, in the case $\alpha = 1$ an ordinary linear differential equation of m–th order with n singular points and when $\alpha > 1$ an ordinary differential equation of m–th order with n super–singular points.

In [6] an integral representation is obtained in $\Gamma_0 = \Gamma\backslash\{0\}, \Gamma = \{-\alpha < x < \alpha\}$, for solutions of the following m–th order linear ordinary differential equation with one singular or super–singular point

$$L_m^\beta(y) = \sum_{j=0}^{m} |x|^{\beta(n-j)} a_j(x) y^{(m-j)}(x) = f(x), \tag{26.2}$$

where $a_0(x) = 1$, depending on the power singularity $\beta(\beta < 1, \beta = 1, \beta > 1)$ and the roots of the characteristic equation

$$\sum_{j=0}^{m} a_j(x)\lambda^{m-j}(x) = 0, \quad a_0(x) = 1, \quad \lambda^0(x) = 1. \tag{26.3}$$

The obtained integral representation is applied for posing correct Cauchy type problems and for their investigation.

Singular ordinary differential equations and their applications are investigated in [1]—[13].

In this work the general solution of equation (26.1) is constructed depending on the positions of the singular ($\alpha = 1$) or super–singular ($\alpha > 1$) points $x = b_k \quad (1 \le k \le n)$ in Γ_0, on the power of the singularities ($\alpha < 1, \alpha = 1, \alpha > 1$) and on the roots of the characteristic equation (26.3). There are two cases: Case 1) $a < b_1 < b_2 <, \dots, b_n < b$; Case 2) $a < b_1 < b_2 <, \dots, b_n = b$. The behaviour of the solution of equation (26.1) in the neighbourhood of the singular or super–singular points $x = b_j \quad (1 \le j \le n)$ is investigated. We have already received the general solution of the equation (26.1), and have applied it for posing Cauchy type boundary value problems with conditions in the singular or super–singular points.

Let $a_j(x) \in C^{(n-1)}(\Gamma) \quad (1 \le j \le n)$. In this case equation (26.1) is represented in the form

$$\prod_{j=1}^{m}\Big(\frac{d}{dx} - \frac{\lambda_{m+1-j}(x)}{\prod_{k=1}^{n} |x - b_k|^\alpha}\Big) y(x) = \frac{f_m(x)}{(\prod_{k=1}^{n} |x - b_k|^\alpha)^m}, \tag{26.4}$$

where $\lambda_j(x) \quad (1 \le j \le m)$ are the roots of the characteristic equation (26.3) and are connected with the coefficients of equation (26.1) by the Vieta formulas,

$$f_m(x) = f(x) + \sum_{j=1}^{m-1} B_j(x) \Big(\prod_{k=1}^{n} |x - b_k|^\alpha\Big)^{j-1} y^{(j-1)}(x), \tag{26.5}$$

$y^{(0)}(x) = 1, \quad B_j(x) \quad (1 \leq j \leq m-1)$ are given functions in Γ.

Specifically, if the coefficients of the equation (26.1) are connected by the equations $B_j(x) = 0 \quad (1 \leq j \leq m-1)$, then we have to find the general solution of the equation

$$\prod_{j=1}^{m}\left(\frac{d}{dx} - \frac{\lambda_{m+1-j}(x)}{\prod_{k=1}^{n}|x-b_k|^{\alpha}}\right)y(x) = \frac{f(x)}{\left(\prod_{k=1}^{n}|x-b_k|^{\alpha}\right)^m}. \tag{26.6}$$

The construction of the general solution of equation (26.1) or (26.6) is dependent on the arrangement of the singular or super–singular points in Γ_0.

In the case, when the left extreme singular or super–singular point coincides with the left boundary point, then the interval Γ_0 breaks up into the n intervals

$$\Gamma_j = \{b_j < x < b_{j+1}\}, \quad 1 \leq j \leq n-1, \quad \Gamma_n = \{b_n < x < b\}, \quad b_1 = a.$$

In this case, equation (26.1) or (26.6) in the intervals $\Gamma_j \quad (1 \leq j \leq n-1)$ has two singular or super–singular point (the left and the right boundary points), in the interval Γ_n it has one singular or super–singular point (the left boundary point). If the solution of equation (26.1) or (26.6) in the interval Γ_j is denoted by $Y_j(x) \quad (1 \leq j \leq n)$, then the function

$$y(x) = \begin{cases} Y_1(x), & x \in \Gamma_1\,, \\ Y_2(x), & x \in \Gamma_2\,, \\ \dots\dots\dots\dots\dots\dots \\ Y_n(x), & x \in \Gamma_n\,, \end{cases} \tag{26.7}$$

gives the solution of equation (26.1) or (26.6) in Γ.

In an analogous way the solution of equation (26.1) or (26.6) is construced in the case, when the right extreme singular or super–singular point coincides with the right boundary point. Similarly the case is handled, when the left extreme and the right extreme singular or super–singular point accordingly coincide with the left and right boundary points and the case, when the left and the right extreme singular or super–singular points do not coincide with the left and right boundary points.

Hence, the problem of finding the general solution to equation (26.1) in Γ_0 reduces to find the general solution of equation (26.1) in intervals with the left and right extreme singular or super–singular points, in intervals with a left extreme singular or super–singular point and in intervals with a right extreme singular or super–singular point.

At first we find the general solution of equation with one singular or super–singular point

$$\sum_{j=0}^{m}|x-c|^{\alpha(m-j)}a_j(x)y^{(m-j)}(x) = f(x) \tag{26.8}$$

in $L_1\backslash\{c\}$, where $L_1 = \{a < x < b\}, \quad a \leq c \leq b$.

For the sake of simplicity below we perform all results for $m = 2$.

In the case $m = 2$, according to formula (26.6) equation (26.1) is represented as

$$(\frac{d}{dx} - \frac{\lambda_2(x)}{|x-c|^\alpha})(\frac{d}{dx} - \frac{\lambda_1(x)}{|x-c|^\alpha})y(x) = \frac{f_2(x)}{|x-c|^{2\alpha}}, \tag{26.9}$$

where

$$f_2(x) = f(x) + B_1(x)y(x), \quad B_1(x) = -|x-c|^{2\alpha}\frac{d}{dx}(\frac{\lambda_1(x)}{|x-c|^\alpha}),$$

$$\lambda_1(x) = -2^{-1}(a_1(x) + \sqrt{D(x)}),$$

$$\lambda_2(x) = 2^{-1}(-a_1(x) + \sqrt{D(x)}), \quad D(x) = a_1^2(x) - 4a_2(x).$$

Introducing the unknown function

$$u(x) = \frac{dy}{dx} - \frac{\lambda_1(x)}{|x-c|^\alpha}y(x) \tag{26.10}$$

we arrive at the equality

$$\frac{du}{dx} - \frac{\lambda_2(x)}{|x-c|^\alpha}u = \frac{f_2(x)}{|x-c|^{2\alpha}}. \tag{26.11}$$

26.2 THE CASE, WHEN THE SINGULAR OR SUPER-SINGULAR POINT COINCIDES WITH THE LEFT BOUNDARY POINT

First assume that the singular point c coincides with the left boundary point, that is $c = a$.

In this case equations (26.10) and (26.11) have the following form

$$\frac{dy}{dx} - \frac{\lambda_1(x)}{(x-a)^\alpha}y = u(x), \tag{26.12}$$

$$\frac{du}{dx} - \frac{\lambda_2(x)}{(x-a)^\alpha}y = \frac{f_2(x)}{(x-a)^{2\alpha}}. \tag{26.13}$$

According to [4]–[6], [12], if the solutions to equations (26.12), (26.13) exist, then they can be represented in the form

$$y(x) = K_a^{\alpha,+}[c_1, \lambda_1, u], \tag{26.14}$$

$$u(x) = K_a^{\alpha,+}[c_2, \lambda_2, \frac{f_2}{(x-a)^{2\alpha}}], \tag{26.15}$$

where

$$K_a^{\alpha,+}[c, \lambda, g] = \exp(\omega_\lambda^\alpha(x))[c + \int_a^x \exp(-\omega_\lambda^\alpha(t))g(t)dt, \quad \text{when} \quad \alpha < 1,$$

$$K_a^{\alpha,+}[c, \lambda, g] = \exp[-\lambda(a)\omega_\alpha(x) + W_\lambda^a(x)]$$

$$\times\left[c+\int_a^x \exp[\lambda(a)\omega_\alpha(t)-W_\lambda^\alpha(t)]g(t)dt\right], \text{ when } \alpha>1,$$

$$K_a^{\alpha,+}[c,\lambda,g]=(x-a)^{\lambda(a)}\exp[W_\lambda^a(x)]$$

$$\times\left[c+\int_a^x \exp[-W_\lambda^a(t)](t-a)^{-\lambda_1(a)}g(t)dt\right], \quad \text{when} \quad \alpha=1,$$

$$\omega_\lambda^\alpha(x)=\int_a^x (t-a)^{-\alpha}\lambda(t)dt,$$

$$\omega_\alpha(x)=[(\alpha-1)(x-a)^{\alpha-1}]^{-1}, \quad W_\lambda^\alpha(x)=\int_a^x \frac{\lambda(t)-\lambda(a)}{(t-a)^\alpha}dt.$$

Here $c_j, j=1,2$, are arbitrary constants.

If the functions $\lambda_j(x), j=1,2$, in the neighbourhood of the point $x=a$ satisfy the conditions

$$|\lambda_j(x)-\lambda_j(a)|\leq H_j^a(x-a)^{\gamma_j} \tag{26.16}$$

where $\gamma_j\geq\alpha-1$, by $\alpha\geq 1, \lambda_j(x)\in C[a,b], \lambda_j(a)<0, j=1,2$, for $\alpha>1, \lambda_1(a)<1, \lambda_2(a)<-1$, for $\alpha=1$, $f_2(x)\in C[a,b], u(x)\in C[a,b]$, then there exists the general solution to equations (26.12), (26.13) in the class $C^1(L_1\backslash\{c\})$ which correspondingly can be represented in the form (26.14), (26.15).

Substituting $u(t)$ from equality (26.15) in the equality (26.14) and introducting the new function

$$y(x)=\exp(\omega_{\lambda_1}^\alpha(x))v(x), \quad \text{for} \quad \alpha<1, \tag{26.17}$$

$$y(x)=(x-a)^{\lambda_1(a)}\exp(W_{\lambda_1}^\alpha(x))v(x), \quad \text{for} \quad \alpha=1 \tag{26.18}$$

and

$$y(x)=\exp[W_{\lambda_1}^\alpha(x)-\lambda_1(a)\omega_\alpha(x)]v(x), \quad \text{for} \quad \alpha>1, \tag{26.19}$$

we arrive at the equality

$$v(x)+\int_a^x K_1(x,\tau)v(\tau)d\tau=F_1^a(c_1,c_2,f), \tag{26.20}$$

where $K_1(x,\tau)$, $F_1^a(c_1,c_2,f)$ are known functions, and c_1,c_2 are arbitrary constants. Hence the next result follows.

Theorem 1 *Let in equation* (26.8), $m=2$, *the singular or super-singular point coincide with the left boundary point. Let the roots* $\lambda_j(x)(j=1,2)$ *of the characteristic equation* (26.3) *be real and different,* $a_j(x)\in C^1(\bar{L}_1), j=1,2, f(x)\in$

$C(\bar{L}_1)$. *Let the functions* $\lambda_j(x), j=1,2$, *for* $\alpha \geq 1$ *satisfy conditions* (26.16). *Moreover, let the conditions*

$$c_1^\alpha(x) = (x-a)^\alpha \lambda_1'(x) - \alpha(x-a)^{\alpha-1}\lambda_1(x) = O((x-a)^{\delta_1}), \quad \delta_1 > 2\alpha - 1,$$

$$f(x) = O((x-a)^{\delta_2}), \quad \delta_2 > 2\alpha - 1, \quad \text{for} \quad x \to a, \quad \alpha < 1;$$

$$\lambda_2(a) < -1, \; c_1^1(x) = O((x-a)^{\delta_3}), \; \delta_3 > 1 + \sqrt{D}, \; \text{for}\, x \to a, \; \alpha = 1;$$

$$\lambda_2(a) < 0, c_1^\alpha(x) = O(\exp[(\lambda_1(a) - \lambda_2(a))\omega_\alpha(x)](x-a)^{\delta_4}), \; \delta_4 > 2\alpha - 1,$$

for $x \to a$, $\alpha > 1$ *be satisfied. Then any solution of equation* (26.8) *for* $m = 2$ *of class* $C^2(L_1)$ *can be represented in the form* (26.17) *for* $\alpha < 1$, (26.18) *for* $\alpha = 1$ *and* (26.19) *for* $\alpha > 1$, *where* $v(x)$ *is the solution of the integral equation* (26.20), c_1, c_2 *are arbitrary constants.*

In the future the solution to equation (26.9) is denoted by $K_{\alpha,a}^{+,1}[c_1, c_2, f]$, when $c = a$ and $\lambda_1(x), \lambda_2(x)$ are real and different, and by $K_{\alpha,a}^{+,2}[c_1, c_2, f]$, when $c = a$ and $\lambda_1(x), \lambda_2(x)$ are real and equal. That is

$$y(x) = K_{\alpha,a}^{+,1}[c_1, c_2, f], \quad \text{if} \quad c = a, \quad \lambda_1(x) \neq \lambda_2(x),$$

$$y(x) = K_{\alpha,a}^{+,2}[c_1, c_2, f], \quad \text{if} \quad c = a, \quad \lambda_1(x) = \lambda_2(x).$$

Remark 1 A similar assertion as in Theorem 1 may be obtained in the case when $\lambda_1(x), \lambda_2(x)$ turn out as complex–conjugated, or as real and equal.

26.3 THE CASE, WHEN THE SINGULAR OR SUPER–SINGULAR POINT COINCIDES WITH THE RIGHT BOUNDARY POINT

Let the singular or super–singular point c coincide with the right boundary point, that is $c = b$. In this case equations (26.10) and (26.11) have the form

$$\frac{dy}{dx} - \frac{\lambda_1(x)}{(b-x)^\alpha} y(x) = u(x),$$

$$\frac{du}{dx} - \frac{\lambda_2(x)}{(b-x)^\alpha} u = \frac{f_1^b(x)}{(b-x)^\alpha},$$

where

$$f_1^b(x) = f_1(x) - (\lambda_1'(x)(b-x)^\alpha + \lambda_1(x)\alpha(b-x)^{\alpha-1})y(x).$$

A similar assertion as in Theorem 1 may be obtained in this case too. For example, when $\lambda_1(x), \lambda_2(x)$ are real and different, we have the following assertion.

Theorem 2 *Let in equation* (26.1) $m = 2$, *and the singular or super–singular point coincide with the right boundary point. Let the roots* $\lambda_j(x)$ $(j = 1,2)$ *of the characteristic equation* (26.3) *be real and different,* $a_j(x) \in C^1(\bar{L}_1)$, $j = 1,2$, $f(x) \in C(\bar{L}_1)$. *Let the functions* $\lambda_j(x)$ $(j = 1,2)$ *for* $\alpha \geq 1$ *satisfy*

$$|\lambda_j(x) - \lambda_j(b)| \leq H_j^b (b-x)^{\delta_j}, \quad \delta_j \geq \alpha - 1, \quad j = 1,2.$$

Moreover, let the conditions

$$c_2^{\alpha,b}(x) = \lambda_1'(x)(b-x)^\alpha + \alpha(b-x)^{\alpha-1}\lambda_1(x) = O\left((b-x)^{h_1}\right),\ h_1 > 2\alpha - 1,$$

$$f(x) = O\left((b-x)^{h_2}\right), \quad h_2 > 2\alpha - 1, \quad for \quad \alpha < 1;$$

$$\lambda_2(b) > 1, \quad \lambda_2(b) - \lambda_1(b) < 1,$$

$$c_2^{1,b}(x) = O\left((b-x)^{h_3}\right),\ h_3 > \lambda_1(b) - \lambda_2(b) + 1, \quad for \quad \alpha = 1;$$

$$\lambda_2(b) < 0, \quad \lambda_1(b) < \lambda_2(b),$$

$$c_2^{\alpha,b}(x) = O(\exp[(\lambda_1(b) - \lambda_2(b))\omega_\alpha^b(x)](b-x)^{h_4}), \omega_\alpha^b(x) = [(\alpha-1)(b-x)^{\alpha-1}]^{-1},$$

$$h_4 > 2\alpha - 1, \quad for \quad \alpha > 1,$$

be satisfied. Then any solution of equation (26.8) with $m = 2$ *of class* $C^2(L_1)$ *can be represented in the form*

$$y(x) \quad = \quad \exp(-\omega_{\lambda_1}^{\alpha,b}(x))v(x) \equiv K_{\alpha,b}^{-,1}[c_1, c_2, f],$$

$$\omega_{\lambda_1}^{\alpha,b}(x) = \int_b^x (b-t)^{-\alpha}\lambda_1(t)dt, \ for \ \alpha < 1,$$

$$y(x) = (b-x)^{-\lambda_1(b)} \exp[W_{\lambda_1}^b(x)]v(x) \equiv K_{1,b}^{-,1}[c_1, c_2, f], \quad for \quad \alpha = 1,$$

$$y(x) = \exp[-W_{\lambda_1}^{\alpha,b}(x) + \lambda_1(b)\omega_\alpha^b(x)]v(x) \equiv K_{\alpha,b}^{-,1}[c_1, c_2, f], \quad for \quad \alpha > 1,$$

where $v(x)$ *is the solution of the second kind Volterra integral equation*

$$v(x) + \int_x^b K_2(x,\tau)v(\tau)d\tau = F_1^b(c_1, c_2, f).$$

Here $K_2(x,\tau)$, $F_1^b(c_1, c_2, f)$ *are known functions, and* c_1, c_2 *are arbitrary constants.*

Remark 2 An assertion similar as in Theorem 2 can be given in the case, when the roots of the characteristic equation (26.3) are real and equal or complex–conjugated.

26.4 THE CASE, WHEN THE SINGULAR OR SUPER–SINGULAR POINT DOES NOT COINCIDE WITH BOUNDARY POINTS

Let the singular or super–singular point $x = c$ not coincide with a boundary point of the interval L_1, that is $a < c < b$. Let $a_j(x) \in C^1(L_1 \backslash \{c\})$, $j = 1, 2$, $f(x) \in C(L_1 \backslash \{c\})$. In the singular or super–singular point $x = c$ they can be have a first kind singularity.

In this case, using the result of Theorem 1,2 one can see, that the general solution to equation (26.9) for $m = 2$ in $L_1\backslash\{c\}$, when the roots of the characteristic equation (26.3) are real and different, is given by the formula

$$y(x) = \begin{cases} K^{-,1}_{\alpha,c}[c_1, c_2, f], \text{for} \quad a < x < c \\ K^{+,1}_{\alpha,c}[c'_1, c'_2, f], \text{for} \quad c < x < b \end{cases} \tag{26.21}$$

From Theorem 1,2 the next result is obtained.

Theorem 3 *Let in equation (26.8), $m = 2$, the singular or super–singular point $x = c$ not coincide with a boundary point, that is $a < c < b$, $a_j(x) \in C^1(L_1\backslash\{c\})$, $j = 1,2$, $f(x) \in C(L_1\backslash\{c\})$ for which the point $x = c$ can be a first kind singularity. The roots $\lambda_j(x)$ $(j = 1,2)$ of the characteristic equation (26.3) are assumed to be real and different. Let the functions $\lambda_j(x), j = 1,2$, satisfy the conditions*

$$|\lambda_j(x) - \lambda_j(c \pm 0)| \leq H_j^{\pm}|x - c|^{\gamma_j^{\pm}}, \quad \gamma_j^{\pm} > \alpha - 1, \quad j = 1,2,$$

for $x \to c \pm 0$, $\alpha \geq 1$. Moreover, let the conditions

$$c_{2,\alpha}(x) = |x - c|^{\alpha}\lambda'_1(x) - \alpha|x - c|^{\alpha}(x - c)^{-1}\lambda_1(x) = O(|x - c|^{\delta_1}),$$

$$\delta_1 > 2\alpha - 1,$$

$$f(x) = O(|x - c|^{\delta_2}),\ \delta_2 > 2\alpha - 1, \quad for \quad x \to c \pm 0, \quad \alpha < 1;$$

$$\lambda_2(c + 0) < -1, \quad \lambda_2(c - 0) > 1, \quad \lambda_1(c + 0) - \lambda_2(c + 0) < 1,$$

$$\lambda_2(c - 0) - \lambda_1(c - 0) < 1,\ c_{2,1}(x) = O((c - x)^{\delta_3}),\ \delta_3 > 1 + \lambda_2(c + 0) - \lambda_1(c + 0),$$

$$for \quad x \to c + 0,$$

$$c_{2,1}(x) = O\left((x - c)^{\delta_4}\right),\ \delta_4 > 1 + \lambda_1(c - 0) - \lambda_2(c - 0), \quad for \quad x \to c - 0, \quad \alpha = 1;$$

$$\lambda_2(c + 0) < 0, \quad \lambda_1(c + 0) > \lambda_2(c + 0),$$

$$c_{2,\alpha}(x) = O[\exp((-\lambda_1(c + 0) + \lambda_2(c + 0))\omega_\alpha^c(x))(x - c)^{\delta_5}],\ \delta_5 > 2\alpha - 1,$$

$$for\ x \to c + 0,$$

$$\lambda_2(c - 0) > 0, \quad \lambda_1(c - 0) - \lambda_2(c - 0) < 0,$$

$$c_{2,\alpha}(x) = O[\exp((\lambda_1(c - 0) - \lambda_2(c - 0))\omega_\alpha^c(x)(c - x)^{\delta_6}],\ \delta_6 > 2\alpha - 1,$$

$$\omega_\alpha^c(x) = [(\alpha - 1)|x - c|^{\alpha - 1}]^{-1} \ for\ x \to c - 0,\ \alpha > 1.$$

Then any solution of equation (26.8) with $m = 2$ of class $C^2(L_1\backslash\{c\})$ can be represented in the form (26.21), where c_j, c'_j $(j = 1,2)$ are arbitrary constants.

In the case, when the roots $\lambda_j(x)$ $(j = 1,2)$ of the characteristic equation (26.3) are real and equal, the following assertion holds.

Theorem 4 *Let in equation (26.8), $m = 2$, the singular or super–singular point not coincide with the boundary points, that is $a < c < b$, let $a_j(x) \in$*

$C^1(L_1\backslash\{c\})$ $j=1,2,$ $f(x)\in C(L_1\backslash\{c\})$ and allow the point $x=c$ to be a first kind singularity for a_j and f. The roots $\lambda_j(x)$ $(j=1,2)$ of the characteristic equation (26.3) are real and equal. Let the function $a_1(x)$ satisfy the conditions

$$|a_1(x)-a_1(c\pm 0)|\leq M^{\pm}|x-c|^{\delta^{\pm}},\quad \delta^{\pm}>\alpha-1,\ for\, x\to c\pm 0,\quad \alpha\geq 1.$$

Moreover, let the conditions

$$c_{2,\alpha}^{a_1}(x)=-|x-c|^{\alpha}2^{-1}(a_1'(x)+\alpha(x-c)^{-1}a_1(x)$$
$$=O\left(|x-c|^{\delta_1}\right),\quad \delta_1>2\alpha-1,$$
$$f(x)=O\left(|x-c|^{\delta_2}\right),\quad \delta_2>2\alpha-1,\ for\, x\to c\pm 0,\quad \alpha<1;$$
$$a_1(c+0)>2,\quad a_1(c-0)<-2,\quad for\quad \alpha=1;$$
$$a_1(c+0)>0,\quad a_1(c-0)<0,\quad for\quad \alpha>1;$$

be satisfied. Then any solution of equation (1.8) with $m=2$ of class $C^2(L_1\backslash\{c\})$ can be represented in the form

$$y(x)=\begin{cases}K_{\alpha,c}^{-,2}[c_1,c_2,f], for & a<x<c,\\ K_{\alpha,c}^{+,2}[c_1',c_2',f], for & c<x<b,\end{cases}$$

where c_j,c_j' $(j=1,2)$ are arbitrary constants.

Remark 3 An assertion similar as in Theorem 3, 4 can be obtained in the case, when the roots of the characteristic equation (26.3) are complex–conjugated.

The solution of equation (26.1) in intervals $\Gamma_j=\{b_j<x<b_{j+1}\}$ with left and right extreme singular or super–singular points is constructed as follows. Fix an arbitrary point $x=x_j^0$ in Γ_j $(x_j^0\neq b_j,b_{j+1})$. Denoting the solution of equation (26.1) in the interval $\Gamma_j^1=\{b_j<x\leq x_j^0\}$ by $Y_j^1(x,c_1,c_2)$, and in the interval $\Gamma_j^2=\{x_j^0\leq x<b_{j+1}\}$ by $Y_j^2(x,c_1,c_2)$, the function

$$Y_j(x)=\begin{cases}Y_j^1(x,c_j^1,c_j^2), & \text{when}\quad x\in\Gamma_j^1,\\ Y_j^2(x,c_j^3,c_j^4), & \text{when}\quad x\in\Gamma_j^2,\end{cases}$$

gives the solution of equation (26.1) in Γ_j $(1\leq j\leq n)$. Moreover, from the continuity of the solution of equation (26.1) in the point $x=x_j^0$

$$Y_j^1(x_j^0,c_j^1,c_j^2)=Y_j^2(x_j^0,c_j^3,c_j^4)$$

and the equality

$$(P_{\lambda_j'}^{\alpha}(y))_{x=x_j^0-0}=(P_{\lambda_j'}^{\alpha}(y))_{x=x_j^0+0},$$

follow with constants c_j^3,c_j^4 expressed through the constants c_j^1,c_j^2.

26.5 CAUCHY PROBLEMS IN THE CASE, WHEN THE SINGULAR OR SUPER–SINGULAR POINT DOES NOT COINCIDE WITH A BOUNDARY POINT

Let the singular or super–singular point c not coincide with a boundary point, that is $a < c < b$.

On the basis of the obtained integral representation for the equation (26.8) the following Cauchy boundary value problem with a set of conditions in the singular or super–singular point $x = c$ are posed.

Problem K_{II}^{c} It is required to find the solution of equation (26.8) of class $C^2(L_1 \backslash \{c\})$, when the roots of the characteristic equation (26.3) in $L_1 \backslash \{c\}$ are real and different, satisfying the boundary conditions

$$\left.\begin{array}{l} y(c-0) = A_1, \quad y(c+0) = B_1, \quad (P_{\lambda_1}^{\alpha}(y))_{x=c-0} = A_2, \\ (P_{\lambda_1}^{\alpha}(y))_{x=c+0} = B_2, \quad P_{\lambda_1}^{\alpha} \equiv d/dx - \lambda_1(x)|x-c|^{-\alpha}, \end{array}\right\} \text{for } \alpha < 1 ;$$

$$\left.\begin{array}{l} [(c-x)^{\lambda_1(c-0)} y(x)]_{x=c-0} = A_1, \\ [(x-c)^{-\lambda_1(c+0)} y(x)]_{x=c+0} = B_1, \\ [(c-x)^{\lambda_2(c-0)} P_{\lambda_1}'(y)]_{x=c-0} = A_2, \\ [(x-c)^{-\lambda_2(c+0)} P_{\lambda_1}'(y)]_{x=c+0} = B_2, \end{array}\right\} \text{for } \alpha = 1 ;$$

$$\left.\begin{array}{l} [\exp(-\lambda_1(c-0)\omega_{\alpha}^{c}(x)) y(x)]_{x=c-0} = A_1, \\ [\exp(\lambda_1(c+0)\omega_{\alpha}^{c}(x)) y(x)]_{x=c+0} = B_1, \\ [\exp(-\lambda_2(c-0)\omega_{\alpha}^{c}(x)) P_{\lambda_1}^{\alpha}(y)]_{x=c-0} = A_2, \\ [\exp(\lambda_2(c+0)\omega_{\alpha}^{c}(x)) P_{\lambda_1}^{\alpha}(y)]_{x=c+0} = B_2, \end{array}\right\} \text{for } \alpha > 1,$$

where $A_j, B_j \quad (j = 1, 2)$ are given constants, $\omega_{\alpha}^{c}(x) = (\alpha-1)^{-1}|x-c|^{1-\alpha}$. The solvability of problem K_{II}^{c} is given in the following assertion.

Theorem 5 *Let in equation* (26.8) $a_j(x) \quad (j = 1, 2)$ *and* $f(x)$ *satisfy the conditions of Theorem 3. Then problem* K_{II}^{c} *has a unique solution, which is given by the formula*

$$y(x) = \begin{cases} K_{\alpha,c}^{-,1}[A_1, A_2, f], \text{for} & a < x < c, \\ K_{\alpha,c}^{+,1}[B_1, B_2, f], \text{for} & c < x < b. \end{cases}$$

Remark 4 A similar result for Problem K_{II}^{c} can be given in the case, when the roots of the characteristic equation (26.3) are real and equal or complex–conjugated.

Remark 5 For equation (26.9), in dependence of the characteristic equation

(26.3) problems of linear conjugated type in the singular or super–singular point c $(a < c < b)$, can be stated and solved.

Remark 6 Problem K_{fI}^{c} can be solved for equation (26.1) in dependence of the location of the singular or super–singular points in Γ_0, the order of the singularities ($\alpha < 1, \quad \alpha = 1, \quad \alpha > 1$) and the roots of the characteristic equation in every interval $\Gamma_j, 1 \leq j \leq n$.

References

[1] Begehr H., Gilbert R.P.: Transformations, transmutations and kernel functions, v. 2, 268 pp. Longman, Harlow, 1993, 268 pp.

[2] Gilbert R.P.: Function theoretic methods in partial differential equations, New York—London, Acad. Press, 1969, 311 pp.

[3] Elschner, J.: Singular ordinary differential operators and pseudodifferential equations. Academie–Verlag, Berlin, 1985, 200 pp.

[4] Rajabov, N.: Integral representations and boundary value problems for some differential equations with singular line or singular surface (Introduction in the second order hyperbolic equation with singular lines). Part 4, Tajik State Univ. Publ., Dushanbe, 1985, 148 pp.

[5] Rajabov, N.: An introduction to the theory of partial differential equations with super–singular coefficients. Dushanbe, 1996, 230 pp.

[6] Rajabov, N.: Ordinary differential equations of higher order with super–singular point, News Acad. Sci. Tajikistan 1–2 (8) (1994), 3–9.

[7] Rajabov, N.: Linear hyperbolic equations of the fourth order with two singular lines. Proc. Asian Math. Conf., Hong–Kong, 1990. World Sci., Singapore, 1992, 387–393.

[8] Rajabov, N.: Integral representation and boundary value problems for the general second order ordinary differential equations with a singular point. Differential and integral equations. Issue 3, Dushanbe 1980, 43–54.

[9] Rajabov, N.: System of two first order ordinary differential equations with n super–singular points. Int. Conf. Differential Equations with Singular Coefficients, Dushanbe, 1996, 69.

[10] Rajabov, N., Mirzoev A.M.: Cauchy type boundary value problems for two first order linear ordinary differential equations with one super–singular point. Int. Conference Differential Equations with Singular Coefficients, Dushanbe, 1996, 71.

[11] Rajabov, N., Chojaev, S.: To the theory of one class of two first order linear ordinary differential equations with one super–singular point. Int. Conf. Differential Equations with Singular Coefficients, Dushanbe, 1996, 74.

[12] Rajabov, N.: Cauchy type boundary value problems for a linear system of ordinary differential equations with singular coefficients. Scientific and

theoretical conference in Tajik State Univ. Ser. Nat. Sci., Dushanbe, 1994, 18.

[13] Rajabov, N., Shevchuk, V.: The theory of ordinary differential equations with super–singular point. Tajik Acad. Sci. Doklady 32 (1989), 506—510.

27 APPLICATIONS OF METHODS OF INTEGRALS AND FUNCTION THEORY TO STATISTICS

Mingzhong Jin

Guizhou College of Nationalities
Guiyang 550025

Xiru Chen

Graduate School, Academia Sinica
Beijing 100080

Zhongyuan Yu

Guizhou University
Guiyang 550000
P.R. China

Abstract: Let $Y_i = x_i'\beta_0 + e_i$, $1 \le i \le n$, $n \ge 1$ be a linear regression model. Denote by $\hat{\beta}_n$ the M-estimate of β_0, using a convex function ρ. In [1], a basic theorem(Theorem A below) concerning the weak consistency of $\hat{\beta}_n$ is established. This theorem raises further problems concerning the consistency of $\hat{\beta}_n$. In this note, some of these questions are considered for the special cases of LAD and LS estimates, and the methods of integrals and real function theory are used.

27.1 INTRODUCTION

Consider a linear regression model

$$Y_i = x_i'\beta_0 + e_i,\ 1 \le i \le n,\ n \ge 1, \tag{27.1}$$

H.G.W. Begehr et al. (eds.), Partial Differential and Integral Equations, 359–369.

where Y_i, x_i, β_0 and e_i are the observation of the target variable, a known p-vector, the unknown parameter vector, and the random error, respectively. Let ρ be a convex function defined on R^1. The M-estimate of β_0, to be denoted by $\hat{\beta}_n$, is defined as a minimizing point of the function $H(\beta) \equiv \sum_{i=1}^n \rho(Y_i - x_i'\beta)$. Denote by the left and right derivative of ρ by ψ_- and ψ_+. About the weak consistency of $\hat{\beta}_n$, Zhao, Rao and one of the present authors established the following result [1].

Theorem A *Let $e_1, e_2, \cdots$ be iid. Suppose that there exists a function ψ, satisfying $\psi_- \leq \psi \leq \psi_+$ and the following conditions:*
1. $E\psi(e_1) = 0$.
2. There exist constants $c > 0$ and $\Delta > 0$, such that

$$|E(\psi(e_1 + u)| \geq c|u|, \ |u| \leq \Delta. \tag{27.2}$$

3. $E\psi^2(e_1 \pm \delta) < \infty$, for some $\delta > 0$.
Write $S_n = x_1x_1' + \cdots + x_nx_n'$. If

$$\lim_{n\to\infty} S_n^{-1} = 0, \tag{27.3}$$

then $\hat{\beta}_n$ is weakly consistent.

This theorem substantially improves early existing results, including those which define the M-estimate as a solution of the equation

$$\sum_{i=1}^{n} \phi(Y_i - x_i'\beta)x_i = 0 \tag{27.4}$$

for some chosen function ψ. In this respect the paper [2] by Yohai and Maronna should be mentioned. In that paper the authors, assuming that $e_1, e_2, \cdots$ are iid., its common distribution and ψ satisfy a set of conditions, prove that (27.3) is sufficient for β_n^*, the solution of (27.4), to be weakly consistent. However, their conditions do not cover some important cases, especially the Least Absolute Deviation(LAD) estimate. More importantly, the conditions improved on ψ do not guarantee the existence of a solution of (27.4), see, for and example, [3]. To ensure the existence of a solution of (27.4), one should add the assumption that ψ is continuous everywhere on R^1 which, as can easily be shown, amounts to require that ρ is continuously differentiable in Theorem A.

At this point two questions can be asked to maintain the consistency of $\hat{\beta}_n$:
1) Under reasonably general conditions on ρ and the common distribution of $\{e_i\}$, can condition (27.3) still be weakened?
2) Can conditions 1–3 be further weakened and (27.3) still remains sufficient?

There are indications that the answer of the first question in "no", see [4] and [5]. To get some insight into the second question, we give a closer look on the conditions 1–3 of Theorem A. First, condition 1, together with the condition

$$E\psi(e_1 + u) \neq 0, \text{ for } |u| \neq 0 \text{ small enough} \tag{27.5}$$

ensure that $h(t) \equiv E(\rho(e_1 + t) - \rho(e_1))$ exists for small $|t|$ and that 0 is the unique minimum of $h(t)$. This, at least for some special $\{x_i\}$, is indispensable to ensure the consistency of $\hat{\beta}_n$. Thus 1 is essential. As for 2, it is much stronger than (27.5). Can it be replaced by (27.5)? The answer is far from trivial.

Take condition 3. It adds a stronger moment condition on $\psi(e_1)$, compared with 1. One may suspect that it is largely technical and not essential in proving the consistency of $\hat{\beta}_n$, just as the existence of finite variance is not essential in proving the classical LLN. Can this condition be got rid of or be replaced by a weaker one? The answer is again far from trivial.

In this paper we shall show that for the two most important special cases of M–estimates – the LS and LAD estimates, the conditions 1–3 turn out to be the best – they cannot be improved any more. Though a conclusive solution remains pending, this gives us the reason to believe that the answer is also "no" for general convex ρ.

27.2 LAD ESTIMATES

For LAD estimates condition 3 has no question, and condition 1 amounts to require $\text{med}(e_1) = 0$. Define

$$c(x) = \begin{cases} P(0 < e_1 < x)/x, & x > 0, \\ P(x < e_1 < 0)/|x|, & x < 0. \end{cases}$$

Then it can easily be seen that condition 2 is equivalent to the following conditions (**Condition** C):

If $P(e_1 \leq 0) = 1/2$, then $\liminf_{x \downarrow 0} c(x) > 0$.
If $P(e_1 \geq 0) = 1/2$, then $\liminf_{x \downarrow 0} c(x) > 0$.

Theorem 1 *Suppose that in model* (27.1) $e_1, e_2, \cdots$ *are iid. If Condition C is not satisfied, then* (27.3), *or even the stronger condition*

$$\max_{1 \leq i \leq n} x_i' S_n^{-1} x_i \to 0 \tag{27.6}$$

is not sufficient to ensure the weak consistency of $\hat{\beta}_n$.

Proof Consider first the special case $p = 1$. Define a random variable ξ with distribution

$$P(\xi = \frac{Y_i}{x_i}) = \frac{|x_i|}{M_n}, \quad 1 \leq i \leq n, \quad M_n = \sum_{i=1}^{n} |x_i|.$$

Then it is easy to see that the LAD estimate $\hat{\beta}_n$ of β_0 is $\text{med}(\xi)$. Hence, if we define $Z_n = \sum_{i=1}^{n} |x_i| I(Y_i/x_i \geq 1)/M_n$, where I is the indicator, then $\{Z_n > 1/2\} \subset \{\hat{\beta}_n \geq 1\}$.

Assume without loss of generality that $\beta_0 = 0$, then $Y_i = e_i$, and from the above it follows that in order to prove the non-consistency of $\hat{\beta}_n$, it suffices to show that

$$\overline{\lim_{n \to \infty}} P(Z_n > \frac{1}{2}) > 0. \tag{27.7}$$

Now suppose that Condition C is not satisfied. For definiteness, assume that $P(e_1 \leq 0) = 1/2$ and $\liminf_{x \downarrow 0} c(x) = 0$. Then given $\varepsilon \in (0, 1/9)$ we can find a sequence $\{t_n\}$ such that

$$t_n \downarrow 0, \ \sum_{n=1}^{\infty} t_n^2 < \frac{\varepsilon}{2}, \ 1 > c(t_1) > c(t_2) > \cdots \to 0. \tag{27.8}$$

and

$$1 \pm u \geq \exp(\pm u - u^2), \ 0 \leq u \leq 2P(0 < e_1 < t_1). \tag{27.9}$$

We prove that a sequence $\{q_n\}$ can be found such that

$$t_{n+1} < q_n \leq t_n, \ n \geq 1, \tag{27.10}$$

$$1 > c(q_1) > c(q_2) > \cdots \to 0, \tag{27.11}$$

$$\sum_{n=1}^{\infty} q_n^2 c^2(q_n) < \frac{\varepsilon}{2}, \tag{27.12}$$

and that for any $n \geq 2$ and rational numbers $r_1, \cdots, r_{n-1}$, we have

$$q_n \neq \sum_{i=1}^{n-1} r_i q_i, \tag{27.13}$$

q_n is defined inductively as follows: Take $q_1 = t_1$. Assume that $q_1, \cdots, q_{n-1}$ have already been determined, satisfying

$$\begin{gathered} t_{i+1} < q_i \leq t_i, \ 1 \leq i \leq n-1, \\ |c(q_i) - c(t_i)| < \min\{c(t_i) - c(t_{i+1}), c(t_{i-1}) - c(t_i)\}, \ 2 \leq i \leq n-1. \end{gathered} \tag{27.14}$$

Put $Q = \{\sum_{r=1}^{n-1} r_i q_i : r_i \text{ is rational}, 1 \leq i \leq n-1\}$. Since Q is denumberable and $c(x)$ is left-continuous, we can find $q_n \notin Q$, $t_{n+1} < q_n \leq t_n$, and (27.14) holds for $i = n$. Obviously, $\{q_n\}$ determined in this way satisfies all required conditions.

The next step is to construct a sequence of positive integers $\{k_i\}$, such that

$$\sum_{n=1}^{\infty} k_n q_n^2 = \infty, \tag{27.15}$$

$$\sum_{n=1}^{\infty} k_n q_n^2 c^2(q_n) \leq \varepsilon. \tag{27.16}$$

For this purpose we note that by (27.8) and (27.10), we have $\sum_{n=1}^{\infty} q_n^2 < \infty$. From this and (27.12), we can define

$$\sum_{i=n}^{\infty} q_i^2 = D(n), \ \sum_{i=n}^{\infty} q_i^2 c^2(q_i) = E(n), \ W(n) = \frac{E(n)}{D(n)}.$$

From (27.11), we get $W(n) \to 0$, hence it can be found a sequence of positive integers $1 = n(0) < n(1) < n(2) < \cdots$, such that

$$\sum_{i=1}^{\infty} W(n) < \frac{\varepsilon}{2}. \tag{27.17}$$

Choose a sequence of positive integers $1 = j_0 < j_1 < \cdots$, such that

$$\sum_{i=n(j_t)}^{n(j_{t+1})-1} q_i^2 > \frac{D(N(j_t))}{2}, \; t = 0, 1, 2, \cdots. \tag{27.18}$$

Now we define

$$k_i = \begin{cases} 1, \; i = 1, \cdots, n(1) - 1, \\ [1/D(n(j_t))], \; n(j_t) \le i < n(j_t + 1), \; t = 0, 1, 2, \cdots, \end{cases}$$

where $[a]$ is the largest integer not exceeding a. Notice that since $D(n) < \varepsilon/2 < 1$ for $n \ge 1$, we have $k_i \ge 1$ for all i. Also, from (27.18) we get

$$\sum_{i=n(j_t)}^{\infty} k_i q_i^2 \ge \sum_{i=n(j_t)}^{n(j_{t+1})-1} k_i q_i^2 \ge \left[\frac{1}{D(n(j_t))}\right] \frac{D(n(j_t))}{2} \ge \frac{1}{4}, \tag{27.19}$$

$$\sum_{i=n(j_t)}^{n(j_{t+1})-1} k_i q_i^2 c^2(q_i) \le \sum_{i=n(j_t)}^{\infty} \frac{q_i^2 c^2(q_i)}{D(n(j_t))} = W(n(j_t)). \tag{27.20}$$

Now (27.15) follows from (27.19), and according to (27.12), (27.17) and (27.20), we have

$$\sum_{n=1}^{\infty} k_n q_n^2 c^2(q_n) = \sum_{i=0}^{n(1)-1} k_i q_i^2 c^2(q_i) = \sum_{t=0}^{\infty} \sum_{i=n(j_t)}^{n(j_{t+1})-1} k_i q_i^2 c^2(q_i)$$

$$\le \sum_{n=1}^{\infty} q_n^2 c^2(q_n) + \sum_{t=0}^{\infty} W(n(j_t)) \le \frac{\varepsilon}{2} + \frac{\varepsilon}{2} = \varepsilon.$$

This proves (27.16).

Now we write $x(i)$ for x_i in model (27.1), and define

$$x(1) = \cdots = x(k_1) = q_1,$$

$$x(k_1 + \cdots + k_i + 1) = \cdots = x(k_1 + \cdots + k_{i+1}) = q_{i+1}, \; i \ge 1.$$

By (27.15), we get $\sum_{i=1}^{\infty} x^2(i) = \infty$. As $\{x(i)\}$ is bounded, (27.6) is true.

We proceed to show that for this choice of $\{x(i)\}$, $\hat{\beta}_n$ does not converge is probability to β_0, which is assumed to be zero. As said earlier, for this purpose we need only to prove (27.7). Since $P(e_1 \leq 0) = 1/2$ and $e_1, e_2, \cdots$ are iid., we have

$$P(\frac{e_i}{x(i)} \geq 1) = \frac{1}{2} - P(0 < e_1 < x(i)) = \frac{1}{2} - x(i)c(x(i)),$$

$$P(\frac{e_i}{x(i)} < 1) = \frac{1}{2} + P(0 < e_1 < x(i)) = \frac{1}{2} + x(i)c(x(i)).$$

Hence, if

$$\{1, 2, \cdots, n\} = \{i_1, \cdots, i_t\} \cup \{i'_1, \cdots, i'_s\}, \ n = t + s, \tag{27.21}$$

we should have

$$G \equiv P(e_i/x(j) \geq 1, \ j = i_1, \cdots, i_t : \ e_j/x(j) < 1, \ j = i'_1, \cdots, i'_s)$$

$$= 2^{-n} \prod_{r=1}^{t} (1 - 2x(i_r)c(x(i_r))) \prod_{l=1}^{s} (1 + 2x(i'_l)c(x(x'_l))).$$

As $x(i)c(x(i)) = P(0 < e_1 < x(i)) \leq P(0 < e_1 < t_1)$, by (27.9) we obtain

$$G \geq 2^{-n} \exp(-2 \sum_{r=1}^{t} x(i_r)c(x(i_r)) + 2 \sum_{l=1}^{s} x(i'_l))) \exp(-4 \sum_{i=1}^{n} x^2(i)c^2(x(i))).$$

Write $\eta = e^{-4\varepsilon}$, $A_n = \sum_{i=1}^{n} x(i)c(x(i))$. From (27.16) we get $\sum_{i=1}^{2} x^2(i)c^2(x(i)) \leq \varepsilon$. Therefore

$$\begin{aligned} G &\geq \eta 2^{-n} \exp(-4|\sum_{r=1}^{t} x(i_r)c(x(i_r)) - \frac{A_n}{2}|) \\ &\geq \eta 2^{-n}(1 - 4|\sum_{r=1}^{t} x(i_r)c(x(i_r)) - \frac{A_n}{2}|) \\ &= \eta 2^{-n}(1 - 4|\sum_{l=1}^{t} x(i'_l)c(x(i'_l)) - \frac{A_n}{2}|). \end{aligned} \tag{27.22}$$

From the definition of $\{x(i)\}$, and by (27.13), we see that for any partition of $\{1, 2, \cdots, n\}$ as (27.21), it is true that

$$\sum_{r=1}^{t} x(i_r) \neq \sum_{l=1}^{s} x(i'_l).$$

Thus one of the two sums must be greater than $M_n/2$. From this, and considering the definition of Z_n, we get

$$\begin{aligned} P(Z_n > \frac{1}{2}) &= \sum{}^* P\left(\frac{e_i}{x(j)} \geq 1, \ j = i_1, \cdots, i_t; \ \frac{e_j}{x(j)} < 1, j = i'_1, \cdots, i'_s\right) \\ &\geq \eta \left(1 - 4 \cdot 2^{-n} \sum{}^* |\sum_{r=1}^{t} x(i_r)c(x(i_r)) - \frac{A_n}{2}|\right), \end{aligned} \tag{27.23}$$

where $\sum^*$ means the summation is taken over all possible disjoining partition of $\{1, 2, \cdots, n\}$ as (27.21), so in $\sum^*$ there are 2^n terms.

We can estimate the right-hand side of (27.22) very easily by the following observation: Define iid. random variables $\xi_1, \cdots, \xi_n$ with common distribution $P(\xi_1 = 0) = P(\xi_1 = 1) = 1/2$. Then obviously

$$2^{-n} \sum{}^* |\sum_{r=1}^{t} s(i_r)c(x(i_r)) - \frac{A_n}{2}| = E|\sum_{i=1}^{n} x(i)c(x(i))(\xi_i - E\xi_i)|. \quad (27.24)$$

From (27.23),(27.24), and remembering that $0 < \varepsilon < 1/9$, we obtain

$$\begin{aligned} P(Z_n > \frac{1}{2}) &\geq \eta(1 - 4(\mathrm{Var}(\sum_{i=1}^{n} x(i)c(x(i))\xi_i))^{1/2}) \\ &= \eta(1 - 2(\sum_{i=1}^{n} x^2(i)c^2(x(i)))^{1/2}) \\ &\geq \eta(1 - 2(\sum_{i=1}^{\infty} k_i q_i^2 c^2(q_i))^{1/2}) \\ &\geq \eta(1 - 2\sqrt{\varepsilon}) \geq \frac{\eta}{3}. \end{aligned}$$

Since $\eta > 0$ does not depend on n, (27.7) follows, concluding the proof for the case of $p = 1$.

If $p > 1$, we define $\{x_i\}$ in (27.1) by

$$x_{(i-1)p+1} = (x(i), 0, \cdots, 0)', \cdots, x_{(i-1)p+p} = (0, \cdots, 0, x(i))', \; i = 1, 2, \cdots,$$

where $\{x(i)\}$ is defined as in the above proof. Then (27.6) is true, but every component of $\hat{\beta}_n$ does not converge in probability to the corresponding component of β_n. Theorem 1 is proved.

27.3 LS ESTIMATES

For LS estimate $p(u) = u^2$. Condition 1 reduces to $E(e_1) = 0$ while condition 2 holds automatically. Condition 3 reads $Ee_1^2 < \infty$.

Theorem 2 *Suppose that in model* (27.1) *the random errors* $e_1, e_2, \cdots$ *are iid., and* $Ee_1^2 = \infty$. *Then* (27.3) *is not sufficient for the weak consistency of the LS estimate* $\hat{\beta}_n$.

For the case $p = 1$, the conclusion of this theorem follows from the following more general result. We state it as a theorem since it is of independent interest.

Theorem 3 *Suppose that* $e_1, e_2, \cdots$ *are iid. If for any real sequence* $\{a_i\}$ *with*

$$A(n) \equiv \sum_{i=1}^{n} a_i^2 \to \infty, \quad (27.25)$$

we have

$$\sum_{i=1}^{n} \frac{a_i e_i}{A^2(n)} \xrightarrow{P} 0 \text{ as } n \to \infty. \tag{27.26}$$

Then

$$Ee_1 = 0, \;\; Ee_1^2 < \infty. \tag{27.27}$$

Proof. From (27.25) we have $\max_{1\le i\le n} |a_i|/A(n) \to 0$, this together with the assumption that $e_1, e_2, \cdots$ are iid, entails that $\{(A(n)^{-1}a_ie_i,\ 1 \le i \le n\},\ n \ge 1$, satisfies the "uan" condition ([6], p.302). Hence the following two are necessary conditions for (27.26) ([6], p.317):

$$b(n) \equiv \sum_{i=1}^{n} \int_{|x|<A(n)/|a_i|} dF(x) \to 0, \;\; n \to \infty, \tag{27.28}$$

$$\begin{aligned}\sigma(n) \equiv \sum_{i=1}^{n} a_i^2 \Big(\int_{|x|<A(n)/|a_i|} x^2 dF(x)\Big) \\ -\Big(\int_{|x|<A(n)/|a_i|} x dF(x)\Big)^2 /A^2(n) \to 0, \; n \to \infty,\end{aligned} \tag{27.29}$$

here F is the common distribution of $e_1, e_2, \cdots$.

Firstly we show that $E|e_1| < \infty$. Denote by $\tilde{F}$ the distribution of $|e_1|$. If $\int_0^\infty x d\tilde{F}(x) = \infty$, define

$$\tilde{F}(c,i) + \tilde{F}(c(i+1)) - \tilde{F}(c\,i), \; i = 0, 1, 2, \cdots,$$

$$G(c,n) = \sum_{i=1}^{n} i\tilde{F}(c,i), \; n = 1, 2, \cdots,$$

we have

$$\begin{aligned}G(c,n) \ge \sum_{i=0}^{n} (i+1)\tilde{F}(c,i) - 2 = \sum_{i=0}^{n} c\tilde{F}(c,i)/c \\ -2 \ge c^{-1} \int_0^{c(n+1)} x d\tilde{F}(x) - 2.\end{aligned}$$

Therefore for any $a, b,\ 0 < a < b < \infty$, we get

$$\lim_{n\to\infty} \big(\inf_{a\le c\le b} G(c,n)\big) = \infty. \tag{27.30}$$

From (27.30), for any positive integer j, a positive integer n_j can be found, such that

$$G(jc\pi^2/6, n_j) \ge 1, \;\; 1/2 \le c \le 1, \tag{27.31}$$

$$T(n_j) \equiv \sum_{i=1}^{n_j} i^{-2} \ge \frac{\pi^2}{12}. \tag{27.32}$$

Write $N(k) = n_1 + \cdots + n_k$, $k \geq 1$. Define a sequence $\{a_i\}$ in the following way:

$$a_i = 1/i, \quad i \leq i \leq N(1),$$
$$a_i = 1/j, \quad i = N(k) + j, \quad 1 \leq j \leq n_{k+1}, \; k \geq 1. \tag{27.33}$$

For this sequence $A(N(k)) = \sum_{r=1}^{k} T(n_r) \geq k\pi^2/12 \to \infty$, so (27.25) is fulfilled. Now prove that (27.28) is false. In fact, by (27.33),

$$b(N(k)) \geq \sum_{i=1}^{n_k} \int_{|x| \geq A(N(k))/|a_i|} dF(x)$$
$$= \sum_{i=1}^{n_k} \int_{|x| \geq A(N(k))} dF(x) \geq G(A(N(k)), n_k). \tag{27.34}$$

From (27.32) we have $1/2 \leq A(N(k))/(k\pi^2/6) \leq 1$. Hence by (27.31) and (27.34), we get $b(N(k)) \geq 1$, for any positive integer k. This shows that (27.28) does not hold, and (27.25) is not true. This contradicts the assumption and thus proves that $E|e_1| = \infty$ is impossible.

Having obtaining $E|e_1| < \infty$, $Ee_1 = 0$ follows by choosing $a_1 = a_2 = \cdots = 1$ on employing the law of large numbers.

Finally we prove $Ee_1^2 < \infty$. If it is not true, we would have $\int_{|x|<\infty} x^2 dF(x) = \infty$. Find a positive integer n_0 such that

$$b_0 = \int_{|x|<n_0} x^2 dF(x) > 0.$$

Find a positive integer $n_1 > n_0$, such that

$$b_1 \equiv \int_{|x|<n_1} x^2 dF(x) \geq 2b_0, \; \sqrt{n_1 b_0} \geq n_0.$$

Assume that $n_0 < n_1 < \cdots < n_k$ have already been determined to satisfy

$$b_1 \equiv \int_{|x|<n_i} x^2 dF(x) \geq 2b_{i-1}, \; \sqrt{n_i b_{i-1}} \geq n_{i-1}, \; i \leq i \leq k. \tag{27.35}$$

Then since $\int_{|x|<\infty} x^2 dF(x) = \infty$, we can find a positive integer $n_{k+1} > n_k$, such that (27.35) holds for $i = k+1$. Thus we determine a sequence of positive integers $\{n_k\}$, $k \geq 0$ for which (27.35) is true for any $i \geq 1$.

Write again $N(k) = n_1 + \cdots + n_k$. Define

$$a_i = \sqrt{b_0/n_1}, \; 1 \leq i \leq n_1,$$
$$a_{N(k)+j} = \sqrt{(b_k - b_{k-1})/n_{k+1}}, \; 1 \leq j \leq n_{k+1}, \; k \geq 1. \tag{27.36}$$

For this sequence $\{a_i\}$ we have $A(N(K+1)) = b_k \geq 2^k b_0 \to \infty$, as $k \to \infty$, so (27.25) is true. Now show that (27.29) is false. In fact, define

$$D_k \equiv \sum_{i=N(k-1)+1}^{N(k)} a_i^2 \int_{|x|<b_{k-1}/|a_i|} x^2 dF(x)/A^2(N(k)).$$

But according to (27.36) for $N(k-1)+1 \leq i \leq N(k)$, we get

$$b_{k-1}/|a_i| = b_{k-1}\sqrt{n_k/(b_{k-1}-b_{k-2})}$$
$$> b_{k-1}\sqrt{n_k/b_{k-1}} = \sqrt{n_k b_{k-1}} \geq n_{k-1}.$$

The last inequality is true by (27.35). Therefore, again by (27.35),

$$D_k \geq b_{k-1}^{-2} n_k \frac{b_{k-1}-b_{k-2}}{n_k} \int_{|x|<n_{k-1}} x^2 dF(x) =$$
$$= b_{k-1}^{-2}(b_{k-1}-b_{k-2})b_{k-1} = (b_{k-1}-b_{k-2})/b_{k-1} \geq \frac{1}{2}. \tag{27.37}$$

Earlier we proved $Ee_1 = 0$. Since $\max_{1\leq i\leq N(k)} |a_i|/A(N(k)) \to 0$, from $Ee_1 = 0$ and $Ee_1^2 = \infty$, we have

$$\sum_{1\leq i\leq M(k)} (\int_{|x|<A(N(k))/|a_i|} x dF(x))^2 \to 0, \ k \to \infty,$$

$$\sum_{1\leq i\leq M(k)} \int_{|x|<A(N(k))/|a_i|} x^2 dF(x) \to \infty, \ k \to \infty.$$

Hence, when k is large, for $1 \leq i \leq N(k)$, we have

$$\int_{|x|<A(N(k))/|a_i|} x^2 dF(x) \geq 2(\int_{|x|<A(N(k))/|a_i|} x^2 dF(x))^2. \tag{27.38}$$

From (27.37) and (27.38), we get

$$\sigma^2(N(k)) \geq \sum_{i=N(k-1)+1}^{N(k)} a_i^2 (\int_{|x|<A(N(k))/|a_i|} x^2 dF(x)$$
$$- \int_{|x|<A(N(k))/|a_i|} x dF(x))^2/A^2(N(k))) \geq \frac{D_k}{2} \geq \frac{1}{4},$$

when k is large. Hence (27.29) fails, and (27.25) is not true. This contradicts the assumption and shows that $Ec_1^2 = \infty$ is impossible, Theorem 5 is proved.

Theorem 3 entials Theorem 2 for the case of $p = 1$. For general $p > 1$, use the argument employed earlier in the proof of Theorem 1.

References

[1] Zhao, L.C., Rao C.R., Chen, X.R.: A note on the consistency of M–estimates in linear models, Stochastic processes. A Festschrift in Honour of Gopinath Kallianpur, ed. S. Cambanis. Springer–Verlag, New York, 1993, 359-367.

[2] Yohai, V.J., Maronna R.A.: Asymptotic behavior of estimators for the linear model. Ann. Statist. 7 (1979), 258-268.

[3] Bai, Z.D., Wu, Y.H., Chen, X.R., Miao B.Q.: On solvability of an equation arising in the theory of M–estimates. Commun. Statist.–Theory and Methods 19 (1990), 363-380.

[4] Li, K.C.: Regression models with infinitely many parameters: Consistency of bounded linear functionals. Ann. Statist. 12(2) (1984), 601-611.

[5] Chen, X.R., Zhao, L.C., Wu, Y.H.: On conditions of consistency of ML estimates. Statistica Sinica, 1993(3), 9-18.

[6] Loève, M.: Probability Theory. van Nostrand, Princeton, N.J., 1960.

International Society for Analysis, Applications and Computation

1. H. Florian et al. (eds.): *Generalized Analytic Functions*. Theory and Applications to Mechanics. 1998 ISBN 0-7923-5043-X
2. H.G.W. Begehr et al. (eds.): *Partial Differential and Integral Equations*. 1999 ISBN 0-7923-5482-6

KLUWER ACADEMIC PUBLISHERS – DORDRECHT / BOSTON / LONDON

MIX
Papier aus verantwortungsvollen Quellen
Paper from responsible sources
FSC® C105338
FSC
www.fsc.org

If you have any concerns about our products,
you can contact us on
ProductSafety@springernature.com

In case Publisher is established outside the EU,
the EU authorized representative is:
Springer Nature Customer Service Center GmbH
Europaplatz 3, 69115 Heidelberg, Germany

Printed by Libri Plureos GmbH
in Hamburg, Germany